STRUCTURE AND FUNCTION OF ROOTS

Developments in Plant and Soil Sciences

VOLUME 58

*The titles published in this series are listed at the end of this volume.*

# Structure and Function of Roots

## Proceedings of the Fourth International Symposium on Structure and Function of Roots, June 20–26, 1993, Stará Lesná, Slovakia

*Edited by*

F. BALUŠKA, M. ČIAMPOROVÁ, O. GAŠPARÍKOVÁ and P. W. BARLOW

Partly reprinted from *Plant and Soil*, Volume 167, No. 1 (1994)

SPRINGER-SCIENCE+BUSINESS MEDIA, B.V.

A C.I.P. Catalogue record for this book is available from
the Library of Congress.

ISBN 978-90-481-4402-0 ISBN 978-94-017-3101-0 (eBook)
DOI 10.1007/978-94-017-3101-0

---

*Printed on acid-free papers*

# Contents



* Chapters indicated with an asterisk are reprinted from *Plant and Soil*, Volume 167, No. 1. (1994).

## ROOTS UNDER STRESS CONDITIONS

*F. Baluška et al. (eds.), Structure and Function of Roots,* IX, 1995.

# Preface

In 1971, the late Dr. J. Kolek of the Institute of Botany, Bratislava, organized the first International Symposium devoted exclusively to plant roots. At that time, perhaps only a few of the participants, gathered together in Tatranska Lomnica, sensed that a new era of root meetings was beginning. Nevertheless, it is now clear that Dr. Kolek's action, undertaken with his characteristic enormous enthusiasm, was rather pioneering, for it started a series a similar meetings. Moreover, what was rather exceptional at the time was the fact that the meeting was devoted to the functioning of just a single organ, the root.

One possible reason for the unexpected success of the original, perhaps naive, idea of a Root Symposium might lie with the fact that plant roots have always been extremely popular as experimental material for cytologists, biochemists and physiologists whishing to probe processes as diverse as cell division and solute transport. Of course, the connection of roots with the rest of the plant is not forgotten either. This wide variety of disciplines is now coupled with the development of increasingly sophisticated experimental techniques to study some of these old problems. These factors undoubtedly contribute to the necessity of continuing the tradition of the root symposia. The common theme of root function gives, in addition, a certain unity to all these diverse activities.

Plant roots have evidently proved to be an extremely suitable theme for a meeting, such as was held in those early days, more than twenty years ago, in Tatranska Lomnica. Not only have the Slovak 'Root Meetings' become a successful tradition, but the subject now attracts a broad international audience with similar meetings being held with increasing frequency throughout the world. This book presents contributions to the Fourth International Symposium on Plant Roots held once again in the High Tatra mountains of Slovakia in June 1993. We hope that these Proceedings will recapture something of the flavour of those days.

Although the Fourth International Root Symposium was itself an ephemeral affair, lasting a mere 5 days, we hope that the contributions will prove useful for a considerable period. Of course, when that period has passed, the time may be ripe for a further Root Symposium in Slovakia!

*The Editors*

STRUCTURAL ASPECTS OF ROOT GROWTH AND DEVELOPMENT

*F. Baluška et al. (eds.), Structure and Function of Roots*, 3–18, 1995.


# Structure and function at the root apex - phylogenetic and ontogenetic perspectives on apical cells and quiescent centres

Peter W. Barlow
*Department of Agricultural Sciences, University of Bristol, Institute of Arable Crops Research, Long Ashton Research Station, Bristol BS18 9AF, UK*

*Key words:* apical cell, ferns, meristem, merophytes, phylogeny, quiescent centre, root

**Abstract**

The summit of roots of various plant species may be occupied by a single, rapidly proliferating tetrahedral apical cell (as in ferns), or by a multicellular and slowly proliferating quiescent centre (as in angiosperms), or by intermediate types of cellular organizations. The present paper attempts to deduce the phylogeny of these various types of cellular patterning at the root apex.

## Introduction

Ever since Bohumil Němec, nearly a hundred years ago, observed a group of non-dividing cells at the summit of the proximal meristem of roots of *Allium cepa*, *Rorippa amphibia* and *Iris germanica* (Němec, 1897), similar observations have been made on numerous other species of angiosperms and gymnosperms. Later, the relative inactivity of these cells in roots of *Allium cepa* was more definitely demonstrated through the use of the metaphase-arresting agent colchicine (D'Amato and Avanzi, 1948; Mangenot, 1942). Prolonged immersion in the drug (up to 7 days) revealed the continuing presence of small diploid nuclei at the summit of the root, whereas highly polyploid nuclei had become plentiful in the surrounding meristematic regions. Additional information on this cellular peculiarilty was provided by Clowes (1956 a, b) through the use of autoradiography of sectioned roots of maize and other species which had been fed radioactive precursors of DNA and RNA. Clowes found a closely defined area of unlabelled cells, which he termed 'quiescent centre', that corresponded to the zone where mitoses had been reported as being infrequent. Interestingly, the quiescent centre (QC) of the maize root also corresponded to a zone which Rabideau and Mericle (1953) had shown three years earlier, also by autoradiography, as failing to incorporate $^{14}C$ translocated from shoots exposed to an atmosphere enriched with $^{14}CO_2$.

More than thirty years since Clowes's illuminating discovery, it is still not clear whether the QC reflects little more than a passive geometrical necessity dictated by the kinematics of cellular growth within the apex. An alternative proposition is that it plays an active and essential role in meristem function (Torrey, 1972). While answers to these questions will probably not be forthcoming until the techniques of molecular biology have more fully unlocked the secrets of the QC (Sabelli et al., 1993), it is nevertheless possible to propose schemes in which the QC has both passive and active attributes in relation to the growth of the root. In this connection, it is useful to recognise the value of comparative studies of cellular structure at the root summit both in a range of species growing under normal conditions (Clowes, 1984; Guttenberg, 1961, 1966, 1968; Pillai, 1966) and in roots of a few species growing under different experimental conditions (Clowes, 1975). Such studies inevitably lead to questions about not only the phylogenetic origin of the QC, but also the relation of the QC to apical growth and how, in an evolutionary context, its properties optimize a plant's response to selective pressures exerted by the environment. From these considerations it can be tentatively concluded that the QC of angiosperms represents a complex fail-safe system (Barlow, 1994a) that ensures uninterrupted development of the root in the face of potentially disruptive environmental conditions (Barlow, 1987). Whether the same can be said of the QC of gymnosperms or of summit cells in roots with other types of apical organization (e.g. those ferns with an 'apical cell') is at present an open question, but could be answered through suitable experimentation. From the phylogenetic point of view, the QC appears to have arisen in parallel with an increasing degree of complexity in the behaviour of the structural initials at the pole of the root. Quiescence and structural complexity may also be aspects of a fail-safe maintenance system that ensures the integrity of the growing root apex.

## Origins

*'...from the fragmentary records at our disposal, . . . no details are to be considered as too trivial provided their value be not overestimated' (Bower, 1889).*

It is impossible to say at what stage during the evolution of the root a QC arose. The fossil record is of little help [though Good and Taylor (1972) have shown that the cellular detail of meristems can be preserved in fossilized material] and, regrettably, surveys of contemporary plant for the presence of a QC in their roots are few. The QC is best known in angiosperm roots (Clowes, 1984), but it has also been demonstrated in gymnosperms (Chouinard, 1959; Peterson and Vermeer, 1980; Wilcox, 1962). Of older phylogeny than either of these groups are the ferns (Filicopsida) and possibly the horsetails (Sphenopsida). Both these last-mentioned groups have root apices with a rather different type of cellular organization from those found in the evolutionarily younger groups. Here, the summit of the root can be occupied by a prominent apical cell, sometimes having a tetrahedral form. Although it is hazardous to deduce phylogenetic progressions of a specific cellular compartment, such as a QC, on the basis of properties exhibited by the root apices of contemporary species, it is, as we shall see, possible to envisage the origin of a multicellular, proliferatively weakly active QC, such as found in present-day angiosperms, from a much simpler ancestral form. The antithesis of the angiosperm QC is the unicellular, proliferatively active apical cell found in the roots of Sphenopsida (e.g. *Equisetum*) (Gifford and Kurth, 1982) and certain present-day Filicopsida (e.g. Salviniales) (Kurth, 1981; Nitayangkura et al., 1980). Since ferns have a longer history than angiosperms, it is clearly of interest to consider whether this last-mentioned type of apical organization can have given rise to the type characteristic of angiosperms, or, alternatively, whether both types have been derived from some third type ancestral to the other two. For example, many of the Lycopsida, which are phylogenetically older than any other plant taxon except, perhaps, the Psilopsida, do not seem to have a prominent apical cell in their roots; nor is it known if they have a QC. However, roots of the lycopod *Selaginella emmeliana* and *S. martensii* are said to have a tetrahedral apical cell (Voronin, 1964), though in roots of many other Selaginellales no such cell is apparent (Guttenberg, 1966). A prominent apical cell is found in the shoot and embryo of some other species of *Selaginella*, whereas such cells do not arise in embryos of the Lycopodiales. Even members of the rootless Psilopsid, *Psilotum nudum*, seem to have an apical cell in the shoot of both the gametophyte and sporophyte generations (Bierhorst, 1954). Thus, the propensity of forming a centrally placed tetrahedral cell may be a very primitive characteristic of apices (Voronin, 1964). Its presence in the superficial cell layer of an apex may be related to the curvature ($\kappa$) of that surface. A high value of $\kappa$ at the summit of the apex of a root (or shoot) may favour a tetrahedral cell at that point, whereas a lower value of $\kappa$ may allow other cell forms to develop (Barlow, unpublished observations).

In comparing the behaviour of root meristems, it is evidently necessary to distinguish between the form of the cells at the apex and their proliferative properties. Reference to Table 1 shows that, within the meristem as a whole, structure can be separated from function. It is supposed that roots of contemporary species within taxonomic divisions which are, in evolutionary terms, in advance of the Lycopsida possess a number of clearly differentiated structural initial cells, or merophytes (i.e. groups of cells of common descent) of zero order designated $M_0$. These cells are polyhedra with a defined number of faces ($F$) that differentiate during embryogenesis and, as an accompaniment of their apical location, collectively serve as a cellular template for the root. An $M_0$ cell divides by mitosis and some, or all, of its faces are sites from which a new daughter merophyte, $M_1$, issues; the other daughter cell of the division retains its identity as a new $M_0$. A succession of $j$ merophytes, all of which are meristematic, can be produced from a given face of $M_0$. The set of merophytes $\{M_0, M_1,...M_j\}$ contains members which may be either proliferatively active (+) or proliferatively quiescent (-). (It is emphasised that 'quiescent' does not necessarily mean inactive; 'quiescent' is a relative term implying slow proliferative activity in contradistinction to rapid activity.) If a subset of merophytes, namely $\{M_0, M_1,.., M_i\}$, is quiescent, but the related subset $\{M_{(i+1)},...,M_j\}$ is active, then the latter subset can be regarded as meristematic, with merophyte

*Table 1.* Relative proliferative activity (+), quiescence (-), or inactivity (o) in three subsets of merophytes constructing the apical root meristem. The meristem is composed of the complete set $\{n(M_0, M_1,..., M_j)\}$, where *n* refers to the number of structural initial ($M_0$) cells

| Merophyte subset | Relative activity | | | | | | | |
|---|---|---|---|---|---|---|---|---|
| $\{M_0\}$ | + | -/o | -/o | + | + | + | -/o | -/o |
| $\{M_1,...,M_i\}$ | + | + | -/o | + | -/o | -/o | + | -/o |
| $\{M_{(i+1)},...,M_j\}$ | + | + | + | -/o | -/o | + | -/o | -/o |
| Type of apical behaviour* | 1 | 2 | 3 | 4 | 5 | 6 | 7 | 8 |

*Examples of apical types: 1 - Young roots of *Azolla pinnata* or *Marsilea vestita*, 2 - Older roots of *Marsilea vestita*, 3 - Angiosperms roots with a quiescent centre, 4,5 - Roots regenerating a functional meristem, 6 - A possibility for a root of *Osmunda regalis* or *Angiopteris evecta* (see Figs. 4 and 5), 7 - Hypothetical and 8 - Inactive root, if o replaces -; older roots of *Azolla pinnata*.

$M_j$ consituting the most proximal meristematic merophyte. Within the context of the root, the subset $\{M_0, M_1,...,M_i\}$ would constitute a quiescent zone. At the proximal border of this zone, one of the component cells of merophyte $M_{(i+1)}$, would have the status of a functional initial cell. The total number of merophytes in each of the subsets $\{M_0\}$, $\{M_1,...,M_i\}$ and $\{M_{(i+1)},...,M_j\}$ depends on *n*, the number of $M_0$ cells and the grade of polyhedron, defined by *F* the number of its faces, to which $M_0$ belongs (the number of faces of $M_0$ actually available for $M_1$ merophyte production is assumed for the moment to equal *F*). As an example: for $\{M_0\}$ in the young root of *Azolla pinnata* $n = 1$ and $F = 4$; $\{M_0\}$ thus consists of a single tetrahedral apical cell (Fig. 1). The constitution of this root meristem can therefore be notated $[\{1(M_0\}, \{4(M_1,...,M_j)\}]$, where $j_{a,b,c} = 7$ for the three root-forming merophytes; for the cap-forming merophyte produced at the fourth face of $M_0$, $j_d = 1$. In angiosperms, a minimal value of *n* = 3 or 4, each with $F > 4$, commonly applies to the $M_0$ cells; for merophytes on the proximal side of $M_0$, *j* may have values up to 7 or more, but *j* may be only 1 or 2 on the distal, cap side.

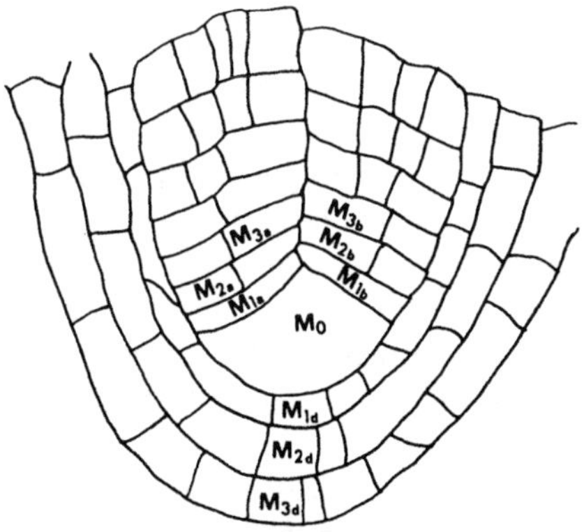


*Fig. 1.* Median longitudinal section through a root apex of the leptosporangiate water-fern *Azolla pinnata* (Salviniales) showing an apical cell $\{M_0\}$ and its immediate derivatives $\{M_{1a}, M_{2a}, M_{3a}, \ldots\}$ and $\{M_{1b}, M_{2b}, M_{3b}, \ldots\}$ (the third set of merophytes $\{M_{1c}, M_{2c}, M_{3c}, \ldots\}$ are absent from this section). Distal to the apical cell are three cap merophytes, $\{M_{1d}, M_{2d}, M_{3d}, \ldots\}$ (Note that while similarly numbered merophytes in the root each belong to the same gyre of merophytes, this is not true of the cap merophytes; these are actually much older than the root merophytes, even though they have the same number.) The form and activity of this root is representative of a Type 1 apex (Table 1). Redrawn from a montage of electron micrographs kindly provided by Prof B E S Gunning.

As indicated in Table 1, the type of meristem structure and behaviour can be designated by numbers 1 through 8 according to the combinations of proliferative activity (+) or quiescence (-) in the three merophytic subsets of the root. Thus, the most extreme phylogenetic transformation leading to the acquisition of a QC from an ancestral type of apex would be from Type 1 where $n = 1$ (+) and $F = 4$, to Type 3 where $n \gg 1$ (-) and $F > 4$.

The roots of ferns (Filicopsida) show many diverse features of apical organization (Guttenberg, 1966; Voronin, 1964) which might enable their forms to be arranged into a series ordered on the basis of progressive modifications to the pattern of cell growth and division, starting from an assumed ancestral apical form. This series might then mirror the phylogenetic sequence. Unfortunately, the corresponding evolutionary history of the ferns is uncertain (Sporne, 1975); ferns may even have a polyphyletic origin. This makes uncertain any conclusions regarding the direction of phylogenetic changes to apical form. One plausible evolutionary sequence is that the Ophioglossales preceded the Marattiales and Filicales, which in turn preceded the Salviniales, the latter order representing the most advanced species of ferns. Each of these orders contains species where the roots possess an apical cell (Table 2). Another categorisation of the ferns regards them as either eusporangiate or leptosporangiate. Leptosporangiate ferns are more delicate and are considered advanced relative to the more robust eusporangiate ferns. The former are characterised by tetrahedral apical cells ($F = 4$) whereas the latter sometimes possess, in their stead, polyhedral cells of a higher grade ($F = \sim 6$ ). Thus, the view of the phylogenetic changes that may have occurred in root apical morphology (which presumably served as a precursor for the development of the angiosperm QC) depends in some measure on whether the possession of a single tetrahedral apical cell is to be regarded as either (a) a primitive character, as it might be in the Ophioglossales, for example, or (b) an advanced character, as in the Salviniales. A third possibility (c) is that such a cell is characteristic of both primitive and advanced forms, and that there was an ancestral type of root from which the type with a single apical cell was derived. This ancestral type may still be reflected in the root organization of the eusporangiate Marattiales and the intermediate Osmundales. Unfortunately, there are no relevant records from apices belonging to the orders Cladoxylales and Coenopteridales (both now extinct) which might have helped clarify this problem. However, ancestral types may still be uncovered in the Lycopsida; the tetrahedral apical cell of certain *Selaginella* species has already been mentioned. In general, the four-sided apical cell may be regarded as a primitive, less complex form of structural initial.

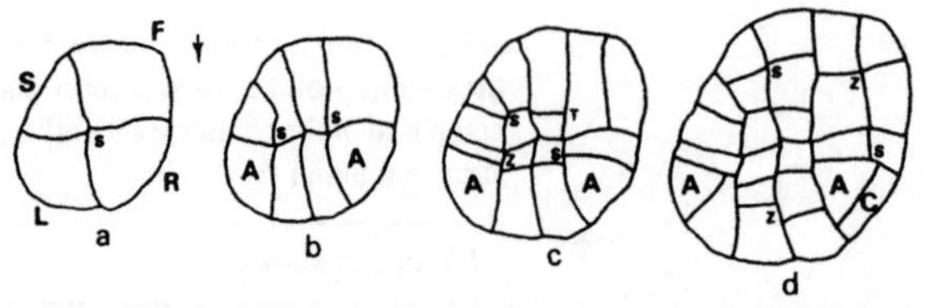


*Fig. 2.* Early divisions of the sporophyte of the leptosporangiate fern *Adiantum concinuum* (Filicales) showing the development of tetrahedral apical cells within the quadrants destined to become root and other organs. The arrow points towards the neck of the archegonium. The sequence of divisions following fertilization is shown from left to right (**a-d**). Synchronous divisions are assumed. Cellular tetrads, which are two pairs of sister cells resulting from two successive divisions of a common grandmother cell, can be classified as either an S, Z or T tetrad (Barlow, 1991). The types of tetrad that have arisen after each division are indicated. All the tetrads resulting from the first four divisions may be shown in the diagram, but some resulting from later divisions may not be represented. The four quadrants of the young sporophyte are: F - foot, S - stem, L - leaf, R - root [shown in (a)]. Note that the apical cell in the root quadrant becomes internal as a result of the fifth division when the cap merophyte (C) is formed. In the leaf quadrant, on the other hand, the apical cell persists in a superficial location (see Fig. 3). Interpreted and redrawn from Bower (1908).

## Fern roots with a tetrahedral apical cell

Taking the embryonic development of *Adiantum concinuum* (Filicales) as typical of a fern whose root contains a single tetrahedral apical cell, it can be easily seen that this cell is created by the earliest divisions of the zygote (Fig. 2). The zygote of the sporophyte is embedded in the thallus of the gametophyte. The first two divisions lead to a tetrad of cells that become the four embryonic quadrants of foot, stem, leaf and root. The two former quadrants face inwards, away from the neck of the archegonium, the two latter quadrants face outwards towards the neck, as is characteristic of leptosporangiate fern embryogeny. Each cell of the tetrad is a tetrahedron, as is inevitable from the orientations of the first two divisions of the spherical zygote (Fig. 2a). Further divisions of the two outer cells - i.e. the quadrants for the future leaf and root - maintain a tetrahedral cell within each set of derivative cells, one of whose walls form part of the outer surface of the respective quadrant (Fig. 2b, c). In the case of the root quadrant, divisions of its tetrahedral cell, $M_0$, create new cell walls that are aligned parallel to each of its original walls, thus forming three basiscopic sectors (subsets) of merophytes, $\{3(M_1,...,M_j)\}$, within which typical root tissues commence to differentiate, and a fourth acroscopic subset of merophytes from which the root cap is derived (Figs. 2d, 3).

This cellular configuration persists within the emerged root of the sporophyte and seems to be fairly general for leptosporangiate ferns. Questions then arise as to whether an apical cell of this type has some special role in the growth and development of the root - for instance 'do mitoses persist in this cell?', or is it, as Gifford (1983) has recently asked, 'no more than a gap in a system of confocal intersecting cell walls?'[1]. It is known from careful studies of the anatomy of roots of *Azolla pinnata* (Salviniales) (Fig. 1) that the apical cell divides rapidly during the early stages of root

[1] Here, Gifford was restating a question originally posed by Sachs (1878) when this latter author considered the system of periclinal and anticlinal walls that apparently both describe and, by their pattern of extension, govern the growth and form of the root.

*Table 2.* Presence of a tetrahedral apical cell (T; $F = 4$) or a higher-grade polyhedral cell (P; $F = 5$ or 6) in root apices of various orders of ferns

| Age of root | Order | | | | | |
|---|---|---|---|---|---|---|
| | Ophioglossales | Marattiales | Osmundales | Filicales | Marsileales | Salviniales |
| Young | T | T | T | T | T | T |
| Mature | T | P | P | T | T | T |
| | Eusporangiate | | | | Leptosporangiate | |
| Status[a] | Primitive | | → | | | Advanced |

[a]According to Sporne (1975).

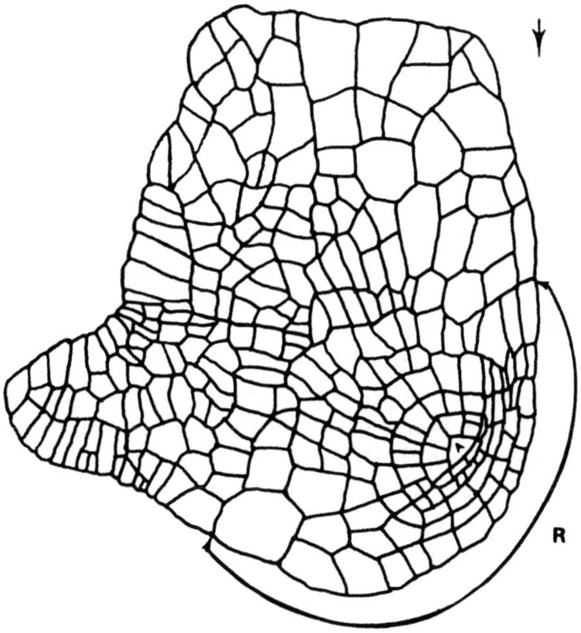


*Fig. 3.* A later stage in sporophyte development (as seen in tissue section) than that shown in Figure 2. The root quadrant (R) with its internal apical cell (A) produces, from its basiscopic faces, merophytes for the root proper (only two, of the three sets of merophytes are visible) and, from its acroscopic face, merophytes for the root cap. Same orientation as in Figure 2 (indicated by the arrow). Redrawn from Bower (1908).

growth (Gunning et al., 1978). Divisions most frequently involve the three basiscopic faces of the tetrahedral apical cell, $M_0$, and hence create merophytes that contribute to the root proper. Less frequent are divisions that involve the fourth, acroscopic, face from which issue the cap merophytes. At this early stage the root shows Type 1 behaviour, according to the criteria of Table 1. However, the roots are determinate in their growth, this process ceasing soon after the apical cell itself has ceased to divide; it conforms to a root of Type 8 in Table 1. Gunning (1978) has suggested that the cessation of proliferation of the apical cell is related to its increasing symplasmic isolation from neighbouring cells by a concomitant decrease in the number of plasmodesmatal interconnections across the common boundary walls. However, this decrease could reflect a general decline in metabolic activity which includes a reduction in the synthesis of tubulin and other proteins required for plasmodesmatal organization. Later, the root is shed from the frond, suggesting that the root is little more than a means by which surplus carbohydrates are 'excreted' from the plant (Barlow, 1994b).

The apical cell of *Azolla* can be taken as an indicator of a relatively advanced type of organization of the summit of the root. Not only is the apical cell (1) a rapidly dividing cell (+), but it is (2) a structural initial cell $M_0$ (i.e. it is the cell to which all other cells within the root can trace their origin) and also (3) a functional initial cell (i.e. it is the cell which, by its division, actively gives rise to all other cells); the continuity of root growth is directly linked to the activity of the apical cell. Possibly, such a regular, even stereotypic, type of root organization is associated with its ephemeral life and its aquatic habitat in which there are relatively few environmental variables that affect anything other than the rate of metabolism.

Roots of other species of leptosporangiate fern, also with an apical cell, show slightly different patterns of behaviour which have led to some confusion concerning the significance of this cell. For example, in some lateral root primordia of *Marsilea strigosa* (Marsileales - see Table 2) studied by Cremonini (1974), and earlier by D'Amato and Avanzi (1965), mitoses were rarely observed in the apical cell, i.e. it is $M_0$(-); the cell

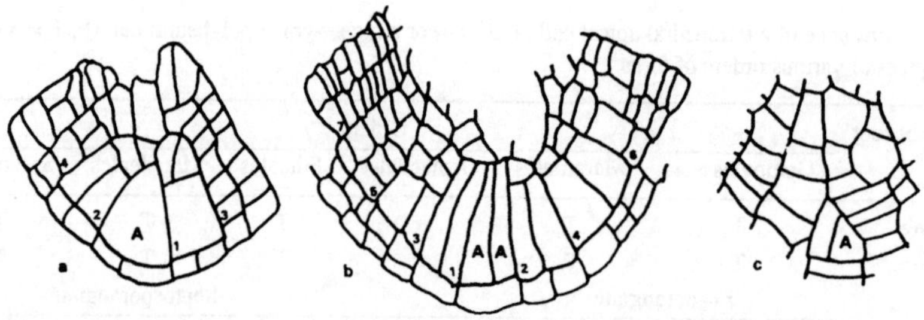


*Fig. 4.* Median longitudinal sections through root apices of *Osmunda regalis* (Osmundales). The sequence (**a-c**) corresponds to roots of increasing age. Merophytes of increasing age are indicated by the numbers in the bottom corner of each (the numbers do not denote the exact sequence of merophyte production; cap merophytes are incomplete and are not numbered. (**a**) A section from a young root with an obvious apical cell (A); periclinal and anticlinal divisions are both frequent in merophytes 3 and 4. (**b**) A section from an older root. Here there is a group of central cells; its apical cells and the youngest merophytes are tentatively identified. Whatever the precise sequence of divisions, periclinal divisions are fewer in the most recent merophytes than they are in the merophytes of the younger root shown in (a). Also, the number of faces of the apical cell(s) has increased from $F = 4$ to $F = 6$. (**c**) A section from a still older root. This has an irregular pattern of cells at the apex (not all cells are shown), but within this group a new near-tetrahedral apical cell has apparently reformed. Sections in (a) and (b) are redrawn from Freeberg and Gifford (1984); (c) is redrawn from Bower (1885, 1889).

was also found to be endopolyploid with its interphase nuclei containing between 8C and 16C units of DNA (the diploid content is 2C-4C). But such a situation was actually characteristic of older root primordia; in younger primordia the apical cell divided frequently and the nucleus was diploid. Therefore, ageing of this root correlates with a declining frequency of division of the apical cell, although this does not necessarily mean that DNA synthesis is impaired since its nucleus participates in an endoreduplication (or endomitotic) cycle; its inability to divide may be due to some type of inhibition of $p34^{cdc2}$ kinase activity (see Nagl, 1993). Here, therefore, such roots switch from Type 1 to Type 2 (see Table 1) during growth. Unfortunately, the precise coupling of endoreduplication cycles in the apical cell with the occasional re-entry of its nucleus to mitosis and its relationship with the proliferative activity of the surrounding cells has not been as carefully investigated as this system seems to warrant. On the other hand, the apical cells of rhizome-borne roots of *M. vestita* have been observed to divide frequently (similar to *Azolla*) and failed to show evidence of becoming endopolyploid (Kurth, 1981). A similar pattern of activity has also been found in *M. diffusa* by Vallade and Bugnon (1979). Roots of these two species seem to continue as Type 1. Nevertheless, it is possible to see that the apical cell of older lateral roots of *M. strigosa*, and also of the parent root axis (D'Amato and Avanzi, 1965), now has features distinct from that of the apical cell of *Azolla* which make it closer in character to a quiescent centre: (1) the apical cell divides less frequently than its neighbours (though it may replicate nuclear DNA at a relatively high rate); (2) it acts as a structural initial, $M_0$(-); (3) it may act less frequently as a functional initial, devolving this role to the three neighbouring basiscopic cells $\{3(M_1)\}$ which are hence the ones that provide for the necessary cell proliferation and continued root growth.

The division of a polyploid apical cell generates similarly polyploid daughter cells or merophytes. These two properties of the apical cell, its division and its ploidy, result in some or all of the sectors of the root becoming polyploid so that the root can ultimately possess a cytochimeric, or mixoploid, constitution. Moreover, lateral roots arising from a merophyte consisting of polyploid cells would inevitably have a polyploid apical cell. While a uniform level of polyploidy need not be deleterious to organ growth and function, a mixture of ploidy levels may be potentially disadvantageous as the cells may grow to different sizes (Reese, 1973) which can lead to an unbalanced physiological system, particularly if environmental conditions are limiting [see, e.g., the results of Hall (1972)]. However, such a situation could be mitigated by the determinate growth of any given root and by the replenishment of the root system with new roots that arise from diploid initial cells located within the extending system of rhizomes (as in *Marsilea*). But for plants which, because of their architecture, have to survive with the aid of a root system more limited in its origins, it is necessary for that system itself to maintain a stock of diploid functional initial cells which can contribute not only to a given root for the whole period of its growth, but which also ensure that any primordia for lateral roots are also diploid.

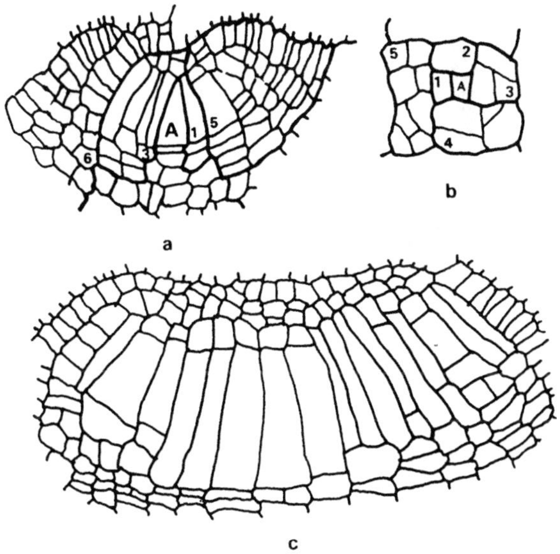


*Fig. 5.* Sections through root apices of *Angiopteris evecta* (Marattiales). (**a**) Median longitudinal section through a young, thin root. It has a tetrahedral apical cell (A), apparently producing merophytes from all its faces. Therefore, the merophytes can be numbered in the sequence in which they could have been produced. (**b**) A transverse section through an older root. The apical cell now has a total of five or six faces but is square when seen in transverse section. The sequence of merophytes is indicated. (**c**) Median longitudinal section through an old, thick root. The central group of six-sided cells is still actively dividing, but the frequency of periclinal divisions has decreased causing the cells to remain large. The cells at the proximal side of this central group contribute to the stele. Since these cells divide more rapidly than the central cells, it is possible that they serve as discrete initials for the stele histogen since they are only slowly replaced as a result of divisions which involve the basiscopic faces of the central group of cells. Redrawn from Voronin (1964).

## Fern roots with polyhedral apical cells

Although a good deal of attention has been paid to the leptosporangiate ferns which possess a tetrahedral apical cell, other members of this group with different types of apical cell, or other groups of ferns (e.g. eusporangiate ferns), have been relatively neglected in recent studies of fern root anatomy. Nevertheless, the summits of their roots show features which are undoubtedly relevant in considering the evolution of the QC. For one thing, a single apical cell is sometimes not so evident because, although there may be such a cell in the early stages of root development, with continued growth of the root its summit becomes occupied a number of cells having a prismatic form (Table 2). One explanation for this would be that the number of faces of the $M_0$ cell which donate walls to a daughter merophyte $M_1$, has increased from $F = 4$ (in the tetrahedral case) to $F > 4$. As a result, the root may not be so clearly structured in sectors when viewed in transverse section, though its merophytic organization is retained (evident when viewed in longitudinal section), particularly in young roots. A good example of this type of apical organization is seen in roots of the leptosporangiate fern, *Osmunda regalis* (Osmundales), studied by Bower ( 1885) and Freeberg and Gifford (1984) (Fig. 4). These latter two authors showed that where a prominent cluster of apical cells ($M_0$, $M_1$, ...) had developed, their nuclei maintained a diploid DNA content. The cluster of apical cells arose because periclinal divisions, which usually occur in the derivatives ($M_1$, $M_2$, ...) of the single, tetrahedral apical cell, $M_0$, failed to occur. The first few divisions in merophytes $M_1$, $M_2$, ... , were anticlinal only, thus increasing the complexity of the apical cluster. At the same time, the orientation of division in $M_0$ changed, as a result of which $F$ became $> 4$. Moreover, in the accompanying illustrations of Freeberg and Gifford (their Figs. 5 and 6) all these apical cells are highly vacuolate. This of itself does not preclude their division - *Azolla* apical cells can divide when vacuolate (Barlow et al., 1982) - but does indicate that synthesis of cytoplasm has lagged behind the expansion of the cell and so may indicate some metabolic differentiation of these apical cells from their neighbours. The lack of periclinal divisions in the youngest merophytes of the root also indicates this, and contrasts with the frequent divisions in the cap merophytes. As a result, the root cap develops its own meristem in which the rate of proliferation outpaces that in the immediately subjacent apical cells. A similar situation has been described in *Angiopteris evecta* (Marattiales) by Koch (1895) and Voronin (1964) and is illustrated in Figure 5. As the root becomes older and thicker, the number of large cells at the summit increases. Exactly how this comes about is not known. The apices of *Osmunda* and *Angiopteris* during the stages described above may initially be of Type 1, but then convert to Types 2 or 3, or even to Type 6 if $M_0$ continues to be proliferatively active and its descendents become quiescent while older merophytes remain proliferative. Eventually all these apices may proceed to Type 8.

The Osmundales and Marattiales, to which *O. regalis* and *A. evecta* respectively belong, are two of the more primitive families in the Filicopsida. Thus, it is not clear whether the geometry and behaviour of their apical cells is of a more primitive character than that described for apical cells of *Azolla* or *Marsilea*. What is certain, however, is that the roots of all species mentioned start development with a tetrahedral cell.

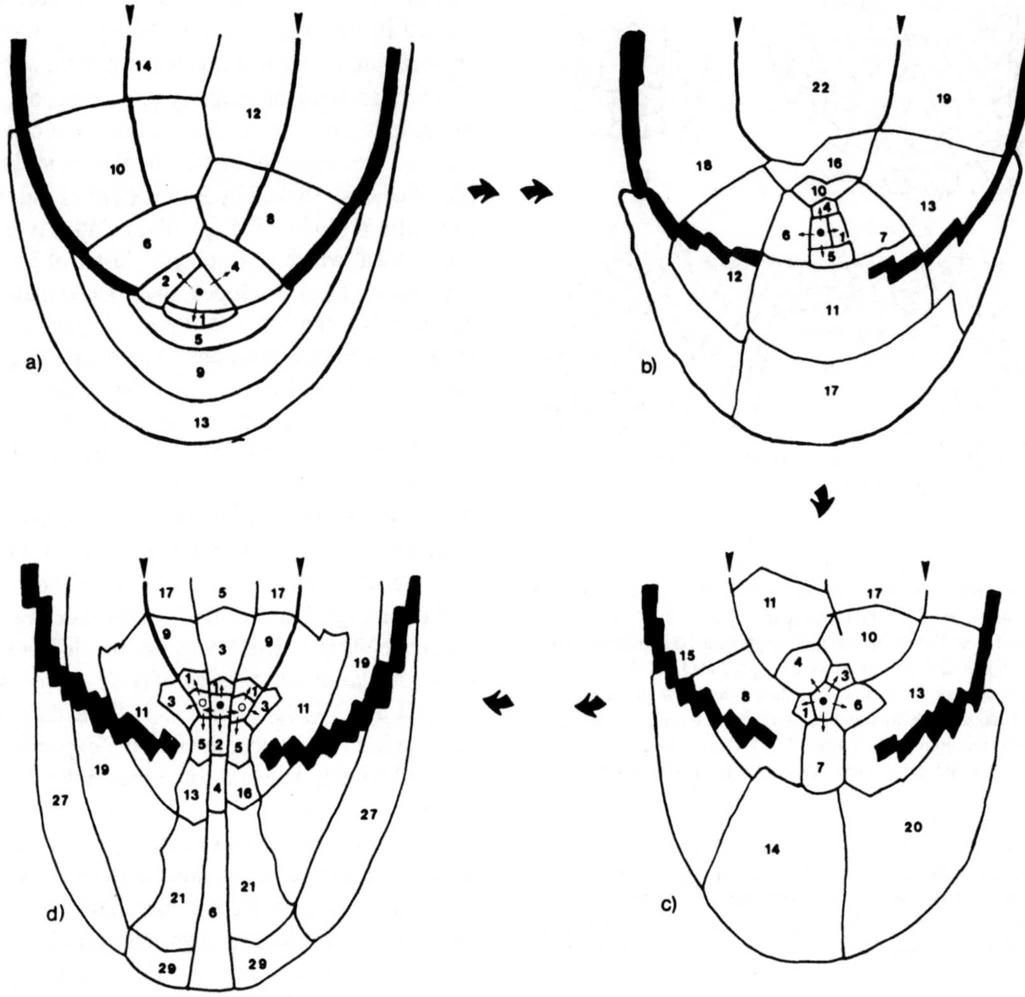


*Fig. 6.* Schematic merophytic patterns, as suggested by median longitudinal sections, in root meristems of (**a**) a leptosporangiate fern, (**b**) an eusporangiate fern, (**c**) a hypothetical intermediate form, and (**d**) a gymnosperm. Their supposed phylogenetic sequence is indicated by the curved arrows. The filled circles in each section mark the structural initials, $M_0$. In (d), open circles denote cells which could also be $M_0$ cells. In this case there might be six $M_0$ cells (○) surrounding a central $M_0$ cell (●), making a set of $\{7(M_0)\}$ cells in all. Arrows extending from the $M_0$ cell(s) indicate the planes in which daughter merophytes, $M_1$, are produced. The $M_0$ cell in (a) is shown with three sides and is tetrahedral ($F = 4$). The $M_0$ cell in (b) is shown with four sides, and represents a cuboid, i.e. $F = 6$, each face being able to contribute to $M_1$. In (c), $F = 7$; thus, the polyhedral grade of this $M_0$, is intermediate between the eusporangiate fern and the gymnosperm. The geometry of this cell is a fusion of a tetrahedron and a prism. In (d), the central $M_0$ cell has eight faces, $F = 8$, all of which can produce new merophytes, some displacing existing surrounding $M_0$ cells. For the sake of simplifying the positioning of the merophyte boundaries, it has been assumed that only two of the faces of the central $M_0$ have been involved in cell production. Each of the surrounding $M_0$ cells also has eight faces but, again for simplicity, it has been assumed that only seven of them have been involved in divisions. Merophytes are numbered according to an assumed sequence of division involving each of the productive faces of the $M_0$ cell(s). Heavy boundary lines represent the dermatogen (boundary between root and cap); this boundary transcends merophyte boundaries since its differentiation may be specified by positional information. Less heavy lines indicated with arrowheads (▼) denote the boundary between stele and cortex. Except for (a), this boundary coincides with merophyte boundaries. Interpreted and redrawn from a diagram of Voronin (1969).

Whether the retention (*Azolla*) or alteration (*Osmunda* and *Angiopteris*) of its shape and that of its immediate descendents represents an evolutionary progression is not clear, though the tendency to form a cellular cluster at the apex is closer to what occurs in the QC of more advanced taxa where the cells are smaller and where there are more structural initial cells of a higher polyhedral grade. Voronin (1969), for example, believes that the regular *Azolla*-type of apical organization, where the tetrahedral apical cell divides in four possible directions ($F = 4$), is derived from a *Marattia*-type of organization, where the apical cell divides in more directions ($F$ = 5 or 6, perhaps) which results in it being less prominent since the number of its faces is similar to that of its daughter merophytes. Nevertheless, what is clear is that the cells in the young merophytes of the root apices of *Angiopteris* and *Osmunda*, and maybe of *Marsilea*, all show the potentiality for slower proliferation than the neighbouring, older merophytes.

## Special properties of apical cells

### *Organogenetic capacity*

The query of Gifford as to whether the apical cell is some sort of geometric 'empty space' possessing no functional significance (Gifford, 1983) can be answered from work on regeneration of apices. First, Kadej (1960) excised portions of the cap from root apices of *Davalia solida* (Filicales); the tetrahedral apical cell was stimulated into frequent periclinal division. Although this resulted in the renewal of the cap merophytes, the form of the apical cell was also maintained. Sometimes a second apical cell came into existence. Second, callus cultures of *Pteris cretica* (Filicales), develop root apices which have a typical tetrahedral cell at their summit (Michaux-Ferrière, 1975, 1980). The presumption was that such cells had been produced by divisions within the callus and that their presence had then led to root organogenesis. Earlier, Bruchmann (1909) had noted tetrahedral cells in stem callus of *Selaginella* which he also presumed were carried into regenerating roots. These examples indicate that, in species which normally possess it, a tetrahedral cell at the summit of a root probably has an inherent capacity to organize new root growth, perhaps as a result of a particular orientation of successive divisions which in turn leads to an ordered set of merophytes. Third, when lateral root primordia are formed in the endodermal lineage of a leptosporangiate fern root axis, one of the first steps is the formation of a tetrahedral initial cell. This requires a particular mode of behaviour in the precursor cell since lateral root initial formation involves a pattern of anisotropic cell growth and oblique divisions that are qualitatively different from the usual type of growth and divisions found in related cells within the same merophyte. Again, the tetrahedral cellular form appears to be one which is favoured in circumstances that are conducive to root organogenesis.

### *DNA segregation*

A little understood feature of fern apical cells is their mode of DNA segregation at mitosis. Interesting experiments with young sporophytes of *Marsilea vestita* indicate that mitosis in the apical cell of the root may be associated with a partial form of conservative DNA segregation. This involves the preferential retention of a particular strand of DNA of the paternal (♂) genome within the apical cell (Kuligowski and Tourte, 1979; Tourte et al., 1980), whereas DNA strand segregation of the maternal (♀) genome of this cell, and of the DNA of both genomes in all other somatic cells, is of the usual (Taylor et al., 1957) semi-conservative type. The evidence for this was based on labelling either the ♂ or the ♀ gametes with $^3$H-thymidine, allowing fertilization to occur, and then counting silver grains over nuclei in autoradiographs of sectioned embryonic sporophytes. When DNA of the ♂ gamete was labelled, nuclear radioactivity was quickly diluted by mitoses throughout the embryo and semi-conservative DNA segregation was therefore presumed. But when the ♀ gamete DNA was labelled, the grain count diminished as expected during the first four cell cycles (producing up to 16 cells), but then nuclear labelling was maintained at a constant level (ca. 50 grains) over nuclei of apical cells beyond this 16-cell stage (cf. Fig. 2). Young sporophytes with ~5000 cells also retained some label over the apical cells, which were assumed still to be mitotically active. Moreover, unequal numbers of grains were found over sister nuclei of recent apical cell divisions. These observations concur with similar unequal grain counts made over sister nuclei in the QC of maize roots (Barlow, 1978). In the fern, the correspondence between the anatomy of the sporophyte and the pattern of autoradiographic grain distribution would support a hypothesis (Cairns, 1975) proposing that the apical cell ($M_0$) retains the same strand of paternal DNA throughout its life and that its mode of conservative segregation partially protects the cell from the effects of mutation by exporting unfaithfully replicated DNA to merophyte $M_1$ and retaining unblemished DNA in $M_0$.

## Gymnosperms and angiosperms with multiple structural initials

### *Division of labour amongst structural initials*

Voronin (1969) proposed that there was a hypothetical form of root meristem organization that linked the form characteristic of roots of eusporangiate ferns with that of gymnosperms (Fig. 6). The distinction is subtle, for it relies upon the structural initial cell of the hypothetical root being a polyhedron of intermediate grade. Thus, increasing the number of faces of $M_0$ from $F = 6$ to $F = 7$ or 8 may be an important step in increasing the diversity of apical forms. The ability to produce new merophytes from each of the increased number of faces of a single cell would require an alteration to the programme controlling the orientation and sequence of mitoses (cf. Barlow, 1991). One solution to this problem, however, would be to increase the number of structural initials (from 1 to $n$) and at the same time reduce the number of division planes of any individual initial cell. Moreover, some anticlinal faces which are shared within the set of initials, $\{n(M_0)\}$, need not be frequently used to generate daughter merophytes, $M_1$, though displacement of

cells by divisions involving these shared faces probably does take place (as shown in Fig. 6d). Infrequent, but precisely oriented, divisions in $M_0$ may underlie the presence of a QC in these roots with more complex organization. Thus, quiescence may be a means of ensuring that the necessary structural complexity of the set of initial cells is preserved.

At present, the physiological basis for the quiescence of the structural initials of spermatophytes - e.g. Types 2 or 3 of Table 1 - is not understood, but intuitively there is no need for the set $\{n(M_0)\}$ to divide, since during the embryogeny of the root they, together with a surrounding set of functional initials, have become already established. The functional initials thus perpetuate the set of cell files derived earlier from the pattern of division dictated by the structural initials. An example of this was given by Vallade et al. (1978) for the root apex of *Petunia hybrida* and is illustrated in Figure 7. Interestingly, in this apex quiescence at first affects only the $n = 4$ cells (Fig. 7a) that are structural initials [i.e. $\{4(M_0)\}$]. But later, quiescence is thought to extend to the proximal group of cells $\{4(M_1)\}$ that had formerly served as functional initials (Fig. 7b). Quiescence may here be an infectious trait, emanating from the $\{4(M_0)\}$ cells, though it could also be an inherited trait, as shown by observations on the growth of the

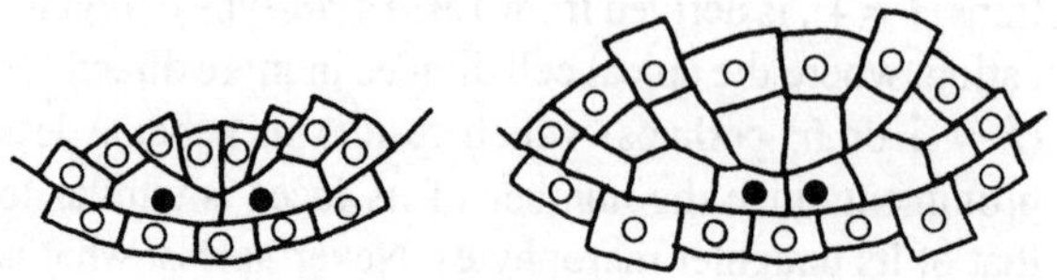


*Fig. 7.* Scheme of the quiescent centre (QC) and initial cells in roots of *Petunia hybrida*. (**a**) Longitudinal section of a young root with a QC consisting of four cells, two of which are seen here in longitudinal section. These, being structural initials, $M_0$, are marked with closed circles. The functional initials immediately overlying them ($M_1$) are marked with an open circle. (**b**) Older root with a larger QC consisting of the marked $M_0$ cells and unmarked cells of merophyte $M_1$. Distal cells in $M_2$ (open circles) are now the functional initials Whether the QC has enlarged due to growth and division of the original four $M_0$ cells, or whether quiescence has affected cells that were previously structural initials is not known. If the latter case applies, a new set of functional initials has been established; if the former case applies, the original functional initials have been displaced basipetally and are continuing to act, still as functional initials, in a new location with respect to the root apex. Adapted from Vallade et al. (1978).

QC in in-vitro-cultured roots of tomato (Barlow, 1992). The cellular set $\{4(M_0)\}$ found in *P. hybrida* seems to correspond to the minimal QC found in angiosperms; it is also found in young roots of *Arabidopsis thaliana* (Dolan et al., 1993). The small number of cells within the QCs of these two species may relate to the small size of their roots (Clowes, 1984).

Quiescence conserves metabolic resources that can be utilized elsewhere in the meristem. It also conserves the structural plan of the root. Structural initials within the QC may be called upon from time to time to re-assert that plan by their division - during germination, for example (Clowes, 1958), but especially when the viability of the root is threatened. It is well known that the QC is sensitive to many kinds of environmental perturbation, all of which have the potentiality to interrupt cell production by the meristem (Clowes, 1975). Thus, although this group of cells is quiescent, it is not inert since it is continually able to receive and respond to signals that lead to its reactivation. The regenerative activity of a multicellular QC need not involve the structural initials, $M_0$, however (see Type 3 roots in Table 1); it is possible that major reactivation of division can occur in the surrounding quiescent group $\{M_1, \ldots, M_i\}$ which has already amplified the pattern of $M_0$. Indeed, this has been observed in roots of maize regenerating their caps (Barlow, unpublished). Moreover, QCs of *Convolvulus arvensis* show a gradation of low proliferative activity with the least active cells being those which include the $\{M_0\}$ group and the rel-

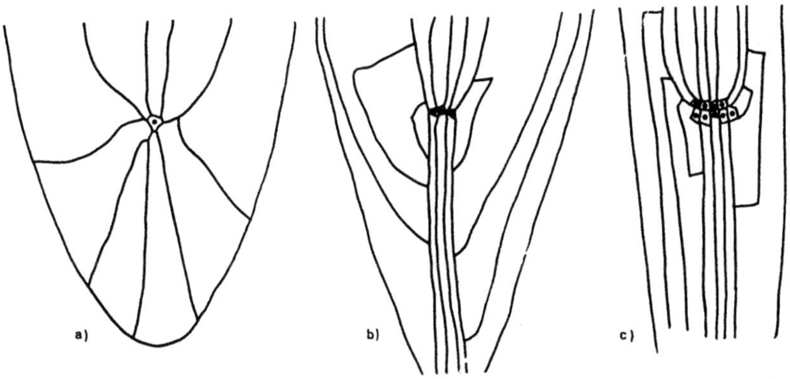


*Fig. 8.* Semi-diagramatic representations of median longitudinal sections through roots of species of different orders of gymnosperms to show a presumed evolutionary progression (**a-c**) in complexity of apical organization. Evolutionary advancement is paralled by an increasing number of groups and tiers of structural initial cells (denoted by filled circles), each group also becoming associated with different histogens. The derivatives of the initials are enclosed within the lines of the diagram. Apparently extraneous lines that do not trace back to an initial [e.g. in (a)] may have traced to a former initial cell that was subsequently displaced by the current initial. (a) *Cycas neocaledonica*. Here, apparently, there is a single structural initial; this is similar to the situation in the hypothetical type of apical organization shown in Fig. 6c. (b) *Gingko biloba*. There is a single tier of initial cells, three of which are shown. (c) *Ephedra foliata*. There are two tiers of initials, the proximal tier giving rise to stelar tissue, the distal tier giving rise to cap and cortex. Redrawn and interpreted from Voronin (1964) (a and b), and Pillai (1966) (c).

atively more active group corresponding to $\{M_1, \ldots, M_i\}$ (Phillips and Torrey, 1972), which would be in accordance with this idea of differential reactivation within the QC.

There is evidently some relationship between the $\{M_0\}$ set and histogenesis that possibly has an evolutionary context. For example, comparative anatomical studies of gymnosperm root apices by Pillai (1963, 1964, 1966) suggest that there is a simple type of organization where the cell-files comprising each of the histogens converge on a common set of histogen initials that correspond with structural initials; a more complex type of organization is supposed in cases where the histogens associate with discrete groups of structural initials (Fig. 8). The various types of organization coincide with the phylogeny of the taxon. Thus, in the Cycadales (Fig. 8a) there is a cup-shaped promeristematic region from which all tissues arise (cf. the situation in ferns described above; it is also similar to the hypothetical apex in Fig. 6c). In the Coniferales and Gingkoales (Fig. 8b), the stele and the root cap columella arise from one set of initials, a second set gives rise to the cortex and the flank of the cap. The Gnetales (*Ephedra* spp.) (Fig. 8c) are yet more advanced in having three sets of initials - say, $\{M_{0a}\}$ for the stele, $\{M_{0b}\}$ for the cortex and cap flank and $\{M_{0c}\}$ for the cap columella. The underlying distinction between these three types of apices resides in the relative rates of cell growth and division in the longitudinal and radial planes in key locations within the whole cellular complex. This in turn permits distinctive walls to be fabricated which can then serve as partitions between the histogens - that is, walls between cortex and stele, and between cortex and cap, in particular. Similar evolutionary considerations apply to angiosperm apices (Voronkina, 1975), especially in relation to the distinction between open and closed types of meristems (Fig. 9), as discussed by Guttenberg (1968) and Clowes (1981). Although division and differentiation are, at the cellular level of organization, often thought of as being rather separate processes, at the organ level (which embraces the root) they are intimately linked since the continual production of new cell division walls is part of an on-going process of patterned histogenesis. Both these aspects of root development perpetuate a course of organogenesis established earlier during embryogenesis (Barlow, 1994c).

It is possible to imagine that discrete initials for the histogens of stele and cortex could also arise in the root apex of eusporangiate ferns. Accompanying the age-related increasing irregularity of division in the apical cell (cf. Fig. 5), a proximal cell in an $M_1$, merophyte could assume the role of initial for the stele, whereas other cells within $M_1$, could become initials for the cortex. In any case, something similar occurs early in development where the root cap is concerned: an autonomous initial for the cap is created as a result of a division of the apical cell in the plane parallel to its acroscopic face (cf. Fig. 1).

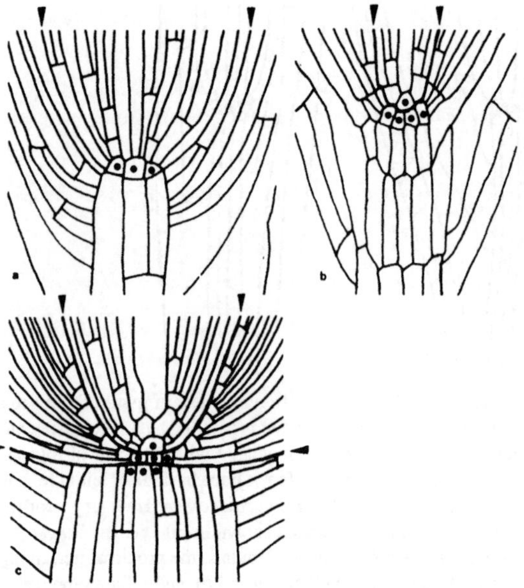

*Fig. 9.* Median longitudinal sections through angiosperm root apices showing cell lineages and varying numbers of tiers or sets of structural initial cells (filled circles). (**a**) *Grevillea robusta* with one tier of initials. (**b**) *Elodea canadensis* with two tiers. (**c**) *Cyperus gracilis* with three tiers. The tiers come into existence because of the distinct planes in which each set of structural initials grows and divides during root ontogeny. This results in certain walls becoming thickened (arrowheads) since they are not violated by a division in one tier forcing a daughter cell into an adjacent tier. The number of structural initials in each tier is not necessarily a minimal number, even bearing in mind that the drawings are two- rather than three-dimensional representations of their disposition. Estimate of the minimal number requires more detailed knowledge of the planes of division in any given tier. Diagrams redrawn from Hejnowicz (1980).

*Selective pressures and the evolution of the QC*

The evolution of the QC of angiosperms and gymnosperms may have been favoured by the increasing probability that roots would emerge from either the seed or the shoot into soil. This environment is a relatively recent one, and is both drier and mechanically more resistant than either water or the swampy silt in which the ancestors of the spermatophytes thrived. The success of the root in the soil environment could well be bound up with the evolution of the root cap. Not only does the cap confer on the root resistance to desiccation (e.g. Leiser, 1968), but it may also contain substances (phenolics, for example) which deter browsing by soil-dwelling animals. Moreover, products of the Golgi apparatus, which include the hydrophilic mucilage secreted from the outer cap cells, have been suggested as regulating the behaviour of the QC (Bennet and Breen, 1989). Mucilage production is probably an adaptation to the drier and more resistant soils in which angiosperms mostly live, although the caps of some aquatic ferns (*Regnellidium diphyllum*, Marsileales) also produce a mucilage (Eastman and Peterson, 1985). On the other hand, root caps of the evolutionarily more ancient Lycopsid, *Isoetes macrospora*, do not share this feature (Peterson et al., 1979). Abscisic acid, a hormone implicated in the drought resistance of roots and synthesized in the cap, has also been suggested as negatively regulating metabolic activities of the QC (Barlow and Pilet, 1984). The loss of distal cap cells activates the QC (Clowes, 1972), again suggesting that the cap regulates the QC. Thus, the QC may owe its presence to chemical regulators synthesised within the root cap itself (Barlow, 1994a), a tissue which in turn permits the root to adjust to environmental conditions such as drought. On the other hand, the presence of a QC enhances the adaptational stratagems of the root apex, for the QC can regenerate a new cap if required; yet its very ability to do this may also be regulated by the cap. Thus, a feedback regulatory loop is established between the cap and QC (Fig. 10). However, as discussed by Feldman (1975), it is possible that the cells surrounding the QC on its proximal face could also be responsible for its quiescent properties.

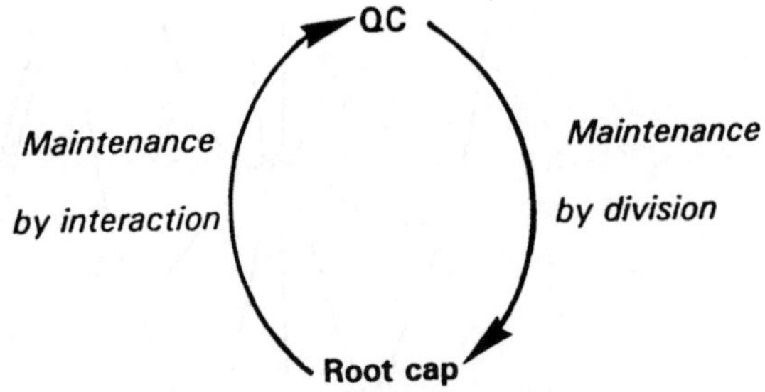


*Fig. 10.* Mutual interaction between cap and QC.

With the development of a QC in Type 3 apices, or indeed with the maintenance of an active apical cell in type 1 apices, one may ask what property of each Type of apex has been selected by evolution? In the case of the QC in a Type 3 apex, it may be its response to environmental variables, transduced by intra- and intercellular signals, that has been selected. But in both types of apices there may be another, deeper aspect of ontogenesis that is subject to selection (not necessarily evolutionary selection), involving a fundamental regulator of root growth and form. In the absence of perturbations to growth, both Type 1 and Type 3 apices consist of particular patterns of cell walls governed by corresponding patterns of cell division (Barlow, 1991). As argued elsewhere (Barlow, 1994c), cell wall patterns probably optimise (in response to evolutionary selection) the support role which walls play in resisting both internal osmotic pressures and external physical pressures from the environment, be it aqueous or solid. These wall patterns are ultimately generated by the sequences and orientations of cytokineses in the apical zone, especially in the $\{M_0, M_1, \ldots, M_i\}$ merophyte set. These, in turn, have as their origin a set of vector fields, or growth tensor, that specifies the three-dimensional pattern of growth within the paraboloidal apex of the root. Hejnowicz (1989) has shown how two different growth tensors can determine contrasting cellular kinematics so that a root can possess either a rapidly proliferating apical cell (Type 1 apex) or a slowly proliferating multicellular QC (Type 3). Later, one tensor with two different sets of parameters was found to suffice for the simulation of the two types of apex (Hejnowicz and Karczewski, 1993). Other patterns of cellular behaviour, such as those that describe open meristems, require other sets of tensor parameters. The problem then becomes one of identifying the cyto-physiological bases of the tensor parameters and of discovering how much scope there is for them to respond to evolutionary selection pressure. Since their basis is akin to a biochemical and/or biophysical topology, they may evolve according to selection of epigenetic properties based on the parameters of reaction-diffusion systems. The tensor fields may alter according to their intrinsic topological properties, and at some point become subject to a 'catastrophe' that alters the topology of the system from one metastable form to another (Thom, 1975). This may account for the quite different parameter sets of the tensor that describe Type 1 and Type 3 apices.

An additional aspect of growth tensors is that, although their parameters may be held constant during growth, the focus of the principal directions of growth (PDG) of the growth tensor may 'wander' within the organ. If the limits of this wandering can be specified, it may be possible to predict a number of developmental events such as the course of apical regeneration (Nakielski, 1992), the alterations to cellular patterning due either to mutation (Nakielski and Barlow, 1993) or to ontogeny, as exemplified by the change from closed to open apical organization (Barlow, 1987; Clowes, 1981), and by the naturally occurring bifurcation of some root apices. The latter situation is seen in roots of ferns where a secondary tetrahedral apical cell can form in proximity to the original apical cell during the normal course of growth (Peterson, 1970). Dichotomy is also a natural feature of roots of Lycopsids (Guttenberg, 1966) and gymnosperm mycorrhiza (Piché et al., 1982). Here, it is as though the focus of the PDG of the growth tensor temporarily shifts beyond the limits of the $\{M_0\}$ set and the effect of this is picked up by a cell in a nearby $M_1$ merophyte. This cell then divides with a frequency and an orientation specified by the tensor, so creating a new apical $M_0$ cell which then becomes the focus of a second system of PDGs which propagates its pattern of growth within the surrounding cellular ensemble.

Once again, it is important to distinguish between the overall form of the cellular pattern and the behaviour of the individual cells. The PDG of the growth tensor specifies the global pattern of cells, whereas the parameters of the tensor equations specify the behaviour of individual cells. With deeper knowledge of root organogenesis it will undoubtedly be possible to explore this aspect of plant biology more extensively and hence come to an understanding of the interplay between the genome, the epigenetic cellular systems (Barlow, 1991), and the environment (Barlow, 1993) in the evolution and behaviour of present-day root apices. This understanding is materially helped also by appreciating the phylogenetic aspect of root organization.

## Conclusion and summary

The evolution of root meristem types (Fig. 11) appears to have been accompanied by an increasing complexity of the geometry of the structural initials - i.e., the cells which specify the basic pattern of cell- files within the root, and termed here $M_0$ cells or merophytes. Although the meristems of ancestral forms of plants, which were the forerunners of the present-day flora, are unavailable for study, it seems probable that their $M_0$ cells could have been tetrahedral, i.e. with four faces or walls, since such cells are well known in roots of present-day ferns (e.g. the leptosporangiate ferns) and plants of more ancient origin (e.g. *Equisetum*). Moreover, all four faces of such cells are used in division to produce new merophytes. It is even possible that $M_0$ cells with three or two merophyte-generating faces could have existed; lens-shaped cells with two curved faces are present as $M_0$ cells at the meristematic apices of fern leaves, for instance (Fig. 11). Later in evolution, the grade of polyhedron for $M_0$ of the root

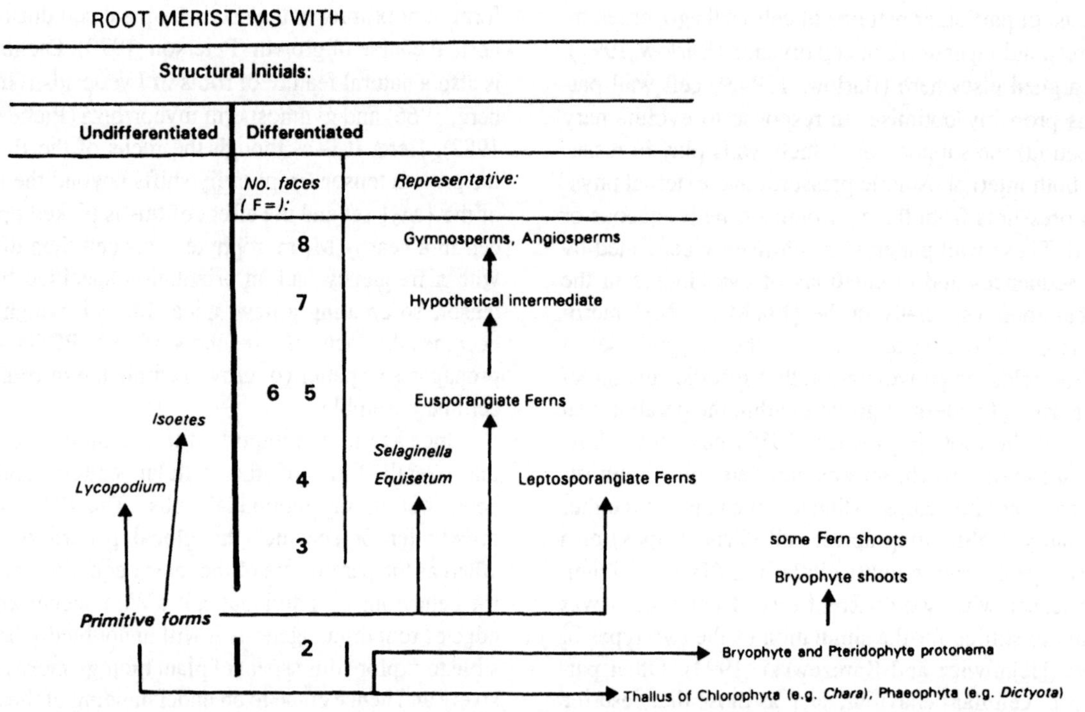


*Fig. 11.* Evolutionary scheme for root meristem development. It is assumed that at some stage early in the phylogeny of plants, a specialized set of structural initials was differentiated in the root apex. These initials are polyhedra of different complexities, denoted by the number of their faces (*F*). However, roots of some pteridophytes do not seem to have specialized initials (to the left of the diagram). The arrows do not strictly reflect the phylogenetic progression of the taxa, but rather indicate the changes in the geometry of the initial cells. Certain other groups of plants are included at the extreme right-hand side to indicate that initials with $F = 3$, and even $F = 2$, can be found, but these seem to occur in plant parts that are not roots. Possibly the root, because of its ability (or necessity) to form a cap, requires initials of greater complexity than other organs.

meristem increased so that six-walled cells were frequent (as in the Marattiales). With the evolution of the Spermatophyta, the $M_0$ group became more complex: their $M_0$ cells basically have eight faces, though up to 14 faces (or more) are possible as a result of the walls of neighbouring cells attaching to, and thus subdividing, the faces of the $M_0$ cells. At the same time, the number of such cells increased, with the consequence that not all the faces of $M_0$ need be used for generating new merophytes. Also, groups of $M_0$ cells became associated with the histogenesis of cortex, stele and root cap. The time of appearance, during evolution, of a proliferatively weakly active QC which includes the $M_0$ population and some of its immediate descendents is not known. In some fern roots it has been reported that the tetrahedral apical cell is inactive. Such a cell may be a prototype of the QC and has some, though not necessarily all, of the latter's biochemical controls (e.g. it cannot control the onset of DNA synthesis). Possibly, the QC of spermatophytes developed concomitantly with the inception of multiple $M_0$ cells and with the development of a root cap which functioned in assisting the transition of the root from an aqueous environment to a drier, soil environment.

A more theoretical approach to meristem generation, also involving a progressive increase in geometric complexity of a structural initial cell, has been taken by Lück and Lück (1993). It seems to arrive at a similar viewpoint to the one presented here, although it was directed more towards an understanding of shoot apices.

Meristems with different numbers of walls (*F*) associated with their $M_0$ cells represent different archetypes of organization. Further work is warranted to reveal whether each archetype is also supported by specific types of physiological or cytological states, especially in relation to the cytoskeletal architecture of the cells that serve as structural initials.

## References


Barlow P W 1978 The concept of the stem cell in the context of plant growth and development. *In* Stem Cells and Tissue Homeostasis. Eds. C S Potten, B I Lord and R J Cole. pp 87–113. Cambridge University Press, Cambridge.

Barlow P W 1987 The cellular organization of roots and its response to the physical environment. *In* Root Development and Function. Eds. P J Gregory, J V Lake and D A Rose. pp 1–26. Cambridge University Press, Cambridge.

Barlow P W 1991 From cell wall networks to algorithms. The simulation and cytology of cell division patterns in plants. Protoplasma 162, 69–85.

Barlow P W 1992 The meristem and quiescent centre in cultured root apices of the *gib-1* mutant of tomato (*Lycopersicon esculentum* Mill.). Ann. Bot. 69, 533–543.

Barlow P W (Ed) 1993 The Impact of the Environment on Roots and Root Systems. Pergamon Press, Oxford. 205 p.

Barlow P W 1994a The cellular and molecular biology of the quiescent centre in relation to root development. *In* Molecular-Genetic Analysis of Plant Metabolism and Development. Eds. P Puigdomènech and G Coruzzi. NATO ASI Series. Springer-Verlag, Berlin. (*In press*).

Barlow P W 1994b The origin, diversity and biology of shoot-borne roots. *In* Biology of Adventitious Root Formation. Eds.T D Davis and B E Haissig. pp 1–23. Plenum Press, New York.

Barlow P W 1994c Cell divisions in meristems and their contribution to organogenesis and plant form. *In* Shape and Form in Plants and Fungi. Ed. D S Ingram. pp. 171–195. Academic Press, London.

Barlow P W and Pilet P-E 1984 The effect of abscisic acid on cell growth, cell division and DNA synthesis in the maize root meristem. Physiol. Plant. 62, 125–132.

Barlow P W, Rost T L and Gunning B E S 1982 Nuclear and cytoplasmic changes during early stages of differentiation in roots of the water-fern, *Azolla pinnata*. Protoplasma 112, 205–216.

Bennet R J and Breen C M 1989 Towards understanding root growth responses to environmental signals: the effect of aluminium on maize. S. Afr. J. Sci. 85, 9–12.

Bierhorst D W 1954 The subterranean sporophytic axes of *Psilotum nudum*. Am. J. Bot. 41, 732–739.

Bower F O 1885 On the apex of the root in *Osmunda* and *Todea*. Quart. J. Microscop. Sci. 25, 75–103.

Bower F O 1889 The comparative examination of the meristems of ferns, as a phylogenetic study. Ann. Bot. 3, 305–392.

Bower F O 1908 The Origin of a Land Flora. A Theory Based Upon the Facts of Alternation. MacMillan, London. 727 p.

Bruchmann H 1909 Von der Wurzelträgern der *Selaginella Kraussiana* A. Br. Flora 95, 150–166.

Cairns J 1975 Mutation selection and the natural history of cancer. Nature 255, 197–200.

Chouinard L 1959 Sur l'existence d'un centre quiescent au niveau de l'apex radiculaire juvénile de *Pinus banksiana* Lamb. Fonds Recherches Forest. Univ. Laval No. 4, 27–31.

Clowes F A L 1956a Nucleic acids in root apical meristems of *Zea*. New Phytol. 55, 29–34.

Clowes F A L 1956b Localization of nucleic acid synthesis in root meristems. J. Exp. Bot. 7, 307–312.

Clowes F A L 1958 Development of quiescent centres in root meristems. New Phytol.57, 85–88.

Clowes F A L 1972 Regulation of mitosis in roots by their caps. Nature New Biol. 235, 143–144.

Clowes F A L 1975 The quiescent centre. *In* The Development and Function of Roots. Eds. J G Torrey and D T Clarkson. pp 3–19. Academic Press, London.

Clowes F A L 1981 The difference between open and closed meristems. Ann. Bot. 48, 761–767.

Clowes F A L 1984 Size and activity of quiescent centres. New Phytol. 96, 13–21.

Cremonini R 1974 Frequenza e localizzazione delle mitosi in primordi di radici laterali di *Marsilea strigosa*. Giorn. Bot. Ital. 108, 155–159.

D'Amato F and Avanzi M G 1948 Reazioni di natura auxinica ed effetti rizogeni in *Allium cepa*. Studio cito-istologico sperimentale. Nuovo Giorn. Bot. Ital. 55, 161–213.

D'Amato F and Avanzi S 1965 DNA content, DNA synthesis and mitosis in the root apical cell of *Marsilea strigosa*. Caryologia 18, 383–394.

Dolan L, Janmaat K, Willemsen V, Linstead P, Poethig S, Roberts K and Scheres B 1993 Cellular organization of the *Arabidopsis thaliana* root. Development 119, 71–84.

Eastman A and Peterson R L 1985 Root apex structure in *Regnellidium diphyllum* (Marsileaceae). Bot. Gaz. 146, 44–55.

Feldman L J 1975 Cytokinins and quiescent center activity in roots of *Zea*. *In* The Development and Function of Roots. Eds. J G Torrey and D T Clarkson. pp 55–72. Academic Press, London.

Freeberg J A and Gifford E M 1984 The root apical meristem of *Osmunda regalis*. Am. J. Bot. 71, 558–563.

Gifford E M Jr 1983 Concept of apical cells in bryophytes and pteridophytes. Ann. Rev. Plant Physiol. 34, 419–440.

Gifford E M Jr and Kurth E 1982 Quantitative studies of the root apical meristem of *Equisetum scirpoides*. Am. J. Bot. 69, 464–473.

Good C W and Taylor T N 1972 The ontogeny of Carboniferous articulates: The apex of *Sphenophyllum*. Am. J. Bot. 59, 617–626.

Gunning B E S 1978 Age-related and origin-related control of the numbers of plasmodesmata in cell walls of developing *Azolla* roots. Planta 143, 181–190.

Gunning B E S, Hughes J E and Hardham A R 1978 Formative and proliferative cell divisions, cell differentiation, and developmental changes in the meristem of *Azolla* roots. Planta 143, 121–144.

Guttenberg H von 1961 Grundzüge der Histogenese höherer Pflanzen II. Die Gymnospermen. Handbuch der Pflanzenanatomie vol. VIII. 4. Gebrüder Borntraeger, Berlin. 172 p.

Guttenberg H von 1966 Histogenese der Pteridophyten. Handbuch der Pflanzenanatomie vol. VII. 2. Gebrüder Borntraeger, Berlin. 312 p.

Guttenberg H von 1968 Der primäre Bau der Angiospermenwurzel. Handbuch der Pflanzenanatomie vol. VIII. 5. Gebrüder Borntraeger, Berlin. 472 p.

Hall O 1972 Oxygen requirement of root meristems in diploid and autotetraploid rye. Hereditas 70, 69–74.

Hejnowicz Z 1980 Anatomia i histogeneza roślin naczyniowych. Państwowe Wydawnictwo Naukowe, Warsaw. 860 p.

Hejnowicz Z 1989 Differential growth resulting in the specification of different types of cellular architecture in root meristems. Env. Exp. Bot. 29, 85–93.

Hejnowicz Z and Karczewski J 1993 Modeling of meristematic growth of root apices in a natural coordinate system. Am. J. Bot. 80, 309–315.

Kadej F 1960 Regeneration der Scheitelzelle bei Farnwurzeln. Acta Soc. Bot. Polon. 29, 363–368.

Koch L 1895 Ueber Bau und Wachstum der Wurzelspitze von *Angiopteris evecta* Hoffm. Jahrb. Wiss. Bot. 27, 369–402.

Kuligowski A and Tourte Y 1979 La determination des cellules apicales chez une ptéridophyte; rôle particulier du génome paternel. Bull. Soc. Fr. Lettres Bot. 126, 491–505.
Kurth E 1981 Mitotic activity in the root apex of the water fern *Marsilea vestita* Hook and Grev. Am. J. Bot. 68, 881–896.
Leiser A T 1968 A mucilaginous root sheath in Ericaceae. Am. J. Bot. 55, 391–398.
Lück J and Lück H B 1993 Paralllel rewriting DW-cellwork systems for plant development. *In* Mathematics Applied to Biology and Medicine (Proc. 1st European Conference on Mathematics Applied to Biology and Medicine, Grenoble, 1991). Eds. J Demongeot and V Capasso. pp 461–466. Wuerz Publishing, Winnipeg.
Mangenot G 1942 Recherches sur l'action des substances mitoclasiques I. Action de la colchicine sur les racines *d'Allium cepa*. Actual. Sci. Ind. 915, 1–120.
Michaux-Ferrière N 1975 Mise en place de cellules apicales sur un cal de *Pteris cretica* L. cultivé in vitro. Etude de l'acquisition de leurs caractères originaux. C. R. Acad. Sci., Paris, Ser. D 281, 783–786.
Michaux-Ferrière N 1980 Formation et evolution de cals chez le *Pteris cretica* L. Z. Pflanzenphysiol. 96, 387–405.
Nagl W 1993 Induction of high polyploidy in *Phaseolus* cell cultures by the protein kinase inhibitor, K-252a. Plant Cell Reps 12, 170–174.
Nakielski J 1992 Regeneration in the root apex; modelling study by means of the growth tensor. *In* Mechanics of Swelling. Ed. T K Karalis. NATO ASI Series Vol. H 64, pp 179–191. Springer-Verlag, Berlin.
Nakielski J and Barlow P W 1993 The growth tensor and specification of the quiescent centre zone in wild-type and mutant tomato roots. Abstracts 4th Internat. Symp. Structure and Function of Roots. Stara Lesna, June 1993, p. 78.
Němec B 1897 Cytologická pozorování na vegetačníh vrcholech rostlin. Sitzungsber. K. Böhm. Ges. Wiss. 1897, 1–26.
Nitayangkura S, Gifford E M Jr and Rost T L 1980 Mitotic activity in the root apical meristem of *Azolla filiculoides* Lam., with special reference to the apical cell. Am. J. Bot. 67, 1484–1492.
Peterson R L 1970 Bud development at the root apex of *Ophioglossum petiolatum*. Phytomorphol. 20, 183–190.
Peterson R L and Vermeer J 1980 Root apex structure in *Ephedra monosperma* and *Ephedra chilensis* (Ephedraceae). Am. J. Bot. 67, 815–823.
Peterson R L, Scott M G and Kott L 1979 Root cap structure in *Isoetes macrospora* Dur. Ann. Bot. 44, 739–744.
Phillips H L and Torrey J G 1972 Duration of cell cycles in cultured roots of *Convolvulus*. Am. J. Bot. 59, 183–188.
Piché Y, Fortin J A, Peterson R L and Posluszny U 1982 Ontogeny of dichotomizing apices in mycorrhizal short roots of *Pinus strobus*. Can. J. Bot. 60, 1523–1528.
Pillai A 1963 Root apical organization in gymnosperms. - Some cycads and *Gingko biloba*. Proc. Indian Acad. Sci. B 57, 211–222.
Pillai A 1964 Root apical organization in gymnosperms. Some conifers. Bull. Torrey Bot. Club 91, 1–13.
Pillai A 1966 Root apical organization in gymnosperms. Root apex of *Ephedra foliata*, with a suggestion on the possible evolutionary trend of root apical structures in gymnosperms. Planta 70, 26–33.
Rabideau G S and Mericle L W 1953 The distribution of $C^{14}$ in the root and shoot apices of young corn plants. Plant Physiol. 28 329–333.
Reese G 1973 Über einen bemerkenswerten Fall von Polysomatie in Wurzelspitzen. Cytologia (Tokyo) 38, 593–601.
Sabelli P, Burgess S R, Carbajosa J V, Parker J S, Halford N G, Shewry P R and Barlow P W 1993 Molecular characterization of cell populations in the maize root apex. *In* Molecular and Cell Biology of the Plant Cell Cycle. Eds. J C Ormrod and D Francis. pp 97-109. Kluwer Academic Publishers, Dordrecht.
Sachs J 1878 Ueber die Anordnung der Zellen in jüngsten Pflanzenteilen. Arb. Bot. Inst. Würzburg 2, 46–104.
Sporne K R 1975 The Morphology of Pteridophytes. 4th ed. Hutchinson, London. 207 p.
Taylor J H, Woods P S and Hughes W L 1957 The organization and duplication of chromosomes as revealed by autoradiographic studies using tritium-labeled thymidine. Proc. Natl Acad. Sci. USA 43, 122–138.
Thom R 1975 Structural Stability and Morphogenesis: An Outline of a General Theory of Models. (English Translation by D H Fowler). Benjamin, Reading, Mass. 348 p.
Torrey J G 1972 On the initiation of organization in the root apex. *In* The Dynamics of Meristem Cell Populations. Eds. M W Miller and C C Kuehnert. pp 1–10. Plenum Press, New York.
Tourte Y, Kuligowski-Andres J and Barbier-Ramond C 1980 Comportement différentiel des chromatines paternelles et maternelles au cours de l'embryogenèse d'une fougère: le *Marsilea*. Europ. J. Cell Biol. 21, 28–36.
Vallade J and Bugnon F 1979 Le rôle de l'apicale dans la croissance de la racine du *Marsilea diffusa*. Rev. Cytol. Biol. Vég. Bot. 2, 293–308.
Vallade J, Alabouvette J and Bugnon F 1978 Apports de l'ontogenèse à l'interprétation structural et fonctionelle du méristème racinaire du *Petunia hybrida*. Rev. Cytol. Biol. Vég. Bot. 1, 23–47.
Voronin N S 1964 Evolution of meristem structure in the roots of plants. Sci. Trans. Fac. Nat. Sci., Kaluga Pedagog. Inst. 13, 3–179 (*In Russian*).
Voronin N S 1969 Apical meristems of the roots in gymnosperms and the principles of their graphical interpretation. Bot. Zhurn. 54, 67–76 (*In Russian*).
Voronkina N V 1975 Histogenesis in root apices of angiospermous plants and possible ways of its evolution. Botan. Zhurn. 60, 170–187 (*In Russian*).
Wilcox H 1962 Growth studies of the root of incense cedar, *Libocedrus decurrens*. The origin and development of primary tissue. Am. J. Bot. 49, 221–236.

*F. Baluška et al. (eds.), Structure and Function of Roots*, 19–26, 1995.


# The origin of the basal part of the embryo root in *Brassica napus* L. and its role in growth of early seedlings

T. Tykarska and M. Kuras
*Department of Plant Morphogenesis, The University of Warsaw, Krakowskie Przedmiescie 26/28, Warszawa, Poland*

*Key words:* axis zonation, *Brassica napus*, germination, root embryogenesis

## Abstract

Observations of growth of zones of a germinating embryo have been carried out: hypocotyl and two parts of the root, the apical one covered with the cap, and basal one lying between the cap and the hypocotyl. Special attention was paid to the basal part of the root.

It has been found that breaking the seed coat is the effect of quick elongation of the hypocotyl. The elongation process then moves to the basal part of the root. When the first root hairs appear, quick growth of the apical part of the embryo root is initiated. The elongation of hypocotyl is restored when elongation of the basal part of the root is extinguished. The length of the basal part of the root steadies on the third day after imbibition of seeds and equals approximately 2mm. Lateral roots appear on the fourth day. The first one usually grows at the border between basal and apical part of the embryo root, others appear further in the apical direction; a little later 1 – 2 roots can appear in the basal part.

At the same time embryological examinations of the cortex have shown that both parts of the root originated from a single mother cell. It divides in the globular embryo into two cells, one of which initiates the apical part of the root, and the other one the basal part of the root and the basal part of the hypocotyl.

## Introduction

One of crucial characteristics in the family of Cruciferae is an unusually regular course of its embryogenesis. This allows identification of borders of the first divisions even in a mature embryo and using them as topographic borders in localization of developmental changes of parts of mature and germinating embryo. For example a cross border of the second division of the cortex mother cell in a globular embryo has been identified in our previous works (Kuras, 1978; Tykarska, 1979) as an anatomical border between hypocotyl and root in a mature embryo. Kuras (1984) has found that meristematic activation of a germinating embryo starts at this level and spreads acropetally in a wave-like fashion. The hypocotyl lying above this border grows because of cell elongation.

However, the border between hypocotyl and root found this way has proven to be unprecise. Root-specific staining with Procion Blue MX-R (Dubrovsky and Tykarska, 1994) carried out after Negbi (1968) and Obroucheva et al. (1984) shows a slightly different course. It goes about 16 dermatogen cells higher than stated before. This newly stated root boundary is strictly correlated with the range of zone where the first root hairs appear and the range of vascular tissue typical for roots. Thus the formerly found root length is supplemented with a 16 cell sector - basal part of the root being an intermediate zone between the meristematic part of the root and the hypocotyl. Its embryological origin and its physiological role in the process of germination and growth of the seedling are the subject of the present work.

## Material and methods

The material of our investigations were developing and germinating embryos of winter rape, *Brassica napus* var. Gorczanski.

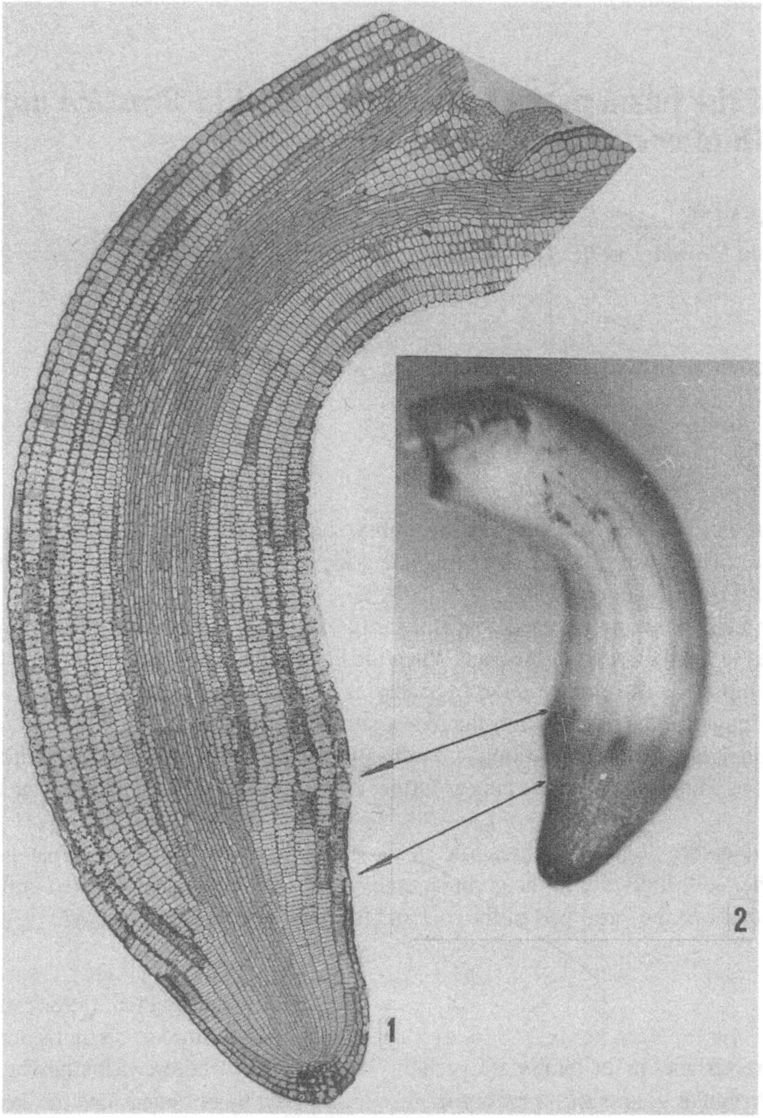


*Fig. 1-2.* The axis of a maturating rape embryo. Fig. 1. longitudinal section, stained by PAS; Fig. 2, an axis stained in toto in 1% Procion Blue MX-R. The arrows show the border between the root and the hypocotyl and the cap base level.

*Investigations of living embryos*

The seeds were put in Petri dishes on a damp blotting paper, in darkness, in temperature of 22° C+/-2. After 24 h the dishes were taken out of the darkness. Seeds with 1mm long germ were stained in toto with 1% water solution of Procion Blue MX-R dye, then rinsed. Cactus needles were carefully pinned into the majority of embryo axes in order to mark the root cap base level. Then the seeds were put on glass panes covered with damp blotting paper in a row in 2 – 3 mm distances. The panes were put slantingly into boxes with water reaching ca 2 cm below the germs. There were 2 – 5 panes in each experiment; 10 germs with a needle and 2 without on every pane. The boxes were covered with glass panes at the top.

Measurements of growth of subsequent zones of axis i.e. two parts of the stained root (from the tip to the root cap base and from the cap base to the border of staining of the axis) and the unstained hypocotyl (to the cotyledon base) were carried out using a microscope (enlargement 3.2×8) with an ocular-micrometer every 3 hours during the first two days. Further observations were carried out with stereomicroscope because

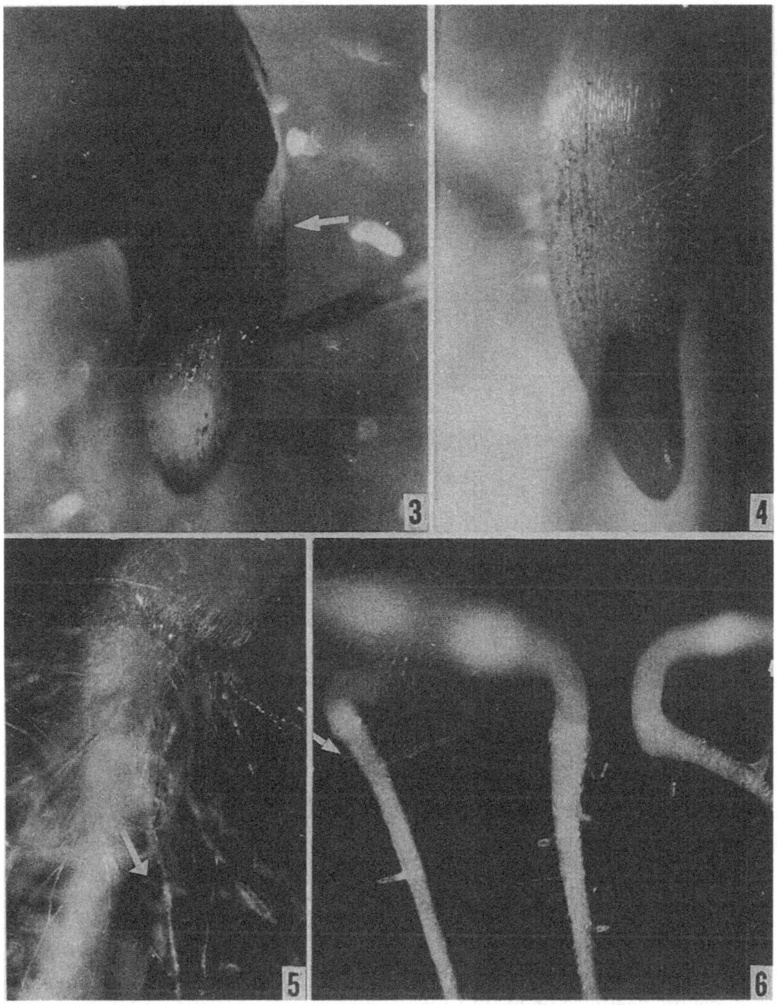


*Fig. 3-6.* Development of a rape seedling. Fig. 3. A seed with a 1mm long germ, stained in toto in 1% water solution of MX-R. The arrow shows the border of axial staining. The cactus needle marks the cap border; Fig. 4. 36 hours from imbibition. The root hairs show the hypocotyl border. There is a cactus needle at the cap border; Fig. 5. A three days old rape seedling. There is a very distinct border between the radical zone with root hairs and the hypocotyl. The arrows shows a needle pinned into the embryo axis at the border between the basal and apical part of the root; Fig. 6. The 4th day of the seedlings development. Lateral roots visible. The arrow shows a small lateral root growing at the basal part of the root.

of increasing length of young seedlings. The experiment was repeated 5 times. Similar measurements were carried out on stained embryo axes taken out from dry seeds, and after 3, 6, 12, 18, 21 hours of imbibition. Photographs of seedlings were made using an Amplival microscope (Carl Zeiss Jena) and light from light pipes (produced by Bratek Co, Poland).

*Examination of fixed material*

Preparates of longitudinal sections of embryos in subsequent phases of embryogenesis, made with a microtome according to the procedure described in earlier works (Tykarska, 1976, 1979) were used for analysis of formation of the root zone. The results of the analysis have been shown in pictures of 12 embryos chosen from the above-mentioned works, with development of the basal part of the root marked in them.

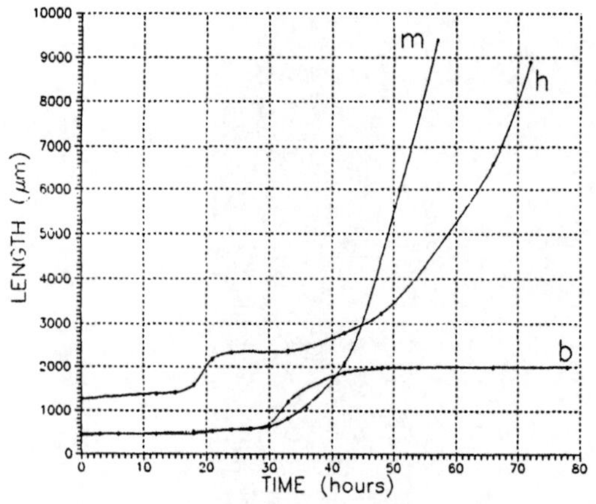


*Fig. 7.* Lengthening of the axial zones in germinating rape embryo. m - meristematic part, b - basal radical part, h - hypocotyl.

## Results

*Observations of living embryos*

Finding three borders in the embryo axis: cap border (marked with a needle), the hypocotyl - root border (the level between stained and non-stained zone of the embryo axis) (Fig. 1–4) and the upper level of the axis at the cotyledon base, allowed to examine the role of all these zones in growth of the embryo axis. Though staining of the root gradually fades during development, root hairs growing precisely from the staining level and then further towards the root tip distinctly show the border between the hypocotyl and the root (Figs. 4, 5). A cactus needle pinned into the axis shows the position of the level of the cap primarily formed in the embryo, although root hairs have already appeared in a large area of quickly lengthening apical zone of the root, i.e. below the needle (Fig. 5).

The measurements show (Fig. 7) that the axis lengthens mainly inside the hypocotyl until breaking the seed coat. Its elongation intensifies during breaking the seed coat and emerging the germ (between 18 – 21h from the imbibition). Then it ceases for about 10 hours. Between the 24 – 27 h (staining, pinning the needles, putting on the glass panes) elongation of the whole embryo axis is hardly visible. From the 30th hour the basal part of the root (adjacent to hypocotyl) undergoes quick elongation. This part, being almost as long as the part covered with the cap in an embryo, now lengthens intensively 12 hours (between 28 and 40h from imbibition). The first root hairs appear about the 36th hour and elongation of the basal zone gets quickly extinguished. During quick growth of the basal part an intensive constant growth of the apical part of the root commences (Fig. 7). The elongational growth of the hypocotyl is restored during termination of growth of the basal part of the root (Fig. 7).

On the third day after imbibition of seeds growth of the basal part of the root is actually finished. The average length of this part is 2mm. On the fourth day, when the root is more than 50mm long, the first lateral roots appear. The first one usually grows at the level of the primarily formed cap (the needle). Other ones grow lower; some of them (1 – 2) may also appear above the first one a little later (Fig. 6). The hypocotyl gets green and starts growing. It straightens vertically, which makes clearer its difference from the neighboring zone (Fig. 2).

*Origin of the staining zone of the root above the cap (on fixed material)*

Before analyzing the origin of the basal root zone the origin of the most important segment borders going across the rape embryo should be reminded and discussed (Tykarska, 1976, 1979) and some new information should be added.

*Origin of the most important segmentational borders in rape embryo axis*

The basic cross borders are formed in proembryo:

- the first one is formed during the division of the zygote (**0** - Fig. 8). In a mature embryo it separates cells originated from the suspensor (i.e. **"iec"** layer containing all initial cells of the root meristem and columella) from the proper embryo originated in development of the apical cell (Figs. 8-19).
- the second one is formed in division of quadrants to octants (**0'** - Fig. 9). In a mature embryo it lies at the base of the cotyledons and it is the upper part of the hypocotyl.

The next group of boundaries are those going through the cortex and the dermatogen. They are formed during development of the proper embryo:

- The first cross border in a forming embryo axis appears in a young globular embryo as a result of division of mother cells of cortex - **mc** (the **hrm** cell is formed at that time - see Figs. 10, 11) and dermatogen of the lower part of the embryo (**I** - Figs. 11, 12). Similarly to the **0'** level, it is not very distinct in longitudinal sections of the embryo, though both borders can be quite precisely identified after a careful analysis of cell families.

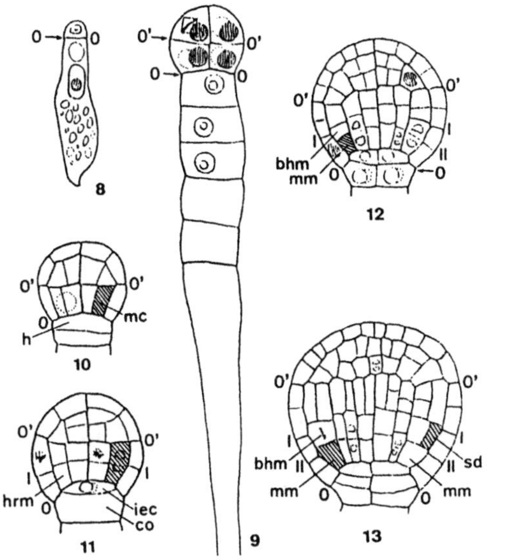


*Fig. 8-13.* Formation of cross borders (0, 0, I', II) in the axis of a developing rape embryo. Fig. 8. A two cell embryo; Fig. 9. The octant stage; Fig. 10. A proembryo with a distinct cortex mother cell (mc) and hypophysis (h); Fig. 11. The early globular stage: the 1st cross division cm initiating the radical part of the embryo. co - the initial cell of columella, iec - which will initiate the layer joining all initial cells of the root meristem, hrm - mother cell of the apical and basal part of the root and basal part of the hypocotyl; Fig. 12. The globular stage: the second cross section mc - initiation of mother cell of the meristematic part of the root cortex (mm) and bhm - mother cell of the basal part of the root and basal part of the hypocotyl; Fig. 13. The late globular stage. Initiation of the subdermatogen mother cell (sd) above mm.

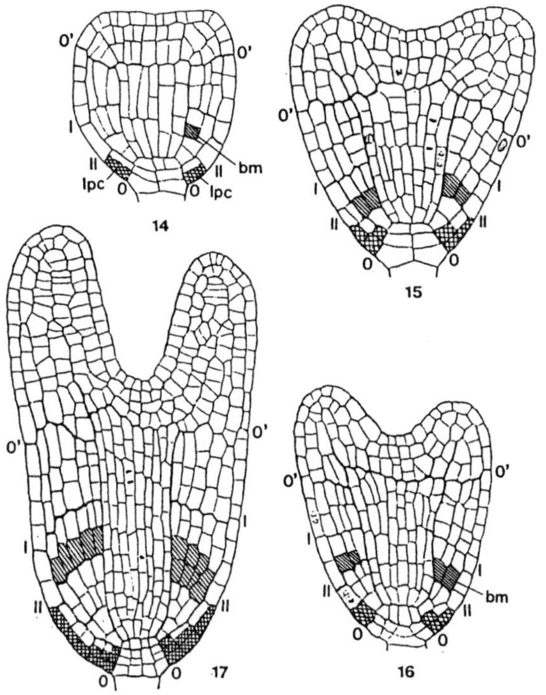


*Fig. 14-17.* Development of the basal part of the root. Fig. 14. The transient stage: initiation of the basal part of the embryo root (bm). Fig. 15, 16. A young heart-shaped embryo: the cortex of the basal part of the root consists of 1 layer of cells (striped area). Lyi.g above it basal part of the hypocotyl (also consisting of 1 layer) has 1 cell more; Fig. 17. Both basal cells, root and hypocotyl one, produce two layers of cortex cells in a late heart stage (the right side of the embryo).

- The second cross border is formed in a late globular embryo. It is initiated in the cortex during division of the lower cell - **hrm** formed from the cortex mother cell - **mc** (Figs.10-12). As a result of this division mother cell of the apical (meristematic) part of the root originates - **mm** (Fig. 12) and the mother cell common for the basal part of the root and the lower = basal part of the hypocotyl - **bhm** (Figs. 12, 13). This is the second generation of the cortex mother cell. The division wall is formed at the level of horizontal wall of dermatogen (dividing also the second generation of cells of their mother cell). Division walls formed at that time will be the border of the upper range of the meristematic part of the root (II).

The border of the basal part of an embryo root can be easily found in a rape embryo axis. In a mature embryo a change of number of cortex cell columns can be seen: there are 5 below it and 6 above it. Besides, all the apical part is distinguished in the axis because of its extremely regular development: the **mm** cell divides across, forming an initial cell of cortex and sharing this direction of division with all subsequent initial cells. Their sister cells will turn into root meristem segments (Kuras, 1978; Tykarska, 1979).

The most distinct point indicating the position of the meristematic border of the embryo root is the cap base. The cap, covering all the meristem, is initiated in two steps due to its double origin: its central part is differentiated at the globular stage during the hypophysis division (Figs. 10-11), and the lateral parts at the transition stage as a result of a longitudinal division of dermatogen cells lying next to the hypophysis (Fig. 14). Usually the whole second generation of the mother cell, lying at the bottom of the axis, between the **0** and **II** border (Fig. 14) undergoes this division differentiating sides of the cap (Tykarska, 1979). For this reason the cap base, distinctly seen at the outer surface of the embryo (Fig. 2), morphologically indicates position of the border of the meristematic part of the root. In order to simplify things we treat the cap base as the border of the meristematic part of all the root (Kuras, 1984).

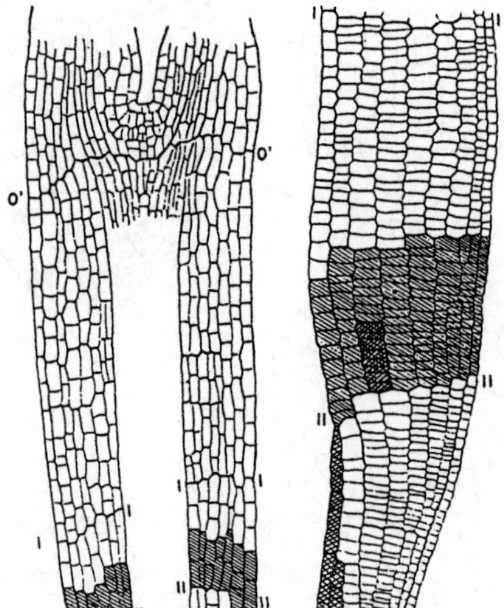

18 19

*Fig. 18-19.* Development of the basal part of an embryo root. Fig. 18. The torpedo-shaped embryo: the cortex of the basal radical part (striped area) consists of two layers of cells. They appear as columns two cells wide; Fig. 19. A fragment of an embryo in the beginning of maturation: the cortex of the basal radical zone consists of 16-celled columns. There are less cells in some columns, e.g. the cell marked with a dot will not divide again. This is a myrosine cell. Families of 8 cells are visible.

*Origin of the basal zone of the root*

The upper border of the differentiating staining used in our investigations lies in the place where (moving from the root-tip to the tip of the shoot) the embryo axis stops widening (Figs. 2, 3). Comparing the stained zone of the root with the anatomical picture of a maturating embryo we can state that the border of staining lies approximately at half of the number of cells lying in cortex cell rows between the **I** and **II** border (Fig. 19). In the lower part, belonging to the root, there are usually 16 or less cells at the inner side of the axis curving (there can be more at the outer side), and in the upper part a little more. The number of these cells in the dermatogen is more or less 16, but it is usually smaller at the concave side of the axis.

In the analyzed zone in 16-cortex cell bands there are usually bundles of 8 cells (Fig. 19) originating from single mother cells. The position of the cells indicates that the mother cell has divided three times giving 2, 4 and finally 8 cells. Sometimes this number is smaller.

This analysis shows that the part of the root above the cap is approximately half of the area between the **I** and **II** border. The second "half" belongs to the hypocotyl, which does not stain. The analysis presented in the chapter "Observations of living embryos" indicates that the mother cell (**bhm**), from which both halves have originated, has been the sister cell of the one from which the meristematic part of the cortex has originated (Figs. 11–13). Both of them were formed at the late globular stage after division of their common mother cell **hrm** (Figs. 11, 12).

Unlike the **mm**, its sister cell's first division is vertical and occurs before the change of the embryo's symmetry. As a result the outer cell is the mother cell of the subdermatogen (**sd** - Fig. 13).

Later the subdermatogen mother cell will divide horizontally. Its sister cell can divide in both directions (Fig. 14). If its division is vertical, the second one will be horizontal. The mitosis go centripetally: from dermatogen to endoderm (Fig. 15).

The horizontal division initiates two cells, the lower one of which will turn into the basal part of the root (Figs. 14,15 - **bm**) and the upper one - the basal part of the hypocotyl. The horizontal wall formed at that time will be a characteristic border of staining and also the border between the radical and hypocotyl part of the embryo root. Its initiation takes place at the transient stage, i.e. during the change of the embryo's axial symmetry (Fig. 14).

In a late heart-shaped embryo, cortex of the basal part of the root consists of 1 or even 2 horizontal layers. There are 4 or 5 cells adjacent to one another in each layer (Fig. 17). Its hypocotyl part is usually 1 cell wider than its root part. The cortex grows thicker due to anticlinal divisions of endoderm. The outer cells of the cortex, lying closer to dermatogen, are formed earlier than the inner ones, so they maturate sooner and start dividing transversally. This leads to forming of two layers of cells both in the radical and hypocotyl "half" (Fig. 17, the right side of the embryo).

At the torpedo stage all cells of the hypocotyl and root part have already undergone cross division (Fig. 18). The endoderm often cuts off periclinally one more cortex cell, initiating the last (6th) column of cortex cells in this section of the axis. However, it occurs more often in the wider hypocotyl "half".

There are three more cross divisions during elongation of the embryo axis, so the radical part above the cap in a maturating embryo has the height of 16 cells, lying one above another in columns. The cortex of a mature embryo consists of 6 columns, as it has been stated in the beginning of this chapter. Number of cells can be smaller in some columns, e.g. a myrosine cell (Tykarska, 1987), which will not undergo the last division, is marked in Figure 19.

## Discussion

Identification of the root-hypocotyl border is difficult in many embryos and it is determined with various probability, mainly according to morphological differentiation of the axis. In rape, this border is assumed to be in the place of sudden change of the axis width, caused by an increase of the number of the cortex cell columns, and thus by the range of the upper border of the 1st layer of the cap. This is the place of initiation of all cytological and cytochemical symptoms of activation, moving apically within the root during germination (Kuras, 1984).

This definition of the border between a hypocotyl and a root has been changed due to the use of differentiating staining of embryo axis with MX-R dye (Dubrovsky and Tykarska, 1994). We have observed that besides the root zone, also the area considered by Kuras (1978, 1984) as basal hypocotyl region has been stained. In addition to typically radical staining properties this zone has root-like vascular tissue (Dubrovsky and Tykarska, 1994). Root hairs have also been found there. Additional proof for radical character of this area is the fact that the both parts of the root origin from a common mother cell, which we have shown in the present work. On the other hand the basal part of the root is genealogically closer related to the lower part of the hypocotyl, because they develop from the same cell.

This means that a rape root in fully mature seeds is a homogeneous structure of uniform vascular tissue and probably consist of homogeneous molecular elements, resulting in similar staining of the cortex due to reaction to dyes (Dubrovsky and Tykarska, 1994). On the other hand it is differentiated into apical part of typically meristematic character (Kuras, 1984) and basal part, the main function of which is growth by elongation, as our research presented in this work has shown; and that is typical for a growing root. Occurence of mitosis in a germinating embryo has not been investigated in our work. They probably occur close to the apical part, but this is a common phenomenon at a border between zones (Rost et al., 1988). The elongation initiated in the hypocotyl moves apically through the basal part of the root (like in isolated *Capsella bursa pastoris* seedlings, Rijven, 1952), according to the thesis of wave-like spreading of activation in germinating embryos (Kuras, 1984). It leads to differentiation into zones, typical for roots (Ivanov, 1973; Schiefelbein and Benfey, 1991).

The investigations indicate that axial direction of the embryo development and simultaneous differentiation leads primarily to the isolation of wholly meristematic root (mitosis occur in both parts of the root, apical and basal one, throughout the embryo development), and then to their physiological differentiation into meristematic and elongational zone. The physiological differentiation probably takes place during seed maturation after extinguishing divisions in the embryo. Similar differentiation of the embryo root has been found in *Echinocactus platyacanthus* (Lux et al., 1993). The process of differentiation goes still further in some plants, leading to formation of lateral root zone in the axis of a mature embryo (Dubrovsky, 1987), or to formation of a secondary seminal root primordia zone (Luxova, 1986).

The presented observations of the rape embryo axis development indicate that the basal part of the root originated as the primary elongational zone maintains its separate character in the axis. This part, placed between two physiologically different zones, i.e. normally developing root and the hypocotyl, acts as a connection, maintaining its transient character due to its origin.

This easy to identify in subsequent stages of germination and growth of seedling zone becomes especially important and deserves further and more precise cytological and anatomical and biochemical investigations.

## Acknowledgements

We would like to thank Dr Joseph Dubrovsky for initiating the subject and its scientific discussion.

## References

Dubrovsky J G 1987 Latent embryo root system of the cucumber. Bot. Zh. 72, 171–178.

Dubrovsky J G and Tykarska T 1994 Visualization of the radicle within the axis of developing and germinating *Brassica napus* L. embryos. Environ. Exp. Bot. (*In press*).

Ivanov V B 1973 Growth and reproduction of cells in roots. *In* Physiology of Roots. Ed. N V Obroucheva. pp 1–40. UNIITI Publishers, Moscow.

Kuras M 1978 Activation of embryo during rape (*Brassica napus* L.) seed germination. I. Structure of embryo and organization of root apical meristem. Acta Soc. Bot. Pol. 47, 65–82.

Kuras M 1984 Activation of rape (*Brassica napus* L. ) embryo during seed germination IV. Germinating embryo. The first zones of mitoses, starch and DNA synthesis and their expansion pattern. Acta Soc. Bot. Pol. 55, 539-563.

Lux A, Mikus M, Hanackova Z and Kristin J 1994 Radicle of *Echinacactus platyacanthus* (Cactaceae). Plant and Soil 167.

Luxova M 1986 The seminal root primordia in barley and the participation of their non-meristematic cells in root construction. Biol. Plant. (Praha) 28, 161–167.

Negbi M 1965 A differential staining for hypocotyl and radicle. Experientia 21, 738–739.

Obroucheva I W Antinova O W Kovalev A G 1984 Sposoby rozgraniczenij kornia i ghipokotylya v zarodishe prorasta yushchego semeni dvudolnych. Bot. 69, 1205–1209.

Rijven A H G C 1952 In vitro studies on the embryo of *Capsella bursa pastoris*. Acta Bot. Neerl. 1, 157–200.

Rost T L, Jones T J and Falk R H 1988 Distribution and relationship of cell division and maturation events in *Pisum sativum* (Fabaceae) seedling roots. Am. J. Bot. 75, 1571–1583.

Schiefelbein J W and Benfey P N 1991 The development of plant roots: New approaches to underground problems. Plant Cell 3, 1147–1154.

Tykarska T 1976 Rape embryogenesis. I. The proembryo development. Acta Soc. Bot. Pol. 45, 3–16,

Tykarska T 1979 Rape embryogenesis. II. Development of embryo proper. Acta Soc. Bot. Pol. 48, 391–421.

Tykarska T 1987 Rape embryogenesis. 6. Formation of the protein bodies. Acta Soc. Bot. Pol. 56, 585–597.

*F. Baluška et al. (eds.), Structure and Function of Roots,* 27–31, 1995.


# Radicle of *Echinocactus platyacanthus* (Cactaceae)

Alexander Lux, Miloš Mikuš, Zora Hanáčková and Jozef Krištín
*Department of Plant Physiology, Comenius University, Mlynská dolina B2, 842 15 Bratislava, Slovakia*

*Key words:* Cactaceae, embryo, radicle, root cap, seed, X-ray microanalysis


## Abstract

Radicle of mature *Echinocactus platyacanthus* embryo is approximately 320 $\mu$m long and represents less then 1/7 of the embryonal axis length. The radicle-hypocotyl boundary can be distinguished according to the striking difference in the size and shape of cells in protoderm and procambium, as well as discontinuity and different number of the cell files in the ground meristem. The root cap is small, consists of 4 layers of cells covering the apex of the radicle. The upper limit of the root cap is approximately 100 $\mu$m closer towards the radicle tip than the radicle-hypocotyl boundary. Ultrastructure of radicle cells showed numerous lipid bodies as is typical for other oily seeds. Protein bodies of variable structure were also present together with other cell structures. Striking differences in protein body structure were found when protoderm and ground meristem were compared. Several small globoid crystals were present in each protein body of the protoderm, while protein bodies in the radicle ground meristem mostly contained one large globoid crystal. X-ray microanalysis revealed presence of P, K and Mg in all analyzed globoid crystals. Fe, Ca and Zn were detected in some of them.


*Abbreviations:* EDX microanalysis – energy-dispersive X-ray microanalysis, GC(s) – globoid crystals, ICP spectroscopy – ion-coupled plasma spectroscopy, LM – light microscopy, PB(s) – protein bodies, SEM – scanning electron microscopy, TEM – transmission electron microscopy.

## Introduction

*Echinocactus platyacanthus* is a barrel cactus known under popular name visnaga. This relatively abundant plant grows in the central and northeastern part of Mexico in semi-arid vegetation (Trujillo, 1984). Succulent stems of *Echinocactus* are used as a source of water and for the preparation of candies. Plants that flower in the spring and autumn, produce fruits in summer and winter respectively. Each fruit contains numerous black seeds approximately 1.3 mm long and 0.9 mm wide.

In genera with highly advanced stem succulence in the family Cactaceae, including the genus *Echinocactus*, the embryo has a well developed hypocotyl but rudimentary cotyledons and a small radicle (Buxbaum, 1955). Mature cactus seeds lack endosperm, but perisperm is present in some primitive species, including those of genus *Echinocactus* (Buxbaum, 1955). Except for an ultrastructural study of *Astrophytum asterias* (Lux, 1990) there are no published structural studies of mature cactus embryo. Reserve substances in this species were located in numerous lipid bodies (LBs) and protein bodies (PBs). Starch grains were very rare.

Protein bodies in seeds vary in size and structure, and considerable variability was observed between species, different tissues and occasionally even within a single cell (e.g. Lott, 1980). Many protein bodies contain dense inclusions called globoid crystals (GCs), which are composed of phytin (a mixture of myo-inositol hexaphosphate salts). The elemental composition of these structures, determined by X-ray microanalysis, showed species- and tissue-specificity for the presence and quantity of individual elements (e.g. Chen and Lott, 1992; Lott et al., 1984; Maldonado and Lott, 1991; Mikuš and Lux, 1990; Mikuš et al., 1992). Globoid crystals contained P, K and Mg, occasionally also Ca, Mn, Fe, Ba, Zn, S, Cl and some other elements (Chen and Lott, 1992; Lott and Buttrose, 1978; Maldonado and Lott, 1991).

The aim of the present study was to characterize the radicle of *Echinocactus platyacanthus* structurally and to compare elemental composition of globoid crystals in protein bodies present in the protoderm and ground meristem of the radicle with other parts of the embryo by use of X-ray microanalysis.

## Materials and methods

Mature dry embryos were isolated from seeds of *Echinocactus platyacanthus* (Link and Otto) collected in vicinity of the village Rayones, Nuevo León, Mexico.

For light microscopy (LM) and X-ray microanalysis the embryos were fixed in 5% glutaraldehyde in 60% ethanol (low water fixation procedure) at 20°C for 5 hours, dehydrated in ethanol and propyleneoxide, and embedded in Durcupan ACM (Fluka). Semi-thin 0.5–1$\mu$m sections were stained with toluidine blue or toluidine blue and basic fuchsin (Lux, 1981) and were examined by LM. For X-ray microanalysis blue sections (approximately 200–230 nm thick) were picked up from redistilled water immediately after sectioning and placed on formvar-carbon-coated copper grids. Our preparation procedure for X-ray microanalysis differs from that published by Lott et al. (1984) since partial leach out of materials must be considered (Skilnyk and Lott, 1992). Unstained sections were analyzed without poststaining in a JEOL 2000 FX electron microscope operating at 80 kV in STEM mode. The elemental composition of globoid crystals was determined with EDX analysis system LINK AN 10/85 S using identical parameters as described before (Mikuš et al., 1992). Minimum 50 spectra acquired from 5 different embryos were used for each value.

For TEM, sections of *Echinocactus* embryos made by a razor blade were fixed in 5% glutaraldehyde in 0.1*M* phosphate buffer pH 7.2 for 5 hours at 20°C, washed in the same buffer and postfixed in 2% osmium tetroxide 2 hours at 20°C. After dehydration in acetone or ethanol and propyleneoxide the samples were embedded in Durcupan ACM (Fluka). Thin sections were stained in uranyl acetate and lead citrate and then viewed in a Tesla 500 electron microscope.

## Results

### *Histology*

The radicle of *Echinocactus platyacanthus* is approximately 320$\mu$m long and represents less then 1/7 of the embryo axis length (about 2100$\mu$m long). The boundary between the hypocotyl and radicle is about 100$\mu$m above the upper limit of the root cap. The boundary can be distinguished easily according to the different size and shape of protodermal cells, which are almost twice as broad in the radial direction in the radicle compared to similar protodermal cells in the hypocotyl (Fig. 1a). Recognition of this boundary in the ground meristem and procambium is more difficult. Discontinuity of cell files in the ground meristem, which can indicate the radicle-hypocotyl boundary was found to occur closer toward the radicle apex. In this level only 4 cortical cell layers are present at each side of the radicle. The level of cell file discontinuity and striking differences in cell length in the procambium (shorter cells in the radicle and longer in hypocotyl) are approximately identical with the radicle-hypocotyl boundary indicated in the protoderm.

The root cap consists of 4 layers of living cells covering the tip of the radicle. Root cap cells are extended on the periphery of the apex approximately 200$\mu$m from the apex of the root cap along the longitudinal axis of the radicle (Fig. 1a).

It can be concluded that radicle of *Echinocactus platyacanthus* mature embryo is relatively very small, simply developed, and the root cap is also small.

### *Cell structure*

Cells of protoderm and ground meristem were observed in TEM. Cells of both analyzed tissues are filled with numerous lipid bodies (LBs, Fig. 1b). The second most frequent structures are protein bodies (PBs), globular or oval in shape (Fig. 1b). The presence and size of globoid crystals (GCs) inside of the protein bodies can be considered as the most apparent difference between protodermal and ground meristem cells. Several small GCs (0.2–0.8$\mu$m in diameter) were present in each PB of protodermal cells, whereas ground meristem cells contained PBs with less numerous, mostly single, large GC (0.25–2.5$\mu$m in diameter) (Fig. 1c, d). Structure of the matrix of PBs showed considerable variability. Some of them were filled by the clumps of transparent matrix with numerous vesicle-like structures, other contained dense homogeneous matrix. In some cases the matrix was located peripherally and in the central part of the PB was separated by one or several light areas. Remnants of GCs were sometimes present inside of these areas. Since GCs were not well preserved after glutaraldehyde-osmium tetroxide fixation, low-water-content fixation

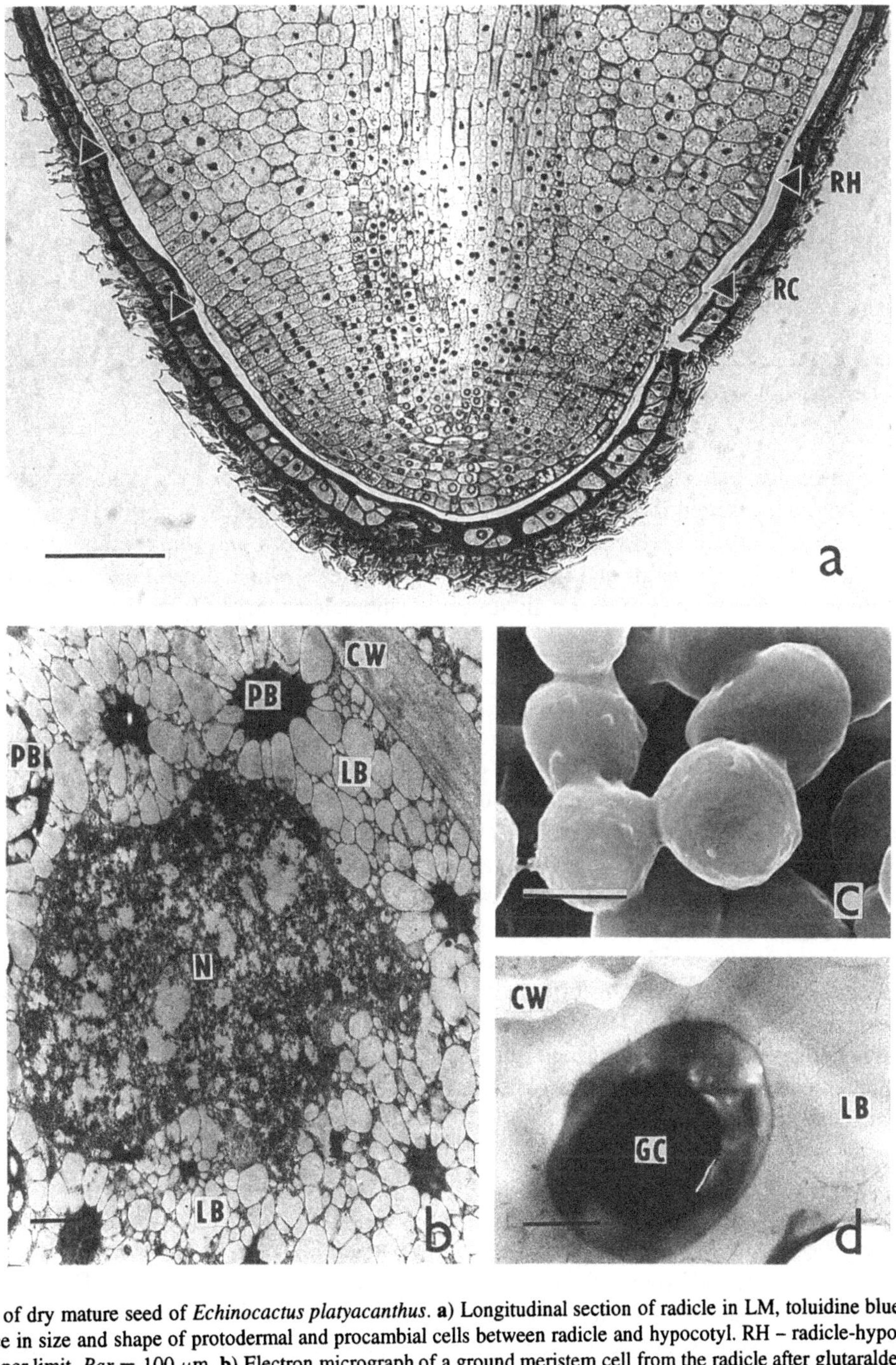


*Fig. 1.* Radicle of dry mature seed of *Echinocactus platyacanthus*. **a)** Longitudinal section of radicle in LM, toluidine blue staining. Note striking difference in size and shape of protodermal and procambial cells between radicle and hypocotyl. RH – radicle-hypocotyl boundary, RC – root cap upper limit. *Bar* = 100 $\mu$m. **b)** Electron micrograph of a ground meristem cell from the radicle after glutaraldehyde – osmium tetroxide fixation showing nucleus (N), part of a cell wall (CW), numerous lipid bodies (LB) and heterogeneous protein bodies (PB). *Bar* = 5$\mu$m. **c)** and **d)** Protein bodies after low-water-content fixation. Bars = 2.5 $\mu$m. c) Protein bodies in protoderm with several small globoid crystals. d) Protein body in ground meristem cell with single large globoid crystal (GC). Lipid bodies (LB) and a part of cell wall (CW) are also shown.

*Table 1.* Mean (±SD) peak-to-background ratios for globoid crystals from protoderm and ground meristem of *Echinocactus platyacanthus* embryos

| Tissue type | n | Peak-to-background ratios* | | | | | |
|---|---|---|---|---|---|---|---|
| | | P | Mg | K | Ca | Fe | Zn |
| **Protoderm** | | | | | | | |
| Radicle | 58 | 3.13 ± 1.19 a | 1.85 ± 0.66 a | 2.56 ± 1.59 ac | 0.27 ± 0.27 a | 0.26 ± 0.11 a | 0.45 ± 0.29 a |
| Hypocotyl | 65 | 3.48 ± 1.36 a | 1.75 ± 0.75 a | 2.22 ± 1.38 a | 0.14 ± 0.10 a | 0.29 ± 0.20 ab | 0.35 ± 0.20 a |
| Cotyledon | 50 | 3.07 ± 1.51 a | 1.23 ± 0.52 b | 1.59 ± 1.20 b | 0.14 ± 0.12 ab | 0.33 ± 0.21 ab | 0.31 ± 0.16 a |
| **Ground meristem** | | | | | | | |
| Radicle | 89 | 3.43 ± 1.36 a | 2.19 ± 0.82 c | 2.86 ± 1.54 cd | 0.06 ± 0.03 b | 0.26 ± 0.12 a | 0.33 ± 0.19 a |
| Hypocotyl | 63 | 4.44 ± 1.30 b | 2.47 ± 0.75 d | 3.26 ± 1.38 d | 0.20 ± 0.33 ab | 0.28 ± 0.19 ab | 0.41 ± 0.36 a |
| Cotyledon | 61 | 4.72 ± 0.66 b | 2.75 ± 0.72 e | 3.83 ± 0.69 e | 0.00 ± 0.00 | 0.35 ± 0.24 b | 0.33 ± 0.17 a |

*For each element, values followed by same letters are not significantly different at $p = 0.05$ or less.

was therefore used for better preservation of GCs which on the contrary resulted in very poor general appearance of the cells structures (Fig. 1c, d).

Centrally located nuclei, plastids and small mitochondria were observed in the cells. Other cells structures were not found.

*X-ray microanalysis of globoid crystals*

Data of EDX analyses of GCs from protodermal and ground meristem cells of *Echinocactus platyacanthus* radicle are shown in Table 1. All analyzed GCs contained P, K and Mg, most of them contained also traces of Fe. Considerable amounts of Zn and Ca were found in GCs of protoderm cells of the radicle. Differences among tissue types were calculated from peak-to-background values (Table 1) showing differences in some elements among radicle, hypocotyls and cotyledons. Gradient of increasing values along the embryonal axis from radicle to cotyledons was found in ground meristem for P, Mg, K and Fe (Table 1).

## Discussion

Boundary between radicle and hypocotyl was assumed after Kuras (1978) Tykarska (1979) and Tykarska and Kuras (1994). These authors studied in detail rape (*Brassica napus*) embryogenesis and found criteria facilitating location of this boundary in the mature embryo. Some of these criteria that also could be applied for *Echinocactus platycanthus* embryos are discontinuity of the ground meristem cell files and different number of ground meristem cell layers in the radicle and hypocotyl. However, tapering of the embryo axis and the neighborhood of the first root cap cells, which usually reached the radicle-hypocotyl boundary in rape (Tykarska, 1979), were not useful for determining this boundary in Echinocactus embryos. In this cactus species the embryo axis is conical up to the boundary of the perisperm junction (not documented here) and the root cap boundary is about 100$\mu$m closer to the root apex than the hypocotyl-radicle boundary. Cactus embryos thus have a more poorly developed root cap than rape embryos. On the other hand we found a new criteria for determination of the radicle – hypocotyl boundary in *Echinocactus*, namely a striking difference in the size and shape of protodermal and procambial cells. These results showed that features locating the radicle-hypocotyl boundary are species specific.

Ultrastructure of *Echinocactus* radicle cells is similar to other cactus species observed by us (Lux, 1990). Considerable variability in PBs structure described for the cells of *Echinocactus* was observed also in other species from different families (e.g. in *Lycopersicon esculentum*, Spitzer and Lott, 1980; *Datura stramonium*, Maldonado and Lott, 1991; *Capsicum annuum*; Chen and Lott, 1992). This PBs variability is caused by differences in the structure of the matrix and the presence of different number of GCs varying in their size. Protodermal cells containing PBs with a simpler structure (lacking GCs) were found also in maize and barley (Lux et al., 1992; Mikuš et al., 1992). Differences in PBs of different tissues in rape embryos, related to developmental changes during embryogenesis were observed by Tykarska (1987), leading to tissue-specific pattern of protein distribution (Diers-Venting and Venting, 1982).

The elemental composition of GCs also tend to be tissue-specific. Our preparation procedure, mainly sectioning using a knife with a water filled boat certainly caused leakage of some materials from GCs as it was shown also by Skilnyk and Lott (1992). However, in similarly prepared samples from maize embryos there was similar ratio of cations when compared with ICP spectroscopy (Mikuš et al., 1994). Moreover, a comparison of elemental composition of GCs from dry- and wet-cut sections showed no complete loss of any element. The procedure used here thus enabled us to show presence of elements and compare their relative amounts between different organs and tissues of *Echinocactus* embryos. The EDX analyses of *Echinocactus* embryos showed gradient of increasing values along the embryonal axis from radicle to cotyledons in ground meristem for the majority of the elements present. The organ and tissue-specific distribution of elements in GCs can reflect some degree of early cell differentiation, which becomes more obvious upon seed germination.

## References

Buxbaum F 1955 Morphology of Cacti. Section III. Fruits and Seeds. pp 174–223. Abbey Garden Press, Pasadena.

Chen P and Lott J N A 1992 Studies of *Capsicum annuum* seeds: structure, storage reserves and mineral nutrients. Can. J. Bot. 70, 518–529.

Dierks-Ventling C and Ventling D 1982 Tissue-specific immunofluorescent localization of zein and globulin in *Zea mays* L. seeds. FEBS Lett. 144, 167–172.

Kuras M 1978 Activation of embryo during rape (*Brassica napus* L.) seed germination. I. Structure of embryo and organization of root apical meristem. Acta Soc. Bot. Pol. 47, 65–82.

Lott J N A 1980 Protein bodies. *In* The Biochemistry of Plants, Vol 1. Eds. P K Stumpf and E E Conn. pp 589–625. Academic Press, Inc, New York, London, Toronto, Sydney, San Francisco.

Lott J N A and Buttrose M S 1978 Thin sectioning, freeze fracturing, energy dispersive X-ray analysis, and chemical analysis in the study of inclusions in seed protein bodies: almond, Brazil nut, and quandong. Can. J. Bot. 56, 2050–2061.

Lott J N A, Goodchild D J and Craig S 1984 Studies of mineral reserves in pea (*Pisum sativum*) cotyledons using low-water-content procedures. Can. J. Bot. 62, 459–469.

Lux A 1981 A rapid method for staining semi-thin sections of plant material. Biológia (Bratislava) 36, 753–757 (*In Slovak*)

Lux A 1990 Ultrastructure of seeds in the family *Cactaceae*. Mature embryo of *Astrophytum asterias* (Zucc.) Lem. Acta F.R.N. Univ. Comen – Physiol. Plant. 26, 59–64.

Lux A, Bobák M and Mikus M 1992 Protein bodies from radicle of maize and barley: their structure and elemental composition of globoid crystals. *In* Root Ecology and its Practical Application, 3.ISRR Symp. Wien, Univ. Bodenkultur, 1991. Eds. L Kutschera, E Hübl, E Lichtenegger, H Persson and M Sobotik. pp 9–12. Verein für Wurzelforschung, Klagenfurt.

Maldonado S and Lott J N A 1991 Protein bodies in *Datura stramonium* seeds: Structure and mineral nutrient composition. Can. J. Bot. 69, 2545–2554.

Mikuš M, Lux A, Crans D C, Shin P K and Krištín J 1994 Comparison of phytate in radicle, plumula, scutellum and endosperm of *Zea mays* and *Zea diploperennis* seeds. Plant and Soil (*In press*).

Mikuš M, Bobák M and Lux A 1992 Structure of protein bodies and elemental composition of phytin from dry germ of maize (*Zea mays* L.). Bot. Acta 105, 26–33.

Mikuš M and Lux A 1990 Heterogeneity of protein bodies in dry and germinating seeds of barley (*Hordeum distichum* L.). Acta F.R.N. Univ. Comen. – Physiol. Plant. 25, 51–60.

Skilnyk H R and Lott J N A 1992 Mineral analysis of storage reserves of *Cucurbita maxima* and *Cucurbita andreana* pollen. Can. J. Bot. 70, 491–495.

Spitzer E and Lott J N A 1980 Thin-sectioning, freeze-fracture, and energy dispersive X-ray analysis studies of the protein bodies of tomato seeds. Can. J. Bot. 58, 699–711.

Trujillo S 1984 Distribución geográfica y ecológica de *Echinocactus platyacanthus*. Un ejemplo de distribución disyunta. Cact. Suc. Mex. 29, 75–80.

Tykarska T 1979 Rape embryogenesis. II. Development of embryo proper. Acta Soc. Bot. Pol. 48, 391–421.

Tykarska T 1987 Rape embryogenesis. VI. Formation of protein bodies. Acta Soc. Bot. Pol. 56, 585–597.

Tykarska T and Kuras M 1994 The origin of the basal part of the embryo root in *Brassica napus* L. and its role in growth of early seedlings. Structure and Function of Roots. Eds. F Baluška, O Gašparíková and M Čiamporavá. Kluwer Academic Publishers, Dordrecht.

*F. Baluška et al. (eds.), Structure and Function of Roots*, 33–39, 1995.


# The relation between cell size, chromosome length and the orientation of chromosomes in dividing root cortex cells

J.L. Oud and N. Nanninga
*BioCentrum Amsterdam, Institute for Molecular Cytology, University of Amsterdam, Plantage Muidergracht 14, NL-1018 TV Amsterdam, The Netherlands*

*Key words:* cell size, chromosome length, chromosome orientation, *Crepis capillaris* root cortex, *Vicia faba*

## Abstract

Cells in the root meristem are organised in longitudinal files. Repeated transverse cell divisions in these files are the prime cause of root growth. Because of the orientation of the cell divisions, we expected to find mitoses with an spindle axis parallel to the file axis. However, we observed in the root cortex of *Vicia faba* large number of oblique chromosome orientations. From metaphase to telophase there was a dramatic increase of the rotation of the spindle axis. Measurements of both the size of the cortex cells and the chromosome configurations indicated that most cells were too small for an orientation of the spindle parallel to the file axis. Space limitation force the spindle into an oblique position. Despite this spindle axis rotation, most daughter cells remained within the original cell file. Only in extremely flat cells did the position of the daughter nuclei forced the cell to set a plane of division parallel to the file axis, which result in side-by-side orientation of the daughter cells. Telophase spindle axis rotations are also observed in *Crepis capillaris* and *Petunia hybrida*.. These species have respectively medium and small sized chromosomes compared to *Vicia*. Since space limitation, which causes the rotation, depends both on cell and chromosome size, the frequency and extent of the phenomenon in former two species is comparatively low.

## Introduction

The number and orientation of cell divisions, as well as cell elongation and differentiation, determines the tissue organisation of a root. Since a root is a tubular structure, three principal orientations of cell divisions can be distinguished: periclinal, radial and transverse (Fig. 1). In the first two classes the axis of the mitotic spindle is oriented perpendicular to the root axis (respectively parallel to the radius and the periphery of the root). Cells which divide transverse show a spindle axis parallel to the root axis. The stem cells for root development are located in the so-called quiescenter. Descendants of the infrequently dividing quiescent center cells ('initial'cells; Barlow, 1978) give rise to the root cap tissue and the various cell layers (for review see Clowes, 1976; Barlow, 1984). According to Luxová (1975) the formation of the layers (epidermis, cortex, stele etc.) is the result of periclinal divisions, whereas anticlinal divisions increase the number of cells in an individual layer, and transverse divisions (followed by cell elongation) are responsible for the longitudinal growth of the root. All three types of cell division can be found in the formative part of the root meristem (Fig. 1). Cell divisions in the proliferative part are almost exclusively transverse. As a result, the cells in the latter part are organised in longitudinal files, parallel to the root axis.

Looking at dividing cells in the proliferative part of the meristem in *Vicia faba* roots, we observed an unexpected large number of oblique anaphases and telophases, apparently caused by a rotation of the spindle axis (Oud and Nanninga, 1992). Oblique chromosome orientations are also observed in the root meristem of *Crepis capillaris* and *Petunia hybrida*. The analysis of three-dimensional confocal microscopy images of root cortex cells, showed that the degree of spindle rotation is related to both the size of the dividing cells and the length of the chromosome arms. The spindle is forced to rotate if the cell height is insufficient for a full separation of the two sets of chromatids during anaphase and telophase. Oblique telophase orien-

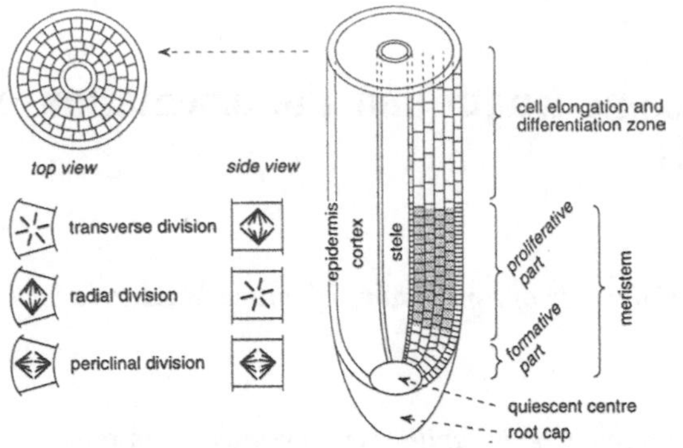


*Fig. 1.* Schematic representation of a root tip with an indication of the three principal orientations of a cell division (drawn as anaphases). The root cap and the different cell layers originate from cells in the quiescent center. The present study is restricted to dividing cells in the proliferative part of the meristematic region of the cortex (grey cells in the side view).

tations influence the position of the nuclei in daughter cells; they are shifted away from the center towards opposite directions. In spite of the spindle axis rotation and the displacement of the nuclei, the relative position of the daughter cells is nearly always unaffected, i.e. on top of each other, contributing to a further longitudinal growth of the cell file.

## Material and methods

*Vicia faba, Crepis capillaris* and *Petunia hybrida* plants were grown under standard conditions in flowerpots in a greenhouse. Lateral roots of 3 to 4 weeks old plants were fixed in ethanol/glacial acid (3:1, v/v) for 30 min. *Vicia* and *Petunia* material was embedded in a 2:1 (v/v) mixture of polyethyleneglycol (PEG) 1500 and PEG 4000. Further details about the embedding technique are given by Oud and Nanninga (1992). 20 to 25 $\mu$m thick sections of the PEG embedded roots were stained in a 50 $\mu$g mL$^{-1}$ McIlvaine (pH 7.0) buffer (+ 5 m*M* $MgCl_2$) solution for 15 min. at room temp., and mounted in glycerine to which 2% DABCO antifade (Sigma D 2522) was added. The *Crepis* roots were macerated and stained according to the standard Feulgen procedure (Oud et al., 1989). The preparations were analysed by confocal microscopy (for technical details see Brakenhoff et al., 1990 and references therein). Depending on the image, voxel sizes of 200 to 400 nm lateral, and 350 to 700 nm axial were used. The 3-D images consisted of 24 to 36 optical sections. For the analysis of the confocal images a HP 425T workstation and Scilimage software (Kate et al., 1990) were used. Solid model like images, as depicted in Figures 2 and 6, were made with the help of a ray-tracing technique, combined with an artificial shadowing to enhance the impression of depth (so-called FSP algorithm; van der Voort, 1989). The IMAP3D program (Houtsmuller et al., 1992; Montijn et al., 1994) was used to visualise simultaneously the XY, XZ and YZ aspect of 3-D images, to measure cell sizes, and to determine the orientation of file and spindle axes. From repeated measurements we estimated an accuracy of approximately $\pm$ 0.5 $\mu$m for length measurements and $\pm$ 2° for angle measurements.

## Results

### *Chromosome orientation in* Vicia faba *cortex cells*

The analyses are based on three-dimensional images (i.e. series of optical sections), obtained by confocal microscopy, of 20–25 $\mu$m thick sections, stained with the DNA specific fluorochrome mithramycin. An example of a computer processed 3-D image is shown in Figure 2. The 3-D images enabled the analysis of 4 samples of 50 cells each, of respectively metaphase, early anaphase, late anaphase and telophase. The discrimination between early and late anaphase was based on the degree of separation of the two sets of chromatids. A cell was classified as late anaphase as soon as all chromatids were fully separated.

To compare the chromosome orientation in relation to the file axis, we constructed on the computer screen a line through each chromosome complement in such a way that it represented the most likely orientation of the mitotic spindle axis (as indicated in Fig. 2). A comparable procedure was used to define the axis of the cell file, whereafter the angle between the spindle axis and the file axis was calculated. There proved to be no preference for the direction of the rotation; on the average there were as many clockwise as counter-clockwise rotations; also in the case of several oblique orientations in the same file of cells. To facilitate the comparison of the subsequent stages of mitosis, we divided all measurements in three classes, comprising of angles of respectively $\leqslant$ 5°, 5° to 20° and $\geqslant$ 20°. The result of the comparison is visualised in Figure 3. These arbitrarily chosen classification gives a good impression of the percentage of cells with a chromosome parallel to the file axis (as expected in a transversally dividing cell), and the cells with moder-

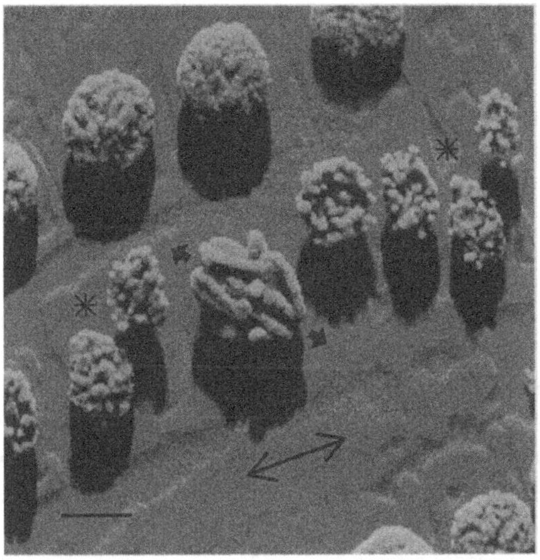

*Fig. 2.* 3-D image of a part of two files of cells in the *Vicia faba* root cortex. The images are based on stacks of optical sections obtained with confocal microscopy and processed with the SFP algorithm. ↔: orientation of the file axes; →: orientation of the anaphase spindle axis; *: pairs of nuclei which are shifted diagonally out of the center of the cell, towards opposite directions. Note the rotation of the spindle axis with respect to the file axis. The *bar* represents 10 $\mu$m.

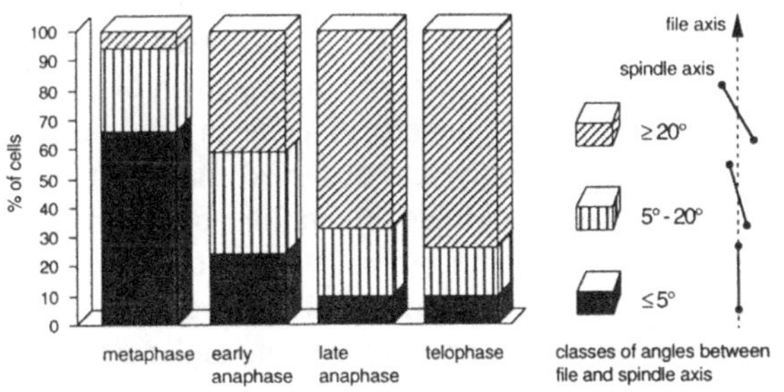


*Fig. 3.* The deviation of the spindle axis with respect to the axis of the cell file. The analysis is based on 4 samples of 50 each, respectively in metaphase, early and late anaphase and telophase. The data are grouped in three classes, which correspond with non-rotated, moderately rotated and severely rotated spindle axes. Note the strong increase in spindle rotation from metaphase to telophase.

ate or severe deviations from the expected orientation. The early anaphase in Figure 2 is characterised by an approximately 40° rotation of the spindle axis. As can be seen in Figure 3, there is a dramatic shift in the percentage of cells with a 20° or more rotation from metaphase to telophase. Two-thirds of all metaphases are aligned (almost) parallel to the file axis, whereas three-quarters of all telophases showed a significant rotation.

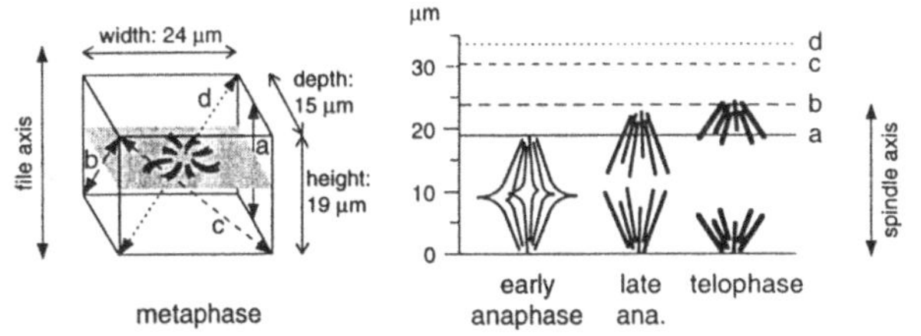


*Fig. 4.* The mean size of a dividing *Vicia faba* root cortex cell, compared with the mean size of early and late anaphase and telophase chromosome orientations. For the sake of clarity, the equator plane is indicated in the schematic representation of the metaphase. In an average sized cell there is only enough space for a full separation of the anaphase and/or telophase chromatids, unless the spindle axis rotates towards an orientation parallel to one of the diagonals (broken lines b, c and d).

Since there proved to be a relation between to the orientation of the chromosomes and the size of the dividing cortex cells (see discussion), we also measured cell sizes. Root cortex cells have a rather rectangular shape; therefore the sizes are measured as cell height, width and depth, corresponding to the edge parallel to, respectively, the longitudinal axis, radius and periphery of the root. As shown in Figure 4, the mean height of dividing cells in the proliferative region of the *Vicia faba* root cortex proved to be 19 $\mu$m (with a range from 8 $\mu$m to 32 $\mu$m. The mean cell width was 24 $\mu$m (20–27 $\mu$m) and the mean cell depth was 15 $\mu$m (12 – 19 $\mu$m. In addition, we also measured the size of the chromosome configurations. The total longitudinal size of the two sets of chromatids of ana- and telophases (see Fig. 4) increased from an average 19 $\mu$m (16 – 22 $\mu$m) in early anaphase to 25 $\mu$m (23 – 26 $\mu$m) at the end of telophase.

The metaphases with a moderately rotated spindle axis (14 cells in a sample of 50 cells) and the 2 cells which showed a more than 20° rotation invariably were cells with a severely reduced space. One of the latter two cells is schematically depicted in Figure 5. This cell was the smallest dividing cell in all samples, only being 8 $\mu$m high. A transverse chromosome orientation is not possible in such a 'flat' metaphase since most of the *Vicia* chromosomes are too long to be fully positioned in the metaphase equator plane. The centromere region of each of the chromosomes is oriented in this plane but a larger part of the remaining chromosome is oriented more or less perpendicular to the equator plane (see the side views in Fig. 5). Besides the observed oblique orientation, also a periclinal or a radial orientation would be possible. However, they

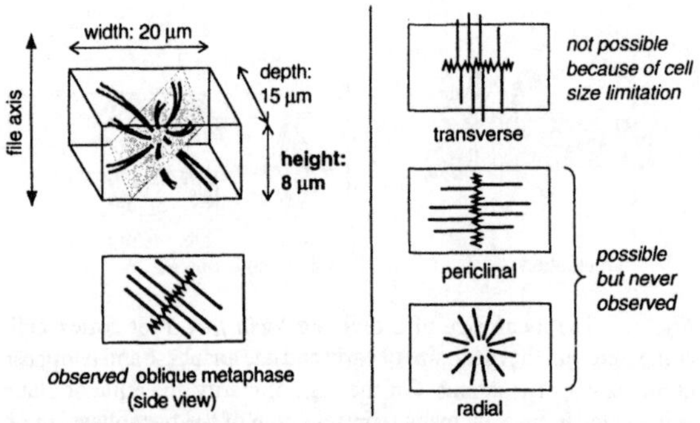


*Fig. 5.* The orientation of the metaphase equator plane in an extreme flat cell (left), and the three principal metaphase chromosome orientations drawn as side views of an equally sized cell (right). Only some of the chromosomes are drawn. The centromeres (black dots) are oriented in the equator plane (grey), whereas larger parts of the chromosome arms stick out of the plane. Also in other, comparable flat, dividing cells in the proliferative part of the cortex meristem, periclinal or radial orientations are never observed.

were never observed in the few examples of extreme flat dividing cells.

*The position of the nuclei in daughter cells*

Besides the chromosome orientation in dividing cells, we also took notice of the position of the nuclei in interphase nuclei. In most cell files pairs of small, somewhat flattened, interphase nuclei can be seen which are often shifted out of the center of the cell, towards opposite directions (in Fig. 2 two of these pairs are indicated by an asterisk). In view of the relatively small size of these cells and their nuclei, they probably represent young $G_1$-phase nuclei. The much larger S- and $G_2$-phase nuclei (Fig. 2: left file of cells) are nearly always positioned in the center of the cell. Apparently, the increase in nuclear size during interphase also results in a less eccentric position. The significance of these 'shifted' $G_1$-phase nuclei will be discussed in the next section.

*Chromosome orientation in* Crepis capillaris *and* Petunia hybrida

As described above, the orientation of the chromosomes in dividing cortex cells has been analysed extensively in *Vicia faba*. Roots of *Crepis capillaris* and *Petunia hybrida* have been studied to verify whether the phenomena found in *Vicia* also occurred in other species. An example of two dividing *Crepis capillaris* root cortex cells is shown in Figure 6. The equator plane of the metaphase has an orientation (as expected) perpendicular to the file axis, whereas the late anaphase shows a 27° rotation of the spindle axis. Among others, the three species are different with respect to the size of the chromosomes. The relevant data for the three species are summarised in Table 1. The mean cell height proved to be almost the same in the three species, in spite of the considerable differences in chromosome size. As expected, the differences in length of the longest chromosome arm (12 $\mu$m for *Vicia* to 5 $\mu$m for *Petunia*) have consequences for the size of the ana- and telophase complements (25 $\mu$m for *Vicia* to 17 $\mu$m for *Petunia*). The estimate of the percentages of cells with an at least moderate spindle axis rotation (90% for *Vicia*, 50% for *Crepis* and 15% for *Petunia*) is roughly in accordance with the differences in cell and chromosome size. Both in *Crepis* and *Petunia* we also observed pairs of daughter cells with shifted interphases nuclei.

## Discussion

*The relation between cell size and spindle axis rotation*

In the proliferative part of the root cortex meristem we expected to find dividing cells with a spindle axis parallel to the file and root axis. Such transverse divisions contribute to the elongation growth of the root by increasing the number of cells in the files (see Fig. 1). Surprisingly, we observed that most dividing cells showed oblique orientations of the anaphase and telophase chromosomes. The relation between cell size (especially cell height) and chromosome orientation proved to be a key factor in the understanding of these oblique orientations, as illustrated for *Vicia faba* in Figure 4. The measurements show a considerable discrepancy between the mean cell height (19 $\mu$m) and the mean size of a telophase configuration (25 $\mu$m). These figures indicate that in an average cell there is not enough space for a full upright (i.e. parallel to the file axis) separation of the chromatids during anaphase and telophase. As a consequence, in a majority of cells (90%) the spindle axis is forced to rotate towards one of the edges during anaphase. To give an impression of the resulting gain in space, the mean size of the different cell diagonals is also shown in Figure 4 (dotted lines b, c and d.). Only cells with a height of at least 25 $\mu$m are large enough to permit the expected non-rotated positioning of the two set of chromatids.

The relation between cell size and spindle axis rotation has been studied in material fixed in 3:1 ethanol

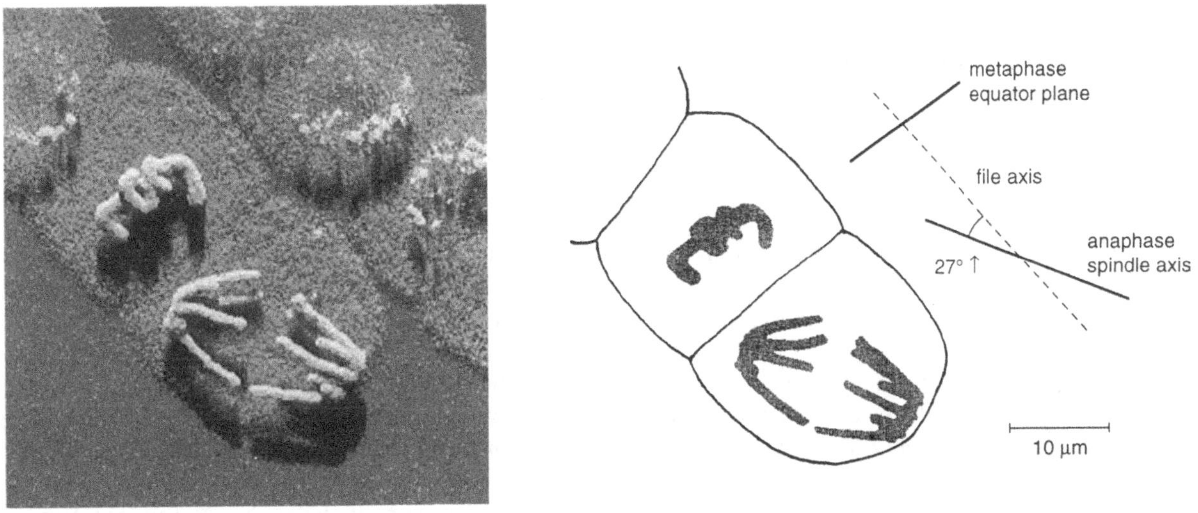


*Fig. 6.* Confocal microscopy based 3-D image of a metaphase and anaphase cell in the root cortex of *Crepis capillaris*. As shown in the accompanying drawing, the spindle axis of the anaphase is 27° rotated compared to the file axis, whereas the metaphase equator plane is oriented in an almost non-rotated position perpendicular to the file axis.

*Table 1.* Mean cell size, chromosome length and percentage of oblique orientations in root cortex cells of three plant species

| Species | Mean cell height | Length of longest chromosome arm[a] | Mean total length of telophase complement[b] | % of telophases with >5° spindle axis rotation[b] |
|---|---|---|---|---|
| *Vicia faba* | 19 $\mu$m | 12 $\mu$m | 25 $\mu$m | 90 |
| *Crepis capillaris* | 19 $\mu$m | 8 $\mu$m | 22 $\mu$m | 50 |
| *Petunia hybrida* | 18 $\mu$m | 5 $\mu$m | 17 $\mu$m | 15 |

[a]measured in 10 metaphases (not treated with a spindle inhibitor);
[b]estimate based on respectively 50, 25 and 25 telophases.

acetic acid. In this respect it is relevant to consider the possibility that the observed phenomena are mainly the result of cell shrinkage by the fixative. There are however several reasons why this is to our opinion rather unlikely. First, if there is a significant shrinkage of the cells one would expect a deformation or compression of the chromosome and spindle configuration rather then a rotation. Second, it is unclear why there is such a difference in the degree of shrinkage, resulting in a considerable range in rotation. And finally, we have recently observed oblique anaphases, telophases and cell plates also in preparations for electron microscopical studies, which were fixed with 3% glutaraldehyde and 2% formaldehyde. One of the consequences of a spindle axis rotation is an eccentric position of the nuclei in the daughter cells. Two examples of $G_1$-phase nuclei which are shifted to opposite directions can be seen in Figure 2. Although, three-quarters of all telophases show a considerable spindle axis rotation (Fig. 3), which results in these 'shifted nuclei', the daughter *cells* are nearly always positioned on top of each other, sharing the same cell file. We observed only rarely that daughter cells are positioned side-by-side (comparable to the result of a radial or periclinal cell division). In other words, even with a spindle axis rotation the cell divisions remain essentially transverse, although the plane of division is often no longer perpendicular to the file axis, but sigmoid shaped or oblique (for details about this aspect see Oud and Nanninga, 1992).

Only in extremely flat dividing cells, as the one depicted in Figure 5, an extreme oblique (i.e. almost vertical) cell wall seems to be inevitable. A more transverse wall will intersect the daughter nuclei. It is expected that the result of the division of an extreme flat cell is a side-by-side positioning of the daughter cells. The low frequency of such flat cells (approximately 2% of all dividing cells) is in accordance with the rather exceptional observation of a file of cells with one pair of cells next to each other instead of on top of each other (Oud and Nanninga, 1992).

The above described model, in which cell size is responsible for the spindle axis rotation and indirectly determines the position of the daughter cells, is in accordance with the concept of the critical aspect ratio as proposed by Barlow and Adam (1989). They found in the root cortex of tomato that 'tall' cells (higher than wide) divide transversally, whereas 'flat' cells (wider than high) divide periclinally. The transition between flat and tall lies in the range of 1.9 to 2.3.

*Oblique chromosome orientation in relation to chromosome size and cell type in various plant species*

Spindle axis rotations occur very frequently, but not exclusively in *Vicia faba* root cortex cells. We observed oblique orientations and pairs of shifted interphase nuclei also in the same tissue of *Crepis capillaris* and *Petuni hybrida*. In these cells, the frequency and degree of rotation seems to be related to the size of the chromosomes of the species. Of the species mentioned above, oblique orientations are most abundant in *Vicia*, are found only occasionally in *Petunia*, whereas *Crepis* takes a position in between. This order roughly corresponds with the mean size of the chromosomes. A remarkable point which emerges from these findings is that, at least for root cortex cells, there is obviously no relationship between the size of the cells and the chromosome. However, there are large differences in cell shape between cell types. Therefore oblique chromosome orientations occur much more frequently in one tissue compared to another. In *Vicia* for example, they are almost absent in the root stele region, in which the cells are considerably longer compared to the more cubical or flat cells in the cortex. The occurrence of spindle axis rotations in other plant tissues is among others reported in stamen and leaf cells of *Tradescantia virginica* (Bělař, 1930) and in guard mother cells of *Allium cepa* (Palevitz and Helper, 1974). Also in these studies it is concluded that cell size limitations are the principal cause of the observed oblique orientations.

*Further research*

Further research with respect to the relation between cell size and spindle axis rotation during cell division is focused on the role of the cortical microtubules in general and that of the pre-prophase band of microtubules (PPB) in particular. It is generally believed that the PPB, which is visible from $G_2$-phase to early prophase, predicts the site of the subsequent cell division (Gunning and Wick, 1985; Williamson, 1991 and references therein). If a cell is able to sense its dimensions prior to cell division, we have to expect the occurrence of oblique PPBs in relative flat cells. On the contrary, fully transverse PPBs in flat cells, should point to the existence of a mechanism to adjust the plane of division from transverse to a sigmoid or oblique shape. In this respect, current studies by De Ruijter (Dept. of Plant Cytology and Morphology, Agricultural University Wageningen, the Netherlands) have not given so far an indication for the occurrence of clear oblique PPBs in flat *Vicia faba* root cortex cells (personal communication; details about this study will be published elsewhere).

## References


Barlow P W 1978 The concept of the stem cell in the context of plant growth and development. *In* Stem Cells and Tissue Homeostasis. Eds. B I Lord, C S Potten and R J Cole. pp 87–113. Cambridge University Press, Cambridge.

Barlow P W 1984 Positional controls in root. *In* Positional Controls in Plant Development. Eds. P W Barlow and D J Carr. pp 281–318. Cambridge University Press, Cambridge.

Barlow P W and Adam J S 1989 Experimental control of cellular patterns in the cortex of tomato roots. *In* Structural and Functional Aspects of Transport in Roots. Eds. B C Loughman, O Gašparikova and J Kolek. pp 21–24. Kluwer Academic Publishers, Amsterdam.

Bělař K 1930 Beiträge zur Kausalanalyse der Mitose. III. Untersuchungen an den Staubfadenhaarzellen und Bladmeristemzellen von *Tradescantia virginica*. Z. Zellforsch. Microsc. Anat. 10, 73–131.

Brakenhoff G J, van der Voort H T M and Oud J L 1990 Three-dimensional image representation in confocal microscopy. *In* Confocal Microscopy. Ed. T Wilson. pp 185–197. Academic Press, London.

Clowes F A L 1976 The root apex. *In* Cell Divisions in Higher Plants. Ed. M M Yeoman. pp 253–284. Academic Press, London.

Gunning B E S and Wick S M 1985 Preprophase bands, phragmoplasts, and spatial control of cytokinesis. J. Cell Sci. Suppl. 2, 157–179.

Houtsmuller A B, Smeulders A W M, van der Voort H T M, Oud J L and Nanninga N 1992 The homing cursor: a tool for three-dimensional chromosome analysis. Cytometry 14, 501–509.

Luxová M 1975 Some aspects of the differentiation of primary root tissues. *In* The Development and Function of Roots. Eds. J G Torrey and D T Clarkson. pp 73–90. Academic Press, London.

Montijn M B, Houtsmuller A B, Oud J L and Nanninga N 1994 The spatial localization of 18S rRNA genes, in relation to the descent of the cells, in the root cortex of *Petunia hybrida*. J. Cell Sci. 107, 457–467.

Oud J L, Mans A, Brakenhoff G J, van der Voort H T M, van Spronsen E A and Nanninga N 1989 Three-dimensional chromosome arrangement of *Crepis capillaris* in mitotic prophase and anaphase as studied by confocal scanning laser microscopy. J. Cell Sci. 92, 329–339.

Oud J L and Nanninga N 1992 Cell shape, chromosome orientation and the position of the plane of division in *Vicia faba* root cortex cells. J. Cell Sci. 103, 847–855.

Palevitz B A and Hepler P K 1974 The control of the plane of division during stomatal differentiation in *Allium*. I. Spindle reorientation. Chromosoma 46, 297–326.

Ten Kate T K, van Balen R, Groen F C A, Smeulders A W M and Den Boer G 1990 SCILAIM: a multilevel interactive image processing system. Pat. Recogn. Let. 10, 429–441.

Van der Voort H T M, Brakenhoff G J and Baarslag M W 1989 Three-dimensional visualization methods for confocal microscopy. J. Microsc. 153, 123–132.

Williamson R E 1991 Orientation of cortical microtubules in interphase plant cells. Int. Rev. Cytol. 129, 135–206.

*F. Baluška et al. (eds.), Structure and Function of Roots,* 41–51, 1995.


# Importance of the post-mitotic isodiametric growth (PIG) region for growth and development of roots

František Baluška[1], Peter W. Barlow[2] and Štefan Kubica[1]
[1]*Institute of Botany, Slovak Academy of Sciences, Dúbravská cesta 14, 842 23 Bratislava, Slovakia and* [2]*University of Bristol, Department of Agricultural Sciences, Institute of Arable Crops Research, Long Ashton Research Station, Bristol BS18 9AF, UK*

*Key words:* aerenchyma, cell growth, gravitropism, maize roots, microtubules, morphogenesis, root cortex, root development, tomato roots, *Zea mays* L.

## Abstract

Plant cells are assumed to embark on rapid elongation immediately after the cessation of their mitotic divisions at the proximal end of the meristem. However, a comprehensive appraisal of most of the data previously published on this matter, as well as several recently obtained findings, has convincingly revealed, at least for root cells, that this belief is incorrect. For example, morphometric analysis of the maize root apex has clearly defined a distinct transitional region between the meristem and the zone of rapid cell elongation. This growth region is operationally defined as the post-mitotic isodiametric growth (PIG) region. In the middle of the cortex, this region may have a length similar to that of the meristem in steady-state growing roots. We believe that PIG is a specific phase in root cell ontogeny and that cells of the PIG region have more in common with the meristematic cells than with the rapidly elongating cells, although many of the metabolic properties of cells in the PIG region are rather unique. In this regard, certain properties of cells in the PIG region, such as the distributions of their microtubules and their sensitivity towards auxin and calcium, were found to be of crucial importance for the induction of differential cell growth patterns essential for the graviresponse of maize roots. Other data indicate that the PIG region is of considerable significance for the overall development of roots because of its characteristic plasticity under unfavourable external conditions.

## Introduction

The generally accepted view, that plant cells embark on rapid cell elongation immediately after the cessation of their mitotic divisions, has been widely propagated in much of the anatomical literature since the beginning of this century. However, a comprehensive appraisal of most of the data previously published on this matter, as well as recently obtained findings, has convincingly revealed, at least for root cells, that this belief is incorrect.

Burström (1942, 1950, 1969, 1971) was the first who not only recognized this feature of growing root cells but, in addition, distinguished between two phases of cell elongation which differed in their pattern of cell growth. In particular, he proposed that there was an initial slower phase of cell elongation which later changed into a rapid phase of cell elongation. We shall be concerned with the initial phase of cell elongation since we believe it to occur in a zone of the root that is especially important in many aspect of root development and in the response of roots to their environment.

## Cell growth patterns in the post-mitotic growth regions

The current interpretation of the sigmoid curve of cell lengths along a file of cells is that the relative rate of cell elongation (Barlow et al., 1991) or the relative elemental rate of elongation (Barlow et al., 1991; Sharp et al., 1988; Silk et al., 1986) increases to a maximum and then declines as cell elongation ceases (Fig. 1). Thus, plant cells elongate throughout the growth region, and their longitudinal walls increase all along their length by the interpolation of new wall material. Elongation can therefore be expressed in units of % $h^{-1}$ for an element of growth (Barlow et al., 1991; Sharp et al.,

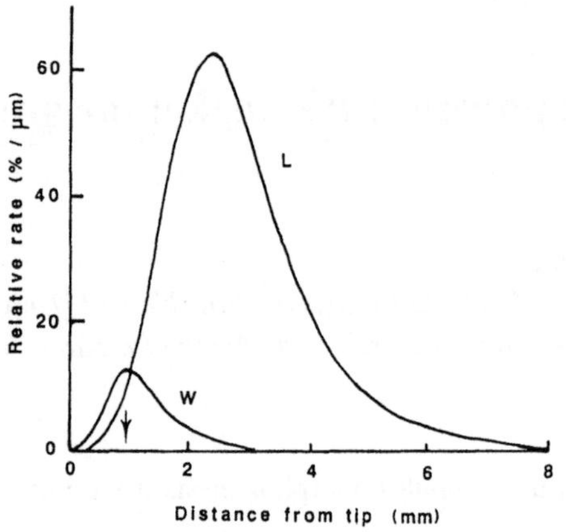


*Fig. 1.* Relative rates (% $\mu m^{-1}$ axial length moved) of cell lengthening (L) and cell widening (W) along the axis (mm distance from root tip) of the cortex of a tomato root. The arrow indicates the proximal limit of the cortical meristem with respect to the tip of the root (which is at 0 mm distance). This position is slightly distal to the point where the two lines intersect; the latter is the point of isodiametric growth. It is also close to where cell widening is proceeding at maximal rate.

1988; Silk et al., 1986) at a given distance from the root tip.

The early phase of cell elongation is characterized by smaller cells and a slower relative rate of elongation. It starts in the proximal part of the apical meristem (Baldovinos de la Pena, 1953; Baluška et al., 1990, 1993a; Barlow, 1969; Bednara, 1974; Jensen and Kavaljian, 1958; Webster, 1980) and continues until after cessation of mitotic divisions (Baluška et al., 1990; Bednara, 1974; Burström, 1942, 1950; Demchenko, 1984; Luxová, 1981). The more rapid phase of cell elongation (Bruström, 1942, 1950) occurs only when the first phase of non-proliferate cell growth has already increased the average cell length at least threefold (Baluška et al., 1990, 1993a; Bendara, 1974; Demchenko, 1984). Thus, the proximal boundary of the apical meristem does not appear to be associated with any profound change in the rate of cell elongation. Such a change occurs only later in cellular ontogeny when the slower relative rate of cell elongation, characteristic of meristematic and immediately post-mitotic cells, transforms into a more rapid one.

This characteristic feature of the cell growth in root apices appears to be unavoidable if we take into consideration the fact that individual cell files of the maize root differ significantly regarding the distances from the root tip at which their divisions cease (Clowes, 1983; Luxová, 1980). Moreover, the symplastic nature of cell growth and its rapid acceleration during the second phase of cell elongation, whereby maize root cells traverse the cell elongation region (a distance of 4–5 mm) in only 2 h (Kubica et al., 1991) imply the existence of a transition region between the meristem and the cell elongation zone simply from spatiotemporal considerations. The post-mitotic transition zone could, therefore, give the cells time to complete mitosis before rapid elongation commences. Rather indirect, but nevertheless substantial support for the independence between the cessation of mitotic divisions and the onset of rapid cell elongation comes from studies of Ivanov (1980, 1981). He found that even the complete suppression of mitotic divisions in the meristem did not affect the time and rate at which cells continued to make their transitions to rapid elongation. Therefore, the beginning of rapid cell elongation is not a direct consequence of the cessation of mitosis and, consequently, these two events must be spatially separated and regulated independently.

It was again Burström (1971) who first pointed out another significant feature of the post-mitotic transitional region of wheat roots. In this case, it concerned the polarity of the cell growth. He reported that both the basal part of the meristem as well as the immediate post-mitotic region exhibited anisotropic cell growth. With the onset of more rapid cell elongation, cell growth then became strictly isotropic, in the longitudinal direction only. Therefore, in addition to an absence of any change in the rate of cell elongation, cells at the proximal border of the meristem lack any profound reorientation of their growth, whereas later in ontogeny this does occur. More recently, this interesting feature of cell growth in plant roots was confirmed, and generalized to other plant species, through the use of more sophisticated morphometric methods (Baluška et al., 1990, 1993a, c; Barlow et al., 1991; Silk et al., 1986).

Cell shapes along the growth region of maize roots can be described using Form Factor values computed, for instance, by the ASBA image analyser (Baluška et al., 1990). Form Factor curves reflect the distinctive course of cell shape changes during root growth (Baluška et al., 1990, 1993a, c). Another approach for characterizing growth patterns, in this instance in tomato root apices, utilized the relative rates of cell elongation and cell widening (Barlow et al., 1991). Both relative rate curves (Fig. 1) and Form Factor curves (Fig. 2B) clearly defined a distinct transitional region between the meristem and the zone of rapid cell elongation in the root apex (for maize see Fig. 2A). This region, which, in the cortex of steady-state growing maize roots can exhibit a length similar to that of

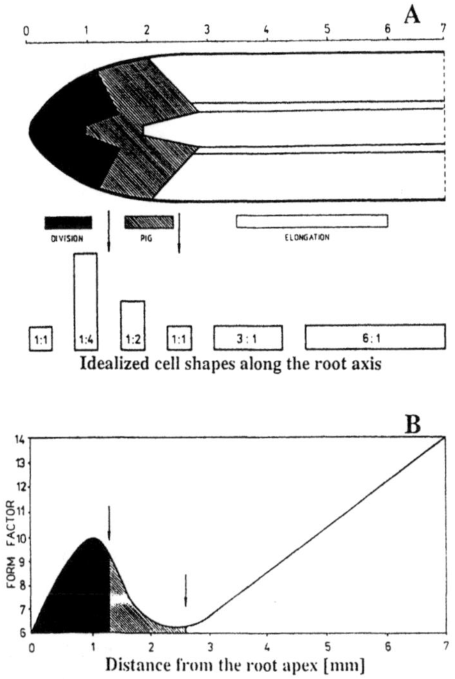


*Fig. 2.* **A.** Schematic illustration showing spatial patterns of cell divisions, postmitotic isodiametric cell growth and cell elongation in the maize root apex. Idealized cell shapes as detected in the middle cortex using Form Factor values (adapted from Baluška et al., 1990, 1993c). Arrows correspond to the proximal boundaries of the meristem and PIG region in the middle cortex. **B.** Form Factor (FF) curve computed from data from the middle cortex cells (adapted from Baluška et al., 1990, 1993c). FF = 6 represents a square; FF = 7 represents a rectangle with a length to width ratio of 2:1; and FF = 12.4 represents a rectangle with a length to width ratio of 5:1. Arrows denote the proximal boundaries of the meristem and the PIG region in the middle cortex.

the meristem, was operationally defined as the post-mitotic isodiametric growth (PIG) region (Baluška et al., 1990; Ishikawa and Evans, 1992).

Interesting relationships between the extent of the meristem and of the PIG region in the root apex were described during germination of maize roots (Deltour et al., 1989). A short PIG region became discernible very early in germination (60 h after soaking), concomitant with the establishment of the meristem. However, it was only after a further 30 h of growth that the size of these two growing regions acquired similar values. On the other hand, an anatomical survey of branch roots of field-grown maize undergoing transition to determinate growth showed prominent PIG regions that appeared larger than the corresponding apical meristems (Varney and McCully, 1991).

Therefore, in contrast to earlier suggestions that the growth of immediately post-mitotic cells represents the first phase of rapid cell elongation (Burström, 1942, 1950, 1971), we prefer the idea that there is an unique PIG phase in root cell ontogeny (Baluška et al., 1990, 1992, 1993c). In fact, as we shall discuss, cells of the PIG region have more in common with cells of the meristem than with the rapidly elongating cells and, furthermore, many properties of cells in the PIG region are not shared by other root regions. We already mentioned the anisotropic mode of cell expansion and the relatively slow rate of cell elongation, both of which are typical of the basal part of the meristem and the PIG region but not of the region exhibiting the second phase of elongation which is characterized by more rapid cell growth. There are several other common features between cells in the meristem and PIG regions regarding their nuclear metabolism - these are summarized in Table 1.

## Cells in the PIG region exhibit metabolic properties similar to those found in the meristem

Firstly, cell growth in both the meristem and the PIG region appears to be accompanied by an increase in cytoplasm, whereas this ceases to be prominent in the elongation zone where vacuolation predominates. This can be inferred not only from the relatively slight development of the vacuome in cells of the PIG region compared to the extreme vacuolation of rapidly elongating cells (quantitatively documented for root cells of *Azolla pinnata* by Barlow et al., 1982), but also from the pattern of protein and dry matter increments as determined in successive 1-mm-long segments of the maize root apex (Gandar, 1980; Pahlavanian and Silk, 1987; Silk and Erickson, 1980; Silk et al., 1986.) A similar conclusion can be reached from inspection of data on content of dry matter or protein published much earlier (Baldovinos de la Pena, 1953; Brown and Broadbent, 1951; Jensen, 1955).

Maize roots growing under low water potential show a significantly shorter growth zone and a smaller final cell length (Fraser et al., 1990; Saab et al., 1992; Sharp et al., 1988) than adequately watered roots. Interestingly, the spatial distribution of cell elongation was unaffected in the meristem and in the PIG region (Ishikawa and Evans, 1993; Pritchard et al., 1993; Saab et al., 1992; Sharp et al., 1988; Spollen and Sharp, 1991). The lack of cell length response in these two regions to water stress was reported to depend on the accumulation of abscisic acid in the cytoplasm (Saab et al., 1992). Abscisic acid may help

*Table 1.* Cell growth patterns and nuclear activities in different growth regions of maize roots

| | Cell division | Cell widening | Cell lengthening | rRNA transcription | DNA synthesis |
|---|---|---|---|---|---|
| Meristem | X | X | X(slow) | X | X |
| PIG region | – | X | X(slow) | X* | X* |
| Elongation region | – | – | X | – | X(slow) |

*rRNA transcription already starts to cease and DNA synthesis starts to slow in the PIG region if the nuclei here contain a high proportion of condensed chromatin (Baluška, 1990; Kubica et al., 1989).

preserve the longitudinal cell growth in the apical part of the growth region due to its stimulation of proline deposition which was found to play an essential role in osmotic adjustment of the cytoplasmic compartment at low water potentials (Voetberg and Sharp, 1991). All these observations support the notion that the cytoplasmic mode of cell growth prevails not only in the meristem but also in the PIG region. An important role in preserving the cell elongation patterns in the apical part of the growth region (first three millimetres) at low water potentials could be played by xyloglucan endotransglycosylase. Activity of this wall-loosening enzyme was at a high level and was unchanged by osmotic stress in cells of the meristem and the PIG region (in contrast to rapidly elongating cells where levels were lower) of maize root apices (Pritchard et al., 1993).

In addition to low water potentials, cytoplasmic cell growth seems also to be sustained under other unfavourable conditions that inhibit turgor-driven cell elongation. For instance, the inhibition of root growth by KCl (10 mol. $m^3$) or cold treatments (4 °C) allowed cells to reach lengths corresponding only to those acquired at the end of PIG region under control conditions (Pritchard et al., 1988, 1990). On the other hand, cells of the PIG region and meristem were found to be sensitive to coumarin, whereas cells that had already embarked on rapid elongation were not affected by such treatment (Rufelt, 1959; Svensson, 1971). These observations suggest that the meristematic and PIG region cells have properties different from rapidly elongating cells. The differential response to KCl, for instance, indicates that, in this environment, meristematic cells can still grow and produce new cells which enter the PIG region, but here they accumulate since they cannot accomplish the transition to rapid elongation owing to a block in the vacuolation process.

In close correlation with protein deposition rates and dry weight increments, increases in nucleic acid content are also more prominent in the meristem and PIG regions than in the rapid elongation region (Baluška, 1990; Holmes et al., 1955). In relation to this, nuclear metabolism was found to change profoundly — not immediately after the cessation of mitotic divisions but later, when the rate of cell elongation transformed from the slower to the more rapid mode. At this crucial transition point of root cell development, nuclear DNA synthesis markedly slows down in all endopolyploid tissues (Kubica, 1986; Kubica et al., 1989), an event apparently closely associated with prominent condensation of the nuclear chromatin complex (Baluška, 1990; Kubica, 1984). Furthermore, intense rRNA transcription was observed to occur only in cells whose nuclei displayed a rather loose chromatin structure — all such cells were either meristematic or mainly stelar cells in the PIG region. The nuclear architecture appears to determine also the tissue-specific differences in cell size in these two apical regions of the maize root (Baluška and Kubica, 1992). In this regard, it is notable that the elemental growth rate and rRNA transcript maturation were found to exhibit the same pattern in individual tissues along the whole root growth region, irrespective of the proportion of condensed chromatin in their nuclei (Kubica et al., 1991).

Cell growth coordinated with the increase of cytoplasm is believed to be controlled by the uptake of organic nutrients, such as sugars and amino acids, from the apoplast (Schobert and Komor, 1987). This process is expected to require an inwardly directed ion current (Weisenseel et al., 1992). On the other hand, turgor-driven rapid cell elongation is expected to be associated with an outwardly directed ion current which should enable the symport/antiport uptake of potassium and sodium ions required for the extensive development of the vacuole. The patterns of ionic currents that were detected in or around diverse root apices (Collings et al., 1992: Iwabuchi et al., 1989; Weisenseel et al., 1992) precisely correspond to suggestions that growth dominated by the increase in cytoplasm is typical not only for meristematic cells but that it continues also in the PIG region (Baluška et al., 1990; Kubica et al., 1989; Weisenseel et al., 1992).

**Cells in the PIG region exhibit several specific metabolic properties**

The outward flow of electric current that is typical of the turgor-driven phase of cell elongation seems to be associated with, and dependent on, a low exogenous calcium concentration. This can be inferred from observation that an artificial lowering of calcium in the medium reversed the ionic current patterns around cress root apices, and even favoured outward flow of current near the root tip (Weisenseel et al., 1992). These authors hypothesized that the high activity of plasma membrane $H^+$-ATPases (Parets-Soler et al., 1990), which is associated with the turgor-driven rapid cell elongation, is possibly controlled by a lowering of the calcium concentration below a certain threshhold level. This event is known to occur naturally when cells are displaced further away from the root tip (Burström, 1968).

If this hypothesis is correct, then an artificial increase of calcium level in cells should prevent the onset of rapid cell elongation. This has now been confirmed experimentally: calcium was reported not only to act as a strong inhibitor of root elongation (Hasenstein and Evans, 1986) but also cells of the PIG region were characterized as the specific target site for calcium action (Ishikawa and Evans, 1992). Unilateral application to the root surface of agar blocks containing calcium elicited positive root curvature (i.e. calcium-mediated inhibition of cell elongation at the site of its application). Importantly, this response could not be triggered by plain agar or by calcium application beyond the PIG region (Ishikawa and Evans, 1992). These observations suggest that high calcium concentration specifically inhibits the transition to the rapid phase of cell elongation. Interestingly, differential growth responses in the PIG region of maize roots can also be induced by local treatment with auxin (Meuwly and Pilet, 1987, 1991; Pilet and Meuwly, 1986). However, auxin application elicited negative curvature (i.e. auxin-mediated stimulation of cell elongation at the site of its application) which was strongest if the site of auxin application was located between 1.5 – 2.5 mm from the root tip (Meuwly and Pilet, 1991). More recently, peculiar properties of maize root cells in the PIG region with regard to their sensitivity towards auxin were found to be of crucial importance for the induction of differential elongation during the graviresponse (Ishikawa et al., 1991; Ishikawa and Evans, 1993). High concentration of exogenous auxin is known to suppress completely cell elongation in the elongation region (Burström, 1957; Goodwin, 1972; Hejnowicz, 1961) but was observed to enhance cell elongation in the PIG region; this then enabled the subsequent recovery of root growth (Ishikawa and Evans, 1993). Particularly interesting is the possibility that cells of the PIG region (Ishikawa and Evans, 1993; Muday, 1994) are responsible for the surprising ability of maize roots to respond to gravity even in the presence of exogenous auxin at concentrations (Katekar and Geissler, 1992) which completely inhibit the growth of vertically oriented roots. This finding indicates that the elongation of cells in the upper part of the gravistimulated PIG region is perhaps enhanced by some mechanism independent of auxin and would thus be inconsistent with the Cholodny-Went theory of gravitropism (cf. Sievers and Zieschang, 1992).

All these findings stress the intimate involvement of calcium ions and auxin molecules in the changing cellular metabolism specifically associated with the PIG region. They suggest that both auxin and calcium are involved in switching root cell development from a mode associated with more or less apolar (anisotropic) and mainly cytoplasmic growth to one that is strictly polar (isotropic) and dependent on the extensive formation of the vacuome, which in turn contributes to a turgor-driven elongation process. Other work indicates that profound reorientations of cortical microtubules (CMTs), a cytoskeletal element whose organization depends on the inherent dynamic instability of MTs (Mitchison and Kirschner, 1984; Schilstra et al., 1991) which in turn is related to the state of cellular metabolism, are of great importance for this developmental switch in the pattern of cellular growth. CMTs in the PIG region were found to accomplish a characteristic shift in arrangement from a more or less random and homogeneous distribution to one with clearly ordered transverse arrays with respect to the root axis. This event appeared to be essential for the transition from the more or less anisotropic growth of the PIG region to the strictly isotropic growth typical of the region of rapid cell elongation (Baluška et al., 1992, 1993a). Stabilization of CMTs in the transverse plane was found to be controlled by gibberellic acid because when the endogenous level of this phytohormone was decreased root cells failed to start their strictly polarized elongation and, instead, continued to widen throughout the elongation region (Barlow et al., 1991; Baluška et al., 1993a). As a consequence of this, cell growth polarity was disturbed, even in the meristem, and the PIG region was shifted proximally, away from the root apex. Recently, several root morphogenesis mutants have been isolated which fail to accomplish the major switch in cellular development from anisotropic expansion to isotropic elongation (Baskin et al., 1992; Benfey et al., 1993) The diversity in their appearances and properties, as well as the tissue-specific expressions of these mutations, all suggest that cells in the PIG region are accomplishing a complicated transformation which are regulated by a large number of genes. Curiously enough, some processes accomplished in cells of the PIG region are insensitive to metabolic inhibitors because they fail, as we have mentioned above, to affect the onset of rapid cell elongation (Ivanov, 1980, 1981). A characteristic feature of cells in the PIG region is the intense metabolism of sugars which is rendered visible in the prominent accumulation of starch grains in the cytoplasm of xylem and cortical cells (Baluška et al., 1993a). This metabolic specificity of the PIG region might result from the unloading of phloem elements since in this region phloem becomes mature and invertase activity reaches its highest level (Hellebust and Forward, 1962). Interestingly, cells of the PIG region are sensitive to the external application of some sugars; these cause selective collapse of cells in the PIG region and lead to a 'physiological decapitation' of the root (Rufelt, 1959; Stenlid, 1957). Although this finding confirms the metabolic uniqueness of the PIG region, the mechanism of 'physiological decapitation' of roots is not known. The process might explain the mysterious loss of apical meristems in certain lateral roots of the maize root system (Varney and McCully, 1991).

**Role of the PIG region in root growth**

Specific morphological and metabolic features of cells in the PIG region seem to provide the growing root apices with properties that enable them to behave as highly dynamic structures that efficiently exploit the soil. Particularly, cells of the PIG region seem to be indispensable for the root penetration of compact soil; and, equally importantly, they appear to be essential for all tropic movements of roots.

The mechanical impedance of growing roots promotes radial growth of post-mitotic cortical cells (Atwell, 1993; Wilson et al., 1977) resulting in the formation of a swelling behind the meristem. This event is associated with a lateral force that helps open up compact soil ahead of a growing tip. We have found that specific properties of the microtubular cytoskeleton in cells of the inner cortex might be involved in the formation of swellings in the PIG region. Auxin and ethylene could be considered to represent possible candidates responsible for the swelling of the PIG region (Lachno et al., 1982; Sarquis et al., 1991) because these two phytohormones were found to alter the polarity of cell growth in the cortex just behind the root meristem (Elliason et al., 1989; Whalen and Feldman, 1986). Interestingly, PIG cells of the inner cortex, which display homogeneously distributed arrays of CMTs that are sensitive to unfavourable external conditions (Baluška et al., 1992, 1993b), seem to act as a specific target site for ethylene (Baluška et al., 1993c). In contrast, cells of the outer cortex show bundled and ethylene-insensitive CMTs, and here the polarity of cell growth is rather stable.

Generally, growing plant organs consist of tissues that vary profoundly with respect to the polarity of their consitutent cells, their cell-wall structures, and their patterns of physical tension. Inevitably, due to this variation, some tissues must limit the growth of other tissues. Although the spatial distribution of tissue stresses in roots is poorly described at present, it is clear that it differs fundamentally from above-ground organs where the situation is better known (Kutschera, 1989). For organs such as coleoptiles, epicotyls or hypocotyls, the epidermal cells have been unequivocally identified as the essential tissue which limits not only the growth rate of the whole organ (Bergfeld et al., 1987; Hodick and Kutschera, 1992; Hohl and Schopfer, 1992; Ishida and Katsumi, 1992; Kutschera et al., 1987; Kutschera, 1992) but also controls its phototropic and gravitropic curvatures (Nick et al., 1990, 1992). Specific properties of outer epidermal walls such as, for example, their multilayered helicoidal texture and conspicuous thickness, seem to play crucial role in the reduction of cell- wall plasticity and growth (Hodick and Kutschera, 1992; Kutschera, 1992). Auxin-mediated reorientation of CMTs was reported to be responsible for altering the properties of this growth-limiting cell wall and allowing a concomitant expression of the growth potential of the underlying cellular layers (Bergfeld et al., 1987; Nick et al., 1990, 1992; Zandomeni and Schopfer, 1993). Thus, in the case of above-ground organs, the epidermis constitutes the primary target tissue for the growth-promoting action of auxin.

Despite the paucity of information regarding roots when compared with the above-ground organs, one thing is clear — the root epidermis is, unlike the situation in the shoot, rather a passive partner in growth control (Barlow, 1982, 1989). In accordance with this conclusion, epidermis cells were reported to bulge extensively in a specific root mutant of *Arabidopsis thaliana* without affecting the expansion of the whole organ (Baskin et al., 1992). Furthermore, an unspecified sub-epidermal tissue, rather than epidermis itself, was hypothesized to act as the auxin-sensitive tissue that could control root growth (MacDonald and Hart, 1987). For maize roots, we propose that PIG-region cells in the outer cortex — i.e. three or four cell files beneath the epidermis (Baluška et al., 1993c) — are of primary importance in root growth control. This notion is in agreement with the hypothesis that all tropic and other movements, which represent a class of morphogenetic processes, are brought about by controlled deformations at organ or tissue boundaries (Barlow, 1994).

Specific features of the outer cortex such as its well organized arrays of CMTs which are resistant to perturbations, the early onset of cell elongation and the absence of intercellular spaces, certainly contribute to the generation of large internal tissue stresses. Stresses were found to be maximal just in the PIG region (Burström, 1971) where the graviresponding cells are localized and where the root curvature arises (Barlow and Rathfelder, 1985; Evans, 1991; Ishikawa et al., 1991; Ishikawa and Evans, 1992, 1993; Meuwly and Pilet, 1991; Sievers and Hejnowicz, 1992; Stenz and Weisenseel, 1993; Zieschang and Sievers, 1991). Furthermore, differential growth of PIG-region cells, located not only in the epidermis (Zieschang and Sievers, 1991) but also in the outer cell layers of the maize root cortex (Shen-Miller et al., 1978), was reported to be of specific significance for the development of gravitropic curvature. Microsurgical removal of epidermal and outer cortical cells provided direct evidence that the outer cells layers of maize root are essential for execution of gravitropic curvature (Yang et al., 1990), though for contrasting results see Barlow (1982) and Björkman and Cleland (1991). These ideas gain support from a recent immunofluorescence study showing that reorientations of CMTs occurred only in cells of the outer four to five cortical layers along the lower side of the PIG region in gravireacting maize roots (Blancaflor and Hasenstein, 1993). CMTs in cells of the epidermis, inner cortex and stelar tissues, and in cells at the upper side of the PIG region, preserved their original orientations.

In this respect it is worth mentioning that the outer cortex has also been found to show peculiar properties regarding the endogenous content of calcium and auxin, two principal mediators of root gravitropism. Post-mitotic growing cells of the outer cortex, but not of the inner cortex, were characterised by a very low content of calcium (Dauwalder et al., 1985; Jaunin et al., 1991). Furthermore, basipetally transported auxin (supplied originally to the root apex by an acropetal stelar flow that subsequently becomes redistributed within the root cap (Young et al., 1990)) is accomplished preferentially through the outer cortex (Ohwaki and Tsumuri, 1976; Tsumuri and Ohwaki, 1978). When these observations are combined with findings that IAA stimulates cell elongation (Burström, 1957; Ishikawa and Evans, 1993; Meuwly and Pilet, 1991) whereas calcium inhibits it (Ishikawa and Evans, 1992), specifically in the PIG region, we are tempted to hypothesize that a relatively high content of auxin and a relatively low content of calcium are the main factors reponsible for the early onset of rapid cell elongation in the outer cortex.

A delicate balance between the internal tissue tensions in straight growing roots might be related to symmetrical ionic current patterns around roots (Collings et al., 1992; Weisenseel et al., 1992). Even a small disturbance to this finely balanced system, elicited for instance by the reorientation of growing roots in a gravitational field (Barlow and Rathfelder, 1985; Shen-Miller et al., 1978; Weisenseel et al., 1992; Zieschang and Sievers, 1991) or by the application of a weak electric field (Stenz and Weisenseel, 1993), causes changes to the cell growth pattern specifically in the PIG region. Interestingly, if a strong electric field was applied to growing maize roots, it failed to induce the characteristic differential growth patterns in the PIG region and, moreover, a reversed curvature developed in the basal part of the elongation zone of roots and stopped their growth. Present data indicate that specific changes in cell growth polarity occurring in the PIG region of gravistimulated roots enable not only the re-establishment of symmetrical ion current fluxes

(Collings et al., 1992; Weisenseel et al., 1992), but also the acquisition of a new equilibrium of internal tissue stresses. Perhaps this is really what gravitropism is for: it is to restore a stable electrical field around the root — and this is actually the self-correcting element for any deviation of the root from its liminal angle. Thus, the reoriention of a root by means of a tropism might be a secondary event: it results from an injured electric field and is the means of its repair (Barlow and Zieschang, 1994).

Several findings indicate that the PIG region is not only the site where the differential growth of gravire-sponding cells occurs but, in addition to the root cap, it also might be a site of gravity perception (see also Poff and Martin, 1989). For instance, the gravitropic responses in the PIG region (i) start very early after receiving the stimulus (Barlow and Rathfelder, 1985; Collings et al., 1992; Evans, 1991; Ishikawa and Evans, 1990; Nantawisarakul and Newman, 1992); (ii) they occur even at concentrations of auxin which completely inhibit the growth of vertically oriented roots (Ishikawa and Evans, 1993; Katekar and Geissler, 1992); (iii) they are independent of the root tip during this initial phase of the gravistimulation (Stinemetz, 1992). Additionally, in accordance with the possible involvement of the PIG region cells in gravity sensing, Shen-Miller and McNitt (1978) found markedly altered intracellular distributions of dictyosomes in cells of the outer cortex located in the upper half of the PIG region in gravistimulated roots. However, existence of a very fast electrical signalling mechanism of unknown nature between the root cap and PIG region (Hejnowicz et al., 1991; Sievers and Hejnowicz, 1992), could overcome the first (i) of the above-listed indications that gravity perception is also in the PIG region.

## Significance of the PIG region for root development

Under poorly-aerated conditions, maize roots develop an extensive network of gas-filled lacunae, or aerenchyma, behind their growth zone (Cambell and Drew, 1983). Ethylene has been shown to play a crucial role in the schizogenous/lysigenous conversion of root cortex into an aerenchyma (Atwell et al., 1988; He et al., 1992). Although the cortical cavities develop only after the cessation of cell growth, a growing root apex appears to be necessary for their formation (Atwell et al., 1988).

Detailed morphometric and immunofluorescence analyses revealed two cortical domains within the PIG and elongation regions of the maize root apex – the outer and inner cortex (Baluška et al., 1993c). Only the inner domain is involved in aerenchyma formation whereas the outer cortical domain preserves its integrity and serves as a mechanical support tissue (Deacon et al., 1986; Konings and Verschuren, 1980). As mentioned earlier, the three to four outermost cell layers beneath the epidermis exhibit several structural features that distinguish them from cells of the inner cortex. First, their cell walls are tightly appressed and consequently there are fewer of the prominent extracellular spaces which are typical for the inner cortex. Due to this, the epidermis, hypodermis and the outer cortical cell layers constitute a coherent sleeve-like structure that can be isolated by enzymatic digestion from the remainder of the cortical tissue (Clarkson et al., 1987). Second, cells of the outer cortex start their rapid phase of elongation closer to the root tip (Baluška et al., 1993c), and their cell walls form collenchyma thickenings (Clarkson et al., 1987) which, in mature regions of the root, are strongly autofluorescent due to a high content of lignin (McCully and Canny, 1985). In addition, spatial analysis of a cDNA clone corresponding to an mRNA putatively involved in suberin biosynthesis showed that this probe was localized preferentially in the three or four cell layers of the outer cortex as well as in endodermis (Held et al., 1993). CMTs of these cells are organized, as we have already mentioned, into prominent bundles at the onset of rapid cell elongation (Baluška et al., 1992).

The bundled CMTs in cells of the outer cortex are resistant not only to low temperature and colchicine treatments but also to ethylene application (Baluška et al., 1993b, c). Ethylene has been found to disorganize and, at high ethylene concentrations, even to cause disintegration of the arrays of CMTs in cells of the inner cortex. In accordance with these effects on CMTs, exogenous ethylene, applied at a concentration that induces the development of a prominent aerenchyma (Atwell et al., 1988), also elicits differential effects on the polarity of cell growth in the PIG and elongation regions (Baluška et al., 1993c). While cells of the outer cortex continued to show a growth pattern more or less typical for control roots, cells of the inner cortex showed an altered mode of cell growth, e.g. cell widening was apparently promoted at the expense of cell elongation. Therefore, the differential sensitivity of CMTs towards ethylene in outer and inner cortical cells of the PIG region appears to be developmentally relevant for the further differentiation of cortical cells into either aerenchymatous cavities (inner cortex) or lignified mechanical support tissue at the root periphery (outer cortex). All these observations suggest that cells of the inner cortex act as a highly specific target for ethylene.

In conclusion, the PIG region seems to be of considerable significance not only for morphogenetic root movements but also, because of its plasticity under unfavourable external conditions, for the overall development of roots and their response to the external environment.

## References


Atwell B J 1993 Response of roots to mechanical impedance. Environ. Exp. Bot. 33, 27–40.

Atwell B J, Drew M C and Jackson M B 1988 The influence of oxygen deficiency on ethylene synthesis, 1-aminocyclopropane-1-carboxylic acid levels and aerenchyma formation in roots of *Zea mays*. Physiol. Plant. 72, 15–22.

Baldovinos de la Pena G 1953 Growth of the root tip. *In* Growth and Differentiation in Plants. Ed. W E Loomis. pp 27–54. Iowa State College Press, Ames, Iowa, USA.

Baluška F 1990 Nuclear size, DNA content and chromatin condensation are different in individual tissues of the maize root apex. Protoplasma 158, 45–52.

Baluška F and Kubica Š 1992 Relationships between the content of basic nuclear proteins, chromatin structure, rDNA transcription and cell size in different tissues of the maize root apex. J. Exp. Bot. 43, 991–996.

Baluška F, Hauskrecht M and Kubica Š 1990 Postmitotic 'isodiametric' cell growth in the maize root apex. Planta 181, 269–274.

Baluška F, Parker J S and Barlow P W 1992 Specific patterns of cortical and endoplasmic microtubules associated with cell growth and tissue differentiation in roots of maize (*Zea mays*). J. Cell Sci. 103, 191–200.

Baluška F, Parker J S and Barlow P W 1993a A role for gibberellic acid in orienting microtubules and regulating cell growth polarity in the maize root cortex. Planta 191, 149–157.

Baluška F, Parker J S and Barlow P W 1993b The microtubular cytoskeleton in cells of cold-treated roots of maize (*Zea mays*) shows tissue-specific responses. Protoplasma 172, 84–96.

Baluška F, Brailsford R W, Hauskrecht M, Jackson M B and Barlow P W 1993c Cellular dimorphism in the maize root cortex: involvement of microtubules, ethylene and gibberellin in the differentiation of cellular behaviour in postmitotic growth zones. Bot. Acta 106, 394–403.

Barlow P W 1969 Cell growth in the absence of division in a root meristem. Planta 88, 215–223.

Barlow P W 1982 Cellular aspects of gravitropism in roots. *In* Plant Growth Substances. Ed. P F Wareing. pp 507–518. Academic Press, London.

Barlow P W 1989 Anatomical controls of root growth. Aspects Appl. Biol. 22, 57–66.

Barlow P W 1994 Cell divisions in meristems and their contribution to organogenesis and plant form. Bot. J. Linn. Soc., Suppl. (*In press*).

Barlow P W and Rathfelder E L 1985 Distribution and redistribution of extension growth along vertical and horizontal gravireacting maize roots. Planta 165, 134–141.

Barlow P W and Zieschang H E 1994 Root movements: towards an understanding through attempts to model the processes involved. Plant and Soil 165.

Barlow P W, Brain P and Parker J S 1991 Cellular growth in roots of a gibberellin-deficient mutant of tomato (*Lycopersicon esculentum* Mill.) and its wild-type. J. Exp. Bot. 42, 339–351.

Barlow P W, Rost T L and Gunning B E S 1982 Nuclear and cytoplasmic changes during early stages of cell differentiation in roots of the water fern, *Azolla pinnata*. Protoplasma 112, 205–216.

Baskin T I, Betzner A S, Hoggart R, Cork A and Williamson R E 1992 Root morphology mutants in *Arabidopsis thaliana*. Aust. J. Plant Physiol. 19, 427–437.

Bednara J 1974 Root growth rate and incorporation of $^3$H-thymidine following partial removal of the root tip in *Zea mays*. *In* Structure and Function of Primary Root Tissues. Ed. J Kolek. pp 23–32. Veda, Bratislava.

Benfey P N, Linstead P J, Roberts K, Schiefelbein J W, Hauser M-T and Aeschbacher R A 1993 Root development in *Arabidopsis*: four mutants with dramatically altered root morphogenesis. Development 119, 57—70.

Bergfeld R, Speth V and Schopfer P 1987 Reorientation of microfibrils and microtubules at the outer epidermal wall of maize coleoptiles during auxin-mediated growth. Bot. Acta 101, 57–67.

Björkman T and Cleland R E 1991 Root growth regulation and gravitropism in maize roots does not require the epidermis. Planta 185, 34–37.

Blancaflor E B and Hasenstein K H 1993 Organization of cortical microtubules in graviresponding maize roots. Planta 191, 231–237.

Brown R and Broadbent D 1951 The development of cells in the growing zones of the root. J. Exp. Bot. 1, 249–263.

Burström H 1942 Die osmotische Verhältnisse während des Streckungswachstum der Wurzel. Lantbr-Högsk. Ann. 10, 1–30.

Burström H 1950 Studies on growth and metabolism of roots. IV. Positive and negative auxin effects on cell elongation. Physiol. Plant. 3, 277–292.

Burström H 1957 On the adaptation of roots to $\beta$-indolylacetic acid. Physiol. Plant. 10, 187–197.

Burström H 1968 Calcium and plant growth. Biol. Rev. 43, 287–316.

Burström H 1969 Influence of the tonic effects of gravitation and auxin on the cell elongation and polarity in roots. Am. J. Bot. 56, 679–684.

Burström H 1971 Tissue tensions during cell elongation in wheat roots and a comparison with contractile roots. Physiol. Plant. 25, 509–513.

Cambell R and Drew M C 1983 Electron microscopy of gas space (aerenchyma) formation in adventitious roots of *Zea mays* L. subjected to oxygen shortage. Planta 147, 350–357.

Clarkson D T, Robards A W, Stephens J E and Stark M 1987 Suberin lamellae in the hypodermis of maize (*Zea mays*) roots, development and factors affecting the permeability of hypodermal layers. Plant Cell Environ. 10, 83–93.

Clowes F A L 1983 Exit from the mitotic cycle in root meristems of *Zea mays* L. Ann. Bot. 51, 385–393.

Collings D A, White R G and Overall R L 1992 Ionic currents changes associated with the gravity-induced bending response in roots of *Zea mays* L. Plant Physiol. 100, 1417–1426.

Dauwalder M, Roux S J and Rabenberg L K 1985 Cellular and subcellular localization of calcium in gravistimulated corn roots. Protoplasma 129, 137–148.

Deacon J W, Drew M C and Darling A 1986 Progressive cortical senescence and formation of lysigenous gas space (aerenchyma) distinguished by nuclear staining in adventitious roots of *Zea mays*. Ann. Bot. 58, 719–727.

Demchenko N P 1984 Mitotic and endoreduplication cycles in the development of the wheat root metaxylem cell lines. Tsitologiya 26, 382–391. (*In Russian*).

Deltour R, Antoine N and Bronchart R 1989 Root ontogenesis during early germination of *Zea mays*. Ann. Bot. 64, 107–116.

Eliasson L, Bertell G and Bolander E 1989 Inhibitory action of auxin on root root elongation not mediated by ethylene. Plant Physiol. 91, 310–314.

Evans M L 1991 Gravitropism: Interaction of sensitivity modulation and effector redistribution. Plant Physiol. 95, 1–5.

Fraser T E, Silk W K and Rost T L 1990 Effects of low water potential on cortical cell length in growing regions of maize roots. Plant Physiol. 93, 648–651.

Gandar P W 1980 The analysis of growth and cell production in root apices. Bot. Gaz. 141, 131–138.

Goodwin R H 1972 Studies on roots. V. Effects of indoleacetic acid on the standard root growth pattern of *Phleum pratense*. Bot. Gaz. 133, 224–229.

Hasenstein H K and Evans M L 1986 The calcium dependence of rapid auxin action in maize roots. Plant Physiol. 81, 439–443.

He C -J, Morgan P W and Drew M C 1992 Enhanced sensitivity to ethylene in nitrogen- or phosphate-starved roots of *Zea mays* L. during aerenchyma formation. Plant Physiol. 98, 137–142.

Hejnowicz Z 1961 The response of the different parts of the cell elongation zone in root to external $\beta$-indolylacetic acid. Acta Soc. Bot. Pol. 30, 25–42.

Hejnowicz Z, Krause E, Glebicki K and Sievers A 1991 Propagated fluctuations of the electric potential in the apoplasm of *Lepidium sativum* L. roots. Planta 186, 127–134.

Held B M, Wang H, John I, Wurtele E S and Colbert J T 1993 An mRNA putatively coding for an *O*-methyltransferase accumulates preferentially in maize roots and is located predominantly in the region of the endodermis. Plant Physiol. 102, 1001–1008.

Hellebust J A and Forward D F 1962 The invertase activity of the corn radicle and its activity in successive stages of growth. Can. J. Bot. 40, 113–126.

Hodick D and Kutschera U 1992 Light-induced inhibition of elongation growth in sunflower hypocotyls. Biophysical and ultrastructural investigations. Protoplasma 168, 7–13.

Hohl M and Schopfer P 1992 Cell-wall tension of the inner tissues of the maize coleoptile and its potential contribution to auxin-mediated organ growth. Planta 188, 340–344.

Holmes B E, Mee L K, Hornsley S and Gray L H 1955 The nucleic acid content of cells in the meristematic, elongating and fully elongated segments of roots of *Vicia faba*. Exp. Cell Res. 8, 101–113.

Ishida K and Katsumi M 1992 Effects of gibberellin and abscisic acid on the cortical microtubule orientation in hypocotyl cells of light-grown cucumber seedlings. Int. J. Plant Sci. 153, 155–163.

Ishikawa H and Evans M L 1990 Gravity-induced changes in intracellular potentials in elongating cortical cells of mung bean roots. Plant Cell Physiol. 31, 457–462.

Ishikawa H and Evans M L 1992 Induction of curvature in maize roots by calcium or by thigmostimulation. Role of the postmitotic isodiametric growth zone. Plant Physiol. 100, 762–768.

Ishikawa H and Evans M L 1993 The role of the distal elongation zone in the response of maize roots to auxin and gravity. Plant Physiol. 102, 1203–1210.

Ishikawa H, Hasenstein K H and Evans M L 1991 Computer-based video digitizer analysis of surface extension in maize roots. Kinetics of growth rate changes during gravitropism. Planta 181, 381–390.

Ivanov V B 1980 Specificity of spatial and time organization of root cell growth in connection with functions of the root. Sov. Plant. Physiol. 26, 720–728.

Ivanov V B 1981 Cellular basis of root growth. *In* Soviet Science Review, Biological Series Vol. 2. Ed. V S Kulachev. pp 365–392. Overseas Publications, New York.

Iwabuchi A, Yano M and Shimizu H 1989 Development of extracellular electric pattern around *Lepidium* roots: its possible role in root growth and gravitropism. Protoplasma 148, 94–100.

Jaunin F, Boutinard Rouelle-Rossier V and Hofer R-M 1991 Cytochemical and X-ray microanalytical studies of calcium in the extending zone, particularly in the rhizodermis of *Zea mays* roots. Physiol. Plant. 81, 15–23.

Jensen W A 1955 A morphological and biochemical analysis of the early phases of cellular growth in the root tip of *Vicia faba*. Exp. Cell Res. 8, 506–522.

Jensen W A and Kavaljian L G 1958 An analysis of cell morphology and the periodicity of division in root tip of *Allium cepa*. Am. J. Bot. 45, 365–372.

Katekar G F and Geissler A E 1992 On the role of the NPA receptor in the root gravitropic response mechanism. *In* Progress in Plant Growth Regulation. Eds. C M Karssen. L C Van Loon and D Vreugdenhil. pp 921–927. Kluwer Academic Publishers, Dordrecht, The Netherlands.

Konings H and Verschuren G 1980 Formation of aerenchyma in roots of *Zea mays* in aerated solutions, and its relation to nutrient supply. Physiol. Plant. 49, 265–270.

Kubica Š 1984 Chromatin condensation during cell cycles of differentiating barley root central metaxylem. Biológia (Bratislava) 39, 1067–1073.

Kubica Š 1986 Duration of the last endocycle in differentiating cells of barley root central metaxylem. Biológia (Bratislava) 41, 1053–1060.

Kubica Š, Baluška F and Gašparíková O 1989 Pattern of nucleic acids synthesis in the root apex of *Zea mays* L. Biológia (Bratislava) 44, 201–207.

Kubica Š, Baluška F and Hauskrecht M 1991 Elemental growth rate and rRNA transcript maturation exhibit the same pattern in individual tissues of the maize root apex. Ann. Bot. 68, 387–391.

Kutschera U 1989 Tissue stresses in growing plant organs. Physiol. Plant. 77, 157–163.

Kutschera U 1992 The role of the epidermis in the control of elongation growth in stems and coleoptiles. Bot. Acta 105, 246–252.

Kutschera U, Bergfeld R and Schopfer P 1987 Cooperation of epidermis and inner tissues in auxin-mediated growth of maize coleoptiles. Planta 170, 168–180.

Lachno D R, Harrison-Murray R S and Audus L J 1982 The effects of mechanical impedance to growth on the levels of ABA and IAA in root tips of *Zea mays* L. J. Exp. Bot. 33, 943-951.

Luxová M 1980 Kinetics of maize root growth. Ukr. Zhurn. Bot. 37, 68-72. (*In Russian*).

Luxová M 1981 Growth region of the primary root of maize (*Zea mays* L.). *In* Structure and Function of Plant Roots. Eds. R Brouwer, O Gašparíková, J Kolek and B C Loughman. pp 9–14. M Nijhoff/Dr W Junk, The Hague.

MacDonald I R and Hart J W 1987 New light on the Cholodny-Went theory. Plant Physiol. 84, 568–570.

McCully M E and Canny M J 1985 Localisation of translocated $^{14}C$ in roots and root exudates of field-grown maize. Physiol. Plant. 65, 380–392.

Meuwly P and Pilet P-E 1987 Maize root growth and localized indol-3yl-acetic acid treatment. Plant Physiol. 84, 1265–1269.

Meuwly P and Pilet P-E 1991 Local treatment with indole-3-acetic acid induces differential growth responses in *Zea mays* L. roots. Planta 185, 58–64.

Mitchison T and Kirschner M W 1984 Dynamic instability of microtubule growth. Nature 312, 237–242.

Muday G K 1994 Tomato root growth, gravitropism, and lateral development. Plant Physiol. Biochem. 32, 193–203.

Nantawisarakul T and Newman I A 1992 Growth and gravitropism of corn roots in solution. Plant Cell Environ. 15, 693–701.

Nick P, Bergfeld R, Schäfer E and Schopfer P 1990 Unilateral reorientation of microtubules at the outer epidermal wall during photo- and gravitropic curvature of maize coleoptiles and sunflower hypocotyls. Planta 181, 162–168.

Nick P, Schäfer E and Furuya M 1992 Auxin redistribution during first positive phototropism in corn coleoptiles. Plant Physiol. 99, 1302–1308.

Ohwaki Y and Tsurumi S 1976 Auxin transport and growth in intact roots of *Vicia faba*. Plant Cell Physiol. 17, 1329–1342.

Pahlavanian A M and Silk W K 1987 Effect of temperature on spatial and temporal aspects of growth in the primary maize root. Plant Physiol. 87, 529–532.

Parets-Soler A, Pardo J M and Serrano R 1990 Immunocytolocalization of plasma membrane H-ATPase. Plant Physiol. 93, 1654–1658.

Pilet P -E and Meuwly P 1986 Local application of indole-3-acetic acid, by resin beads to intact growing maize roots. Planta 169, 16–22.

Poff K L and Martin H V 1989 Site of graviperception in roots: a re-examination. Physiol. Plant. 76, 451–455.

Pritchard J, Wyn Jones R G and Tomos A D 1988 Control of wheat root growth. The effects of excision on growth, wall rheology and root anatomy. Planta 176, 399–405.

Pritchard J, Barlow P W, Adam J S and Tomos A D 1990 Biophysics of the inhibition of the growth of maize roots by lowered temperature. Plant Physiol. 93, 222–230.

Pritchard J, Hetherington P R, Fry S C and Tomos A D 1993 Xyloglucan endotransglycolase activity, microfibril orientation and the profiles of cell wall properties along growing regions of maize roots. J. Exp. Bot. 44, 1281–1289.

Rufelt H 1959 Influence of coumarin on geotropism and growth of wheat roots. Svensk Bot. Tidskr. 53, 312–318.

Saab I N, Sharp R E and Pritchard J 1992 Effect of inhibition of abscisic acid accumulation on the spatial distribution of elongation in the primary root and mesocotyl of maize at low water potentials. Plant Physiol. 99, 26–33.

Sarquis J I, Jordan W R and Morgan P W 1991 Ethylene evolution from maize (*Zea mays* L.) seedling roots and shoots in response to mechanical impedance. Plant Physiol. 96, 1171–1177.

Schilstra M J, Bayley P M and Martin S R 1991 The effect of solution composition on microtubule dynamic instability. Biochem. J. 277, 839–847.

Schobert C and Komor E 1987 Amino acids uptake by *Ricinus communis* roots: characterization and physiological significance. Plant Cell Environ. 10, 493-500.

Sharp R E, Silk W K and Hsiao T C 1988 Growth of the maize primary root at low water potentials. I. Spatial distribution of expansive growth. Plant Physiol. 87, 50–57.

Shen-Miller J and McNitt R E 1978 Quantitative assessment of the ultrastructural changes in primary roots of corn (*Zea mays* L.) after geotropic stimulation. Plant Physiol. 61, 649–653.

Shen-Miller J, McNitt R E and Wojciechowski M 1978 Regions of differential cell elongation and mitosis, and root meristem morphology in different tissues of geotropically stimulated maize root apices. Plant Physiol. 61, 7–12.

Sievers A and Hejnowicz Z 1992 The gravitropic response chain of roots: new perspectives. *In* Plant Cell Walls as Biopolymers with Physiological Functions. Ed. Y Masuda. pp 13-18. Yamada Science Foundation, Osaka, Japan.

Sievers A and Zieschang H E 1992 What remains of the Choldny-Went theory? It does not fit the growth pattern of cells during bending of a root. Plant Cell Environ. 15, 789–790.

Silk W K and Erickson R O 1980 Local biosynthetic rates of cytoplasmic constituents in growing tissue. J. Theoret. Biol. 83, 701–703.

Silk W K, Hsiao T C, Diedenhofen U and Matson C 1986 Spatial distributions of potassium, solutes, and their deposition rates in the growth zone of the primary corn root. Plant Physiol. 82, 853–858.

Spollen W G and Sharp R E 1991 Spatial distribution of turgor and root growth at low water potentials. Plant Physiol. 96, 438–443.

Stenlid G 1957 A comparison of the toxic effects of some sugars upon growth and chloride accumulation in young wheat roots. Physiol. Plant. 10, 807–823.

Stenz H-G and Weisenseel M H 1993 Electrotropism of maize (*Zea mays* L.) roots. Facts and artifacts. Plant Physiol. 101, 1107–1111.

Stinemetz C 1992 Altered IAA responsiveness in the elongation region of graviresponding roots is independent of the root tip. Plant Physiol. Suppl. 99, 367.

Svensson S-B 1971 The effect of coumarin on root growth and root histology. Physiol. Plant. 24, 446–470.

Tsurumi S and Ohwaki Y 1978 Transport of $^{14}C$-labeled indoleacetic acid in *Vicia faba* root segments. Plant Cell Physiol. 19, 1195–1206.

Varney G T and McCully M E 1991 The branch roots of *Zea*. II. Developmental loss of the apical meristem in field-grown roots. New Phytol. 118, 535–546.

Voetberg G S and Sharp R E 1991 Growth of the maize primary root at low water potentials. III. Role of increased proline deposition in osmotic adjustment. Plant Physiol. 96, 1125–1130.

Webster P L 1980 Analysis of heterogeneity of relative division rates in root apical meristems. Bot. Gaz. 141, 353–359.

Weisenseel M H, Becker H F and Ehlgötz J G 1992 Growth, gravitropism, and endogenous ion currents of cress roots (*Lepidium sativum* L.). Plant Physiol. 100, 16–25.

Whalen M C and Feldman L J 1986 The effect of ethylene on root growth of *Zea mays* seedlings. Can. J. Bot. 66, 719–723.

Wilson A J, Robards A W and Goss M J 1977 Effects of mechanical impedance on root growth in barley, *Hordeum vulgare* L. II. Effects on cell development in seminal roots. J. Exp. Bot. 28, 1216–1227.

Yang R L, Evans M L and Moore R 1990 Microsurgical removal of epidermal and cortical cells: evidence that the gravitropic signal moves through the outer cell layers in primary roots of maize. Plant Physiol. 180, 530–536.

Young L M, Evans M L and Hertel R 1990 Correlations between gravitropic curvature and auxin movements across gravistimulated roots of *Zea mays*. Plant Physiol. 92, 792–796.

Zandomeni K and Schopfer P 1993 Reorientation of microtubules at the outer epidermal wall of maize coleoptiles by phytohormone, blue-light photoreceptor, and auxin. Protoplasma 173, 103–112.

Zieschang H E and Sievers A 1991 Graviresponse and the localization of its initiating cells in roots of *Phleum pratense* L. Planta 184, 468–477.

F. Baluška et al. (eds.), Structure and Function of Roots, 53–59, 1995.


# The completion of cell proliferation and growth in wheat radicle

N.P. Demchenko
*Komarov Botanical Institute, Russian Academy of Sciences, Prof. Popov str. 2, 197376 Sanct-Petersburg, Russia*

*Key words:* DNA contents, DNA synthesis, growth, proliferation, radicle, wheat

## Abstract

The pattern of proliferation and growth of cortical and central metaxylem cells in a radicle and the transitional zone of a wheat embryo was studied during the final stages of embryogenesis. Cell divisions finished nearer the root tip in the central metaxylem than was the case in the cortex. After divisions ceased the cells of both tissues maintained the ability to synthesize DNA and the cells began DNA endoreduplication. The maximal levels of endoreduplication were 4C and 8C in cortical and central metaxylem cells, respectively. As a result of nonsimultaneous cessation of divisions, the metaxylem cells were two or three times longer than cortical cells. The proportion of cells with the maximal DNA content was smaller in the transitional zone than in the radicle. During the final embryonal stages cell growth rate was decreased. It was established that the transition of cells to DNA synthesis was inhibited in all sites of the radicle during the completion of embryogenesis. The cell growth was stopped in proximal sites of the radicle. In the division zone the cells which had already begun DNA synthesis were able to complete it and divided. Cell growth stopped simultaneously with completion of proliferation in this zone.

## Introduction

The processes of growth and cell proliferation during embryogenesis have not yet been studied well enough. The characters of these processes have been usually estimated by the changes of the dimensions of embryo or embryonal axis as well as of the cell number (Danovich, 1982). Changes in cell dimensions were studied using morphometric method in developing roots of in vitro cultured immature maize embryos (Van Zeist et al., 1990). There are several reviews of works in which a DNA content has been determined in the quiescent embryonic cells of various plants (Barlow, 1973; Deltour, 1985; Obrucheva, 1982; Olszewska, 1977; Troyan et al., 1983). In these works it was established that the majority of cells had a 2C DNA content. However, considerable a proportion of cells had a 4C DNA content was discovered in radicles of *Triticum durum, Zea mays, Vicia faba* and *Pisum sativum.* It was shown that this proportion of 4C cells depended on the environmental conditions at the time of seed maturation (Bogdanov et al., 1967). In dormant embryos of *T. durum* the proportion of 4C cells also differed according to the type of apex (primary root, seminal roots of the first and second pairs) (Avanzi et al., 1969). In radicles cells having a 4C DNA content can be induced by an arrest during the short time interval between DNA synthesis and mitosis (Avanzi et al., 1969; Brunori et al., 1966; Brunori, 1967). Alternatively in some tissues, some of the cells that have completed a DNA synthesis have already passed to differentiation (Avanzi et al., 1963; Brunori et al., 1966). Brunori and Ancora (1968) established that in the radicle of dry seeds of *Allium cepa* the majority of nuclei with a 4C DNA content belonged to differentiated cells. However, in the above mentioned works the DNA content of the embryonic nuclei was determined on squash preparations. This technique does not allow the exact the location of the cells having a 4C DNA content to be determined. The proportion of cells having a 2C or 4C nuclear DNA content in different sites and tissues of a radicle has only been investigated on the sections of embryos of *Pisum sativum* (Jones, 1977) and *Zea mays* (Barlow, 1987). However, these authors did not observe the connection between the occurrence of the cells having the nucleus with a 4C DNA content and the stage of the tissue development during embryogenesis. Luxova (1986) has given her opinion that cells of the proximal part of the radicle are more advanced in the development at the completion of barley embryogenesis in comparison with cells of distal part. Upon germination the cells of the proximal part begin elongation and progressive differentiation and do not proliferate.

Embryos in the caryopses of *Triticum aestivum* are highly organized. The cell-files of all root tissues can be easily observed in longitudinal sections. In the present study the changes of the growth rate of radicular cells in the course of development, and lengths of division zones in cortical and central metaxylem files along with the localization of the cells having a nucleus with a 4C DNA content were established. Besides, the sequence of inhibitions of growth, DNA synthesis and division of radicular cells was established at the completion of embryogenesis.

## Materials and methods

Caryopses of winter wheat *Triticum aestivum* ssp *lutescens* cv. "Bezostaya-1" were germinated in darkness at 25±1°C on moistened filter paper. Under such conditions DNA synthesis in root cells begins after 15 hr (Mory et al., 1972). Radicle growth and cell divisions begin later (Grif, 1958). Therefore, the caryopses were fixed after 11 hr of germination. Caryopses were fixed in Navashin's fixative (chromic acid, formalin, glacial acetic acid, 10:4:1 v/v/v) and were embedded in paraffin wax. Longitudinal sections (10 $\mu$m) were stained by Feulgen's reaction for 4 hr, hydrolysis being accomplished for 30 min. in 6*N* HC1 solution at 22°C. The sections were rinsed in freshly prepared $SO_2$ water. Cell walls were stained for 40 min. with a 0,5% solution of Alcian blue in 3% acetic acid. Sections were dehydrated, and mounted under a coverslip in Canada Balsam.

One or two files of outer cortical layers and the central metaxylem files were investigated, working from the initial cell (IC) to the most proximal cell of each file in medium longitudinal sections (see Fig. 1A-D). In each file, cell lengths were measured with an ocular micrometer calibrated to the nearest micron. Twelve cortical cell-files from 7 embryos and eleven central metaxylem cell-files were studied.

The relative DNA content of the nucleus of each cell of a file was determined using a Opton's SMP-01 microscope-photometer with the wavelength of the scanning light beam set at 550 nm. The total relative DNA content in the nucleus was estimated from all sections of the nucleus.

## Results

The changes in cell length and nuclear DNA content in a cortical file of an unemerged wheat radicle is represented in Figure 2A. The minimum and maximum lengths of cortical cells changed little along the radicular axis. The initial and the most distal cells of a file, as well as some cells located in the root base and the transitional zone, were longer than cells located in others sites of a file. In the root base and the transitional zone these cells were formed as a result of periclinal divisions. Therefore, the radicular cells did not pass to elongation in the course of embryogenesis. In the nuclei of the IC and the following 41 cells of a file, DNA content was constant, although the lengths of these cells varied. In the nucleus of cell number 42 the DNA content was double that was of the nuclei of the preceding cells. The latter were considered to have a 2C DNA content. Cells with a 4C DNA content more often occurred towards the root base. The longest cells always had a 4C DNA content. The minimal lengths of cells with a 2C and 4C nuclear DNA content were 5 $\mu$m and 7 $\mu$m , respectively. In each file studied, cells at the apical zone had a 2C nuclear DNA content only. The length of this zone varied from 185 $\mu$m to 444 $\mu$m (mean 347±20 $\mu$m) in various files. The number of cells in this zone varied from 19 to 60 cells (mean 43±3 cells).

In the central metaxylem file the initial and several distal cells also had a nucleus with a 2C DNA content (Fig. 2B). However, further away from the IC cell length and nuclear DNA content increased, the maximal DNA content being 8C. The length of the apical zone where the distal cells of a central metaxylem file had only a 2C DNA content varied from 33 $\mu$m to 165 $\mu$m (mean 97±14 $\mu$m). The number of cells in this zone varied from 3 to 23 cells (mean 13±2 cells).

The changes of cell length and mean nuclear DNA content in all files studied in the cortex and in the central metaxylem along the axis of the unemerged wheat radicle are represented in the Tables 1 and 2. The mean length of the IC of the cortical and metaxylem files was larger than that of the distal cells. Cell length changed little at distances 101–500 $\mu$m and 1–200 $\mu$m from the IC in the cortex and the metaxylem, respectively. In cortical files the mean cell length increased slightly at a distance > 501–600 $\mu$m from the IC. In central metaxylem files cell length gradually increased at distances more 200 $\mu$m from the IC and reached the maximal level 501–600 $\mu$m from the IC. In latter site the mean cell length was three times larger than at sites 1–

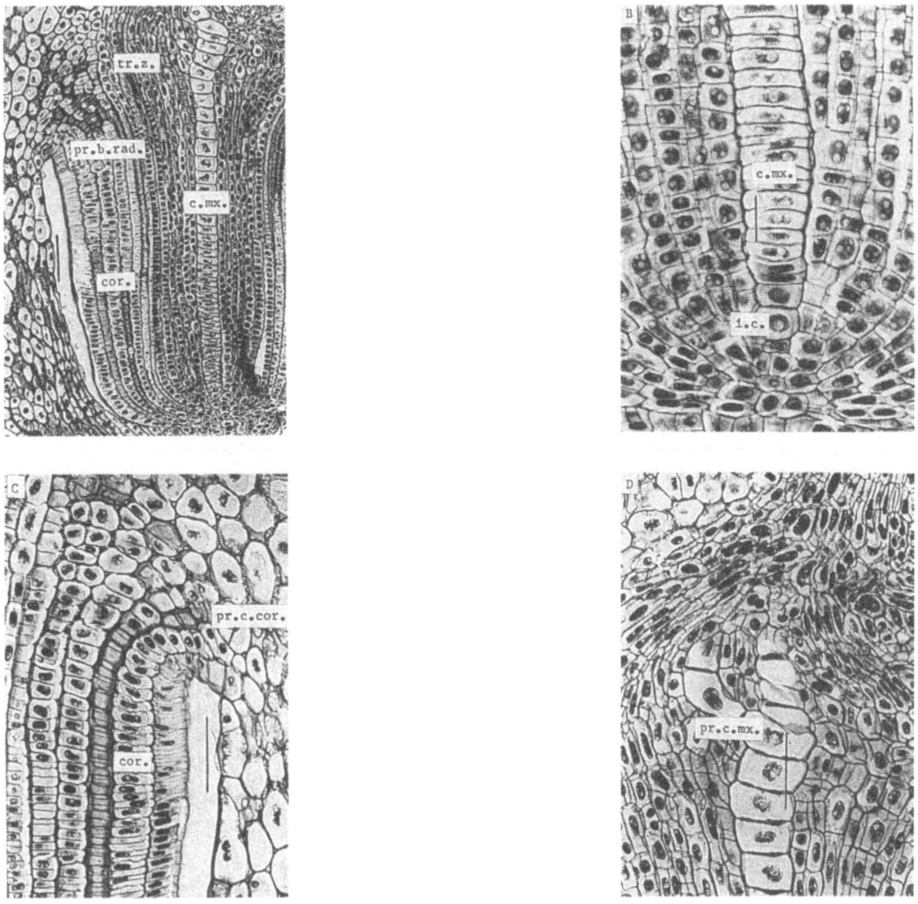


*Fig. 1.* Longitudinal sections of embryos in caryopses of *Triticum aestivum* L. after 11 hr of the germination. **A**, a radicle. rad. — a radicle, c.mx. — a central metaxylem cell-file, cor. — a cortical cell-file, pr.b.rad. — the proximal boundary of a radicle, tr.z. — the tranzitional zone. The *bar* represents 100 μm. **B**, the distal part of a radicle. c.mx. — a central metaxylem cell-file, i.c. — the initial cell of the central metaxylem file. The *bar* represents 20 μm. **C**, the proximal part of a radicle. cor. — a cortical cell-file, pr.c.cor. — the proximal cell of the cortical file. The *bar* represents 50 μm. **D**, the transitional zone. pr.c.mx. — the proximal cell of the central metaxylem file. The *bar* represents 50 μm.

200 μm from the IC. In sites located 601–700 μm and further away from the IC cell length was unchanged or slightly decreased. The proximal boundary of the radicle occurred at distances 490–665 μm (mean 575±14 μm) and 410–625 μm (mean 506±20 μm) from the IC in cortical and metaxylem files, respectively. Files located beyond this boundary occupy the transitional zone.

In the cortex the mean nuclear DNA content was slightly different in the initial and the following cells of a file up to about 400 μm from the IC. In more proximal sites it gradually increased and reached a maximal level in sites 501–600 μm from the IC. In sites located 601–800 μm from the IC the mean nuclear DNA content changed little. In central metaxylem it was little different in the initial and the following cells up to a distance of no more than 100 μm from the IC. In proximal sites the mean DNA content then gradually increased and reached a maximum level at a distance of 501–600 μm from the IC. In the latter site the mean DNA content was three times larger than that found at a distance 1–200 μm from the IC. In sites located at the distances 601–700 μm and 701–800 μm from the IC it decreased.

The mean nuclear DNA content of cortical and central metaxylem cells changed along radicle axis as a result of the increase in the proportion of cells in which DNA endoreduplication occurred (Tables 1, 2; Fig. 2A, B). The maximal level of DNA endoreduplication was 4C and 8C DNA content in cortical and central metaxylem cells, respectively. Cells that had stopped in S-phase of the mitotic or the endoreduplication cycles

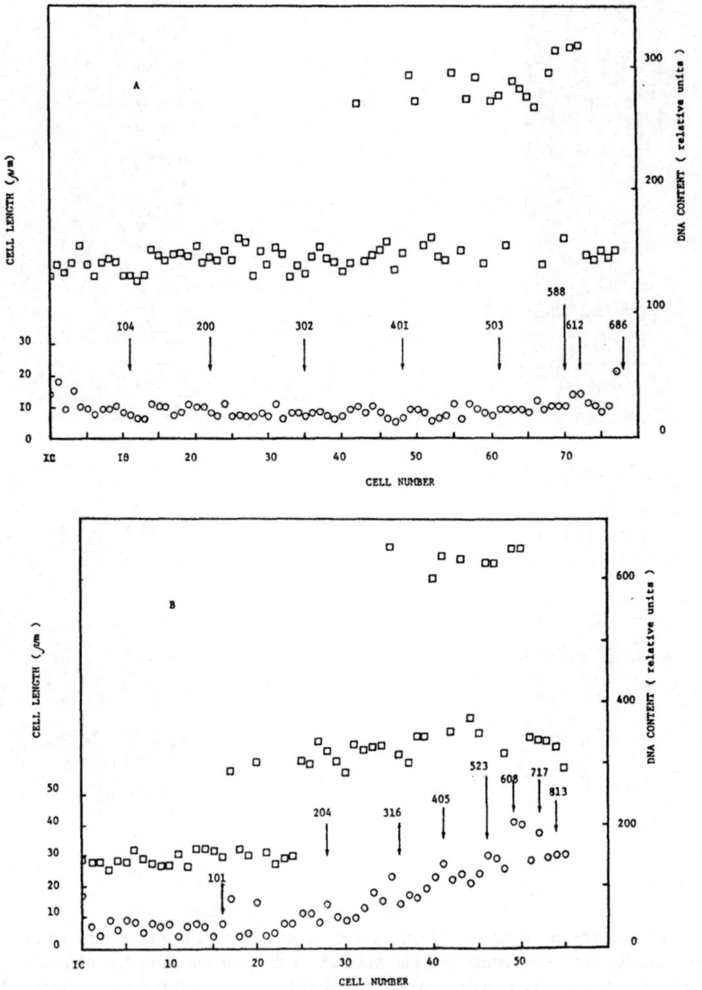


*Fig. 2.* Changes of cell length (○) and DNA content in the nucleus (□) in cortical (**A**) and central metaxylem (**B**) files of the unemerged radicle in the caryopsis of *Triticum aestivum* L. after 11 hr of the germination. IC - the initial cell of a file. The distances of the cells from the IC are marked by arrows (μm); the distance of the proximal boundary of a radicle from the IC is marked by longer arrow (μm).

were not found. Most cortical cells had a 4C DNA content and the central metaxylem cells had an 8C DNA content in sites located at a distance 501–600 μm from the IC where the root boundary is usually found. In the transitional zone the proportion of these metaxylem cells decreased.

## Discussion

The results of the present study show that the majority of wheat radicular cells having a nucleus with a 4C DNA content was formed as a result of the sequence of cessation of cell proliferation. The inhibition of the transition to division occurred before that of the transition to DNA synthesis. The distances of the sites (or the boundaries) from the IC in which the inhibition of the transition to division occurred are different in the various tissues. The transition to division was inhibited nearer the root tip in the central metaxylem than it was in the cortex. Therefore, some cortical cells had time to complete one endoreduplication cycle while some central metaxylem cells had time to complete two. However, in the growing wheat root some central metaxylem cells completed three endoreduplication cycles and the proportion of cortical cells having the nuclei with a 4C DNA content was greater than in the radicle (Demchenko, 1981, 1982, 1984a). In the latter, some of the cells did not have time to begin either the first DNA endoreduplication (cortex) or the next round of it (cen-

*Table 1.* Changes of cell length, nuclear DNA content and the distribution of DNA contents of nuclei in the cortex along of the longitudinal axis of the unemerged radicle in the caryopsis of *Triticum aestivum* L. after 11 hr of the germination

| Distance from the initial cell of a file, μm | The number of cells investigated | Mean cell length ± SE μm | Mean nuclear DNA content ± SE relative units | The distribution of DNA contents of nuclei, % ± SE | |
|---|---|---|---|---|---|
| | | | | 2C | 4C |
| Initial cell | 12 | 17,6±1,6 | 142±4 | 100 | 0 |
| 1–100 | 118 | 10,4±0,4 | 142±1 | 100 | 0 |
| 101–200 | 155 | 7,6±0,2 | 146±1 | 99±1 | 1±1 |
| 201–300 | 168 | 7,2±0,1 | 146±2 | 98±1 | 2±1 |
| 301–400 | 162 | 7,3±0,2 | 154±3 | 93±2 | 7±2 |
| 401–500 | 145 | 8,0±0,2 | 177±5 | 77±4 | 23±4 |
| 501–600 | 129 | 9,7±0,3 | 207±6 | 52±8 | 48±8 |
| 601–700 | 80 | 11,8±0,4 | 197±8 | 61±9 | 39±9 |
| 701–800 | 11 | 13,7±1,2 | 198±23 | 61±20 | 39±20 |

*Table 2.* Changes of cell length, mean nuclear DNA content and the distribution of DNA contents of nuclei in the central metaxylem along of the longitudinal axis of the unemerged radicle in the caryopsis of *Triticum aestivum* L. after 11 hr of the germination

| Distance from the initial cell of a file, μm | The number of cells investigated | Mean cell length ± SE μm | Mean nuclear DNA content ± SE relative units | The distribution of DNA contents of nuclei, % ± SE | | |
|---|---|---|---|---|---|---|
| | | | | 2C | 4C | 8C |
| Initial cell | 11 | 16,7±0,8 | 160±4 | 100 | 0 | 0 |
| 1–100 | 142 | 8,0±0,2 | 170±4 | 92±3 | 8±3 | 0 |
| 101–200 | 131 | 8,5±0,2 | 217±7 | 62±5 | 38±5 | 0 |
| 201–300 | 98 | 11,5±0,3 | 300±9 | 11±6 | 85±6 | 4±2 |
| 301–400 | 70 | 15,6±0,4 | 386±15 | 0 | 73±9 | 27±9 |
| 401–500 | 55 | 21,9±0,5 | 444±20 | 2±2 | 50±9 | 47±9 |
| 501–600 | 42 | 28,1±0,8 | 492±22 | 0 | 39±9 | 61±9 |
| 601–700 | 37 | 28,5±1,2 | 419±23 | 0 | 58±12 | 42±12 |
| 701–800 | 29 | 24,7±1,3 | 316±8 | 0 | 97±3 | 3±3 |
| 801–900 | 15 | 26,6±1,7 | 356±26 | 0 | 86±9 | 14±9 |

tral metaxylem) because the completion of radicular development had begun. This started with the inhibition of the transition of cells to DNA synthesis in all sites and the cessation of cell growth in the root base. In proximal sites of the radicle the cells that had begun DNA synthesis completed it. The remaining cells finished proliferation. Most cells with a 4C and 8C nuclear DNA content occurred in the root base (Tables 1, 2). However, cell length did not increase in this site. In the remaining sites cells continued to grow and divide. Cells which began DNA synthesis were able to complete it and divided. Probably, some cells did not have time to divide because the transition to mitosis was inhibited (Avanzi et al., 1969; Brunori et al., 1966; Brunori, 1967). However, there were very few of these cells in the radicle. In my work such cells could have occurred in the central metaxylem where the most distal cell of a file had a nucleus with a 4C DNA content (cell number 4; see Results). Probably, the cell growth rate had decreased during the final stages of radicle development. The minimal lengths of cortical and central metaxylem cells in the radicle (Fig. 2A,

B) were less than those of growing roots (Demchenko, 1984a, b). Cell growth had been stopped together with the cessation of cell proliferation in these sites.

In the transition zone a small proportion of the cells had the maximal DNA content. Probably, cell growth and proliferation stopped earlier in the transitional zone than in the radicle itself.

My data are confirmed by other researches (Avanzi et al., 1963, 1969; Brunori, 1967; Brunori and Ancora, 1968) that at a late stage of embryo development DNA synthesis ceases before mitosis. However, the causes of such a sequence of inhibitions are obscure. It is likely that DNA synthesis is more sensitive to the dehydration of the seed or embryo than mitosis. However, data demonstrating the relationship between DNA synthesis and mitosis in the radicle and the water content of the seed or embryo are lacking (Deltour, 1985).

During the germination of rye, wheat and barley caryopses the first mitoses in the roots were observed at certain distances from the cap junctions (Luxova, 1986; Pukhal'skaya, 1949). This fact has been explained that the basal sites could be made up of a greater proportion of cells with a 4C nuclear DNA content which resume proliferation before those with a 2C DNA content (Barlow, 1987). Not enough work in which the direct observation of this sequence during germination has been done. On germination of *Allium cepa* seeds the proportion of these cells (with a 4C DNA content) in the radicle was only 0,18% (Brunori and Ancora, 1968). I believe that where inhibition of the transition to mitosis has occurred the cell is not capable of starting proliferation during germination. The zone in a file where this occurs is between the most distal cell with a 4C DNA content and the cell located at the root boundary (see Fig. 2A, B). Radicular cells located in an apical zone of a file where the proliferation has been stopped at the transition to DNA synthesis (2C nuclei) may be capable of resuming proliferation (see Fig. 2A, B). Upon germination the first mitoses occur at a certain distance from the cap junction as a result of the transfer of metabolites necessary for the cell transitions to DNA synthesis and mitosis. These metabolites reach the proximal cells which retain the ability to proliferate. Upon germination, the beginning of cell elongation takes place in the same order, cells located at the root base are beginning to elongate before the others (Nikitina, 1991).

## References


Avanzi S, Brunori A, Nuti-Ronchi V and Scarascia-Mugnzza G T 1963 Occurrence of 2C (G1) and 4C (G2) nuclei in the radicle meristems of dry seeds in *Triticum durum* Desf. Its implication in studies on chromosome breakage and development processes. Caryologia 16, 553–558.

Avanzi S, Brunori A and D'Amato F 1969 Sequential development of meristems in the embryo of *Triticum durum*. A DNA autoradigraphic and cytophotometric analysis. Dev. Biol. 20, 368–377.

Barlow P W 1973 Mitotic cycles in root meristems. *In* The Cell Cycle in Development and Differentiation. Eds. M Balls and F S Billett. pp 133–165. Cambridge University Press.

Barlow P W 1987 Cellular packets, cell division and morphogenesis in the primary root meristem of *Zea mays* L. New Phytol. 105, 27–56.

Bogdanov Yu F, Liapunova N A and Sherudilo A I 1967 Cell population in pea embryos and root tip meristem. Microphotometric and autoradiographic studies. Tsitologiya 9, 569–576 *(In Russian)*.

Brunori A 1967 Relationship between DNA synthesis and water content during ripening of *Vicia faba* seed. Caryologia 20, 333–338.

Brunori A and Acora G 1968 The DNA content of nuclei in the embryonic root apices of dry seeds of *Allium cepa* and their radiation response. Caryologia 21, 261–269.

Brunori A, Avanzi S and D'Amato F 1966 Chromatid and chromosome aberrations in irradiated dry seeds of *Vicia faba*. Mutat. Res. 3, 305–313.

Danovich K N 1982 The structure and the formation of a seed. *In* Seed Physiology. Ed. A A Prokof'ev. pp 5–47. Nauka, Moscow *(In Russian)*.

Deltour R 1985 Nuclear activation during early germination of the higher plant embryo. J. Cell Sci. 75, 43–83.

Demchenko N P 1981 Cytophotometrical and autoradiographical study of distribution in mitotic cycle of primary cortex cells in roots of *Triticum aestivum* L. 1. Zone of cell division. Tsitologiya 23, 1247–1255 *(In Russian)*.

Demchenko N P 1982 Cytophotometrical and autoradiographical study of distribution in mitotic cycle of primary cortex cells in roots of *Triticum aestivum* L. 2. A zone of DNA synthesis in cells without their division. Tsitologiya 24, 17–25 *(In Russian)*.

Demchenko N P 1984a Mitotic and endoreduplication cycles in the development of the wheat root metaxylem cell lines. Tsitologiya 26, 382–391 *(In Russian)*.

Demchenko N P 1984b Passage of sister cells of wheat root cortex through mitotic cycle phases. Tsitologiya 26, 552–559 *(In Russian)*.

Grif V G 1958 Study of growth processes in winter rye vernalized at temperature below 0°C. Fiziol Rast 5, 524–529 *(In Russian)*.

Jones P A 1977 Nuclear DNA content of embryonic radicles and cultured stationary phase root tips of pea *(Pisum sativum* L.). Am. J. Bot. 64, 455–460.

Luxova M 1986 The seminal root primordia in barley and the participation of their non-meristematic cells in root construction. Biol. Plant. (Praha) 28, 161–167.

Mory Y Y, Chen D and Sarid S 1972 Onset of deoxyribonucleic acid synthesis in germinating wheat embryos. Plant Physiol. 49, 20–23.

Nikitina L I 1991 Structural peculiarities of germinated barley caryopses formed under different temperature conditions. Fiziol. Biokhim. Kult. Rast. 23, 490–499 *(In Russian)*.

Pukhal'skaya E C 1949 The localization of the first mitoses in root meristems of higher plants. Dokl. Akad. Nauk. SSSR 68, 915–917 *(In Russian)*.

Obrucheva N V 1982 Seed germination. *In* Seed Physiology. Ed. A A Prokof'ev. pp 223–274. Nauka, Moscow *(In Russian)*.

Olszewska M 1977 Mechanizmy regulacji cyklu komorkowego u roslin. Post. Biol. Komorki 4, 141–183.

Troyan V M, Yaworskaya V K and Kalinin F L 1983 Biochemical and temporal parameters of cell as a basis of the regulation of growth and plant development. *In* Biochemistry of the Regulation of Plant Cell Ontogenesis. Ed. F L Kalinin. pp 7–149. Naukova Dumka, Kiev *(In Russian)*.

Van Zeist C M, Van Lammeren A A and Schel J H N 1990 Changes in cell dimensions in developing roots of in vitro cultured immature maize embryos *(Zea mays* L.). Bot. Gaz. 151, 158–166.

*F. Baluška et al. (eds.), Structure and Function of Roots*, 61–67, 1995.


# Structural basis of root hair formation: Early development of trichoblasts and atrichoblasts

Dieter Volkmann and Petra Peters
*Botanisches Institut, Universität Bonn, Venusbergweg 22, D-53115 Bonn, Germany*

*Key words:* cress root, development, root hairs

## Abstract

The development of rhizodermis cells — trichoblasts and atrichoblasts — was investigated in cress radicles by light and electron microscopy. In serial transverse sections — progressing acropetally from the area of root hair formation to the apical meristem prospective trichoblasts could be identified by the following characteristics already at the meristematic level:

1. their localization above the anticlinal cell wall of the cortex cells, 2. their wedge-like shape, 3. their remarkable dense cytoplasm, and 4. their cell length. These characteristics indicated that a large number of trichoblasts was alreadey determined during embryogenesis. The development of dermatogen cells into prospective trichoblasts commenced when the subjacent cells of the outer cortex divided longitudinally in anticlinal direction, a process which caused changes in the arrangement of the intercellular system. Dermatogen cells than grew into these newly developed intercellular space and assumed characteristics of prospective trichoblasts.

Thin sectioning electron microscopy demonstrated that prospective trichoblasts possess more ribosomal particles and smaller vacuoles indicating a more meristematic character of this cell type. The cell wall between trichoblasts and cortex cells was characterized by numerous plasmodesmata mainly grouped in large pit fields. In prospective trichoblasts and atrichoblasts a large number of cortical microtubules was observed running perpendicular to the longitudinal axis of the cells.

*Abbreviations:* pats – prospective atrichoblasts, pts – prospective trichoblasts

## Introduction

Histogenesis of higher plants originates during embyogenesis. The further development of specialized tissues starts with the germination of the seed. The rhizodermis of many higher plants is characterized by a pattern of cell differentiations where rows of hair cells (trichoblasts) alternate with rows of non-hair cells (atrichoblasts) (Guttenberg, 1960). Whereas a considerable number of investigations are concerned with the elongation of unipolar growing root hairs (for a review see Sievers and Schnepf, 1981), detailed information on the differentiation process of trichoblasts and atrichoblasts is rare (compare Robards, 1983; Hofer, 1991). Bünning (1951), using roots from white mustard, was able to distinguish between prospective trichoblasts and atrichoblasts under the light microscope. He argued that the relative isolation of dermatogen cells due to the system of intercellular spaces around the outer cortical cells might cause the differentiation of dermatogen cells into trichoblasts. We investigated the development of dermatogen cells into trichoblasts and atrichoblasts by serial sections under the light and electron microscope. Additionally, we used freeze-fracture electron microscopy to get more information concerning the communication via plasmodesmata between prospective trichoblasts and their neighbouring cells.

## Material and methods

Dry seeds from cress (*Lepidium sativum* L., Chrysanth Bonn, FRG) were soaked in tap water for 30 min. The seeds were germinated on filter paper in a chamber under high humidity for 24 h in the dark. For light,

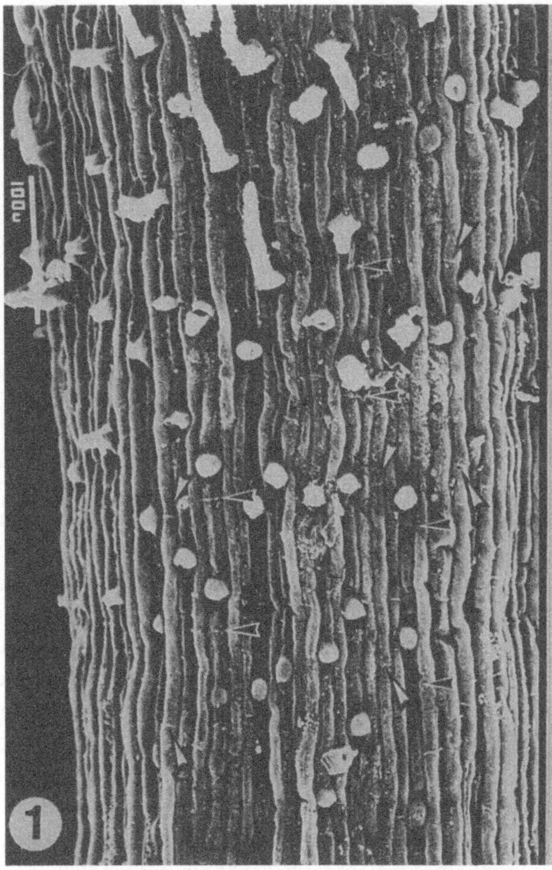


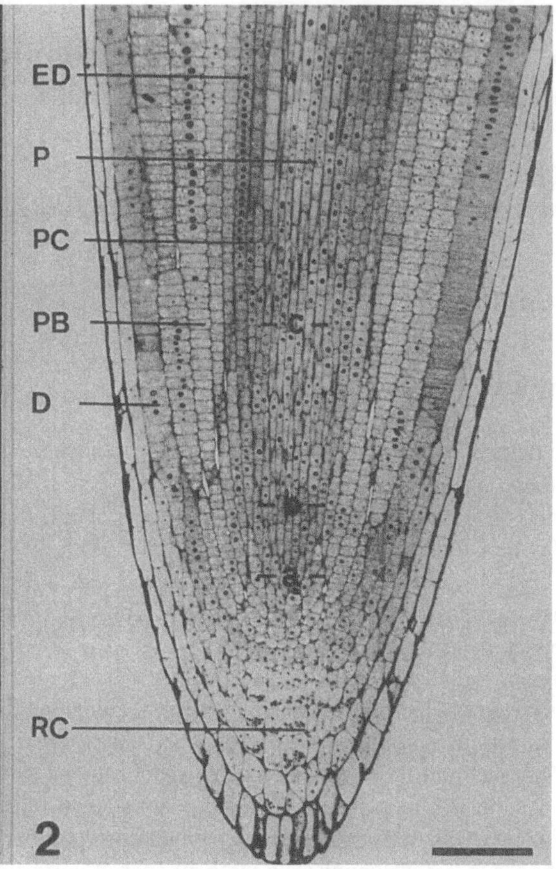


*Fig. 1.* Rhizodermis of a cress root after scanning electron microscopy showing the pattern of root hair formation. One longitudinal row of trichoblasts alternates with one to three rows of atrichoblasts. Transversal cell walls of trichoblasts and atrichoblasts are indicated by small and large arrowheads respectively. *Bar* 100$\mu$m. Magnification 220×.

*Fig. 2.* Median longitudinal section through the root tip of a 24-h-old radicle from cress. The root cap (RC) and histogens, the dermatogen (D), periblem (PB) with endodermis (ED), pericambium (PC) and plerome (P) are visible. Cross sections of the following figures 3a, b, c are indicated by a, b, c. *Bar* 100$\mu$m. Magnification 150×.

electron and freeze-fracture electron microscopy roots were treated as described elsewhere (Volkmann,1981). For scanning electron microscopy the same fixation procedure was performed. After fixation in glutaraldehyde the samples were dehydrated in alcohol series, critical point dried and sputtered with gold. Light microscopy was done with a Zeiss photomicroscope III, electron microscopy with a Zeiss EM 10 at 60 kV, scanning electron microscopy with a Leitz AMR 1200 at 40 kV. Freeze fracturing was performed in Balzers freeze-fracture apparatus BAF 300.

## Results

The region of root hair initiation in cress coincided with the end of the root cap tissue. In general, it was the region of extensive elongation, i.e. approximately 2000 to 2500 $\mu$m from the tip. The rhizodermis of cress was characterized by a relatively strict pattern (Fig. 1) where normally one longitudinal row of trichoblasts alternated with one or two rows of atrichoblasts. The cell length of atrichoblasts was always greater than that of trichoblasts. In the region of hair initiation, the mean length of trichoblasts was 100 $\mu$m, that of atrichoblasts 150 $\mu$m.

Initiation of root hairs was not strictly synchronized between different rows of trichoblasts. It occurred in a wide band of about 400 $\mu$m in the axial direction. At the cellular level, the site of hair outgrowth was mainly

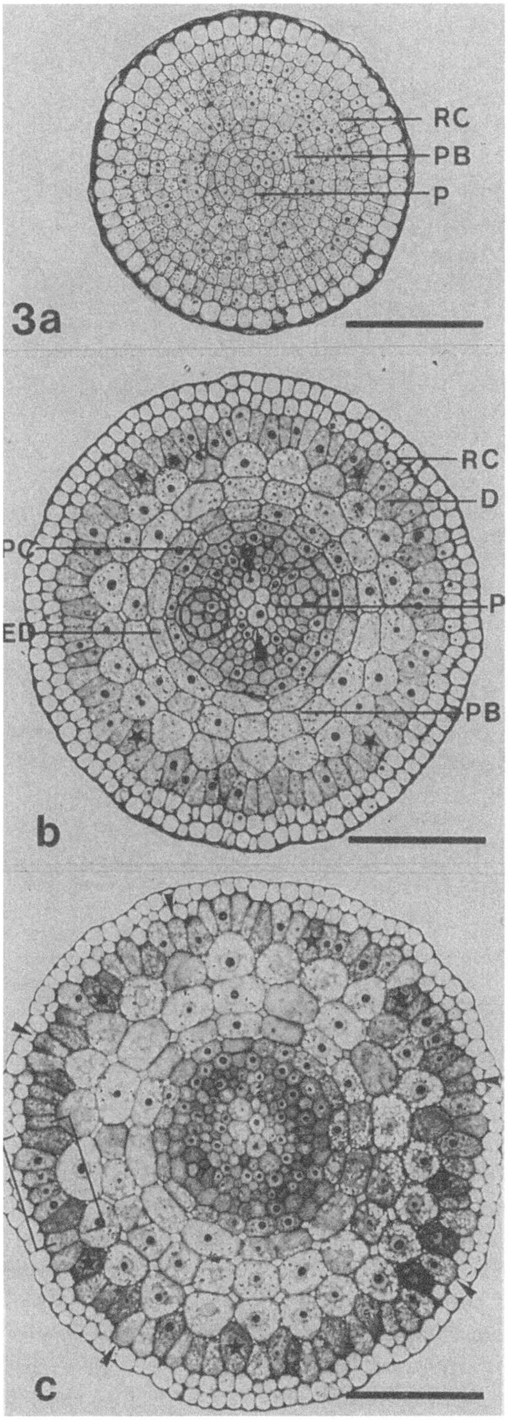


*Fig. 3.* Serial cross sections from the tip of a 24-h-old radicle of cress (for localization in longitudinal direction compare Fig. 2). At 240 μm from the tip (**a**) layers of root cap cells (RC), the periblem (PB), and the plerome (P) are clearly visible. The root periphery is protected by slime. At 360 μm from the tip (**b**), two cell layers of root cap cells (RC) and the histogens dermatogen (D), periblem (PB, three layers of prospective cortex cells, endodermis ED)), pericambium (PC) and plerome (P) with the first elements of protoxylem (arrowheads) and protophloem (circle) are clearly evident. Between prospective cortex cells an intercellular system is already developed. In the dermatogen, prospective trichoblasts are indicated by their location above anticlinal cell walls of the cortex cells and by their wedge-like shape (asterisks). At 535 m from the tip (**c**) cytological differences in prospective trichoblasts (asterisks) and atrichoblasts (arrowheads) become visible, i.e. the number of vacuoles and the density of the cytoplasm. Brackets indicate the beginning of a developmental process depicted in Figure 4. *Bar* 100 μm. Magnification 220×.

localized in the apical half of the trichoblast cell. Longitudinal and transversal sections through the tip (Figs. 2, 3a–c) demonstrated the clear histological situation in the cress root, which at this stage of its development was mainly established in the radicle of the embryo (compare Bünning, 1951; Guttenberg, 1960 for other Brassicaceae). Root cap tissue and the histogens, i.e. dermatogen, periblem, and plerome were clearly visible even within the meristem. 360 μm from the tip (Fig. 3b), two layers of the root cap tissue protected the root. In the dermatogen (prospective rhizodermis) prospective trichoblasts (pts) and prospective atrichoblasts (pats) could be distinguished by three characteristics, their location, cell shape and cytoplasm. Pts were always localized over the anticlinal longitudinal cell wall of the outer cortex cells. Their cell shape was characterized by a larger inner periclinal wall and a smaller outer one thus making a wedge-like shape in cross section. The cytoplasm was extremely dense. By contrast, pats were localized over the periclinal walls of outer cortex cells; their cell shape was exactly the inverse of that of the pts. The cytoplasm was remarkably translucent. Thus, the number of pts and the pattern of alternating trichoblasts and atrichoblasts could be determined by the number of underlying outer cortex cells. This situation became particularly obvious between 440 and 2000 μm from the tip, where the number of outer cortex cells increased by anticlinal longitudinal cell divisions (Fig. 3c). The consequences of such cell divisions could be followed by serial sections (Fig. 4a–d, localized between 535 and 660 μm from the tip (brackets indicate a region of cell division in an outer cortex cell)). Four dermatogen cells were localized over the periclinal cell wall of a large outer cortex cell (Fig. 4a). In the next sections, further from the tip, the place of one cortex cell had been taken by two cells which apparently caused a remarkable change of the cell shape of the cortex cell and of the intercellular system in this area (Fig. 4b). Now two dermatogen cells

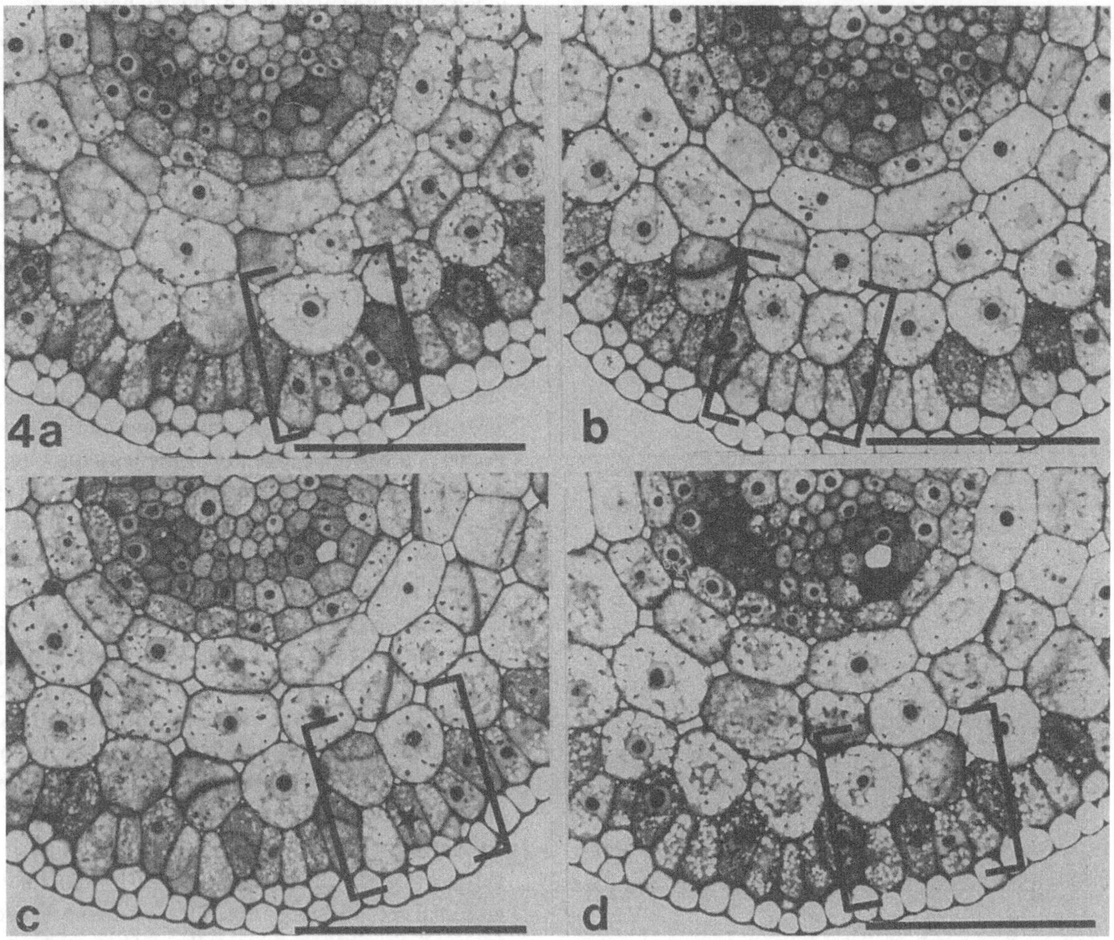


*Fig. 4.* Details from serial cross sections at 535 $\mu$m (**a**), 585 $\mu$m (**b**), 610 $\mu$m (**c**) and 660 $\mu$m (**d**) from the tip. Brackets indicate an area of presumed division of an outer cortex cell. a) Four dermatogen cells are localized over the periclinal cell wall of an outer cortex cell, two prospective trichoblasts follow in lateral direction. b) A large intercellular space is visible between the daughter cells and two dermatogen cells. None of these four dermatogen cells show characteristics of prospective trichoblasts like dense cytoplasm, lesser vacuolation or wedge-like shape. c) One of the dermatogen cells (asterisk) has enlarged into the intercellular space. The vacuolation in this cell is now less pronounced than in the neighbouring cells. d) A prospective trichoblast (asterisk) is located over the anticlinal cell wall of the daughter cells. It is characterized by its dense cytoplasm, its wedge-like shape and by smaller vacuoles. *Bar* 100 $\mu$m. Magnification 350×.

were localized over a large intercellular space near the anticlinal cell wall of the two daughter cells. In a next step (Fig. 4c), one of the dermatogen cells grew into the newly developed intercellular space. In the fourth cross section (Fig. 4d), this development was completed. Now the dermatogen cell, originally distinguished by the characterisitics of a pat showed the characteristics of a pt. This development was finished after three inferred steps of cell divisions in a dermatogen cell.

Electron micrographs from pts and pats (Fig. 5) showed the characteristic differences between both cell types in a more pronounced manner. Particularly, the inversion of the cell shape resulting from the arrangement of the intercellular system became obvious. Further differences were documented by larger vacuoles and a smaller number of ribosomal particles in the cytoplasm of pats. Areas with a large number of plasmodesmata were visible in cell walls between cortex and dermatogen cells and between pats and pts. However, plasmodesmata were never observed between dermatogen cells (pats and pts) and secretory cap cells. Additionally, the periclinal cell walls between dermatogen cells and cap cells were remarkably thick, being up to ten-times thicker than either the anticlinal cell walls between dermatogen cells or the walls between dermatogen and cortex cells. Freeze-

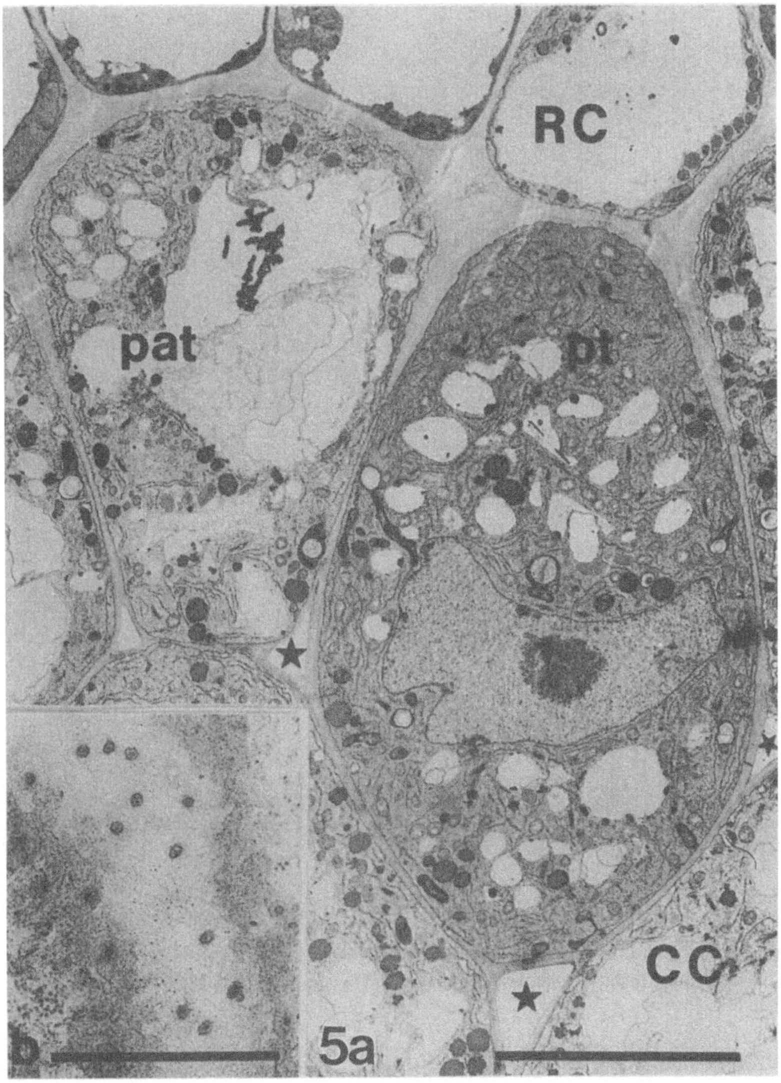


*Fig. 5.* Electron micrographs (cross sections) from two dermatogen cells 800 μm from the tip. Due to the arrangement of intercellular spaces (asterisks) prospective atrichoblasts show a wedge-like shape. Additionally, there are visible differences in the density of the cytoplasm and in the progress of vacuolation. The cell walls between dermatogen cells are extremely thin and the same is true for the wall between dermatogen and cortex cells (CC). By contrast, the cell walls between dermatogen and root cap cells (RC) are very thick. Here plasmodesmata were never observed, whereas large fields of plasmodesmata occur between dermatogen cells (prospective trichoblasts, pt and atrichoblasts, pat) and dermatogen and cortex cells. *Bar* 5**a** 10 μm. Magnification 3.400×. *Bar* 5**b** 1 μm. Magnification 32.000×.

fracture micrographs of the plasma membrane (PF) from pts (Fig. 6a) and pats (Fig. 6b) demonstrated that both cell types possessed a remarkable number of plasmodesmata which was similar in both cell types. Furthermore, in tangential sections, a large number of cortical microtubules became visible in pats and pts. These microtubules were running more or less perpendicular to the longitudinal axis of these cells, often in groups of 4 to 6 or more (data not shown).

## Discussion

Root hair formation in cress occurred in the region of dermatogen approximately 2500 μm from the tip. This coincided almost exactly with the end of the root cap tissue. The differentiation of dermatogen cells into trichoblasts and atrichoblasts, however, takes place much earlier. As Bünning (1951) pointed out, pts in white mustard (*Sinapis alba*) are always localized over anticlinal cell walls of the outer cortex cells. In *Lepidium* a

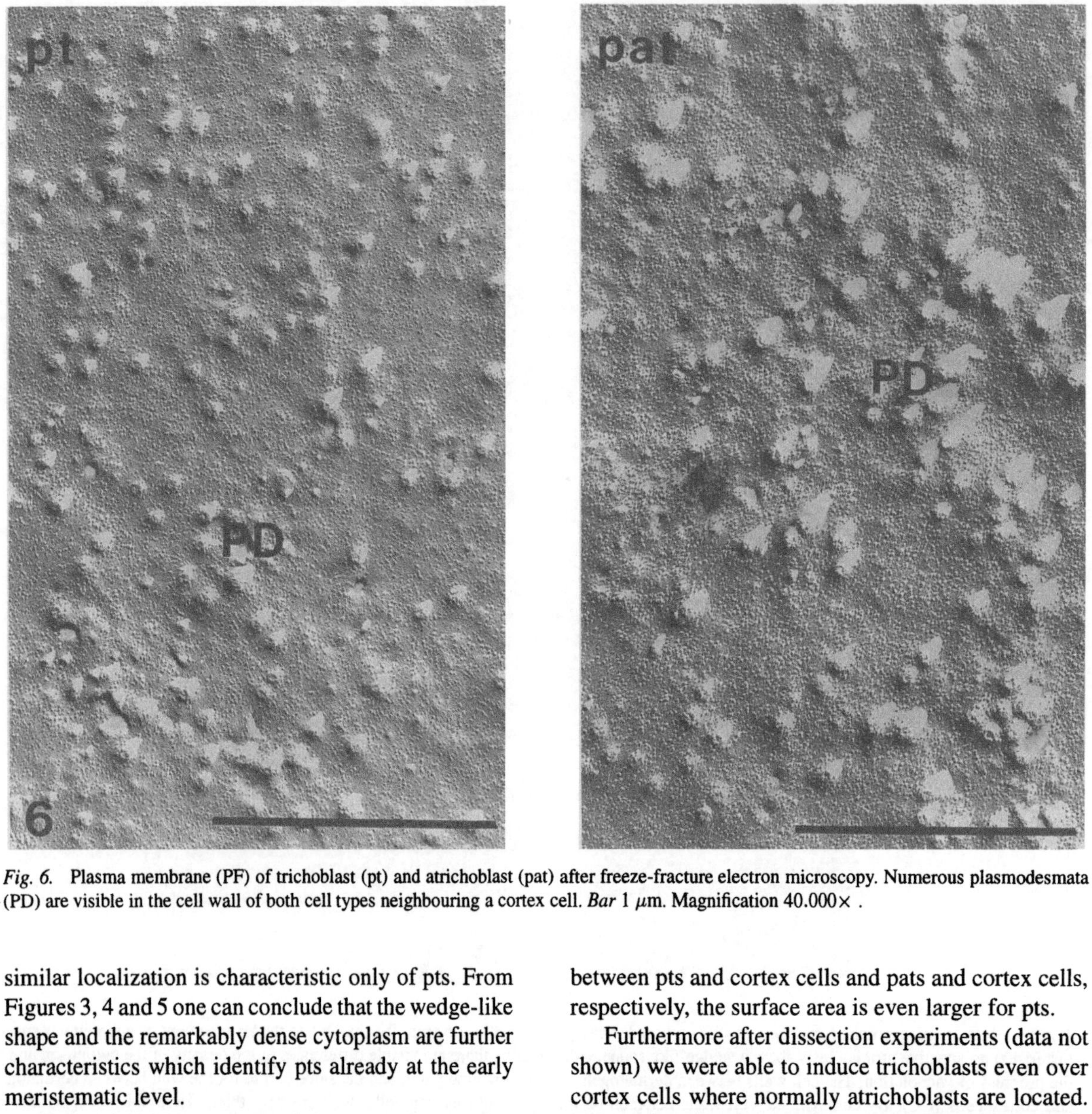


*Fig. 6.* Plasma membrane (PF) of trichoblast (pt) and atrichoblast (pat) after freeze-fracture electron microscopy. Numerous plasmodesmata (PD) are visible in the cell wall of both cell types neighbouring a cortex cell. *Bar* 1 $\mu$m. Magnification 40.000× .

similar localization is characteristic only of pts. From Figures 3, 4 and 5 one can conclude that the wedge-like shape and the remarkably dense cytoplasm are further characteristics which identify pts already at the early meristematic level.

Bünning (1951) considered the particular location of dermatogen cells as the reason why these cells differentiate into pts, suggesting it was due to the relative isolation of dermatogen cells by the intercellular space above the outer cortex cells. He suggested that the cortex cells produce an inhibitory substance which prevents the formation of root hairs in those dermatogen cells which are in intimate contact with cortex tissue. Our freeze-fracture investigations demonstrate a large number of plasmodesmata running between pats and cortex cells but this was also found between pts and cortex cells. Thus, pts are obviously not isolated from their neighbouring cells. Calculating the contact area between pts and cortex cells and pats and cortex cells, respectively, the surface area is even larger for pts.

Furthermore after dissection experiments (data not shown) we were able to induce trichoblasts even over cortex cells where normally atrichoblasts are located. This again indicates that the isolation of trichoblasts by the intercellular space has not that importance for root hair formation, as was claimed by Bünning (1951).

The meristematic character of pts is another important point. Here we are in full agreement with Bünning. Our results indicate that sometimes processes of a dedifferentiation occurred during the development of rhizodermis cells. Figures 4a–d show four rhizodermis cells. So far these cells are located above cortex cells they are in a similar developmental stage. After the longitudinal division of the cortex cell and three transversal division steps of these rhizodermis cells, one of the rhizodermis cells occupies a new position close to a newly formed intercellular space. In this position the rhizodermis cell dedifferentiates into a more meristematic cell. Successive serial sections show that this cell becomes a root hair. Thus, pts seem to be distinguished by retarded differentiation in comparison to their neighbouring cells, the pats. The development of trichoblasts therefore may represent another example for postmitotic cytoplasmic growth (compare Baluska et al., 1990, 1994; Weisenseel et al.,1992).

One of the most important question remains, however: what are the factors which control the differentiation of trichoblasts and atrichoblasts. Is it the position of the cell within the histogen (Barlow, 1984), or is there an epigenetic stress factor (Lintilhac, 1987)?

## References


Baluška F, Kubica S and Hauskrecht M 1990 Postmitotic "isodiametric" cell growth in the maize root apex. Planta 181, 269–274.

Baluška F, Barlow P W and Kubica S 1994 Importance of the postmitotic isodiametric growth (PIG) region for growth and development of roots. Plant and Soil 167.

Barlow P W 1984 Positional controls in root development. *In* Positional Controls in Plant Development. Eds. P W Barlow and D J Carr. pp 281–318. Cambridge University Press, Cambridge, UK.

Bünning E 1951 Über die Differenzierungsvorgänge in der Cruciferenwurzel. Planta 39, 126–153.

Guttenberg H v 1960 Der primäre Bau der Angiospermenwurzel. *In* Encyclopedia of Plant Anatomy Vol, VIII/3. Eds. W Zimmermann, P G Ozenda and H D Wulff. pp 112–137. Gebrüder Borntraeger, Berlin, Germany.

Hofer R M 1991 Root hairs. *In* Plant Roots — The Hidden Half. Eds. Y Waisel, A Eshel and U Kafkafi. pp 129–149. Marcel Dekker Inc., New York, USA.

Lintilhac P M 1987 Plant cytomechanics and its relationship to the development of form. *In* Cytomechanics. The Mechanical Basis of Cell Form and Structure. Eds. J Bereiter-Hahn, O R Anderson and W-E Reif. pp 230-241. Springer Verlag, Berlin, Germany.

Robards A W 1983 General and molecular cytology. Prog. Bot. 45, 1–17.

Sievers A and Schnepf E 1981 Morphogenesis and polarity of tubular cells with tip growth. *In* Cytomorphogenesis in Plants. Ed. O Kiermayer. pp 265–299. Springer Verlag, Wien, Austria.

Volkmann D 1981 Structural differentiation of membranes involved in the secretion of polysaccaride slime by root cap cells of cress (*Lepidium sativum* L.). Planta 151, 180–188.

Weisenseel M H, Becker H F and Ehlgötz J G 1992 Growth, gravitropism, and endogenous ion currents of cress roots (*Lepidium sativum* L.). Plant Physiol. 100, 16–25.

*F. Baluška et al. (eds.), Structure and Function of Roots*, 69–74, 1995.


# The relationship between growth, nucleus migration and cytoskeleton in root hairs of radish

S. Sato, Y. Ogasawara and S. Sakuragi
*Department of Biology, Faculty of Science, Ehime University, Bunkyo-cho 2–5, Matsuyama 790, Japan*

*Key words:* actin filaments, microtubules, nucleus migration, radish, root hairs


## Abstract

The aim of this study was to investigate whether and how the cytoskeleton is involved in the mechanism keeping the nucleus in the sub-apical region (ca. 50–150 $\mu$m, from the tip) of root hairs of radish. In actively-growing root hairs of 2 days-old plants grown at 23°C the nucleus was always localized in the sub-apical region, whereas the nucleus was frequently located toward the base in non-growing root hairs of 3 days-old plants. The root-hair tip grew at a constant rate while the nucleus sometimes changed its migration speed within the sub-apical region. Colchicine liberated the nucleus from the sub-apical region. This phenomenon also occured in plants exposed to 0–1°C for a long time. When the cold-treated plants were transferred to a 23°C incubator, some root hairs began to grow again after a lag phase (1–2 hrs). During this lag phase, the nucleus always returned to the sub-apical region. The root hairs whose nucleus did not recover its position in the sub-apical region never grew again. According to the immunofluorescence pattern of microtubules (MTs), actively-growing root hairs were discriminated into three parts: (1) an amorphous zone located at the tip and which emits uniform strong fluorescence, (2) an Mt-containing ($MT^+$) zone immediately behind the amorphous zone and is occupied by MTs running parallel to the root-hair axis, and (3) a zone where MTs were absent which succeeds the $MT^+$ zone. The nucleus was always localized at the basal end of the $MT^+$ zone. The ends of MTs were anchored at the surface of the nucleus. In elongated root hairs, the $MT^+$ zone shortened and the nucleus was frequently free. Non-growing root hairs differentiated neither the amorphous nor the $MT^+$ zones. In these hairs the nucleus was most frequently located toward the base. Fluorescence microscopy using rhodamine-phalloidin clearly revealed well-developed bundles of actin filaments, especially in actively-growing root hairs, which ran the entire axis of the hairs. In conclusion, the present study reconfirmed that the nucleus is always kept in the sub-apical region while root hairs are actively growing, but it begins to be liberated from this region when the hairs weaken or cease their tip-growth. MTs are responsible for the maintenance of the nucleus in the sub-apical region.


## Introduction

The nucleus has long been known to be localized toward the tip of root hairs. In tip growing plant cells the nucleus migrates towards the tip as the cells extend (Schmiedel and Schnepf, 1980; Schnepf, 1982). But a reliable explanation has not been given until recently concerning how the nucleus is kept in the region close to the tip. Indirect immunofluorescence using monoclonal anti-tubulin antibodies has shown that internal bundles of MTs connect the nucleus to the tip of the root hair in *Vicia hirsuta* (Lloyd et al., 1987). This technique has also been applied to root hairs of several higher plants and results have shown that profiles of MTs in root hairs are not always the same (Lloyd and Wells, 1985; Traas et al., 1985). We have also noticed that the nucleus is not always present in the sub-apical region and is sometimes localized in a region far from the tip. Thus, it seems that further investigations are required to establish a role of MTs in keeping the nucleus in the sub-apical region and to explain the relationship between tip growth, nucleus migration toward the tip or liberation from the sub-apical region, and the cytoskeleton. In this context, this study describes results from root hairs of radish.

## Materials and methods

### *Plant cultivation*

Seeds of radish (*Raphanus sativus* L.) purchased from commercial sources were placed in water in Petri dishes and germinated in a constant-temperature incubator at 23°C. They germinated after 1 day and had roots with many root hairs after 2 days.

### *Determination of tip growth and nucleus migration*

A plant was placed between two pieces of vinyl tape stuck on a clean glass slide. The root of the plant was immersed in water and covered with a coverslip. Then the glass slide with the plant was set on a microscope. Tip growth and nucleus migration were monitored by phase-contrast microscopy. Both were determined by means of a micrometer fixed in an eye-piece. Tip-to-nucleus distance was defined as the distance between the tip of a root hair and the tip-side edge of the nucleus.

### *Detection of MTs and actin filaments*

Plants were fixed in 8.0% freshly prepared paraformaldehyde in MT stabilizing buffer (MTSB: 50 m*M* PIPES, 5m*M* EGTA, 5m*M* $MgSO_4$, 5% DMSO pH 7.0). The following procedures are basically the same as those of Doonan and Clayton (1986). Briefly, plants were macerated at a room temperature for 20 min in a mixture of 2% cellulase (Onozuka RS, Yakult Honsha Co., Japan) and 0.5% pectolyase Y-23 (Seishin Pharmaceutical Co., Japan) in MTSB. Indirect immunofluorescence was carried out using monoclonal anti alpha-tubulin (BioMakor, Israel) and fluorescein isothyocyanate (FITC)-conjugated anti-mouse immunoglobulin G. They were diluted 1:1000 and 1:20 with PBS (phosphate buffered saline), respectively. Then, specimens were stained with 1 $\mu$g mL$^{-1}$ DAPI (4'-6-diamino-2-phenylindole) in PBS and sealed with a mixture of glycerin : PBS (9:1) containing 5% n-propyl gallate.

For actin filaments, plants were fixed at room temperature for 1 hr in 8% formaldehyde in MTSB (pH 7.0). They were rinsed in 1/15 *M* phosphate buffer containing 5 m*M* EGTA (pH 6.9) (PPB) and were macerated in enzyme mixture of 1% cellulase and 1% macerozyme (Onozuka R-10, Yakult Honsha Co., Japan). The specimens were rinsed in PPB, placed on a glass slide and air-dried. They were stained with rhodamine-phalloidin (Wako Pure Chemistry industries Co., Japan) diluted 1:20 with PBS for 1 hr at 37°C after a brief permeabilization with 0.1% Triton X-100 in PBS, then stained with 1 $\mu$g mL$^{-1}$ DAPI in PBS.

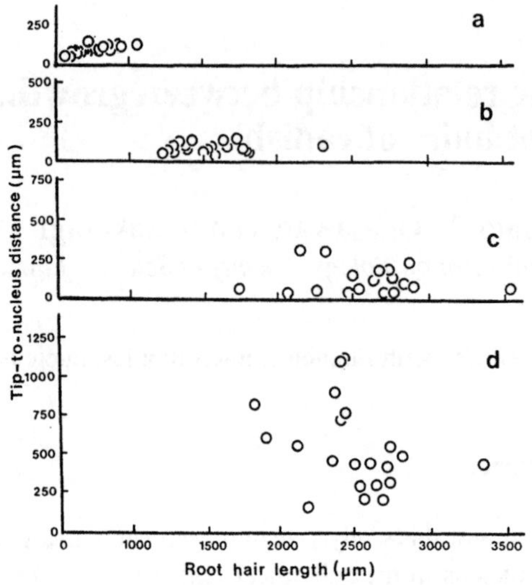


*Fig. 1.* Tip-to-nucleus distance in a root hair in four 5 mm-length segments of the root of a 3 days-old plant. **a**: 0–5 mm; **b**: 5–10 mm; **c**: 10–15 mm; **d**: 15–20 mm.

## Results

### *Relationship between tip growth and nucleus position*

To investigate nucleus position in different growth stages of root hairs, the roots were cut into 5 mm-length segments, and nuclear position was determined in the root hairs of each segment. In 2 days-old plants, almost all root hairs had the nucleus in the sub-apical region (ca. 50–150 $\mu$m from the tip). This is, however, not the case in 3 days-old plants. A representative result in a 3-day-old plant is shown in Figure 1. Most root hairs in the 0–5 mm and 5-10 mm segment had the nucleus in the sub-apical region. On the other hand, the nuclear position very frequently shifted from the sub-apical region, especially in the 15–20 mm segment. A similar result was obtained in three other plants investigated.

Tip growth and nuclear position were plotted every five minutes in an actively growing root hair (Fig. 2). The tip grew at a rather constant rate while the nucleus sometimes changed its migrating speed. The nucleus, however, was always within the sub-apical region. This was basically the case in five other actively-growing root hairs examined. Contrary to this, no tip-growth

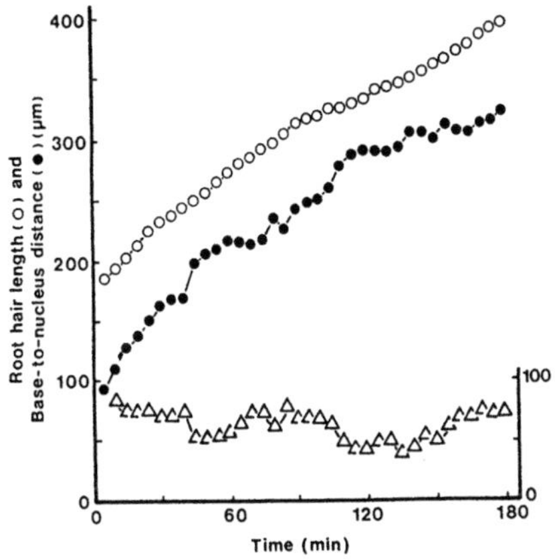


*Fig. 2.* Relationship between tip growth and nucleus migration. Tip-to-nucleus distance varies a little during tip growth.

was seen in root hairs whose nucleus was liberated toward the base from the sub-apical region. Out of four cases examined, three showed that the nucleus gradually moved towards the base. Thus, the nucleus is always kept in the sub-apical region in actively-growing root hairs, but tends to be liberated from this region in non-growing root hairs.

### *An effect of colchicine and cold treatment on nucleus position*

In order to investigate whether MTs are responsible for keeping the nucleus in the sub-apical region, plants were treated with colchicine (Fig. 3). The frequency with which the nucleus was liberated from the sub-apical region gradually increased with increasing concentration of colchicine. Lengthy treatment also induced a high frequency of liberated nuclei compared with shorter treatment, except in the case of the $5 \times 10^{-5}\%$ colchicine treatment.

When plants were exposed to cold (0 – 1°C), the liberated nucleus also increased in frequency as the treatment period increased (Fig. 4). In plants exposed to cold for 17 hrs, more than half of the root hairs had a liberated nucleus. The nucleus recovered its position in the most distal zone of the sub-apical region when plants were transferred back to 23°C. A typical example of this is shown in Figure 5. Plotting tip growth and nucleus migration against time clearly indicated that the tip began to grow after the nucleus recovered its position in the sub-apical region (Fig. 6). However, the chances were rather small that the liberated nucleus could recover its position in the sub-apical region. The root hairs with the liberated nucleus located far from

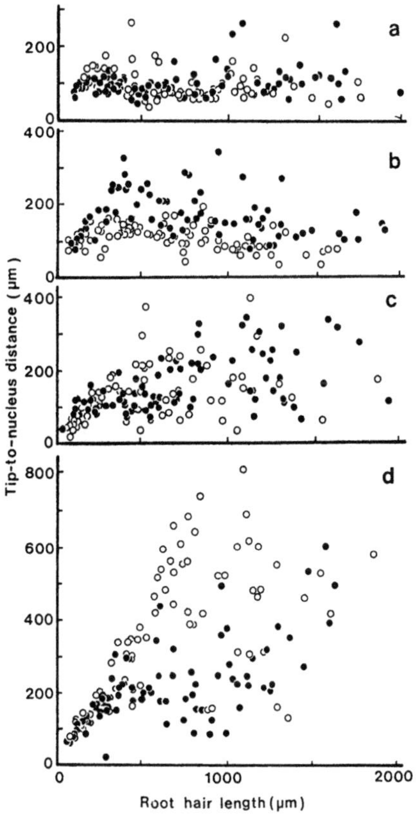


*Fig. 3.* Effect of colchicine treatments for 3 hrs (●) and 5 hrs (○) on tip-to-nucleus distance in root hairs of 2 days-old-plants. **a:** $5 \times 10^{-5}\%$; **b:** $5 \times 10^{-4}\%$ **c:** $5 \times 10^{-3}\%$; **d:** $5 \times 10^{-2}\%$.

the sub-apical region never began to grow. Thus it is probable that the nucleus in the growing hair is held in position by the MTs.

### *MT distribution and nucleus position in root hairs*

The distribution of MTs in root hairs was investigated by means of indirect immunolocalization. Depending on, its immunofluorescence pattern, an actively-growing root hair was clearly differentiated into three regions (Fig. 7). The apical dome emitted strong and uniform fluorescence (amorphous zone). More proximally was a region filled with conspicuous MTs mostly running parallel with the root hair axis ($MT^+$ zone). The basal end of the $MT^+$ zone was clearly delimited by a dark zone where no MTs were seen ($MT^-$ zone). In actively-growing root hairs the $MT^+$ zones were well developed and the maximum span of the $MT^+$ zone was about 150 $\mu$m, which corresponded with the maximum value of tip-to-nucleus distance (Fig. 7a). The nucleus was always localized at the basal end of

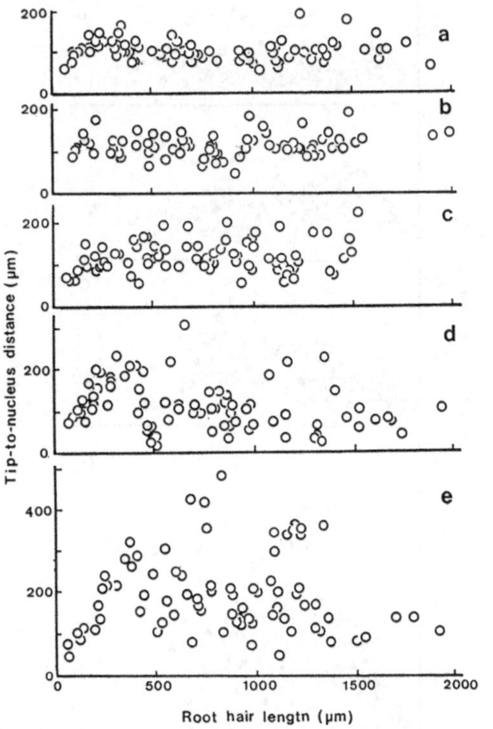


*Fig. 4.* An effect of different time of low temperature (0–1°C) treatment on tip-to-nucleus distance in root hairs of 2 days-old plants. **a**: control; **b**: 3 hrs; **c**: 6 hrs; **d**: 14 hrs; **e**: 17 hrs.

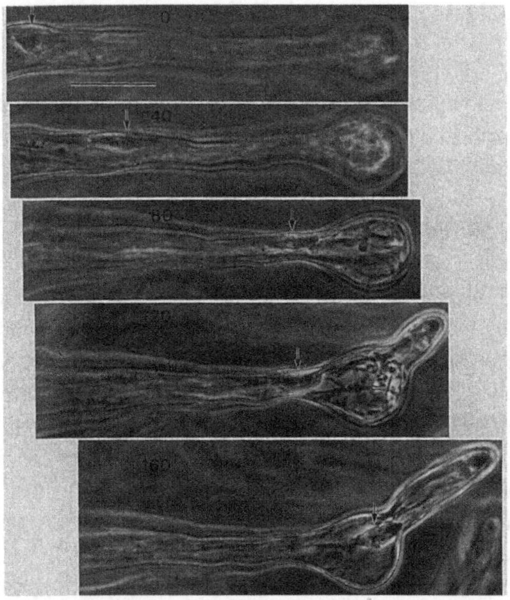


*Fig. 5.* A series of pictures showing the migration of a liberated nucleus toward the most distal position of the sub-apical region prior to tip growth. The plant was transferred back to 23°C after growing for 12.5 hrs at 0–1°C. Number in each picture indicates time after the transfer. Arrows indicate nucleus position. *Bar* equals 40 μm.

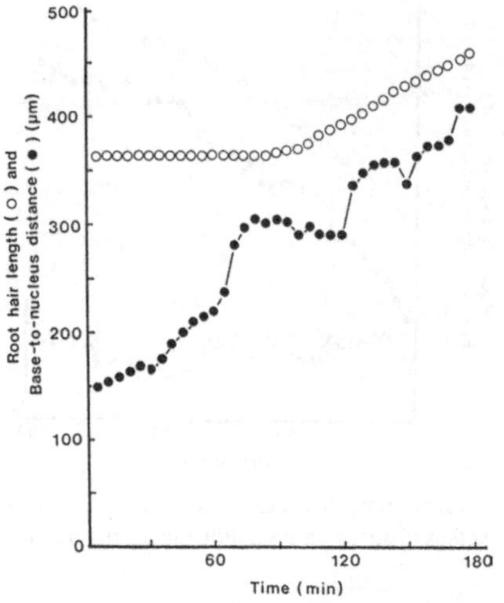


*Fig. 6.* Relationship between tip growth and nucleus migration when a plant was transferred back to 23°C after exposed to 0–1°C for 12.5 hrs. The data was obtained from the root hair shown in Figure 5.

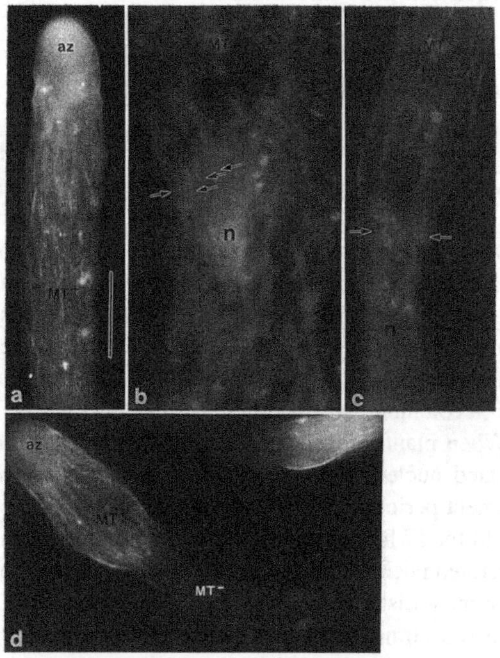


*Fig. 7.* Immunofluorescence microscopy of root hairs. An actively growing root hair with a well developed $MT^+$ zone ($MT^+$) **a**: MTs (arrows) are anchored at the nucleus (n); **b**, **c**: An elongated root hair reduces the $MT^+$ zone and no longer has its nucleus at the basal end of the sub-apical region. az; amorphous zone, $MT^-$; $MT^-$ zone. *Bar* equals 10 μm.

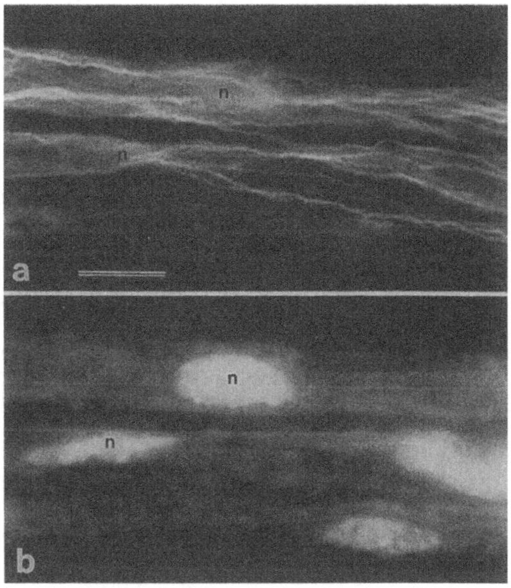


*Fig. 8.* Actin filaments in an actively growing root hair. **a**: The nucleus (n) is visualized with DAPI. *Bar* equals 10 $\mu$m.

the $MT^+$ zone and MTs were anchored at the surface of the nucleus (Fig. 7b, c).

In an elongated root hair which is expected to decrease or stop tip-growth, the $MT^+$ zone extremely reduced or disappeared (Fig. 7d). The nucleus was no longer localized at the basal end of the $MT^+$ zone and was seen in the $MT^-$ zone.

Actively growing root-hairs were treated for 3 hr at 23°C with $5 \times 10^{-2}$% colchicine. Profiles of MTs were very similar to those seen in root hairs grown without colchicine.

### *Distribution of actin filaments*

Thick cables of actin filaments were seen in an actively-growing root hair (Fig. 8). They ran throughout the root hair and very frequently converged or diverged. The nucleus was usually seen at these crossed sites. The cables of actin filaments in actively-growing root hairs were much thicker than those in elongated root hairs.

## Discussion

Computerized deblurring of immunofluorescence profiles of MTs has indicated that the nucleus is connected to the tip by internal bundles of MTs in root hairs of *V. hirsuta* (Lloyd et al., 1987). This was basically the case in root hairs of radish. No clear bundle of MTs, however, was discerned associating with the nucleus in radish. Instead, some fine clusters of MTs were diffusely anchored at the tip-side surface of the nucleus.

MTs have been reported to be distributed throughout the root hairs in several plants (Lloyd and Wells, 1985; Traas et al., 1985). On the other hand, the present study indicated that radish root hairs are differentiated into three regions, the amorphous, $MT^+$ and $MT^-$ zones. These regions are expected to have their own specialized features. Previous electron microscopy has reported that short MTs are more abundant in the tip than in other regions and the maximum number of MTs is attained at 25 $\mu$m from the tip (Seagull and Heath, 1980). Thus, it is suggested that the uniform fluorescence in this region may reflect incomplete assembly of MTs. It has been generally accepted that the cortical MTs play an important role in depositing cellulose microfibrils on the cell wall (Gunning and Hardham, 1982; Heath and Seagull, 1982). In the $MT^+$ zone, therefore, cellulose microfibrils may be actively deposited on the newly formed wall. Thus, the invasion of the nucleus into the $MT^+$ zone may hinder the cortical MTs from depositing, cellulose microfibrils. Concurrent with the weakening or cessation of tip growth, the $MT^+$ zone appeared to reduce. The present study suggests that the nucleus is disconnected from the MTs at this time. The migration of the disconnected nuclei toward the base is probably due to actin filaments (Lloyd et al., 1987).

The previous study, however, has shown that MTs go into the tip in root hairs (Lloyd and Wells, 1985; Traas et al., 1985). This means that the fixation method used in the present study may have been inappropriate to protect MTs in the tip. In other words, MTs in the tip may be unstable compared to those in the $MT^+$ zone.

Profiles of MTs were not changed after colchicine treatment even though the nucleus was very frequently liberated from the sub-apical region by this treatment. This is superficially incompatible with the present conclusion that MTs connect the nucleus to the tip. The explanation for this contradiction may be that MTs connected to the nucleus are sensitive to colchicine while the cortical MTs are not. Two types of MTs, cortical and cytoplasmic MTs, are seen in non-dividing cells of maize (Baluška et al., 1992) and cytoplasmic MTs actually appear to be more sensitive to antimicrotubular treatments than cortical MTs (Baluška et al., 1993; Baluška and Barlow, 1993). Application of indirect immmofluorescence to sections of root hairs may give some useful information concerning this supposition.

No decisive explanation has been given with respect to how the nucleus, once moved to the base by cold treatment, recovers its position again in the sub-apical region when the plant is transferred from cold to 23°C. It would be interesting to investigate whether MTs develop from the nucleus or from the tip. However, we do not deny the possibility that the nucleus exposed to cold could still be connected to the tip by MTs.

## References

Baluška F and Barlow P W 1993 The role of the microtubular cytoskeleton in determining nuclear chromatin structure and passage of maize root cells through the cell cycle. Eur. J. Cell Biol. 61, 160–167.

Baluška F, Parker J S and Barlow P W 1992 Specific patterns of cortical and endoplasmic microtubules associated with cell growth and tissue differentiation in root of maize (*Zea mays* L.). J. Cell Sci. 103, 191–200.

Baluška F, Parker J S and Barlaw P W 1993 The microtubular cytoskeleton in cells of cold-treated roots of maize shows tissue-specific responses. Protoplasma 172, 84–96.

Doonan J H and Clayton L 1986 Immunofluorescent studies and the plant cytoskeleton. *In* Immunology in Plant Science. Ed. T L Wang. pp 11–136. Cambridge University Press, Cambridge, London, New York, New Rochelle, Melbourne, Sydney.

Gunning B E S and Hardham A R 1982 Microtubules. Ann. Rev. Plant Physiol. 33, 651–698.

Lloyd C W, Pearce K J, Rawlins D J, Ridge R W and Shaw P J 1987 Endoplasmic microtubules connect the advancing nucleus to the tip of Legume root hairs, but F-actin is involved in basipetal migration. Cell Motil. Cytoskel. 8, 27–36.

Lloyd C W and Wells B 1985 Microtubules are at the tips of root hairs and form helical patterns corresponding to inner wall fibrils. J. Cell Sci. 75, 225–238.

Seagull R W and Heath I B 1980 The organization of cortical microtubule arrays in the radish root hair. Protoplasma 103, 205–229.

Schmiedel G and Schnepf E 1980 Polarity and growth of caulonema tip cells of the moss *Funaria hygrometrica*. Planta 147, 405–413.

Schnepf E 1982 Morphogenesis in moss protonemata. *In* The Cytoskeleton in Plant Growth and Development. Ed. C W Lloyd. pp 321–344. London, New York Academic Press

Traas J A, Braat P, Emons A M C, Meekes H and Derksen J 1985 Microtubules in root hairs. J. Cell Sci. 76, 303–320.

*F. Baluška et al. (eds.), Structure and Function of Roots*, 75–86, 1995.


# Morphology and anatomy of rice roots with special reference to coordination in organo- and histogenesis

Shigenori Morita and Keisuke Nemoto
*Faculty of Agriculture, The University of Tokyo, 1-1-1 Yayoi, Bunkyo-ku, Tokyo 113, Japan*

*Key words:* anatomy, histogenesis, morphology, morphometry, organogenesis, *Oryza sativa* L. rice, root

## Abstract

Rice belongs to monocotyledon which is characterized by having a so-called fibrous root system. Such root system is built up with seminal and nodal roots with numerous lateral roots. Morphology and anatomy of rice roots which is fundamentally the same as other cereal crops, has been relatively well-described. Rice roots, however, have their own characteristics including well-developed air space in the matured cortex. In addition, many examples of coordination were shown in root organo- and histogenesis as well as root function. Morphometric analysis on anatomy of matured rice roots including vascularization suggests that differentiation and maturation of each tissue in roots proceed quite integratedly. Because vascular system is quite important with reference to root function, many studies have been done. However, following aspects of vascularization have rarely been examined; (a) connections between vascular systems of stem and nodal roots, (b) changing patterns of vascular system along the root axes, (c) connections between vascular systems of the parental roots and their lateral roots and (d) connections among different kinds of vascular elements in the same roots. Morphology and anatomy with reference to coordination in organo- and histogenesis should be further studied in order to understand both structure and function of rice roots.

## Introduction

Rice is an important food crop in the world, especially in Asia. Therefore, numerous agronomic studies have been done for understanding and controlling development and growth of rice plants. However, information on rice roots is relatively limited compared with the shoot, primarily because it is quite time-consuming to examine rice roots in soil under flooded conditions. Because most important agricultural practices in growing rice, for example fertilization and water management, affect the growth of the root system through soil, knowledge of the structure and development of an individual root as well as the whole root system is essential for selecting effective management practices in rice cultivation.

In this article there will be first a brief description on the structure and development of the root system in rice. Then, studies on morphology and anatomy of rice roots will be reviewed from the viewpoint of comparison with other plant species. General morphology and anatomy of rice roots are quite similar to other monocots including cereal crops (for example, Esau, 1965). A rice root, however, has features different from others, partly because it is usually grown under flooded conditions. In addition, coordination in organo- and histogenesis of roots will be shown in rice, which may help to study further the relationships between their structure and function.

## Structure and development of the root system

A rice plant is characterized by having a relatively shallow and compact root system (Fig. 1A; Mawaki et al., 1990; Morita et al., 1988) compared with other upland cereal crops, for example maize (Kaeriyama and Yamazaki, 1983; Weaver, 1926; Yamazaki and Kaeriyama, 1982) and wheat (Gregory et al., 1978; Morita et al., 1993; Weaver, 1926), probably because rice is usually grown under flooded conditions. The structure of the root system in rice, however, is fundamentally the same as other cereals. The rice root system consists of seminal, nodal and lateral roots (Figs. 1B-D; Kawata et al., 1963) which correspond to pole-borne, stem-borne and root-borne roots, respectively

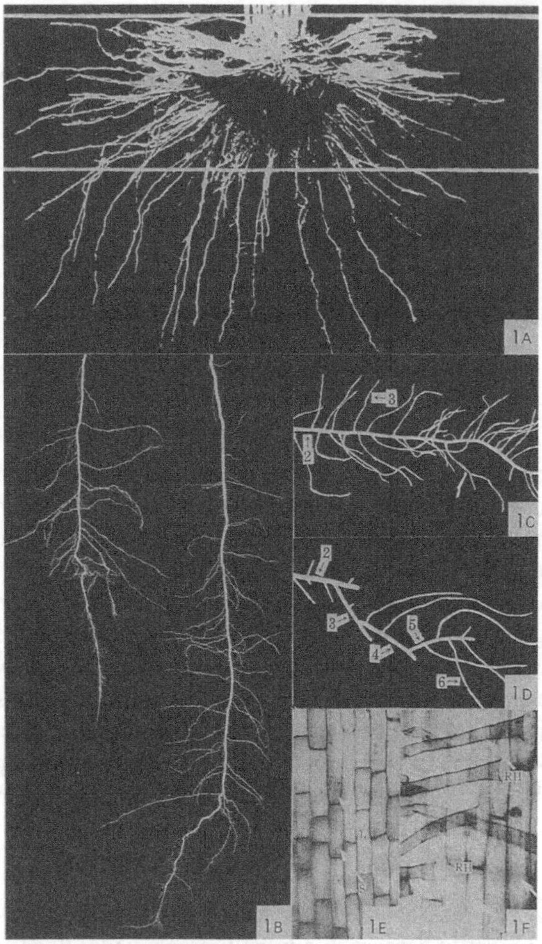


*Fig. 1.* Rice root system and its components. **A**: Root system at the harvest stage (Kawata et al., 1963). The upper and lower white lines indicate the soil surface and plow pan, respectively, and the distance between them is about 15cm. **B**: Nodal roots with lateral roots (Kawata and Soejima, 1974). **C-D**: Branching patterns of nodal roots (Kawata and Soejima, 1974). 2: secondary root (1st-order lateral root), 3: tertiary root (2nd-order lateral root), and so on. **E-F**: Epidermal cells and root hairs of nodal roots. S: short cell, L: long cell, RH: root hair. (Kawata and Ishihara, 1964).

(Barlow, 1994). A rice plant has only one seminal root (Fujii, 1961; Kawata et al., 1963), though the number of seminals varies in other species (Fujii, 1961; Percival, 1921; Yamazaki and Kaeriyama, 1982; Yamazaki and Nakamoto, 1983). The root system of rice, on the contrary, is characterized by numerous nodal roots, partly because rice plants have many tillers. Both seminal and nodal roots (primary roots) form 1st-order lateral roots (secondary roots), 1st-order lateral roots produce 2nd-order laterals (tertiary roots), and so on (Figs. 1B-C). Sometimes higher-order lateral roots are developed in rice: 5th-order lateral roots were found on nodal roots which emerged at later growth stages (Fig. 1D; Kawata and Soejima, 1974). In addition, seminal and nodal roots, as well as most lateral roots (at least up to 4th-order lateral roots), bear root hairs which also are an important component of the root system (Fig. 1F; Kawata and Ishihara, 1959; Kawata and Chung, 1976).

Emergence of nodal roots proceeds acropetally along both the main stem and tillers keeping pace with successive leaf unfolding (Fujii, 1961; Kawata et al., 1963). Lateral roots are also developed in synchrony with shoot growth. When the Nth leaf (for example, 10th leaf) is unfolding, nodal roots are emerging at the (N–3) node (7th node). At the same time, 1st- and 2nd-order lateral roots are developing on nodal roots at (N–4) and (N–5) nodes (6th and 5th nodes), respectively. Such coordination in organogenesis is the reason why the phyllochron concept is quite useful to understand and predict development and growth of rice plants (Nemoto et al., 1993).

Thus, the root system of the rice plant is composed of several different kinds of roots with complicated age distribution.

## Anatomy and histogenesis of roots

### *Root anatomy*

Anatomy of both seminal and nodal roots in rice is quite similar to other cereals. The root apical meristem is surrounded by a root cap (Fig. 2A) and all tissues in nodal roots originate from the root apical meristem which consists of three histogens, calyptrogen, dermatogen-periblem complex and plerome (Fig. 5A; Kawata and Lai, 1965b; Shimabuku, 1960). Although histological observations on the root apical meristem in rice are limited (Mori, 1972; Shimabuku, 1960), it is suggested that there should be a quiescent center (Kawata and Matsui, 1977d), as in other species (Clowes, 1958, 1967). The root proper has three tissue systems, epidermis, cortex and stele (Figs. 2A-C).

## Root formation

A seminal root originates as a radicle during embryogenesis (Hoshikawa, 1989). Primordia of nodal roots, on the other hand, are initiated from cells of the ground meristem which are just adjacent to the peripheral vascular cylinder in the stem (Fig. 3A; Kaufman, 1959;

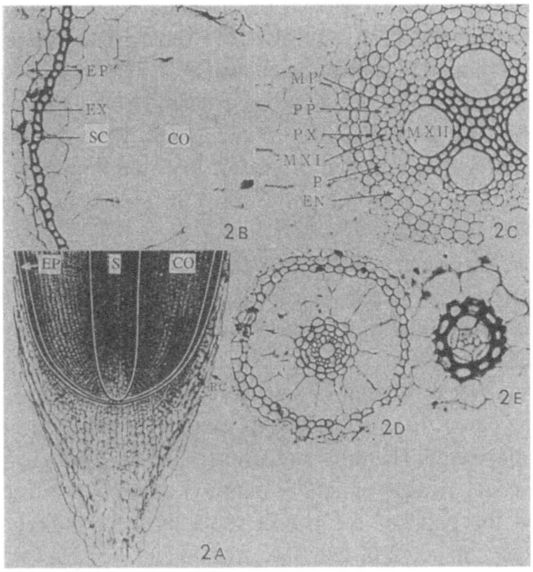


*Fig. 2.* Anatomy of nodal and lateral roots in rice. **A**: Longitudinal section of nodal root (Kawata et al., 1977b). RC: root cap, Cl: columella, T: meristematic tissue in the peripheral zone, EP: epidermis, CO: cortex, S: stele. **B-C**: Cross section of nodal root (Kawata et al., 1977e). EP: epidermis, EX: exodermis, SC: sclerenchyma, CO: cortex, EN: endodermis, P: pericycle, PX: protoxylem vessel, MXI: early metaxylem vessel, MXII: late metaxylem vessel, PP: protophloem sieve tube, MP: metaphloem sieve tube. **D**: Cross section of thick lateral root (Kawata et al., 1977e). **E**: Cross section of thin lateral root (Kawata et al., 1977e).

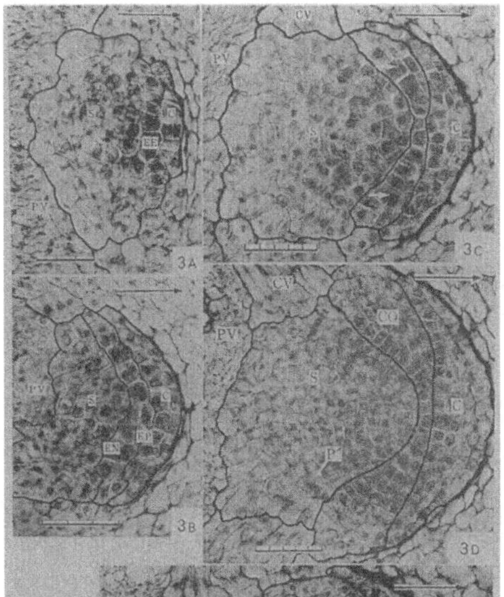


*Fig. 3.* Successive stages in the development of rice nodal root primordium in stem (Kawata and Harada, 1975). PV: peripheral vascular tissue in stem, CV: connecting vascular system, C: root cap, EE: epidermis-endodermis complex, CO: cortex, S: stele, P': connecting phloem, MXII: late metaxylem vessel. *Bars* in A, B and in C-D is 30$\mu$m and 50$\mu$m, respectively.

Kawata et al., 1972b). The organization (Kaufman, 1959; Kawata and Harada, 1975) and diameter (Kawata and Harada, 1977e) of root primordia are established before their emergence from the stem (Figs. 3A-E). In the case of rice, root primordia are, in principle, initiated at internodes as well as at nodes of the stem, which may be an another reason why the root system of rice has numerous nodal roots. Determination of the number and diameter of root primordia is another important problem because it should strongly influence the dimension and distribution of a whole root system. Usually, a positive correlation is found between the number of root primordia and the surface area of the vascular cylinder in the stem where root primordia can be initiated (Kawata et al., 1978b; Kawata and Harada, 1980; Yamazaki and Nemoto, 1986). There is also a significant correlation between the diameter of the root and the diameter of the vascular cylinder in the stem (Yamazaki and Nemoto, 1986). These facts indicate that formation and development of root primordia should keep pace with growth of the stem, which may be examples of a coordination in organo- and histogenesis of rice roots.

## Root cap

Because the root cap is thought to be a site of graviperception as well as a tissue producing plant growth regulators in relation to gravitropism (Jackson and Barlow, 1981; Juniper, 1976; Moore and Evans, 1986), there have been many studies of the root cap (Moore and McClelen, 1983a, b). Its developmental morphology, however, has rarely been examined. In rice, a primary root cap, i.e. a root cap of a nodal root just emerged from the stem, is characterized by stratified cell layers parallel to the outer surface of the root cap (Fig. 4A). Several days after emergence of a nodal root, the primary root cap detaches and is sloughed off, and an ordinary root cap with a well-developed columella is

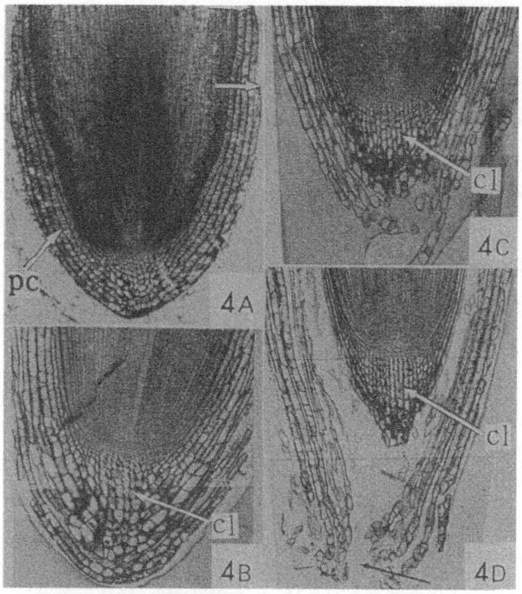


*Fig. 4.* Stages in the development of root cap in rice nodal root (Kawata et al., 1979b). pc: primary root cap, cl: columella.

produced (Figs. 4B-D; Kawata et al., 1979c). Establishment of a columella in the root cap might possibly be coordinated with the onset of graviperception, because the growth direction of each nodal root is usually determined soon after root emergence.

Recently, the size of the root cap or columella has been shown to be important regarding root plagiogravitropism (Moore and Pasieniuk, 1984a, b; Pilet, 1982; Ransom and Moore, 1985). There is wide variation in the size of root caps depending on the position of root emergence along the main stem and tillers in rice. Strong correlations have been found among diameter of the root proper, length and diameter of the root cap, the number of cell files and statocytes (columella cells containing starch grains) in columella (Kawata et al., 1977c). Such morphological correlations suggest a coordination in histogenesis of the root, including the root cap.

## Epidermis and root hair

The epidermis of rice roots is differentiated from the dermatogen-periblem complex at the root tip by the first periclinal division (Fig. 5B; Kawata and Matsui, 1977d). Although there are no significant differences in length of epidermal cells in the meristematic area, two types of epidermal cells, short and long cells, can be distinguished at the basal part of nodal roots (Fig. 1E). The root hair is an outgrowth of a short cell at its apical end (Fig. 1F; Kawata and Ishihara, 1959). The root hairs of lateral roots develop in a similar way (Kawata and Chung, 1976). The oxidizing ability of the short cells with a root hair is greater than the long cells at the basal part of the nodal roots in rice (Kawata and Ishihara, 1964).

There are positive correlations between both frequency and length of root hairs and diameter of the parental nodal root, respectively (Kawata and Chung, 1977a, 1979a). Intimate correlations are also found between formation of root hairs and development of the parent root, including its elongation rate and vascularization (Kawata and Chung, 1977a, 1979a). Initiation, elongation and maturation of root hairs occur on the parts of the parent roots where protoxylem, early metaxylem and late metaxylem vessels begin to mature, respectively (Kawata and Chung, 1979a). Such morphogenetical coordination suggests the interrelationships between the functions of root hairs and the parent root.

## Cortex

Endodermis, the innermost cell layer of cortex, is differentiated from the dermatogen-periblem complex by the first periclinal division (Fig. 5B). All the cell layers of cortex including exodermis and sclerenchyma (Fig. 2B) originate from the endodermis by successive periclinal divisions (Fig. 5C; Kawata and Lai, 1965b). A similar developmental pattern in the cortex has been found in many other plant species (Williams, 1947), though several additional periclinal divisions sometimes occur in the outer cortex (Seago and Wolniak, 1976; Seago and Marsh, 1989). Most researchers believe that only the periclinal divisions of the endodermis are important in determining the number of cortex cell layers. However, it has been realized in rice roots that relationships between periclinal and transverse divisions of endodermis may control the number of cell layers in the cortex (Kawata and Lai, 1965b; Kawata and Matsui, 1977d).

Because the endodermis plays a possible role in water relation of roots (Kawata and Lai, 1968), the endodermis has been relatively well studied in rice (Clark and Harris, 1981; Kawata and Lai, 1967) as well as in other species (Bonnett, 1968; Bryant, 1934; Falk and Sitte, 1961; Ledbetter, 1967; Nagahashi et al., 1974). It has been shown that there is histogenetical

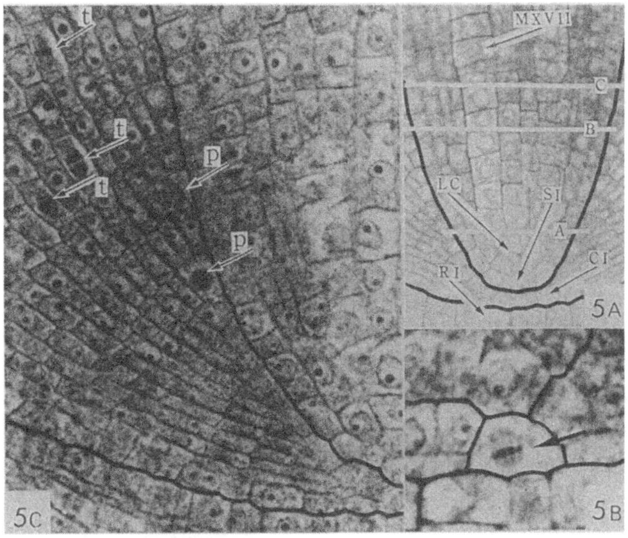


*Fig. 5.* Apical meristem of rice nodal roots. **A**: apical meristem (Kawata et al., 1978a). RI: calyptrogen, CI: dermatogen-periblem complex, SI: plerome, LC: large cells, MXVII: late metaxylem vessel. White lines A, B and C correspond to the approximate levels of cross sections in Fig. 6A-C. **B**: The first periclinal division of dermatogen-periblem complex to differentiate epidermis and endodermis (Kawata and Matsui, 1977c). C: Periclinal and transverse divisions in cortex. p: periclinal division. t: transverse division (Kawata and Matsui, 1977c).

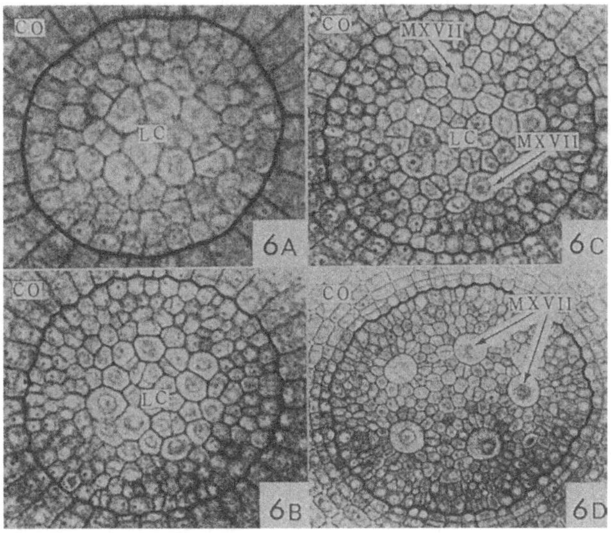


*Fig. 6.* Successive stages in the differentiation of late metaxylem vessel in rice nodal root (Kawata et al., 1978b). CO: cortex, LC: large cells, MXVII: late metaxylem vessel.

coordination between the development of the endodermis, vascularization and lateral root formation in rice roots (Kawata and Lai, 1968). For example, where early and late metaxylem vessels are beginning to mature, primordia of lateral roots are also starting to develop and the Casparian strips are fully matured. In addition, water absorption is most active where the Casparian strips are developed (Kawata and Lai, 1968).

In the later developmental stage of the root, cortical cells are broken down and an air-space is well-developed, except around the lateral roots, this being is the most conspicuous feature in anatomy of rice roots grown under flooded conditions (Figs. 2B-C; Katayama, 1966; Mori, 1972). The air-space in the cortex has been thought to be aerenchyma (Fujii, 1961; Katayama, 1966). In fact, it has already been proved that the air-space is able to function as a pathway for air. Some researchers, however, suspect that the development of the air-space in the cortex might have a possible relation to lateral root formation (Kawata, 1956; Kono, 1968; Kono and Yamada, 1969, 1972b; Kono et al., 1972a). In maize roots, a similar air-space is poorly developed under the conditions of oxygen shortage (Campbell and Drew, 1983). Recently, it has been shown that ethylene has a possible relation to development of the air-space in maize roots (Drew et al., 1979; Konings, 1982)

## Vascularization in roots

Because vascular elements in roots are quite important with reference to the absorption and translocation of water and nutrients, numerous studies on vascularization have been done in many plant species (Bunning, 1951; Heimsch, 1951; Peterson, 1967), including rice (Mori, 1972; Shimabuku, 1960). However, detailed observations on the patterns of both differentiation and maturation of all the vascular elements in any given species are relatively limited.

Both seminal and nodal roots of rice have three kinds of vessels, protoxylem, early metaxylem and late metaxylem vessels, as well as two kinds of sieve tubes, protophloem and metaphloem sieve tubes (Fig. 2C; Kawata et al., 1977f). In the stele, the pericycle, the outermost cell layer of stele, differentiates first. Then, of all the vascular elements, the late metaxylem vessels are the first to differentiate from the large cells near the cap junction (Figs. 5A, 6A-D). Just after the differentiation of these vessels, mother cells of protophloem sieve tubes (PPMC) with metaphloem sieve tubes are differentiated in the peripheral area of the stele. Each PPMC is a hexagonal (sometimes pentagonal) cell adjacent to the pericycle cell layer; a metaphloem sieve tube is located at the inner side of a PPMC (Fig. 7A). Following two oblique divisions in the PPMC, one protophloem sieve tube and two companion cells are formed (Figs. 7B-D), which means that both protophloem sieve tube and its companion cells originate from the same PPMC. Then, differentiation of early metaxylem vessels takes place. Each early metaxylem vessel adjacent to the pericycle is located between two neighboring phloem poles of a PPMC with its associated metaphloem sieve tubes (Fig. 7A).

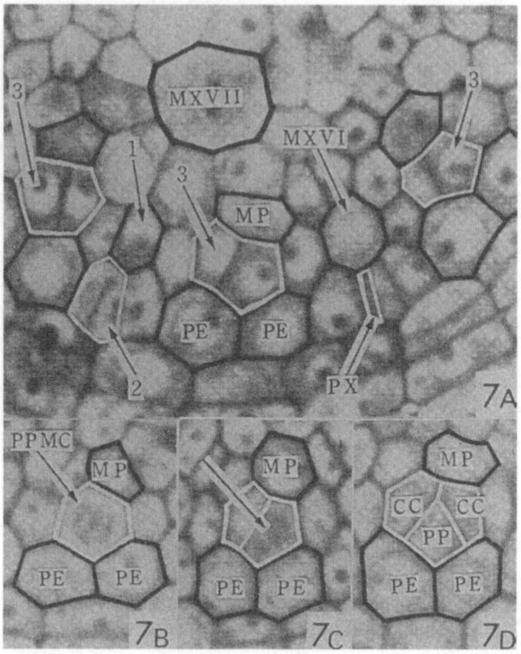


*Fig. 7.* Differentiation of early metaxylem and protoxylem vessels as well as metaphloem and protophloem sieve tubes in rice nodal root (Kawata et al., 1978b). 1: differentiating early metaxylem vessel. 2: radial division of pericycle cell to differentiate protoxylem vessel. 3: protophloem mother cell after the first oblique division. PX protoxylem vessel, MXVI: early metaxylem vessel, MXVII: late metaxylem vessel, PE: pericycle, PPMC: protophloem mother cell, MP: metaphloem sieve tube, PP: protophloem sieve tube, CC: companion cells.

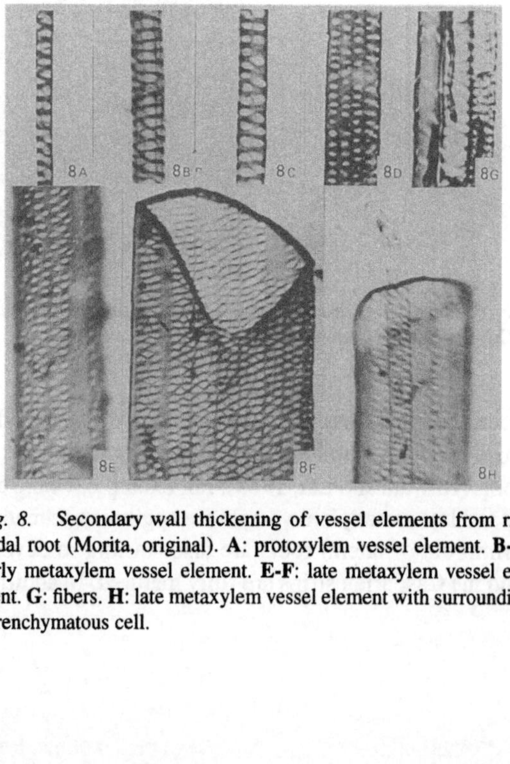


*Fig. 8.* Secondary wall thickening of vessel elements from rice nodal root (Morita, original). **A**: protoxylem vessel element. **B-D**: early metaxylem vessel element. **E-F**: late metaxylem vessel element. **G**: fibers. **H**: late metaxylem vessel element with surrounding parenchymatous cell.

After the differentiation of an early metaxylem vessel, radial cell divisions occur in the pericycle cells adjacent to early metaxylem vessels to differentiate protoxylem vessels last (Fig. 7A). Thus, the pericycle cell layer is interrupted by protoxylem vessels. Differentiation of both vessels and sieve tubes proceeds centrifugally (Kawata et al., 1978a). However, detailed examinations of the secondary wall thickening of vessel elements after maceration (Figs. 8A-F) show that maturation of vessels proceeds centripetally (Morita, unpublished data). Protoxylem vessels mature first, followed by early metaxylem vessels. Maturation of late metaxylem vessels occurs last. The patterns of the secondary wall thickening of vessel elements might be regulated by both the diameter of the vessel elements and the influence of neighboring cells (Figs. 8G-H, Morita, unpublished data). Recently, it was shown in maize roots that there still are end walls of late metaxylem vessel elements in the proximal part of roots and that maturation of these vessel elements takes place far from the root tip (McCully and Canny, 1989; Wenzel et al., 1989). Apparently, maturation is highly correlated with the presence of a root sheath at that time. Histological observations on vascularization in rice as well as in maize suggest that protoxylem and early metaxylem vessels function in the proximal part of roots, whereas late metaxylem vessels translocate water and nutrients mainly in the basal parts. An experimental study using aqueous dye has partly confirmed this hypothesis in maize roots (Mistrikova and Kozinka, 1989). Connections between different vascular elements will be discussed in a later section.

In the case of sieve tubes, intracellular contents of protophloem sieve tubes disappear first and then the metaphloem sieve tube matures which means a centripetal maturation occurs in the phloem, as in the vessel elements (Morita, unpublished data).

## Morphometric analysis on root anatomy

It is thought that the dimensions of the vascular system in the root are an quite important factor regarding their function. Accordingly, the numbers of late metaxylem vessels and protophloem sieve tubes in rice roots were examined. Because, as described above, the number of xylem poles (protoxylem and early metaxylem) is the same as that of phloem poles (protophloem and metaphloem), in principle it is enough to examine the number of late metaxylem vessels and protophloem sieve tubes (or PPMC) as representatives of the "inner and outer vascular elements" in the rice root, respectively (Kawata et al., 1979b).

The numbers of late metaxylem vessels and protophloem sieve tubes in the basal part of nodal roots depend on their location on the stem. There are intimate correlations between numbers of vascular elements and the root diameter, the stele diameter, and the pericycle cell number, respectively (Kawata et al., 1979b; Nemoto et al., 1993). A principal component analysis for the same correlation matrix showed that the first principal component (PC1), interpretable as a general size factor, accounted for 81% of the total variation (Nemoto et al., 1993). Such high explanatory value of PC1 suggests that each tissue in rice roots is tightly integrated in terms of mophogenetic control. Similar close correlations between number of vascular elements and root thickness were found at the root tip where vessels and sieve tubes differentiate (Kawata et al., 1979b).

A R-mode cluster analysis for the matrix suggests that such concordance among the morphological characters might be a manifestation of the history of the histogenetical processes. In the resultant dendrogram, the examined characters were chained in the order that exactly reflected their histogenetical sequence at the root tip. It is most probable that the numbers of vessels and sieve tubes may primarily be controlled by the dimensions of the fields where these tissues differentiate, and that the later development of each tissue keeps pace with each other (Nemoto et al., 1993). Although there are quite few morphometric analyses on root structure (Barlow and Rathfelder, 1984), they must be quite effective for understanding morphogenesis at the root tip.

## Lateral root formation

In rice roots, lateral roots originate in the pericycle (Kawata and Shibayama, 1965a), as in many other plant species (Bell and McCully, 1970; Byrne, 1972; Seago, 1973). Primordia of lateral roots are initiated opposite phloem poles and between two neighboring xylem poles (Fig. 9A; Kawata and Shibayama, 1965a), though the location of lateral root initiation varies with plant species, being opposite to either xylem or phloem in most cases. The formation of lateral root primordia starts by anticlinal cell divisions followed by periclinal cell divisions in pericycle and endodermis (Figs. 9B-G; Kawata and Shibayama, 1965a). Usually, most of the lateral roots are derived from both pericycle and

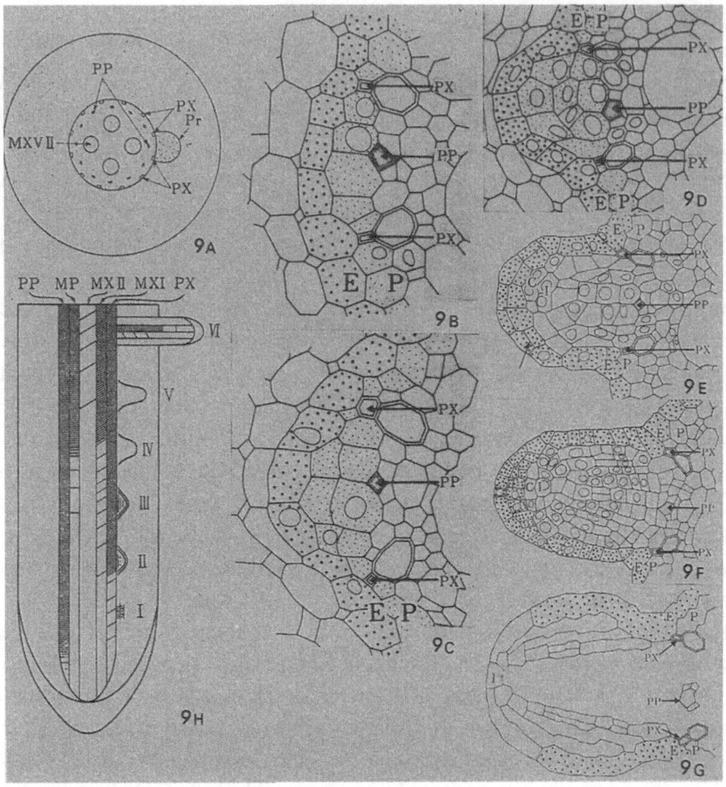


*Fig. 9.* Lateral root formation. **A**: Location of lateral root initiation in cross section of root (Kawata and Shibayama, 1965a). **B-G**: development of lateral root primordium in cross section of root (Kawata and Shibayama, 1965a). **H**: correlations between lateral root formation and vascular system development in longitudinal section of root (Kawata and Takada, 1972). Pr: primordium of lateral root, PX: protoxylem vessel, MXI: early metaxylem vessels, MXII: late metaxylem vessel, PP: protophloem sieve tube, MP: metaphloem sieve tube. EN: endodermis, P: pericycle.

endodermis (Bell and McCully, 1970; Byrne, 1972; Seago, 1973). Observations on longitudinal sections of rice nodal roots have shown that there are mature protophloem sieve tubes and maturing protoxylem vessels in the region where the first primordia of lateral roots are initiated (Fig. 9H; Kawata and Takada, 1972a). Similar correlations between lateral root formation and vascularization of the parental root were also found in some plant species (Crooks, 1933; Hayward, 1942). It is known in rice roots that the distance of the first lateral root primordium from the root tip depends on the diameter and elongation rate of the parent root (Kawata and Ishihara, 1977b). On a nodal root with a larger diameter which is elongating faster, the first lateral root will be formed and will emerge further from the root tip, which indicates an another example of coordination in organo- and histogenesis in rice roots.

In rice, there are two types of lateral roots, thick and thin ones, which differ from each other in morphology and anatomy (Kawata et al., 1962; Kawata and Shibayama, 1965a; Kawata and Soejima, 1974; Kono et al., 1972a). A thick lateral root with a relatively large diameter bears higher-order of lateral roots, while a thin lateral root with a small diameter does not branch (Figs. 1C-D). In a thick lateral root, the cortex is well-developed compared to a thin one. The most striking difference in anatomy is that a thick lateral root has only one late metaxylem vessel in the center of the stele, but this is missing in a thin lateral root (Figs. 2D-E). Such remarkable differences in morphology and anatomy between two types of lateral roots suggest possible differences in their function, but details are still unknown (Yamauchi, 1988). Two such kinds of lateral roots are found in other cereal crops as well (Yamauchi et al., 1987)

## Connecting patterns of vascular tissues in and between roots

Another important aspect of root structure with regard their function of translocating water and nutrients is the patterns of connections between vascular elements, though this has rarely been studied. Some problems regarding these connections are summarized as follows: (a) connections between vascular systems of stem and nodal roots, (b) changing patterns of vascular system along the root axes, (c) connections between vascular systems of the parental roots and their lateral roots, (d) connections among different kinds of vascular elements in the same root.

In rice plants the connecting tissues between the vascular systems of stem and nodal roots are differentiated from the cells of the ground meristem which are adjacent to the peripheral vascular cylinder in the stem (Kawata et al., 1972b). Emergence of nodal roots in rice may correlate with the formation of the connecting tissues between vascular elements of stem and root primordia (Kawata and Harada, 1975). Also, connecting patterns between the vascular system of the stem and the nodal roots may be important with reference to conductivity. Recently, a special structure of the vascular system at the base of barely roots has been found and named as the "hydraulic safety zone" (Luxova, 1991).

Studies on vascular connections between nodal roots and their lateral roots are quite limited (Byrne et al., 1977, 1982). In rice roots, connections between the vessels of nodal roots and their lateral roots, as well as between the vessels of lateral roots and their higher-order lateral roots have already been described in detail (Figs. 11A-D; Kawata et al., 1977f).

In rice nodal roots, there are no connecting tissues among late metaxylem, early metaxylem and protoxylem vessels, excluding the locations of lateral root formation where the tracheary elements differentiated from stelar parenchyma make connections between the different kinds of vessels (Figs. 11E-H; Kawata et al., 1977f). Such findings, as well as the order of maturation of the three different vessels, suggest that "inner" and "outer vascular elements", in principle, are independent of each other, and that water absorbed by the root will enter into the protoxylem and early metaxylem vessels first and then be transferred into late metaxylem vessels through the connecting tissues in the basal part of the root where lateral roots are formed. Recently, such a hypothesis on the pathway of water and nutrients has been partly demonstrated

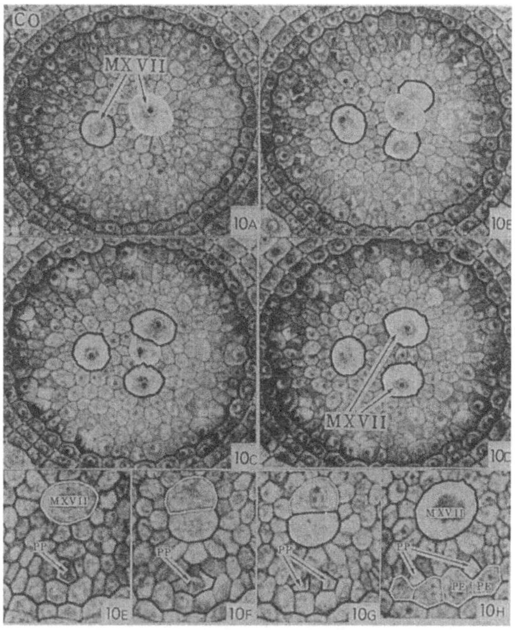


*Fig. 10.* Connecting patterns of vascular systems in rice nodal roots. **A-D**: An example of the connecting pattern of inner vascular system (late metaxylem vessels) (Morita, original). LMX: late metaxylem vessel. **E-H**: connecting pattern of outer vascular system (protophloem sieve tubes). CO: cortex, PE: pericycle, MXVII: late metaxylem vessels, PP: protophloem sieve tubes. (Kawata et al., 1978a).

in maize roots (McCully and Canny, 1989; Mistrikova and Kozinka, 1989).

In the nodal root of rice, the number of vascular elements (pole number) usually decreases along the root axis with the decreases of stele and root diameter (Kawata et al., 1979b). In such a case, the connecting patterns in each vascular elements where their number decrease are also quite important with reference to the translocation of water and nutrients. Preliminary observations at the root tip (Kawata et al., 1978a), suggest that the xylem pole of protoxylem and early metaxylem vessels end blindly and the two neighboring phloem poles fuse into one phloem pole, and the root will, in principle, lose one pole (Figs. 10E-H; Kawata et al., 1978a). There also are several connecting patterns where the number of late metaxylem vessel decreases (Figs. 10A-D). Recently a similar study has been done in barley roots (Luxova, 1991).

Thus, rice is quite a rare and important plant species, in which the network of vascular elements in and between roots, with special reference to the connecting patterns, has been considerably clarified.

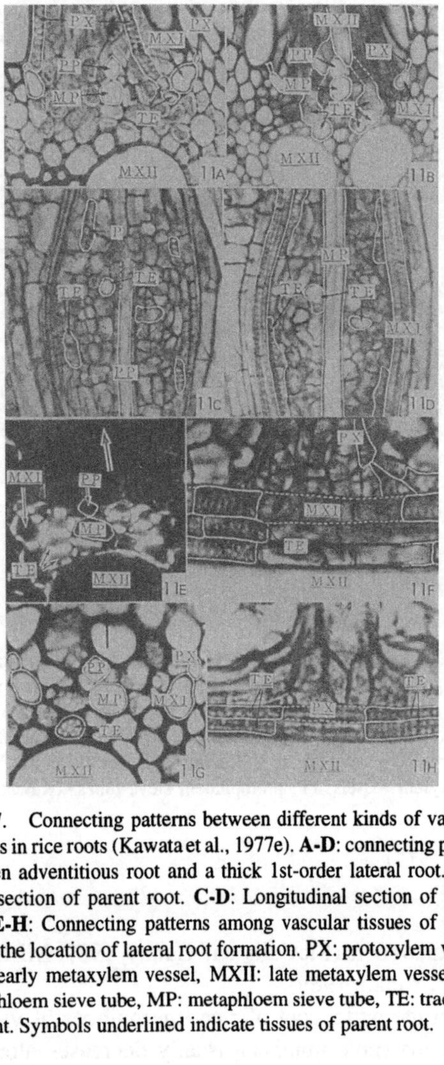

*Fig. 11.* Connecting patterns between different kinds of vascular systems in rice roots (Kawata et al., 1977e). **A-D**: connecting pattern between adventitious root and a thick 1st-order lateral root. **A-B**: Cross section of parent root. **C-D**: Longitudinal section of parent root. **E-H**: Connecting patterns among vascular tissues of parent root at the location of lateral root formation. PX: protoxylem vessel, MXI: early metaxylem vessel, MXII: late metaxylem vessel, PP: protophloem sieve tube, MP: metaphloem sieve tube, TE: tracheary element. Symbols underlined indicate tissues of parent root.

## Concluding remarks

In rice roots there are many examples of coordination in organo- and histogenesis, including root hair formation, endodermis development, lateral root formation and vascularization. These facts suggest that morphogenesis of each tissue keeps pace with one another quite integratedly. At this time, it is thought that late metaxylem vessels could play the role of a kind of "organizer" in root morphogenesis for the following reasons: (a) at the site quite near the cap junction late metaxylem is the first of all tissues to be differentiated in the stele, excluding the pericycle (Fig. 5A), (b) a special arrangement of cells surrounding late metaxylem vessels is found during the process of vascular differentiation (Fig. 6), (c) differentiation of vessel elements proceeds centrifugally, (d) lateral roots without a late metaxylem vessel cannot initiate higher-order laterals, (e) the structure of lateral roots lacking late metaxylem elements is poorly developed (Figs. 2D-E). There also are close correlations between root morphogenesis and water absorption or root elongation, respectively, which indicate that root function might be based on root structure and developmental stages.

## References


Barlow P W and Rathfelder E L 1984 Correlations between the dimensions of different zones of grass root apices, and their implications for morphogenesis and differentiation in roots. Ann. Bot. 53, 249–260.

Barlow P W 1994 Origin, biology and diversity of adventitious roots. *In* Biology of Adventitious Root Formation. Eds. T D Davis and B E Haissig. pp 1–23. Plenum, New York, USA.

Bell J K and McCully M E 1970 A histogenetical study of lateral roots initiation and development in *Zea mays*. Protoplasma 70, 179–205.

Bonnett H T 1968 The root endodermis: fine structure and function. J. Cell Biol. 37, 199–205.

Bryant A E 1934 A demonstration of the connection of the protoplasts of the endodermal cells with the Casparian strips in the root of barley. New Phytol. 33, 231.

Bunning E 1951 Uber die Differenzierungsvorgange in der Cruciferenwurzel. Planta 39, 126–153.

Byrne J M 1972 Lateral root development in *Malva*. Am. J. Bot. 59, 678.

Byrne J M, Pesacreta T C and Fox J A 1977 Development and structure of the vascular connection between the primary and secondary root of *Glycine max* (L.) Merr. Am. J. Bot. 64, 960–965.

Byrne J M, Byrne J M and Emmitt D P 1982 Development and structure of the vascular connection between the primary and lateral root of *Lycopersicon esculentum*. Am. J. Bot. 69, 287–297.

Campbell R and Drew M C 1983 Electron microscopy of gas space (aerenchyma) formation in adventitious roots of *Zea mays* L. subjected to oxygen shortage. Planta 157, 350–357.

Clark L H and Harris W H 1981 Observation on the root anatomy of rice (*Oryza sativa* L.). Am. J. Bot. 68, 154–161.

Clowes F A L 1958 Development of quiescent centres in root meristems. New Phytol. 57, 85–88.

Clowes F W L 1967 The quiescent centre. Phytomorph. 17, 132–140.

Crooks D M 1933 Histological and regenerative studies on the flax seedling. Bot. Gaz. 95, 209–239.

Drew M C, Jackson M B and Gifford S 1979 Ethylene-promoted adventitious rooting and development of cortical air spaces (aerenchyma) in roots may be adaptive responses to flooding in *Zea mays* L. Planta 147, 83–88.

Esau K 1965 Plant Anatomy 2nd ed. John Wiley and Sons, New York.

Falk H and Sitte P 1961 Untersuchungen am Caspary-streifen. Proc. Europ. Region. Conf. Electron Microscopy, Delft 2, 1063–1066.

Fujii Y 1961 Studies on the regular growth of the roots in rice plants and wheats. Agric. Bull. Saga Univ. 12, 1–117.*

Gregory P J, Mcgowan M, Biscoe P V and Hunter B 1978 Water relations of winter wheat. 1. Growth of the root system. J. Agric. Sci. 91, 91–102.

Hayward H E 1942 The anatomy of the seedling and roots of the Valencia orange. USDA Tech. Bull. 786, 1–31.

Heimsch C 1951 Development of vascular tissues in barley roots. Am. J. Bot. 38, 523–538.

Hoshikawa K 1989 The Growing Rice Plant, an Anatomical Monograph. Nobunkyo, Tokyo, Japan. 310p.

Jackson M B and Barlow P W 1981 Root geotropism and the role of growth regulators from the cap: a re-examination. Plant Cell Environ. 4, 107–123.

Juniper B E 1976 Geotropism. Ann. Rev. Plant Physiol. 27, 385–406.

Kaeriyama N and Yamazaki K 1983 The development of rooting zone in soil in relation to the growth direction and the elongation rate of the primary roots in corn plants. Jpn. J. Crop Sci. 52, 508–514.*

Katayama T C 1966 Anatomy of the root of *Oryza sativa* with emphasis on developmental processes. Seiken Ziho 18, 6–22.

Kaufman P B 1959 Development of the shoot of *Oryza sativa* L. III. Early stages in histogenesis of the stem and ontogeny of the adventitious root. Phytomorph. 9, 382–404.

Kawata S 1956 Studies on root formation in certain cultivated plants. III. Relation between primary root formation and canal development in crown roots of rice plants. Proc. Crop Sci. Soc. Japan 24, 232–236.*

Kawata S and Ishihara K 1959 Studies on the root hairs in rice plant. Proc. Crop Sci. Soc. Japan 27, 341–348.*

Kawata S, Kamata E and Yamazaki K 1962 Studies on elements of the wood in rice plants. Proc. Crop Sci. Soc. Japan 30, 266–278.*

Kawata S, Yamazaki K, Ishihara K, Shibayama H and Lai K-L 1963 Studies on root system formation in rice plants in a paddy. Proc. Crop Sci. Soc. Japan 32, 163–180.*

Kawata S and Ishihara K 1964 Studies on the oxidizing ability of root hairs in rice plants. Proc. Crop Sci. Soc. Japan 33, 168–173.*

Kawata S and Shibayama H 1965a On the lateral root primordia formation in the crown roots of rice plants. Proc. Crop Sci. Soc. Japan 33, 423–431.*

Kawata S and Lai K-L 1965b On the meristematic state of the endodermis in the crown roots of rice plants. Proc. Crop Sci. Soc. Japan 34, 210–216.*

Kawata S and Lai K-L 1967 On the differentiation of Casparian dots of the endodermis in the crown roots of rice plant. Proc. Crop Sci. Soc. Japan 36, 75–84.*

Kawata S and Lai K-L 1968 On the correlation between the endodermis differentiation and the water absorption in the crown roots of rice plants. Proc. Crop Sci. Soc. Japan 37, 631–639.*

Kawata S and Takada H 1972a On the first lateral root primordium formation closer to crown root apex of rice plants. Proc. Crop Sci. Soc. Japan 41, 111–119.*

Kawata S, Harada J and Yamazaki K 1972b On the crown root primordium formation in the stem of rice plants. Proc. Crop Sci. Soc. Japan 41, 296–309.*

Kawata S and Soejima M 1974 On superficial root formation in rice plants. Proc. Crop Sci. Soc. Japan 43, 354–374.*

Kawata S and Harada J 1975 On the development of the crown root primordia in rice plants. Proc. Crop Sci. Soc. Japan 44, 438–457.*

Kawata S and Chung W 1976 Formation of root hair in lateral roots of rice plants. Proc. Crop. Sci. Soc. Japan 45, 436–442.*

Kawata S and Chung W-I 1977a Formation of root hair in crown roots of rice plants and nitrogen fertilizer. Jpn. J. Crop Sci. 46, 199–204.*

Kawata S and Ishihara A 1977b Correlations among size of root apex, root apex-lateral distance, and elongation rate in rice roots. Jpn. J. Crop Sci. 46, 228–238.*

Kawata S, Nishimaki K and Yamazaki K 1977c The apical structure of crown roots in rice plants. Jpn. J. Crop Sci. 46, 393–402.*

Kawata S and Matsui S 1977d On the cortex formation in the apical meristems of the elongating crown roots of rice plants. Jpn. J. Crop Sci. 46, 403–413.*

Kawata S and Harada J 1977e On the diameter determination of crown root primordia in rice plants. Jpn. J. Crop Sci. 46, 423–430.*

Kawata S, Sasaki O and Yamazaki K 1977f On the structure of the crown root and the lateral root, and the vessel connection between them, in rice plants. Jpn. J. Crop Sci. 46, 569–579.*

Kawata S, Morita S and Yamazaki K 1978a On the differentiation of vessels and sieve tubes at the root tips of rice plants. Jpn. J. Crop Sci. 47, 101–110.*

Kawata S, Harada J and Yamazaki K 1978b On the number and the diameter of crown root primordia in rice plants. Jpn. J. Crop Sci. 47, 644–654.*

Kawata S and Chung W-I 1979a On the elongation of root hairs of rice plants. Jpn. J. Crop Sci. 48, 123–130.*

Kawata S, Morita S and Yamazaki K 1979b On the numbers of vessels and sieve tubes in crown roots of rice plants. Jpn. J. Crop Sci. 48, 502–509.*

Kawata S, Suzuki S and Yamazaki K 1979c The detachment of the primary root caps in rice plants. Jpn. J. Crop Sci. 48, 303–310.*

Kawata S and Harada J 1980 The effect of defoliating tiller leaves on the crown root formation in the main stem of rice plants. Jpn. J. Crop Sci. 49, 587–592.*

Konings H 1982 Ethylene-promoted formation of aerenchyma in seedling roots of *Zea mays* L. under aerated and non-aerated conditions. Physiol. Plant. 54, 119–124.*

Kono Y 1968 Histogenetical studies on cortical disintegration in rice roots. Proc. Crop Sci. Soc. Japan 37, 235–243.*

Kono Y and Yamada N 1969 Studies on the developmental physiology of the cortical disintegration in rice roots. Proc. Crop Sci. Soc. Japan 38, 477–488.*

Kono Y, Igeta M and Yamada N 1972a Studies on the developmental physiology of the lateral roots in rice seminal roots. Proc. Crop Sci. Soc. Japan 41, 192–204.*

Kono Y and Yamada M 1972b Studies on the developmental physiology of the relationship between the cortical disintegration and lateral root growth in rice seminal roots. Proc. Crop Sci. Soc. Japan 41, 256–266.*

Ledbetter M C 1967 The Casparian strip, a site of a functional tight junction in plant roots. J. Cell Biol. 35, 79A.

Luxova M 1986 The hydraulic safety zone at the base of barley roots. Planta 169, 465–470.

Luxova M 1991 Mechanisms of reduction of the stelar pattern along barley roots. Bot. Acta 104, 163–168.

Mawaki M, Morita S, Suga T, Iwata T and Yamazaki K 1990 Effect of shading on root system morphology and grain yield of rice plants (*Oryza sativa* L.). Jpn. J. Crop Sci. 59, 89–94.*

McCully M E and Canny M L 1989 Pathway and process of water and nutrient movement in root. *In* Structural and Functional Aspects of Transport in Root. Eds. B C Loughman, O Gasparikova and J Kolek. pp 3–14. Kluwer Academic Publishers, Dordrecht.

Mistrikova I and Kozinka V 1989 Patterns of long-distance movement of water in roots. *In* Structural and Functional Aspects of Transport in Roots. Eds. B C Loughman, O Gasparikova and J Kolek. pp 165–168. Kluwer Academic Publishers, Dordrecht.

Moore R and McClelen C E 1983a A morphometric analysis of cellular differentiation in the root cap of *Zea mays*. Am. J. Bot. 70, 611–617.

Moore R and McClelen C E 1983b Ultrastructural aspects of cellular differentiation in the root cap of *Zea mays*. Can. J. Bot. 61, 1566–1572.

Moore R and Pasieniuk J 1984a Structure of columella cells in primary and lateral roots of *Ricinus communis* (Euphorbiaceae). Ann. Bot. 53, 715–726.

Moore R and Pasieniuk J 1984b Graviresponsiveness and the development of columella tissue in primary and lateral roots of *Ricinus communis*. Plant Physiol. 74, 529–533.

Moore R and Evans M L 1986 How roots perceive and respond to gravity. Am. J. Bot. 73, 574–587.

Mori T 1972 Tissue differentiation and development in rice roots. Bull. Fac. Agric. Hirosaki Univ. 18, 236–304.

Morita S, Suga T and Yamazaki K 1988 The relationship between root length density and yield in rice plants. Jpn. J. Crop Sci. 57, 438–443.*

Morita S, Okuda H and Abe J 1993 Spatial distribution and structure of wheat root system. *In* Low-input Sustainable Crop Production Systems in Asia. Ed. Korean Society of Crop Science. pp 399–404.

Nagahashi G, Thompson W W and Leonard R T 1974 The Casparian strip as a barrier to the movement of lanthanum in corn roots. Science 183, 670–671.

Nemoto K, Baba T and Morita S 1993 Shoot and root development in rice related to the phyllochron. 1993 Agronomy Abstract, 119.

Nemoto K, Morita S and Yamada S 1993 Morphometric analysis on the anatomy of rice roots. Jpn. J. Crop. Sci. 62(Extra Issue 1), 250–251.**

Percival J 1921 The Wheat Plant, a Monograph. Duckworth, London.

Peterson R L 1967 Differentiation and maturation of primary tissues in white mustard root tips. Can. J. Bot. 45, 319–331.

Pilet P E 1982 Importance of the cap cells in maize root gravireaction. Planta 156, 95–96.

Ransom J S and Moore R 1985 Geoperception in primary and lateral roots of *Phaseolus vulgaris* (Fabaceae). III. A model to explain the differential georesponsiveness of primary and lateral roots. Can. J. Bot. 63, 21–24.

Seago J L 1973 Developmental anatomy in roots of *Ipomoea purpure*. II. Initiation and development of secondary roots. Am. J. Bot. 60, 607–618.

Seago J L Jr. and Wolniak S M 1976 Cortical ontogeny in roots. I. *Zea mays*. Am. J. Bot. 63, 220–225.

Seago J L Jr. and Marsh L C 1989 Adventitious root development in *Typha glauca* with emphasis on the cortex. Am. J. Bot. 76, 909–923.

Shimabuku K 1960 Observations on the apical meristem of rice roots. Bot. Mag. Tokyo 73, 22–29.

Weaver J E 1926 Root Development of Field Crops. McGraw-Hill, New York.

Wenzel C L, McCully M E and Canny M J 1989 Development of water conducting capacity in the root systems of young plants of corn and some other C4 grasses. Plant Physiol. 89, 1094–1101.

Williams B C 1947 The structure of the meristematic root tip and origin of the primary tissues in the roots of vascular plants. Am. J. Bot. 34, 455–462.

Yamauchi A, Kono Y and Tatsumi J 1987 Comparison of root system structures of 13 species of cereals. Jpn. J. Crop Sci. 56, 618–631.

Yamauchi A 1993 Root type 2. Root Research 2, 51–54.**

Yamazaki K and Kaeriyama N 1982 The morphological characters and the growing directions of primary roots of corn plants. Jpn. J. Crop Sci. 51, 584–590*.

Yamazaki K and Nakamoto T 1983 Primary root formation in the growing shoots of several cereal crops. Jpn. J. Crop Sci. 52, 342–348.*

Yamazaki K and Nemoto K 1986 Morphological trends and interrelationships of leaves stem parts and roots along the main axes of rice plants. Jpn. J. Crop Sci. 55, 236–243.*

* in Japanese with English summary or abstract;

** in Japanese.

*F. Baluška et al. (eds.), Structure and Function of Roots*, 87–89, 1995.


# The localization of $\beta$-glucuronidase in roots of transgenic sugar beet by means of azocoupling procedure

K. Beneš[1], S. Vitha[1], K.M.A. Garland[2] and M.C. Elliott[3]
[1]*Faculty of Biology, University of South Bohemia, Č. Budějovice, Czech Republic,* [2]*Plant Biotechnology Group, Dundee Institute of Technology, Dundee, Scotland and* [3]*Department of Applied Biology, DeMontfort University, Leicester, UK*

*Key words:* azocoupling technique, $\beta$-glucuronidase, indigogenic technique, root tip, transgenosis

## Abstract

Using in toto incubation of formalin-fixed material, a positive azocoupling reaction was obtained in roots of transgenic sugar beet bearing the $\beta$-glucuronidase reporter gene. In the two clones tested, the post-coupling procedure with naphthol ASBI $\beta$-D-glucuronide as a substrate and Fast Blue BB salt as a coupler led to the same localization patterns as observed with the indigogenic procedure. New approaches in studies of root growth and differentiation are mentioned, resulting from the use on transgenic material.

*Abbreviations:* DMF – N,N-dimethylformamide, GUS – $\beta$-glucuronidase

## Introduction

The use of traditional methods of plant anatomy can hardly bring any significant contribution to the study of root differentiation. Therefore, new approaches have been searched for with which to advance in this field. Histochemical studies represent one such attempt. Work of this type has been done also in our laboratory (Beneš, 1974; Beneš et al., 1981, 1989). New stimuli may come now from the utilization of the results of gene engineering. With this approach, we intend to evaluate, from the point of view of tissue differentiation, the patterns generated in root tips of transgenic plants expressing reporter genes driven by different 'tissue specific' promoters. During the last several years, the use of the $\beta$-glucuronidase (GUS) marker gene has become very popular. Many technical problems, however, are encountered here. Concerning the localization of GUS in transgenic plants, it is usually not possible to utilize the usual processing and enzyme reactions that are common for work with animal material (Stoward and Pearse 1991) and which have proved competent for revealing glycosidases in seedling roots (Beneš et al., 1973, 1981). In the present work we intend to focus on one of the methodical points concerned here. We aim to assess the localization of GUS marker enzyme in root tips of transgenic sugar beet by means of an azocoupling procedure. Although methods of this type have been included in methodological papers (Jefferson, 1987; Jefferson et al., 1987), we are not aware of any work concerning transgenic plants where azocoupling procedures were used successfully (Gallagher, 1992).

## Materials and methods

### *Plant material*

Hairy roots of transgenic sugar beet (*Beta vulgaris* L. cv. Salohill) were grown in vitro in liquid Gamborg's B5 medium, supplemented with sucrose 2% w/v and kanamycin sulphate 20 mg $L^{-1}$. For our experiments , two clones were selected (Vitha et al., 1992). Both clones result from *Agrobacterium tumefaciens*-mediated transformation, where *Agrobacterium* acts as a binary vector, carrying a recombinant plasmid (with genes for GUS and for resistance to kanamycin) and a helper plasmid pRi A4b. The helper plasmid is of *A. rhizogenes* origin and is responsible for the hairy root phenotype. In clone F19, the GUS gene is under the control of the rolC promoter, whereas in clone J10 the GUS gene is driven by the CaMV 35S promoter (Phillips et al., 1992).

*Tissue processing*

Root tips about 5 mm long were used. After some preliminary experiments, in toto incubation was performed on material fixed in ice-cold 4% formaldehyde (10% formalin) in phosphate buffer pH 7.0 for 30 minutes. Before incubation, the roots were washed in several changes of cold $H_2O$.

*Incubation media*

The incubation medium for the indigogenic assay was prepared according to Jefferson (1987); 1 mg of 5-Br-4-Cl-3-indolyl $\beta$-glucuronide (X-glucuronide; Clontech) was dissolved in 200 $\mu$L of N,N-dimethylformamide (DMF), and made up to 10 mL with sodium phosphate buffer 50m*M*, pH 7.0, containing 0.5 m*M* potassium ferricyanide and 0.5 m*M* postassium ferricyanide. The material was incubated overnight at room temperature.

For the azocoupling technique, both the simultaneous and post-coupling procedures were tried. In the former case, the medium contained 2.5 mg of a substrate dissolved in 0.2 mL DMF or methanol, 9.8 mL phosphate buffer pH 7.0 and 10 mg of Fast Blue BB salt (purified grade, 1/2 $ZnCl_4$; Sigma). The following substrates were tried: 1-naphthyl-$\beta$-D-glucuronide, 6-bromo-2-naphthyl-$\beta$-D-glucuronide, and naphthol ASBI $\beta$-D-glucuronide (all from Sigma). During overnight incubation the material was transferred to fresh medium every 3 or 4 hours. In the post-coupling procedure the same substrates and the same salt were used. After overnight incubation in a medium with the substrate (the diazonium salt omitted) and a brief rinse in $H_2O$, the material was transferred for 10 minutes into a medium containing the diazonium salt only (substrate and DMF omitted).

Without-substrate controls were run both in indigogenic and azocoupling assays.

*Microscopic examination*

After the incubation, the material was washed in $H_2O$ and observed as water-mounted, fresh preparations under the microscope.

**Results and discussion**

The use of **indigogenic procedure** led to the following results: In J10 clone activity was remarkably high in promeristem and in both plerome and cortex of the primary meristem, whereas in clone F19 the promeristem was negative as was the cortex, while in plerome the phloem poles were uniformly stained regardless of the distance from the tip. The localization patterns observed correspond with our previous results (Vitha et al., 1992).

The localization of Gus activity, as revealed by the indigogenic procedure, was the basis for the evaluation of results obtained in parallel experiments by **azocoupling methods.**

After **simultaneous** coupling no positive readout could be done with any of the substrates used, since the non-specific staining in without-substrate controls was very intensive.

Using a **post-coupling** procedure, a positive reaction was obtained with naphthol ASBI $\beta$-D-glucuronide but not with the other substrates tested. Though the positive reaction appeared at the same sites as in the case of indigogenic procedure (Figs. 1 to 4), the localization patterns were somewhat blurred due to diffusion – apparently – of the primary reaction product. Nevertheless, this is – to our best knowledge – the first case of positive results in transgenic, GUS bearing plants using the azocoupling procedure.

Our previous experiments (unpublished results) proved that in hairy toots of transgenic sugar beet without the GUS gene, no intrinsic GUS activity appears at pH 7.0.

As far as 1-naphthyl- and 6-bromo-2-naphthyl-$\beta$-D-glucuronides are concerned, further in vitro enzymological work is needed to discern whether those substrates were not hydrolysed in our objects, or whether the primary reaction product diffused out of the tissue.

The material processing, successful when assaying intrinsic glycosidases in plants, i.e., frozen sections from Ca-formol-fixed material (Beneš et al., 1973, 1981), often fails in the work with transgenic objects bearing GUS activity. The in toto incubation must often be used, though it is well known that in toto incubation may lead to false localization (e.g., Beneš et al., 1992). In many cases it is not possible to distinguish between the tissue-specific gene expression and the differential penetration of the substrate into the tissue. This seems not to be the case in the present work, since distinct differences in GUS localization were achieved in root clones bearing constructs with different promoters.

We must keep in mind that, concerning GUS activity in transgenic plants, the prokaryotic gene is acting in a eukaryotic cell, which could influence the process of glycosylation of the enzyme formed, its targeting to

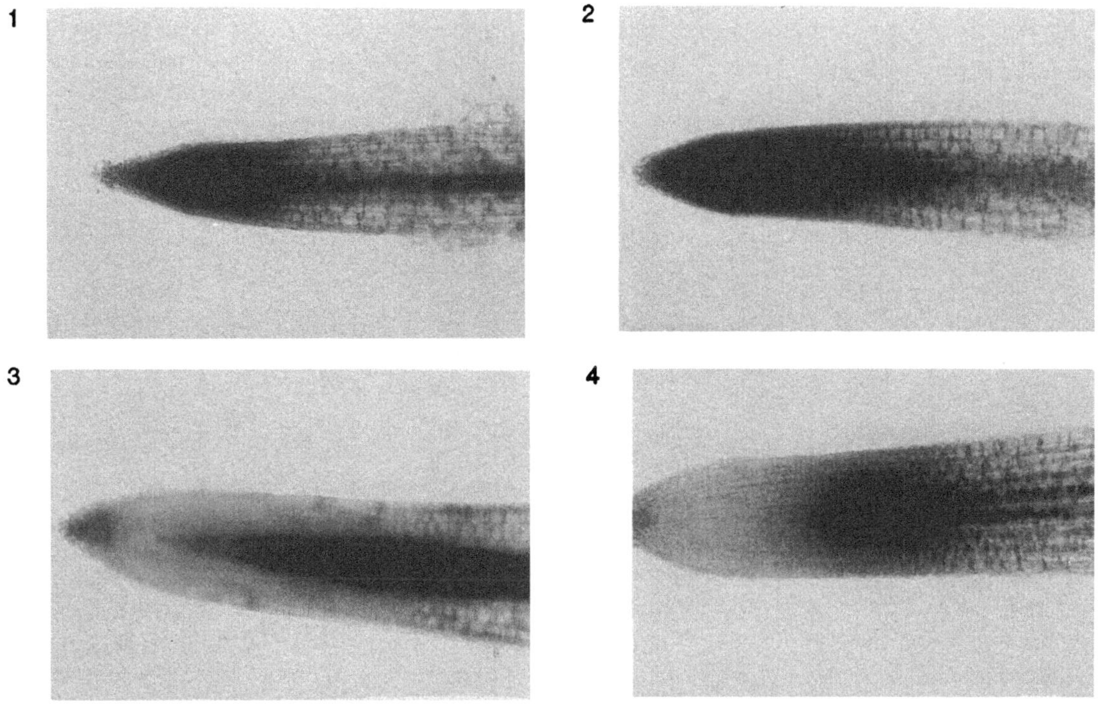


*Fig. 1–4.* A comparison of the indigogenic and postcoupling $\beta$-glucuronidase assays in root tip of in vitro cultivated transgenic sugar beet. *Fig. 1.* Clone J 10. Indigogenic assay. *Fig. 2.* Clone J 10. Post coupling assay. *Fig. 3.* Clone F19. Indigogenic assay. *Fig. 4.* Clone F19. Post coupling assay.

the particular cell compartments, etc. This may be the source of problems for the work with tissue sections. Special studies are in progress in our laboratory in this respect.

## Acknowledgement

The present study was supported by Tempus mobility grant (JEP No. 2216–91).

## References

Beneš K 1974 Histogenesis and non-specific esterase localization in the root tip. *In* Structure and Function of Primary Root Tissues. Ed. J Kolek. pp 261–264. Veda, Bratislava, Czechoslovakia.

Beneš K, Georgieva Y D and Poláčková D 1973 The presence and distribution of alpha and beta glucosidase in root tip. Biol. Plant. 15, 88–94.

Beneš K, Hadaočová J and Hlaráček J 1989 The heterogenity of root tip cells and tissues as revealed by in situ enzyme localization and isoenzyme patterns: A comparison of proteo- and esterolytic activities. *In* Structural and Functional Aspects of Transport in Roots. Eds. B C Loughman et al. pp 29–33. Kluwer, Dordrecht, The Netherlands.

Beneš K, Ivanov V B and Hadačová V 1981 Glycosidases in the root tip. *In* Structure and Function of Plant Roots. Eds. R. Brouwer et al. pp 137–139. Nijhoff, The Hague.

Beneš K, Voldánovà-Michalová M and Vitha S 1992 Localization artifacts caused by in toto incubation to reveal GUS activity in transgenic plants. *In* Progress in Basic, Applied and Diagnostic Histochemistry. p 2. Czechoslovak Histo- and Cytochemical Society, Hradec Králové, Czechoslovakia.

Gallagher S R 1992 GUS Protocols: Using the GUS Gene as a Reporter of Gene Expression. Academic Press, San Diego, CA, USA. 213 p.

Jefferson R A 1987 Assaying chimeric genes in plants: The GUS gene fusion system. Plant Mol. Biol. Rep. 5, 387–405.

Jefferson R A, Kavanagh T A and Bevan M W 1987 GUS fusions: $\beta$ -glucuronidase as a sensitive and versatile gene fusion marker in higher plants. EMBO J. 6, 3901–3907.

Phillips J P, Xing T, Gartland J S, Gartland K M A and Elliott M C 1992 Variation in $\beta$-glucuronidase activity of clones of transformed sugar beet roots. Plant Growth Regul. 11, 319–326.

Steward P J and Pearse A G E 1991 Histochemistry, Vol. 3. Churchill Livingstone, London, UK.

Vitha S, Brada J, Vlachová M, Phillips J P, Gartland K M A and Elliot M C 1992 Histochemical analysis of $\beta$-glucuronidase expression patterns in root tips of trasgenic sugar beet. *In* Progress in Basic, Applied and Diagnostic Histochemistry. p 64, Czechoslovak Histo- and Cytochemical Society, Hradec Králové, Czechoslovakia.

*F. Baluška et al. (eds.), Structure and Function of Roots,* 91–98, 1995.


# Effect of gibberellin and ancymidol on the growth and cell wall components of pea (*Pisum sativum* L.) roots

Eiichi Tanimoto
*Biology Laboratory, College of General Education, Nagoya City University, Mizuho-ku, Mizuho-cho, Nagoya 467, Japan*

*Key words:* ancymidol, cell wall, gibberellin, pea (*Pisum sativum* L.), polysaccharide, root growth

**Abstract**

Externally applied gibberellin $A_3$ ($GA_3$) is known to show little effect on root elongation whereas it strongly promotes stem elongation in peas. $GA_3$ was, however, found to enhance root elongation when root elongation was suppressed by ancymidol (Anc), an inhibitor of gibberellin (GA) biosynthesis. In a series of dose-response experiments, Anc was found to inhibit elongation and induced expansion of roots. These Anc effects were reversed when $GA_3$ was given together with Anc.

In this report, these effects of $GA_3$ and Anc on the growth and morphology of roots were re-confirmed. In addition, the effects of these growth regulators on acid-induced elongation and cell-wall components were investigated. Anc-treated roots showed little elongation response to acid solution, whereas Anc plus $GA_3$-treated roots exhibited a rapid response. Although GA is known to increase the osmotic concentration of stem cells to enhance elongation growth, $GA_3$ did not increase the osmotic concentration in root cells. Since cell elongation is regulated by cell-wall extensibility and osmotic concentration for water uptake, it was suggested that $GA_3$ -enhanced elongation of roots takes place by the $GA_3$-effect on cell walls.

$GA_3$ modified sugar composition and the molecular mass of cell-wall polysaccharides. The relative amounts of arabinose and galactose in the elongating cell walls of $GA_3$ plus Anc-treated roots was different from that of Anc-treated roots. $GA_3$ plus Anc-treated roots also contained larger molecular-mass components of pectin and hemicellulose than Anc-treated roots. These results suggest that the $GA_3$ regulation of elongation and expansion of roots is accompanied by a modification of matrix polysaccharides in the cell walls. Since Anc-treated roots were thicker than $GA_3$ plus Anc-treated roots, these changes in cell walls may reflect the increase in horizontal cell walls due to the thickening.

## Introduction

The role of gibberellin (GA) in stem elongation has been extensively studied in dwarf and normal pea plants (Brian and Hemming, 1957; Ross et al., 1989; Tanimoto et al., 1967). However, we have limited information on the relationship between gibberellin and root growth, since externally applied GA shows little effect on root growth (Tanimoto, 1988, 1990). Gibberellin $A_3$ ($GA_3$) was found to promote root elongation of lettuce (Tanimoto, 1987) and peas (Tanimoto 1988, 1990) when the root growth was simultaneously suppressed by ancymidol (Anc), an inhibitor of GA biosynthesis (Coolbaugh et al., 1982; Rademacher, 1990). Such an interaction between $GA_3$ and Anc has been reported on the shoot growth of *d-5* dwarf maize (Britz and Saftner, 1987), but little is known about the mode of growth regulation by these growth regulators in the root system.

Cell elongation is controlled by cell-wall extensibility and water uptake driven by the osmotic potential of the cells. Although $GA_3$ has been reported to increase cellular osmotic potential to promote stem elongation of cucumber (Katsumi et al., 1980) and of pea (Miyamoto and Kamisaka, 1988), no such effect has been reported on roots. Since acid-induced growth response is one of the characteristics of elongating roots (Edwards and Scott, 1976), effects of Anc and $GA_3$ on acid-growth and osmotic concentration of the root segments were examined in the present study.

The role of matrix polysaccharides in the growing cell walls has been extensively studied in stems and coleoptiles (Fry, 1986; Hayashi, 1989; Hoson and Masuda, 1991; Masuda, 1990; Nishitani and Masuda, 1981; Sakurai, 1991). Although studies on matrix sugars in growing cell walls of roots are few, Masuda and Pilet (1983) reported that changes in sugar composition of cell walls occur along the root axis of maize roots. Thus, growth regulators such as $GA_3$ and Anc may possibly affect these polysaccharides (Katsumi et al., 1992; Tanimoto, 1992).

This paper reports the effects of $GA_3$ and ancymidol on the growth, osmotic concentration, acid-induced growth response and cell-wall polysaccharides of pea roots.

## Materials and methods

*Preparation of roots*

Pea roots from *Pisum sativum* L. cv. Alaska were prepared as described elsewhere (Tanimoto, 1988). Seeds were surface sterilized in 0.1% sodium hypochlorite solution for 30 min, rinsed and imbibed in running tap water for 20 h at 23°C. Germinating seeds were placed on notches in the top edge of a frame of acrylic resin which was designed to prepare straight roots (Tanimoto, 1988). Seeds were incubated for 20 h in a humid box at 23°C in the dark. Straight roots, 20 ± 5 mm long, were selected and used for the experiments.

*Hydroculture of seedling roots*

Twenty seedlings were supported on a stainless steel net (5-mm mesh) fixed in a polystyrene cup (150 mL). The roots were dipped in the $GA_3$ and/or Anc solution containing inorganic nutrients (Tanimoto, 1988). Seedling roots were grown in room light at 23°C.

*Acid-growth of root segments*

Seedling roots were treated with Anc with or without $GA_3$ for 24 h and then apical 10-mm root segments were excised from seedling roots with the 2-mm root tip removed. Acid-growth was measured by the method previously reported by Tanimoto et al. (1989). Sixteen root segments were vertically held in the measuring box of a 'Rhizometer' (Uchida Trading Co. Tokyo, Japan). The rhizometer simultaneously measured each elongation of 16 roots every 10 min by a microcomputer-aided hydroculture equipment (Tanimoto and Watanabe, 1986).

*Measurement of osmotic concentration*

Osmotic concentrations of root segments and buds were measured by vapor pressure osmometry as described by Miyamoto and Kamisaka (1988). One to two hundred sections excised every 1-mm distance from root tip were collected in air-tight vials and frozen at -25°C. Frozen sections were transferred to a funnel with a stainless steel net (140 mesh) fixed in the bottom. The funnel was centrifuged in a vial at 4°C for 10 min at $1000 \times g$. The osmotic concentrations of the filtrates were determined by a vapor pressure osmometer (Wescor Model 5100 C. Utah, USA) with 5 measurements on each sample. The experiments were duplicated and one set of similar results was presented.

*Analysis of cell wall sugar*

Cell-wall sugars were analyzed by the method of Nishitani and Masuda (1981). Seedling roots were treated with the growth solution containing 10 $\mu M$ ancymidol with or without 0.1 $\mu M$ $GA_3$ for 6 days. Consecutive 1-mm long segments were excised from the tip to 8-mm behind the tip of the root by a multi-bladed guillotine. A hundred segments of each zone were collected in a small vial and frozen at -20°C. Frozen-thawed segments were ground in a centrifuge tube by a glass rod with purified quartz powder added. Ground tissue of each root zone was sequentially washed by centrifugation ($1000 \times g$) with cold water, ethanol, acetone, a mixture of methanol and chloroform (1:1, v:v), acetone, and ethanol. Ethanol-washed cell-walls were treated with 200 mg $L^{-1}$ Pronase (Kaken Kagaku Co., Tokyo Japan) in 50 m$M$ phosphate buffer pH 7.0 for 18 h at 30°C to remove proteins. Cell walls were then suspended in 50 m$M$ Hepes buffer pH 7.0, heated at 70°C for 5 min and treated with 10 units $mL^{-1}$ pancreatic $\alpha$ -amylase at 37°C for 3 h. After washing the cell walls with water, cell-wall polysaccharides were hydrolyzed by 2 $N$ trifluoroacetic acid (TFA) at 120°C for 1 h. Neutral sugars liberated by TFA-hydrolysis were acetylated and quantitatively determined by gas-chromatography using inositol as an internal standard.

*Gel-filtration analysis of cell-wall polysaccharides*

Seedling roots were treated with the growth solution containing 10 $\mu M$ Anc with or without $GA_3$ (0.1 $\mu M$) for 4 days. The elongation zone, 2–6 mm behind the root tip, was excised from seedling roots and kept frozen at -20°C. One to two hundred segments were homogenized with a mixture of ice-cold 50 m$M$ Tris-HCl buffer pH 7.4 and Tris-buffer-saturated phenol by Biomixer (Model BM-1, Nihon Seiki Kaisha Ltd., Tokyo, Japan). The advantage of using Tris-buffer-phenol to inactivate pectin-hydrolases was reported by Huber (1991). The homogenate was centrifuged at 2,000×*g* for 20 min at 4°C and cell-wall fraction was separated. The cell walls were then purified by solvent extractions and Pronase treatment as described above. Purified cell walls were then treated with 50 m*M* CDTA (trans-1,2-cyclohexanediamine-*N,N,N',N'*-tetraacetic acid) in 50 m*M* acetate buffer pH 6.5 to solubilize pectins at 23°C (Huber, 1991). The extraction was repeated 5 times, 4, 4, 14, 4 and 4 h each. The CDTA extracts were combined, concentrated by vacuum evaporation at 40°C and then applied to a column bed (41 cm high, 1.6 cm wide) of sepharose CL-2B-300 equilibrated in acetate-EDTA buffer (30 m*M* Na-acetate buffer pH 4.5 containing 30 m*M* NaCl and 3 m*M* $Na_2$-EDTA). Pectins eluted with acetate-EDTA buffer were determined in fractions of 1.53 mL according to Blumenkrantz and Asboe-Hansen (1973). Cell walls were further treated 3 times with hot CDTA solution (Hot CDTA fraction) at 98°C for 1 h each and then 3 times with hot 20 m*M* ammonium oxalate solution pH 4.0 (Hot Amm Ox fraction) at 98°C for 1 h each.

Hemicellulose polysaccharides were extracted 5 times with 24% KOH solution under nitrogen gas for 4, 4, 14, 4, 4 h each. Hemicellulose fraction (HC fraction) were neutralized with acetic acid, dialyzed against distilled water at 4°C, concentrated at 40°C and applied to a column bed (41 cm high, 1.6 cm wide) of sepharose CL-4B equilibrated in 0.1 *M* KOH. The HC fraction was eluted with 0.1 *M* KOH and total sugars in 1.53 mL fractions were determined according to Dubois et al. (1956).

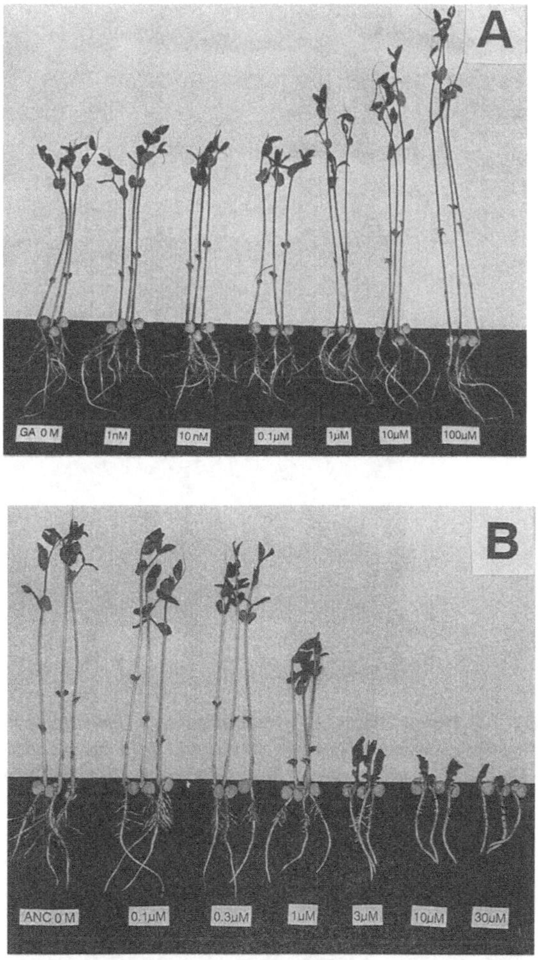


*Fig. 1.* Typical effects of $GA_3$ and Anc on the growth of pea seedlings. Seedlings were treated with $GA_3$ (**A**) or with Anc (**B**) for 6 days. $GA_3$ and Anc were applied to hydroculture solution. A: $GA_3$ concentrations are, from left to right, 0, 1 n*M*, 10 n*M*, 0.1 $\mu M$, 1 $\mu M$, 10 $\mu M$ and 100 $\mu M$. B: Anc concentrations are 0, 0.1 $\mu M$, 0.3 $\mu M$, 1 $\mu M$, 3 $\mu M$, 10 $\mu M$ and 30 $\mu M$.

## Results

### *Growth regulation by $GA_3$ and ancymidol*

Typical effects of $GA_3$ and Anc on the growth of pea seedlings are shown in Figures 1–4. Externally applied $GA_3$ showed little effect on root elongation whereas it promoted stem elongation (Fig. 1A). The inhibitory effect of Anc is shown in Figure 1B. Stem elongation was apparently suppressed at 1 $\mu M$, while the inhibition of main root elongation appeared at 10 $\mu M$. Effects of $GA_3$ on the recovery from Anc-inhibition are shown in Figure 2. Stem and root elongations were significantly reduced by 10 $\mu M$ Anc (Fig. 2A left two). Stem elongation was reversed by increasing $GA_3$ concentrations and it was completely restored at 10 $\mu M$ $GA_3$ (Fig. 2A). Root elongation, however, showed a different feature of recovery. It was partially recovered at 1 n*M* and completely recovered at 0.1 $\mu M$ $GA_3$ (Fig. 2B). The effects of $GA_3$ and Anc on the growth of pea seedlings observed in these photographs were compatible with the previous report of quantitative measurements of the seedling growth (Tanimoto, 1988). Since the previous report (Tanimoto, 1988, 1990) also elucidated that the elongation and the morphology of pea roots treated with $GA_3$ alone or with $GA_3$ plus Anc were similar to those of control roots (no growth regulator

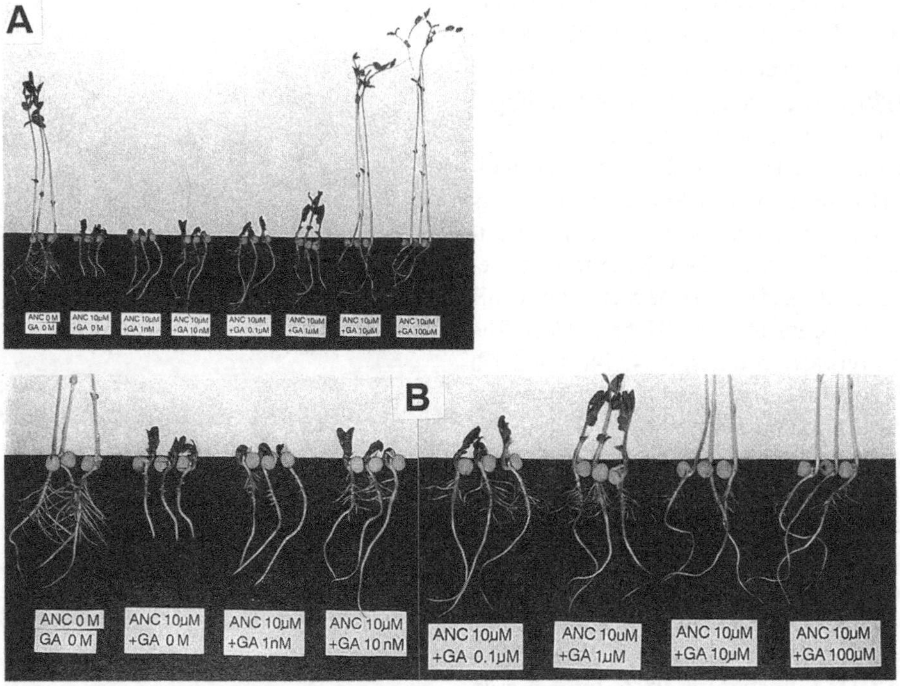


*Fig. 2.* Typical effect of $GA_3$ concentrations on the elongation recovery of shoots (**A**) and roots (**B**) from Anc suppression. Pea seedlings were treated with different concentrations of $GA_3$ in the presence of 10 $\mu M$ Anc. The left-most seedlings are control with no growth regulators added. $GA_3$ concentrations are, from left to right, 0, 0, 1 $nM$, 10 $nM$, 0.1 $\mu M$, 1 $\mu M$, 10 $\mu M$ and 100 $\mu M$.

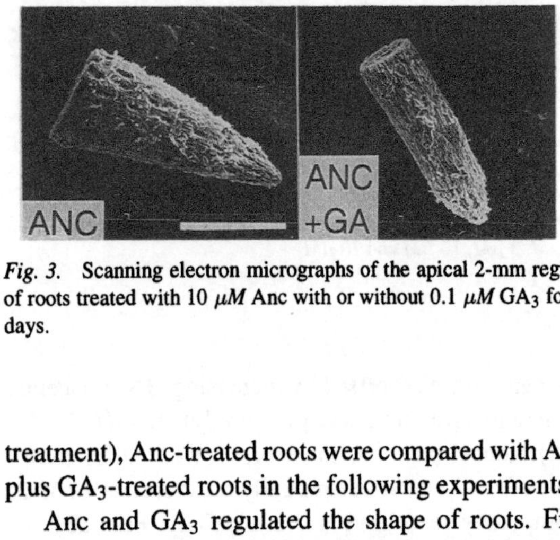


*Fig. 3.* Scanning electron micrographs of the apical 2-mm region of roots treated with 10 $\mu M$ Anc with or without 0.1 $\mu M$ $GA_3$ for 4 days.

treatment), Anc-treated roots were compared with Anc plus $GA_3$-treated roots in the following experiments.

Anc and $GA_3$ regulated the shape of roots. Figures 3 and 4 show scanning electron micrographs of roots treated with 10 $\mu M$ Anc with or without 0.1 $\mu M$ $GA_3$ for 4 days. Anc made roots thick whereas $GA_3$ prevented it. Anc induced the thickening of an apical 2-mm segments (Fig. 3) through to 10-mm behind the tip (Fig. 4).

Anc-thickened roots had expanded cortex cells, while $GA_3$-treated slender roots contained slender cortex cells (Fig. 4). $GA_3$ completely cancelled the Anc-induced morphological change in the cortex cells.

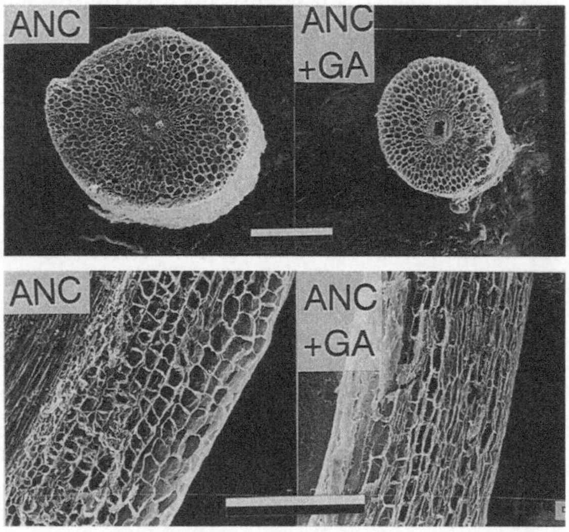


*Fig. 4.* Scanning electron micrographs of cross and longitudinal section of root segments excised 8–10 mm behind the root tip. Treatments as in Figure 3.

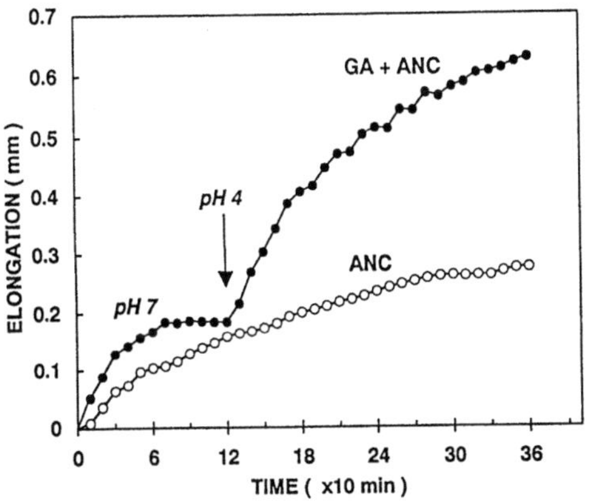


*Fig. 5.* Effect of Anc- and $GA_3$-pretreatment on the acid-induced elongation of root segments. Seedling roots were pretreated with Anc with or without $GA_3$ and the elongation of apical root segments was recorded every 10 min in a Rhizometer as described in Materials and methods. The root segments were grown in a growth solution of pH 7 for 120 min and then pH value was changed to pH 4 as indicated by the arrow. Means of 12 roots were indicated.

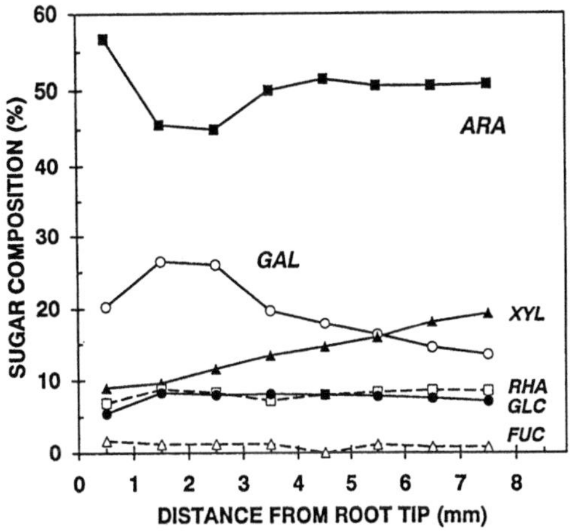


*Fig. 6.* Sugar composition of the cell walls as determined in successive 1-mm segments from the tip to 8-mm behind the tip. Roots were treated for 6 days with 10 $\mu M$ Anc. Relative amounts (%) of six major sugars to total amounts of six neutral sugars are indicated. ARA=arabinose; GAL=galactose; XYL=xylose; RHA=rhamnose; GLC=glucose; FUC=fucose.

The effect of Anc and $GA_3$ on acid-growth response of the elongation zone was investigated using a Rhizometer. The time course of acid-induced rapid elongation is shown in Figure 5. The elongation zone of roots lost its responsiveness to acid solution when the seedling roots were pretreated with Anc for 24 h, while Anc plus $GA_3$-pretreated roots showed a rapid response to acid solution (Fig. 5). The effect of $GA_3$ on osmotic concentration of root segments was investigated in Anc-treated roots (Table 1). $GA_3$ showed little effect on osmotic concentration of any part of the root during a 2-day treatment.

Sugar composition of the cell walls was measured along the root axis to obtain the characteristics of growing cell walls. Figure 6 indicates neutral sugar composition (% of total sugars) of every 1-mm segments prepared from Anc-treated roots for 6 days. Sugar composition changed from the tip to the base of the root. Arabinose decreased and galactose increased at the elongation zone. Xylose increased towards the base. Figure 7 shows the effects of $GA_3$ and Anc on sugar composition after 6-days of treatment. $GA_3$ plus Anc-treated roots contained a lower arabinose and higher galactose content than Anc-treated roots. Arabinose and galactose occupied about 70% of the neutral sugars in the cell-wall matrix. Major components of cell-wall neutral sugars were affected by $GA_3$ -treatment.

The effect of $GA_3$ and Anc on the molecular-mass distribution of pectins is shown in Figure 8. Molecular mass distribution of pectins on a Sepharose-2B column indicated that $GA_3$ plus Anc-treated roots had a higher molecular-mass population than Anc-treated roots. Although the difference of the two profiles seems to be small, it was reconfirmed by duplicate experiments. It should be noted that the abscissa of the graph has a logarithmic scale. Considerable amounts of pectins were found in Hot CDTA and Hot Amm Ox fractions. Molecular-mass of these pectins was, however, considerably smaller than those of the CDTA fraction obtained at 23°C. The effects of $GA_3$ and Anc on these pectins solubilized at high temperature are under

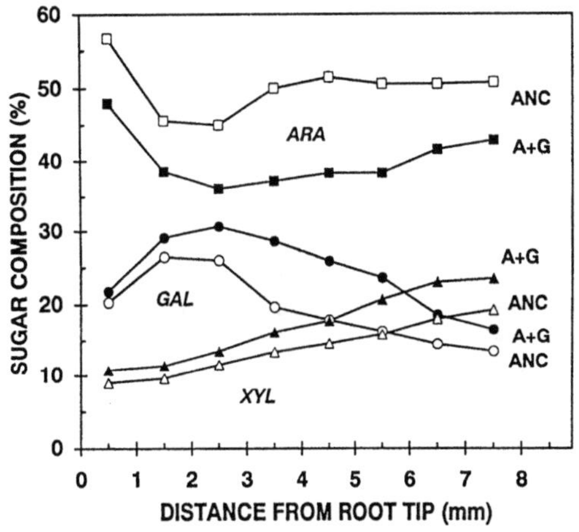


*Fig. 7.* Effect of $GA_3$ on the relative amount of arabinose (ARA, squares), galactose (GAL, circles) and xylose (XYL, triangles). Seedling roots were treated for 6 days with 10 $\mu M$ Anc with or without 0.1 $\mu M$ $GA_3$. Anc (open symbols)=ancymidol-treated roots, A+G (closed symbols)=$GA_3$ plus Anc-treated roots. See legend to Figure 6 for further description.

*Table 1.* Effects of $GA_3$ on osmotic concentration of root segments

| Treatments | | Osmotic concentration(mmole $L^{-1}$) Root segments (Distance from tip, mm) | | | | | | | |
|---|---|---|---|---|---|---|---|---|---|
| | | 1–2 | 2–3 | 3–4 | 4–5 | 5–6 | 6–7 | 7–8 | Bud |
| 1 Day | ANC | 372 | 390 | 360 | 340 | 338 | 388 | 409 | – |
| | ANC+GA | 381 | 372 | 362 | 349 | 378 | 386 | 413 | – |
| | (GA effect %) | (102) | (95) | (101) | (103) | (112) | (99) | (101) | |
| 2 Days | ANC | 311 | 324 | 310 | 302 | 309 | 322 | 329 | 436 |
| | ANC+GA | 324 | 324 | 325 | 313 | 308 | 320 | 337 | 425 |
| | (GA effect %) | (104) | (100) | (105) | (104) | (100) | (99) | (102) | (97) |

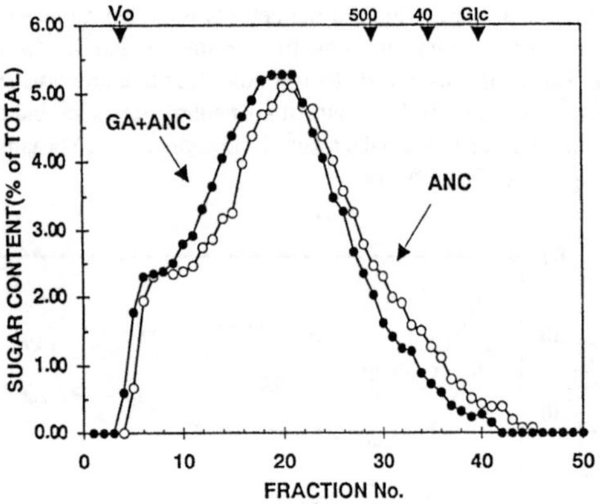


*Fig. 8.* Sepharose CL-2B column chromatography of pectins obtained from the elongation zone of roots treated with 10 $\mu M$ Anc with or without 0.1 $\mu M$ $GA_3$ for 4 days. Relative content (% of total eluate) of uronic acids in each fraction is indicated. Vo=void volume, 500, 40 and Glc indicate the elution points of dextran with mol wt. 500k and 40k, and glucose, respectively.

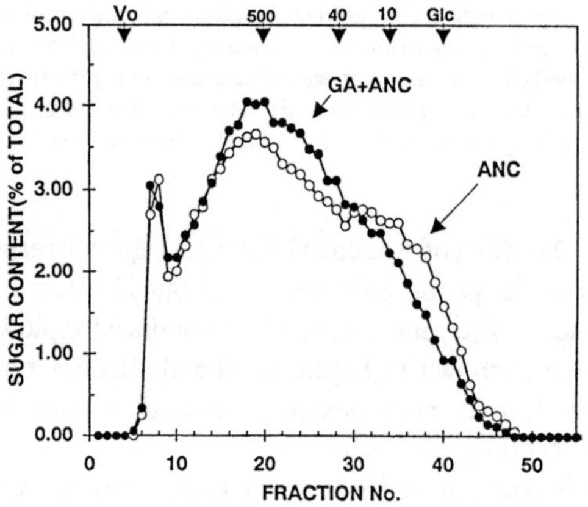


*Fig. 9.* Sepharose CL-4B column chromatography of hemicellulose. See legend to Figure 8 for details.

investigation. Since the pectins eluted at high molecular mass fractions were found to be disintegrated by high temperature treatment at 98°C, molecular mass of the polymer extracted at high temperature should be carefully evaluated.

The effects of $GA_3$ on molecular-mass distributions of hemicelluloses are shown in Figure 9. $GA_3$ plus Anc-treated roots contained higher molecular mass polymers than Anc-treated roots which showed a shoulder peak at lower molecular mass.

## Discussion

A growth-promoting effect of $GA_3$ was confirmed in the root system, but only when root growth was suppressed by Anc. The present dose-response relationships observed by photographs of $GA_3$ and/or Anc-treated seedlings were comparable with those previously reported. These results together suggested that root elongation is regulated by a lower $GA_3$-concentration than stem elongation. Although quantitative determination of GA is necessary, the concept is compatible with the results which have been reported previously (Tanimoto, 1987, 1988, 1990).

Growth promotion of the shoot by GA is interpreted by two mechanisms, i.e. GA increases osmotic concentration of stem cells to enhance water uptake and/or GA increases cell-wall extensibility. The regulation of osmotic concentration may not be involved in $GA_3$-enhanced elongation growth of the root, since a $GA_3$-induced increase in the osmotic concentration was not found in the present experiment (Table 1). In the present experimental conditions, water uptake may not limit the root elongation since roots were dipped in culture solution throughout the experiments.

Acid-induced elongation was significantly suppressed by Anc whereas $GA_3$ prevented the Anc suppression (Fig. 5). Since acid-induced elongation is also known to occur in isolated cell walls, such a rapid elongation response seems to be brought about by an increase in cell-wall extensibility. Although physical measurement of cell-wall extensibility of this species of pea roots is under investigation, $GA_3$ has been reported to increase the cell-wall extensibility of a dwarf pea root by a method using a tensile tester (Tanimoto, 1992). Taking these results together, the interaction between Anc and $GA_3$ in the regulation of root growth seems to be closely related to the cell-wall extensibility of the elongation zone of the roots.

The relationship between cell-wall extension and changes in cell-wall components and/or molecular-mass distribution of cell-wall polysaccharides has been extensively studied on auxin-induced growth of stems (Fry, 1986; Goldberg et al., 1986; Hayashi, 1989; Hoson and Masuda, 1991; Masuda, 1990; Nishitani and Masuda, 1981; Sakurai, 1991; Yamaoka and Chiba, 1983). We have, however, less information on $GA_3$-induced elongation growth (Katsumi et al., 1992; Tanimoto, 1992). Present data indicated that cell walls of Anc-treated thick roots had a different sugar composition to those from $GA_3$ plus Anc-treated slender roots (Fig. 7). The molecular mass of cell-wall polymers was also modified by $GA_3$. $GA_3$ plus Anc-treated slender roots contained larger polymers than Anc-treated thick roots (Figs. 8, 9). These results suggest that $GA_3$-treated roots make new cell-wall polymers with different sugar composition from Anc-treated thick roots in which partial disintegration of hemicelluloses are taking place. Another possiblity is that the change in cell-wall components by $GA_3$-treatment may reflect the increase in the relative amount of transverse cell wall by the thickening, since thick roots contain a larger proportion of such walls.

## Acknowledgements

The author thanks Drs R Yamamoto, T Fujii, S Nakayama, Y Masuda of Tezukayama College, Dr K Miyamoto of Univ Osaka Prefecture, Drs S Kamisaka and T Hoson of Osaka City Univ for their kind help to use an osmometer, Dr K Nishitani of Kagoshima Univ and Dr D J Huber of Univ Florida for their technical advice on analyzing cell-wall polysaccharides.

## References

Blumenkrantz N and Asboe-Hansen G 1973 New method for quantitative determination of uronic acids. Anal. Biochem. 54, 484–489.

Brian P W and Hemming H G 1957 Effect of gibberellic acid and indolylacetic acid on plant cell extension. Nature 179, 417.

Britz S J and Saftner R A 1987 Inhibition of growth by ancymidol and tetcyclacis in the gibberellin-deficient dwarf-5 mutant of *Zea mays* L. and its prevention by exogenous gibberellin. J. Plant Growth Regul. 6, 215–219.

Coolbaugh R C, Swanson D I and West C A 1982 Comparative effects of ancymidol and its analogs on growth of peas and ent-kaurene oxidation in cell-free extracts of immature *Marahmacrocarpus* endosperm. Physiol. 69, 707–711.

Dubois M, Giles K A, Hamilton J K, Reburs P A and Smith F 1956 Colorimetric method for determination of sugars and related substances. Anal. Chem. 28, 350–356.

Edwards K I and Scott T K 1976 Rapid growth responses of corn root segments: Effect of citrate-phosphate buffer on elongation. Planta 129, 229–233.

Fry S C 1986 Cross-linking of matrix polymers in the growing cell walls of angiosperms. Annu. Rev. Plant Physiol. 37, 165–186.

Goldberg R, Morvan C and Roland J C 1986 Composition, properties and localization of pectins in young and mature cells of the mung bean hypocotyl. Plant Cell Physiol. 27, 417–429.

Hayashi T 1989 Xyloglucans in the primary cell wall. Annu. Rev. Plant Physiol. 40, 139–168.

Hoson T and Masuda Y 1991 Inhibition of auxin-induced elongation and xyloglucan breakdown in azuki bean epicotyl segments by fucose-binding lectins. Physiol. Plant. 82, 41–47.

Huber D J 1991 Acidified phenol alters tomato cell wall pectin solubility and calcium content. Phytochemistry 30, 2523–2527.

Katsumi M, Kazama H and Kawamura N 1980 Osmotic potential of the epidermal cells of cucumber hypocotyls as affected by gibberellin and cotyledons. Plant Cell Physiol. 21, 933–937.

Katsumi M, Hara (Kawahara) M and Sato T 1992 Growth and cell walls in a gibberellin mutant of *Zea mays* L. *In* Plant Cell Walls as Biopolymers with Physiological Functions. Ed. Y Masuda. pp 99–104. Yamada Science Foundation.

Masuda Y 1990 Auxin-induced cell elongation and cell wall changes. Bot. Mag. Tokyo 103, 345–370.

Masuda Y and Pilet P-E 1983 Mechanical properties and polysaccharide nature of the cell walls of maize root. Physiol. Plant. 59, 512–517.

Miyamoto K and Kamisaka S 1988 Stimulation of *Pisum sativum* epicotyl elongation by gibberellin and auxin. - Different effects of two hormones on osmoregulation and cell walls. Physiol. Plant. 74, 457–466.

Nishitani K and Masuda Y 1981 Auxin-induced changes in the cell wall structure: Changes in the sugar composition, intrinsic viscosity and molecular weight distributions of matrix polysaccharides of the epicotyl cell wall of *Vigna angularis*. Physiol. Plant. 52, 482–494.

Sakurai N 1991 Cell wall functions in growth and development, A physical chemical point of view. Bot. Mag. Tokyo 104, 235–251.

Ross J J, Reid J B, Gaskin P and MacMillan J 1989 Internode length in *Pisum*. Estimation of $GA_1$ levels in genotypes *Le*, *le* and $le^d$. Physiol. Plant. 76, 173–176.

Tanimoto E 1987 Gibberellin-dependent root elongation in *Lactuca sativa*: Recovery from growth retardant-suppressed elongation with thickening by low concentration of $GA_3$. Plant Cell Physiol. 28, 963–973.

Tanimoto E 1988 Gibberellin regulation of root growth with change in galactose content of cell walls in *Pisum sativum*. Plant Cell Physiol. 29, 269–280.

Tanimoto E 1990 Gibberellin requirement for the normal growth of roots. *In* Gibberellins. Eds. N Takahashi et al. pp 229–240. Springer-Verlag, New York.

Tanimoto E 1992 Modification of cell-wall extensibility and sugar composition by gibberellin $A_3$ in pea (*Pisum sativum*) roots. *In* Plant Cell Walls as Biopolymers with Physiological Functions. Ed. Y Masuda. pp 105–110. Yamada Science Foundation.

Tanimoto E and Watanabe J 1986 Automated recording of lettuce root elongation as affected by indole-3-acetic acid and acid pH by a new rhizometer with minimum mechanical contact to root. Plant Cell Physiol. 27, 1475–1487.

Tanimoto E, Yanagishima N and Masuda Y 1967 Effect of gibberellic acid on dwarf and normal pea plants. Physiol. Plant. 20, 291–298.

Tanimoto E, Scott T K and Masuda Y 1989 Inhibition of acid-enhanced elongation of *Zea mays* root segments by galactose. Plant Physiol. 90, 440–444.

Yamaoka T and Chiba N 1983 Changes in the coagulating ability of pectin during the growth of soybean hypocotyls. Plant Cell Physiol. 24, 1281–1290.

*F. Baluška et al. (eds.), Structure and Function of Roots*, 99–105, 1995.


# Structural and physiological characteristics of the tap root of *Smyrnium perfoliatum* L. (Apiaceae)

Alexander Lux, Elena Masarovicova and Roman Olah
*Department of Plant Physiology, Comenius University, Bratislava, Slovak Republic*

*Key words:* Apiaceae, growth analysis, root anatomy, root respiration, shoot:root ratio, *Smyrnium*

**Abstract**

The thick fleshy tap roots of the biennial forest herb *Smyrnium perfoliatum* L. (Apiaceae) are characterized by excessive development of vascular parenchyma. Numerous starch grains are deposited mainly in phloem parenchyma cells and in rays. Schizogenous secretory cavities, surrounded by epithelial cells were observed in the phelloderm and in secondary phloem. High values of shoot:root ratio (6.78), specific leaf area (56.83 $m^2$ $kg^{-1}$), leaf weight ratio (0.221 g $g^{-1}$) and quite high root respiration rate (0.444 mg $CO_2$ $kg^{-1}$ $s^{-1}$) during growing season confirm that *Smyrnium perfoliatum* is a fast growing species. Decrease of the content of sucrose, proteins, and starch and an increase of reducing sugars in the root during ontogenesis were observed.

*Abbreviations:* $R_D$ — root respiration rate, SLA — specific leaf area, LAR — leaf area ratio, LWR — leaf weight ratio, RWR — root weight ratio.

## Introduction

*Smyrnium perfoliatum* L. is a biennial submediterranean species from the family Apiaceae (Umbelliferae). Slovakia represents the northern boundary of *Smyrnium perfoliatum*'s distribution in Europe. It is considered to be an endangered species in Slovakia, with only one locality where it is an autochthonous species (Devinska Kobyla in the Little Carpathians). *Smyrnium perfoliatum*, an aromatic species, is used in natural medicine and in homeopathy, mainly in southern Europe. In the past it was often cultivated in the gardens of central Europe (Kaspar, 1903). It forms a storage tap root and a rosette of compound leaves in the first year. In the second year a shoot and inflorescence appear. Although the anatomy of the storage tap roots in various species of the Apiaceae has been well studied, information about the genus *Smyrnium* is scarce. In other genera the roots are often thick and fleshy (e.g. Esau, 1940; Hayward, 1938). As shown by Deutschmann (1969), who analyzed root anatomy of 21 species from the family Apiaceae, a characteristic feature of roots in this family is the presence of internal secretory structures, canals and cavities with variable structure and occurrence in different primary and secondary tissues.

According to our results, *Smyrnium perfoliatum* is likely to be a fast growing plant species. Fast growing plants rapidly increase in size and may occupy a large space, both below and the above ground due to their high growth activity. Consequently, these plants have the opportunity to acquire a large share in limiting resources (e.g. light, nutrients or water) in comparison with slower growing individuals. The completion of their life cycle may be facilitated by a fast vegetative growth phase (c.f. Chapin, 1980; Grime, 1979; Grime and Hunt, 1975; Lambers et al., 1990). Fast growing species tend to invest relatively less biomass in roots when grown at an optimum supply of nutrients than do slow growing species. However, at the same time, they have a greater capacity to adjust their investment in root biomass and associated structures in comparison with the slow growing species (Lambers and Poorter, 1992).

The aim of our study was to characterize the structure, respiration rate, growth activity and chemical composition of *Smyrnium perfoliatum* roots. We analyzed the growth parameters to determine the life strategy (slow or fast growing plant) of this species.

## Materials and methods

One year old plants of *Smyrnium perfoliatum* L. were collected from Devinska Kobyla, Little Carpathians region in SW Slovakia, near Bratislava, 450 m above sea level, 16° 59′ 50″ E, 48° 11′ 20″ N. Samples for a structural study were taken in the middle of the growing season, whereas material for the physiological and chemical analyses was collected over the whole growing season.

During the second season, *Smyrnium perfoliatum*, as a biennial plant, started to grow in the middle of March. Compound leaves (2–3) of the rosette grew first, then the main stem was formed with 2 basal compound leaves, 2 upper simple amplexicaul leaves and inflorescence with bracts (Fig. 1a). Optimum of flowering was in the middle of May and fruits were ripe in June–July. The growing season is approximately three month's long and is hot and dry. The maximal air temperature in May 1993 reached 24.7°C and in July 25.6°C. Mean monthly precipitation during the whole growing season was 41 mm (Institute of Hydrometeorology, Slovak Academy of Sciences, Bratislava).

### *Structural study*

Small sections from various parts of tap roots were cut with a razor blade. The samples were fixed in 5% glutaraldehyde in 0.1 mol. $L^{-1}$ phosphate buffer for 5 hours, washed in the same buffer and postfixed in 2% osmium tetroxide. After dehydration in acetone the samples were embedded in Durcupan ACM (Fluka). Approximately 1 $\mu$m thick sections were cut with a glass knife in a LKB Nova ultramicrotome and stained with toluidine blue and basic fuchsin (Lux, 1981).

### *Root respiration rate, growth analysis and chemical composition analysis*

Plants of *Smyrnium perfoliatum* L. were sampled from the above-mentioned plant population during growing season. The individuals, together with their soil monolith (not to disturb water supply), were transferred to laboratory conditions in the early morning. After 1–2 days of acclimatization, root respiration, growth analysis and chemical composition analysis were performed. Before root respiration measurements and chemical analysis, the roots were washed and detached. Root respiration rate was measured under controlled conditions by the gasometric method in a closed system (Infralyt 4, VEB Junkalor Dessau, Germany) at air temperature of 15 °C (earlier phases of ontogenesis — March and April) or 22 °C (later phases of ontogenesis — May and June). During the root respiration measurements the detached roots were exposed in a thermostabilized assimilation chamber (Masarovicova and Elias, 1986). Quantitative parameters of growth analysis were estimated according to Sestak et al. (1971). Chemical analysis of the root was done in the Research Institute of Agroecology, Michalovce, Slovak Republic. Content of the sucrose, reducing sugars, proteins, starch and ascorbic acid was established according to Pribela (1978) and Davidek (1981).

## Results

### *General morphology of the root*

The root of plant in the second year of its growth consisted of a tap root and laterals. The tap root is very well developed as a thick fleshy organ, approximately 40 mm long and 20–30 mm thick at its base, conical, whitish or yellowish brown (Figs. 1b–d). At the beginning of the second growing season the laterals were poorly developed (Fig. 1b), however, during later growth the laterals became numerous and well developed (Fig. 1c). During fructification the tap root was wrinkled, and the laterals were prominent (Fig. 1d).

### *Anatomy of the tap root*

The root was covered by a periderm (Fig. 2a). The phellem was several cell layers thick and the phelloderm was organized in radial rows of cells, with schizogenous secretory cavities of irregular shape, surrounded by an inconspicuous epithelium (Fig. 2a). The secondary vascular tissues was produced by a single cambial layer (Fig. 2b). The tap root of *Smyrnium* was fleshy due to the excessive development of parenchyma in the secondary vascular tissues, especially in the secondary phloem (Fig. 2b). The secondary phloem formed a broad zone, consisting of groups of sieve tubes, companion cells and well developed parenchyma (Fig. 2c). This tissue was the main place of storage deposition, with numerous starch grains visible in the parenchyma cells. Schizogenous secretory cavities of irregular shape were also present in the secondary phloem, being surrounded by epithelium (Fig. 2a).

Secondary xylem was well developed and contained small vessels with diameters up to 50 $\mu$m (Fig. 2d). However, the proportion of parenchyma was lower

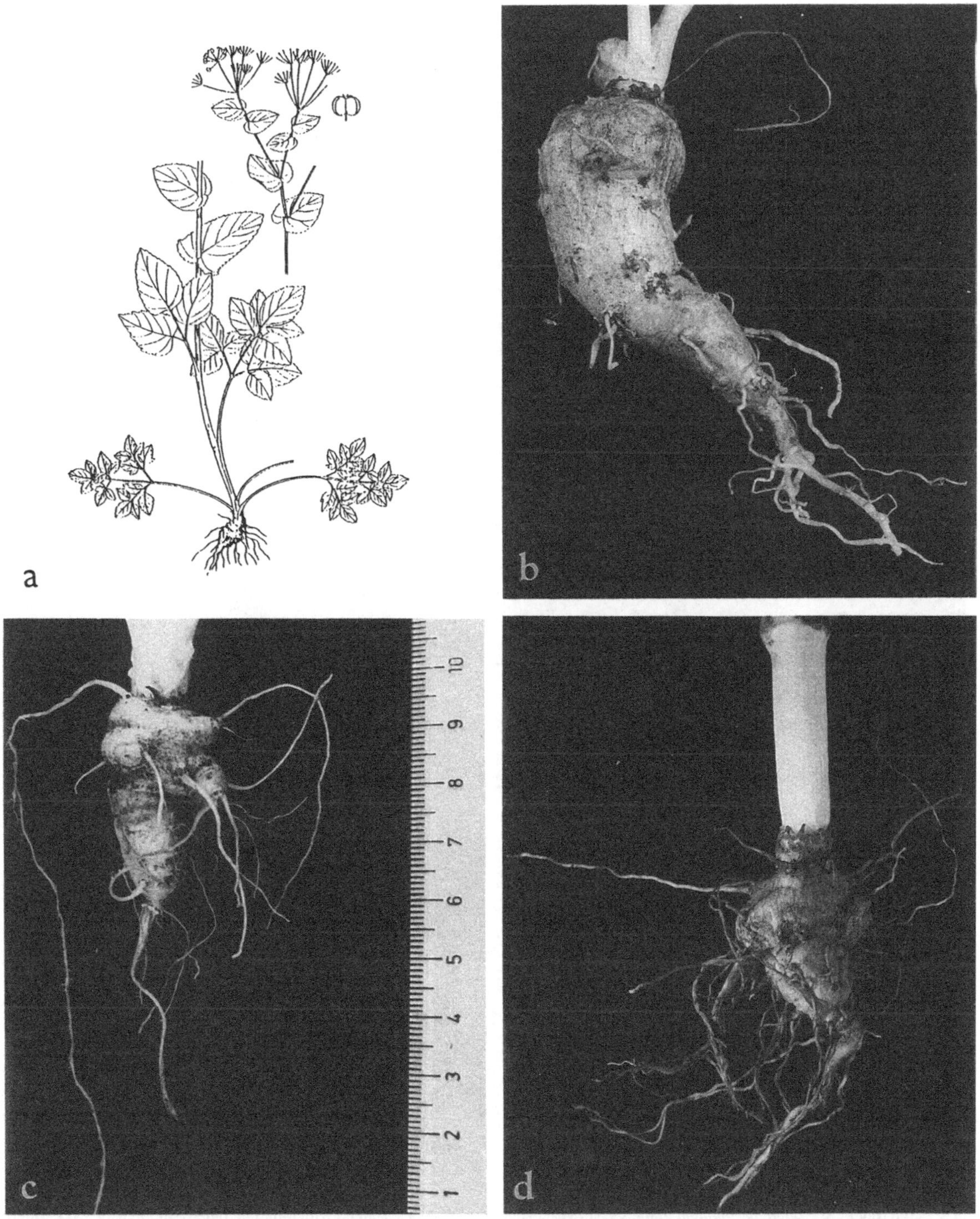


*Fig. 1.* *Smyrnium perfoliatum* in the second year of growth. (a) A plant in flower with a thick fleshy tap root, dimorphic leaves and inflorescence. (b) Storage tap root with poorly developed laterals at the beginning of the growing season. Tap root with numerous well developed laterals in the stage of main stem growth (c) and at the end of the growing season in the stage of fructification (d).

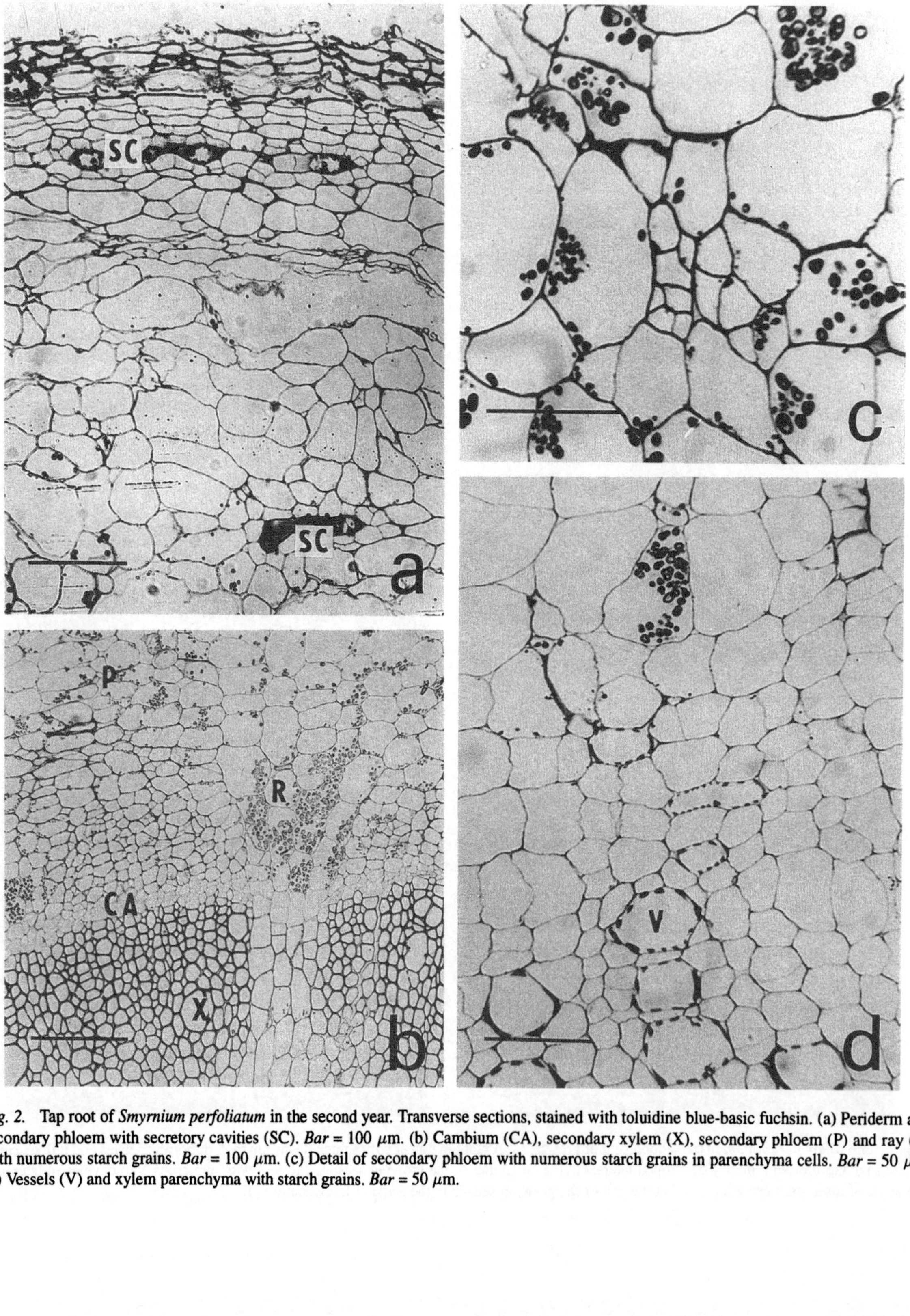


*Fig. 2.* Tap root of *Smyrnium perfoliatum* in the second year. Transverse sections, stained with toluidine blue-basic fuchsin. (a) Periderm and secondary phloem with secretory cavities (SC). *Bar* = 100 $\mu$m. (b) Cambium (CA), secondary xylem (X), secondary phloem (P) and ray (R) with numerous starch grains. *Bar* = 100 $\mu$m. (c) Detail of secondary phloem with numerous starch grains in parenchyma cells. *Bar* = 50 $\mu$m. (d) Vessels (V) and xylem parenchyma with starch grains. *Bar* = 50 $\mu$m.

*Table 1.* Respiration rate of individual roots (mean ± SE) in the phenophase of compound leaf of the rosette, basal compound leaf, upper simple amplexicaul leaf and bracts

| Phenophase | | Root respiration rate ($mg\ CO_2\ kg^{-1}$ (d.m.) $s^{-1}$) |
|---|---|---|
| Compound leaf of the rosette | | 0.4630 |
| | | 0.5559 |
| | | 0.3834 |
| | | 0.3913 |
| | | 0.4239 |
| | $\bar{x}$ | 0.4435 |
| | $s_{\bar{x}}$ | 0.0313 |
| Basal compound leaf | | 0.2099 |
| | | 0.1905 |
| | | 0.1760 |
| | | 0.3790 |
| | | 0.4558 |
| | | 0.4541 |
| | | 0.2556 |
| | $\bar{x}$ | 0.3029[a] |
| | $s_{\bar{x}}$ | 0.0466 |
| Upper simple amplexicaul leaf and bracts | | 0.2803 |
| | | 0.2702 |
| | | 0.3893 |
| | | 0.3137 |
| | | 0.1626 |
| | | 0.3018 |
| | $\bar{x}$ | 0.2863[ns] |
| | $s_{\bar{x}}$ | 0.0300 |

[a] Significant difference at $p = 0.05$, ns, not significant difference.

in this tissue compared with the phloem. Some starch was present in xylem parenchyma cells. The concentration of starch grains was greatest in the rays (Fig. 2b).

*Root respiration, growth analysis and chemical composition*

The highest values of root respiration were found when the compound leaves of the rosette were present (March–April). Later, when the basal compound leaves occurred (May), the root respiration significantly decreased. In the phenophase of fruit formation and fruit ripening (May–June), when the upper simple amplexicaul leaves and bracts were present, the rate of root respiration was lower (Table 1).

*Table 2.* Mean dry mass (± SE) of shout, root and whole plant, shoot-root ratio, total leaf area and total leaf dry mass, some of growth characteristics of *Smyrnium perfoliatum*

| | Dry mass (g) | | | Shoot-root | Leaf area |
|---|---|---|---|---|---|
| | Shoot | Root | Whole plant | ratio | ($m^2$) |
| $\bar{x}$ | 3.732 | 0.5560 | 4.2880 | 6.781 | 0.0539 |
| $s_{\bar{x}}$ | 0.183 | 0.0650 | 0.1890 | 0.786 | 0.0070 |
| Specific leaf area | | | | 56.830 $m^2\ kg^{-1}$ | |
| Total leaf dry mass | | | | 0.948 g | |
| Leaf area ratio | | | | 12.570 $m^2\ kg^{-1}$ | |
| Leaf weight ratio | | | | 0.221 $g\ g^{-1}$ | |
| Root weight ratio | | | | 0.130 $g\ g^{-1}$ | |

*Smyrnium perfoliatum*, showed high values of total leaf area (0.0539 $m^2$), specific leaf area (56.83 $m^2\ kg^{-1}$), leaf area ratio (12.57 $m^2\ kg^{-1}$) and leaf weight ratio (0.221 $g\ g^{-1}$) but lower value of root weight ratio (0.129 $g\ g^{-1}$) and very high value of shoot:root ratio (6.78) (Table 2).

The content of some fundamental organic compounds in the root was determined. It was found that content of the starch, sucrose, proteins and ascorbic acid was higher in the phenophase when growth of the shoots terminated. At the end of May and in June when ontogenesis was finishing and fruits were ripening, the concentration of sucrose, starch and proteins decreased, while the content of reducing sugars in this same period increased (Table 3). We suppose that reducing sugars were allocated from the shoot to the root and they were utilized in respiration. The high value of the shoot:root ratio confirms this hypothesis. Therefore assimilate allocation was in favour of the root. Disproportion of the shoot:root ratio was compensated with equable root respiration during the main period of the growth.

## Discussion

The morphology and anatomy of the storage tap root of *Smyrnium* resembles that of some other species from the family Apiaceae, especially that of *Apium graveolens* (Hayward, 1938). A high proportion of vascular parenchyma, which in the case of *Smyrnium* is mainly formed by phloem parenchyma, is typical. This tissue is responsible for the fleshy character of the root and is a site for deposition of storage substances. Another characteristic of the *Smyrnium* root is the presence of

*Table 3.* Content of sucrose, reducing sugars, proteins, starch and ascorbic acid in individual roots at the beginning and at the end of the growing season, together with their mean (+SE) values

| Organic compounds ($mg\ g^{-1}$ (d.m.)) | | Beginning of growing season | End of growing season |
|---|---|---|---|
| Sucrose | | 76.00 | 67.40 |
| | | 76.30 | 69.30 |
| | | 76.50 | 68.50 |
| | $\bar{x}$ | 76.27 | 68.40[a] |
| | $s_{\bar{x}}$ | 0.14 | 0.55 |
| Reducing sugars | | 56.80 | 80.30 |
| | | 55.10 | 80.30 |
| | | 55.90 | 81.20 |
| | $\bar{x}$ | 55.93 | 80.60[a] |
| | $s_{\bar{x}}$ | 0.49 | 0.30 |
| Proteins | | 45.60 | 38.50 |
| | | 44.90 | 38.50 |
| | | 45.50 | 37.80 |
| | $\bar{x}$ | 45.33 | 38.26[a] |
| | $s_{\bar{x}}$ | 0.21 | 0.23 |
| Starch | | 4.85 | 3.66 |
| | | 4.84 | 3.31 |
| | | 4.80 | 3.44 |
| | $\bar{x}$ | 4.83 | 3.47[a] |
| | $s_{\bar{x}}$ | 0.01 | 0.10 |
| Ascorbic acid | | 0.0152 | 0.0121 |
| | | 0.0145 | 0.0121 |
| | | 0.0146 | 0.0112 |
| | $\bar{x}$ | 0.0149 | 0.0118[ns] |
| | $s_{\bar{x}}$ | 0.0003 | 0.0003 |

[a] Significant difference at $p = 0.01$.
ns, not significant difference.

broad rays with numerous starch grains. The usual kind of secondary growth with a single layer of cambium is common in this family (Esau, 1977).

The family Apiaceae is characterized by a well developed secretory system in all plant parts, including the root. These structures are important places of deposition of various aromatic substances (Deutschmann, 1969). Schizogenous secretory cavities in the secondary phloem and the phelloderm of *Smyrnium* differ from the oil ducts described in secondary phloem of *Apium* (Hayward, 1938) as well as from secretory cavities found in the periderm of the same species (Deutschmann, 1969). Our findings are consistent with previous observations of Deutschmann (1969), that the presence, distribution and character of the secretory structures is species-specific in Apiaceae. This helps the identification of materials for pharmaceutical purposes. Secretory cavities of peridermal origin have rarely been described. However, Deutschmann (1969) after careful observations concluded that secretory cavities are quite frequent in roots of the Apiaceae.

In optimal productive environments, a high leaf weight ratio and a high specific leaf area will be of paramount importance to survive in a dense vegetation. Instead of the relative growth rate potential, factors determining specific leaf area and carbon allocation of a species will determine its occurrence (Poorter, 1990). Accordingly, high values of leaf weight ratio, specific leaf area and shoot:root ratio were found for *Smyrnium perfoliatum* growing on nutrient-rich soils in the large herbaceous layer of the forest. This finding, as well as the other quantitative parameters (net assimilation rate, relative growth rate of shoot and root, unpublished) confirm that *Smyrnium perfoliatum* should be characterized as a fast growing species.

Biomass allocation and morphology appear to have a large impact on the carbon economy of the plant (Poorter, 1990). However, the shoot:root ratio can change dramatically during plant growth, which was also documented for *Smyrnium perfoliatum*. The disproportion of the shoot:root ratio is therefore compensated by increase in physiological activity of the root (c.f. Masarovicova and Navara, 1992).

Poorter et al. (1990) found for twenty four wild species, grown in a growth chamber, root respiration rate ranged from 37 nmol $O_2\ g^{-1}\ s^{-1}$ (1.62 mg $CO_2\ kg^{-1}\ s^{-1}$) to 85 nmol $O_2\ g^{-1}\ s^{-1}$ (2.20 mg $CO_2\ kg^{-1}\ s^{-1}$). These values are higher than the values found for *Smyrnium perfoliatum* growing in natural forest conditions. Analyses of another fast growing species, *Dactylis glomerata*, by van der Werf et al. (1993) revealed even lower root respiration values: 0.92 $\mu$mol $CO_2\ m^{-2}\ s^{-1}$ (0.041 mg $CO_2\ m^{-2}\ s^{-1}$) then we found for *Smyrnium perfoliatum*. Comparing other studies on *Dactylis glomerata* (Wilson, 1982; Wilson and Jones, 1982), with our results for *Smyrnium perfoliatum*, we can conclude that *Smyrnium* has a lower root biomass and higher root respiration than *Dactylis*.

Storage material deposited in the root of a biennial plant during the first year is used for the growth of above-ground part of the plant during the second year (Stehlik, 1982). Therefore the chemical composition of the root during ontogenesis of the plant is variable. The content of dry mass, starch and other organic substances decreases, but the amount of pulp and ash substance increases. At the beginning of the growing season, the rate of decrease of dry mass and sugars is very rapid. Later, the rate of decrease is more moderate and at the end of the ontogenesis an increase of dry mass and sugars occurs. This tendency was observed in many biennial plants (c.f. details for sugar beet see Stehlik, 1982). Qualitative changes in the organic substances (starch, reducing sugars, proteins) in the root of *Smyrnium perfoliatum* showed a similar tendency.

## Acknowledgements

The authors wish to express their gratitude to the Ministry of Environmental Protection for the permission to collect experimental material. The work was supported by grant from the Slovak Grant Agency.

## References

Chapin F S 1980 The mineral nutrition of wild plants. Annu. Rev. Ecol. Sys. 11, 233–260.

Davidek J 1981 Laboratory Manual of Food Analysis. SNTL, Praha. 335 p. (*In Czech*).

Deutschmann F 1969 Anatomische Studien ueber die Exkretgaenge in Umbelliferenwurzeln. Beitr. Biol. Pflanz. 45, 409–440.

Esau K 1940 Developmental anatomy of the fleshy storage organ of *Daucus carota*. Hilgardia 13, 175–226.

Esau K 1977 Anatomy of Seed Plants. John Wiley and Sons, New York, Santa Barbara, London, Sydney, Toronto. 550 p.

Grime J P 1979 Plant Strategies and Vegetation Processes. John Wiley and Sons, Chichester. 421p.

Grime J P and Hunt R 1975 Relative growth rate: Its range and adaptive significance in a local flora. J. Ecol. 63, 393–422.

Hayward H E 1938 The Structure of Economic Plants. The Macmillan Company. New York. 674 p.

Kaspar A 1903 Two introduced plants in vicinity of Prague. Vesmir 32, 84. (*In Czech*).

Lambers H, Cambridge M L, Konings H and Pons T L (Eds) 1990 Causes and Consequences of Variation in Growth Rate and Productivity of Higher Plants. SPB Academic Publishing, The Hague. 363 p.

Lambers H and Poorter H 1992 Inherent variation in growth rate between higher plants: A search for physiological causes and ecological consequences. Adv. Ecol. Res. 23, 187–261.

Lux A 1981 A rapid method for staining semi-thin sections of plant material. Biologia (Bratislava) 36, 753–756. (*In Slovak*).

Masarovicova E and Elias P 1986 Photosynthetic rate and water relations in some forest herbs in spring and summer. Photosynthetica 20, 187–195.

Masarovicova E and Navara J 1992 Ecophysiological processes in fruiting and deblossomed apple trees. Biol. Plant. 34 (Suppl.), 493–494.

Poorter H 1990 Interspecific variation in relative growth rate: On ecological causes and physiological consequences. *In* Causes and Consequences of Variation in Growth Rate and Productivity. Eds. H Lambers, M L Cambridge, H Konings and T L Pons. pp 45–68. SPB Academic Publishing, The Hague.

Poorter H, Remkes C and Lambers H 1990 Carbon and nitrogen economy of 24 wild species differing in relative growth rate. Plant. Physiol. 94, 621–627.

Pribela A 1978 Methods of Food Analysis. Alfa, Bratislava. 255 p. (*In Slovak*).

Sestak Z, Catsky J and Jarvis P G (Eds) 1971 Plant Photosynthetic Production. Manual of Methods. Dr W Junk, The Hague. 818 p.

Stehlik V 1982 The Biology of Species, Varieties and Forms of the Beet of the Genus *Beta* L. Academia, Prague. 444 p.

Van der Werf A, Visser A J, Schieving F and Lambers H 1993 Evidence for optimal partitioning of biomass and nitrogen at a range of nitrogen availabilities for a fast- and slow-growing species. Funct. Biol. 7, 63–74.

Wilson D 1982 Response to selection for dark respiration rate of mature leaves in *Lolium perenne* and its effects on growth of young plants and simulated swards. Ann. Bot. 49, 303–312.

Wilson D and Jones J G 1982 Effect of selection for dark respiration rate of mature leaves on crop yields of *Lolium perenne* cv. S23. Ann. Bot. 49, 313–320.

*F. Baluška et al. (eds.), Structure and Function of Roots*, 107–112, 1995.


# Increased root length and branching in cotton by soil application of the plant growth regulator PGR-IV

D. M. Oosterhuis and D. Zhao
*Department of Agronomy, University of Arkansas, Fayetteville, AR 72703, USA*

*Key words:* Cotton, gibberellic acid, indolebutyric acid, *Gossypium hirsutum* L., nutrient uptake, plant growth regulator, roots, root length

## Abstract

Field and growth chamber studies were used to determine the effect of in-furrow application of PGR-IV on root and shoot development, and yield of cotton. In the field study, an in-furrow application of PGR-IV @ 73 mL $ha^{-1}$ at planting increased yield by 18% compared to the untreated control, and by 11% compared to 2-foliar applications of 292 mL/$ha^{-1}$ each at pinhead square stage of flower development and at first flower appearance. Growth chamber studies revealed that the in-furrow applications of PGR-IV @ 1.131$\mu$L/plant dramatically increased root length (+47%), root dry weight (+29%), number of lateral roots per plant (+75%), and nutrient uptake one week after planting. These differences were still apparent five weeks later at pinhead square but to a lesser degree. The yield enhancement from the foliar applications was associated with increases in leaf growth, nutrient uptake, and boll number, whereas the yield enhancement from the soil application was associated with enhanced root growth and nutrient uptake. The positive effect of PGR-IV on root growth and accelerated early-season growth could have very substantial benefits in cotton production.

## Introduction

Cotton is a perennial with an indeterminate growth habit which is very responsive to management and the environment (Oosterhuis, 1990). Consequently producers and researchers have long been interested in the possibility of controlling plant growth and enhancing yield through the use of plant growth regulators. Much research has been devoted to understanding how these plant hormones work (Guinn, 1982). In the last two decades many synthetic compounds have been developed and tested on cotton often with disappointing results. Part of the reason for this has been due to the lack of basic research with the particular compound to understand the physiology of its response and how to incorporate application of the chemical into existing cotton production systems.

There is a wealth of literature on the role of endogenous plant hormones in shoot growth and development (e.g. Davies, 1987), but their role in controlling root growth is less clear. Generally, root growth is controlled by indoleacetic acid and abscisic acid, with their action being complimentary (Pillet, 1983). Natural control mechanisms involve acropetal movement of endogenous auxin from the shoot, and inhibitors from the root tip, with the interaction of stimulators and inhibitors providing the control (Charlton, 1983). Various secondary metabolites have also been implicated in root development. Rooting of stem cuttings was one of the first commercial uses of synthetic auxins (Scott, 1978). There have been reports of various other chemicals such as indolebutyric acid improving rooting of cuttings (Howard, 1983) and increasing root growth of cotton seedlings (Urwiler and Oosterhuis, 1986). Jackson (1983) listed seventeen chemicals that promoted root extension in intact plants or seedlings. However, the effects of plant growth regulators in promoting root growth, and particularly on increasing lateral root development, have generally shown inconsistent (Charlton, 1983).

PGR-IV is a new plant growth regulator from Microflo, Lakeland, Florida, consisting mainly of gibberellic acid, indolebutyric acid and a proprietary fermentation broth. Research on foliar application of PGR-IV to cotton during for the past seven years at the University of Arkansas has shown consistent increases

*Table 1.* Effects of PGR-IV soil-applied at planting on root growth of cotton grown in 2-L pots in a growth chamber

| Time after planting (days) | Taproot length | | Root dry weight | | Number of lateral roots | | Total root lenght | |
|---|---|---|---|---|---|---|---|---|
| | Control | PGR-IV | Control | PGR-IV | Control | PGR-IV | Control | PGR-IV |
| | ------ cm ----- | | ----- g ----- | | ----- No. ----- | | ----- No. ----- | |
| 1 | 0 | 0 | 0 | 0 | -- | -- | -- | -- |
| 2 | 1.89a[1] | 2.16a | 0.0040b | 0.0054a | -- | -- | -- | -- |
| 3 | 4.12a | 5.09a | 0.0099a | 0.0164b | -- | -- | -- | -- |
| 4 | 5.50a | 7.41b | 0.0123a | 0.0191b | -- | -- | -- | -- |
| 5 | 7.36a | 10.87b | 0.0220a | 0.0254a | 2.5[2]b | 3.9a | 8.5b | 12.6a |
| 6 | 12.60b | 15.51a | 0.0325a | 0.0365a | 10.4a | 11.9a | 19.3b | 26.4a |

[1] Numbers within a row under each heading followed by the same letter are not significantly different ($p = 0.05$).
Lateral root development started on day 5.

in yield averaging about 110 kg lint hectare$^{-1}$ (Hickey and Atkins, 1992; Oosterhuis and Zhao, 1993; Urwiler et al., 1987, 1988). Recent research has documented improved plant growth as reflected in increased shoot dry matter, nutrient uptake, and number of floral buds (squares) retained (Oosterhuis and Zhao, 1993).

Successful and timely cotton seedling establishment plays an important role in profitable cotton production. The early growth of cotton roots directly affects the absorption of water and nutrients, the growth and development of plants, and final yield. However, poor cotton root growth is a major limiting factor in cotton production in the Mississippi Delta. There have been unsubstantiated reports that PGR-IV increased root mass and root activity (Atkins, 1992; Clark et al., 1992). It has been suggested that in-furrow application of PGR-IV may promote early emergence and growth, and enhance earlier boll retention and crop development. However, the evidence for this is lacking. The objective of our research was to investigate the effects of in-furrow applications of PGR-IV at planting on cotton seed germination, emergence, root growth, seedling development, and yield.

## Materials and methods

### *Effects of in-furrow PGR-IV on early root and shoot growth (Growth chamber study)*

Two types of experiments were conducted in a growth chamber at the Altheimer laboratory, University of Arkansas, Fayetteville, AR. Cotton (*Gossypium hirsutum* L.) cultivar Stoneville 506 was planted in washed sand in 2-L pots (10 cm diameter and 30 cm deep) between July and December, 1992. The temperature in the growth chamber was maintained at 30/25°C (day/night), the relative humidity was about 60-80%, and the photoperiod was 12 hours, and light intensity was 600 $\mu$mol m$^{-2}$s$^{-1}$. Plants were watered twice daily with half-strength Hoagland's solution in order to maintain well-watered plants with a sufficient supply of nutrients.

Treatments consisted of: (1) a control sprayed only with water in-furrow, and (2) PGR-IV sprayed in-furrow at a rate of 1.131 $\mu$L pot$^{-1}$ (equivalent to 72 mL ha$^{-1}$) in 20 L water. Applications were made using a $CO_2$ pressurized backpack sprayer. In order to emulate the field practice of in-furrow chemical application, the pots for each treatment were lined up, a 2.5 cm furrow was opened in each and the PGR-IV was sprayed down the "row". One seed per pot was then planted in the furrow and covered with sand. PGR-IV consists of 0.001% (w/v) gibberellic acid, 0.001% indolebutyric acid, and a proprietary fermentation broth with trace amounts of Mg, Cu, Zn, Fe, B, Mo and Co.

In the first experiment, measurements of seed germination, taproot length, number of lateral roots, total root length, and root dry weight were made at daily intervals for the first 7 days after planting. In the second experiment, measurements of plant height, leaf area, taproot length, total root length, and the dry weights of roots, stems and leaves were made at 5 weekly intervals after planting. Nutrient analysis was conduct-

*Table 2.* Effects of soil application of PGR-IV at planting on root growth at weekly intervals after planting for cotton grown in 2-L pots in a growth chamber

| Time | Control | PGR-IV | $LSD_{(0.05)}$ | Increase |
|---|---|---|---|---|
| (weeks) | | | | |
| | | **Taproot length** | | |
| | ----- cm ----- | | | --- % --- |
| 1 | 20.92 | 21.95 | NS[a] | 4.9 |
| 2 | 21.94 | 22.53 | NS | 4.8 |
| 3 | 23.68 | 24.18 | NS | 2.1 |
| 4 | 25.40 | 24.60 | NS | -3.2 |
| 5 | 24.40 | 25.20 | NS | 3.3 |
| | | **Lateral roots** | | |
| | ----- No. ----- | | | --- % --- |
| 1 | 24.3 | 42.6 | 9.1 | 75.3 |
| 2 | 47.1 | 54.0 | NS | 14.7 |
| 3 | 60.2 | 63.6 | NS | 5.7 |
| 4 | 62.8 | 72.4 | 8.7 | 15.3 |
| 5 | 60.4 | 69.4 | NS | 14.9 |
| | | **Total root length** | | |
| | ----- cm ----- | | | --- % --- |
| 1 | 112.1 | 165.4 | 39.3 | 47.6 |
| 2 | 489.6 | 644.1 | 102.1 | 31.6 |
| 3 | 1415.4 | 1652.2 | 226.8 | 16.7 |
| 4 | 2181.7 | 2597.3 | 389.1 | 19.1 |
| 5 | 3753.9 | 4167.8 | NS | 11.1 |
| | | **Root dry weight** | | |
| | ----- g ----- | | | --- % --- |
| 1 | 0.034 | 0.044 | 0.008 | 29.4 |
| 2 | 0.075 | 0.091 | 0.012 | 21.1 |
| 3 | 0.191 | 0.229 | 0.036 | 19.9 |
| 4 | 0.304 | 0.340 | NS | 11.8 |
| 5 | 0.526 | 0.584 | NS | 11.0 |

[a] NS = non significant.

ed at the Agricultural Services Laboratory, University of Arkansas, Fayetteville, using the entire plant sample for each treatment. In all experiments a randomized block design was used with six replications. Each experiment was repeated twice. A sample size of 6 pots for each treatment were used in all measurements, and the data presented are based on the mean of 6 plants.

Total root length was measured using the method of Tennant (1976) by placing the roots of each sample on a glass sheet marked with a 1 × 1 cm grid. Records were then made of the number of intercepts of the roots with the vertical and horizontal grid lines. The total root length was calculated as the number of intercepts × the length conversion factor (0.7857).

*The effect of in-furrow application of PGR-IV on cotton field (Field study)*

A field trial was planted on May 11, 1992 at the Cotton Branch Station in Marianna, Arkansas on a Captina silt loam (Typic Fragiudult). Production practices were in keeping with those recommended by the Cooperative Extension Service (Bonner, 1992). A randomized complete block design with five replications was used. Plot size was four 15-m rows spaced 0.9 m apart. Treatments consisted of (1) a control sprayed only with water in-furrow, (2) PGR-IV sprayed in-furrow at planting at a rate of 73 mL/ha$^{-1}$ followed by 2 applications of 292 mL/ha$^{-1}$ at pinhead square and first flower, and (3) PGR-IV at 292 mL/ha$^{-1}$ at pinhead square and first flower. Applications were made using a $CO_2$ pressurized backpack sprayer using 38 L water/ha. Measurements were made of boll weight, boll number and yield at final harvest in mid-September. The middle two rows of each plot were harvested using an International Mechanical picker. Boll weight and boll number were recorded on 2 × 1 mL lengths of row from the center of each plot.

## Results and discussion

*Germination and emergence*

In Experiment 1, in the growth chamber, the time of radicle emergence from the seed occurred about 2 days after planting and appeared to be unaffected by the PGR-IV treatment. However, the date of 50 percent seedling emergence from the soil occurred one day earlier in the PGR-IV treated plants (data not shown). The average taproot length, root dry weight, number of lateral branches, and total root length were greater in the PGR-IV treatment compared to the unsprayed control during the first week of growth (Table 1). Increases in taproot length, root dry weight, number of lateral branches, and total root length in the PGR-IV treatment 6 days after planting were 23.1%, 12.3%, 14.4% and 36.8%, respectively. The increases in the number of lateral roots and total root length are particularly important for providing a larger root surface for water and nutrient absorption as well as a greater exploration of the soil profile. These results suggested that PGR-IV applied in-furrow at planting could dramatically enhance early root growth during the critical stage of cotton emergence.

*Table 3.* Percentage increase in root, stem, leaf and total dry weight, and root/shoot ratio of growth chamber grown cotton plants treated with PGR-IV at planting compared to untreated control plants

| Time | Dry weight increase (%) | | | | Root/Shoot[1] | |
|---|---|---|---|---|---|---|
| (weeks) | Total | Root | Stem | Leaf | Control | PGR-IV |
| 1 | 16.7 | 29.4 | 10.0 | 8.7 | 0.616b | 0.713a |
| 2 | 12.7 | 21.1 | 17.5 | 11.8 | 0.426b | 0.456b |
| 3 | 14.1 | 19.9 | 13.3 | 11.5 | 0.359b | 0.384b |
| 4 | 11.9 | 11.8 | 10.6 | 12.3 | 0.306b | 0.306b |
| 5 | 3.3 | 11.0 | -1.0 | 2.0 | 0.281b | 0.309a |

[1] Numbers within a row followed by the same letter are not significantly different ($p = 0.05$).

*Table 4.* Mineral nutrient concentration in cotton seedlings grown in a growth chamber and treated with PGR-IV at planting compared to untreated control plants

| Weeks after planting | Treatment | Mineral nutrient concentration | | | | | | | | | |
|---|---|---|---|---|---|---|---|---|---|---|---|
| | | N | P | K | Ca | Mg | Na | Mn | Fe | Zn | Cu |
| | | % | | | | | | g kg$^{-1}$ | | | |
| 1 | Control | 4.52 | 0.67 | 1.75 | 1.03 | 0.54 | 0.16 | 27 | 644 | 88.4 | 43.6 |
| | PGR-IV | 4.71 | 0.70 | 1.54 | 1.14 | 0.58 | 0.18 | 41 | 745 | 65.8 | 47.2 |
| | $LSD_{0.05}$ | NS[a] | NS | NS | NS | NS | NS | 7 | 63 | NS | NS |
| 3 | Control | 3.72 | 0.15 | 3.12 | 2.58 | 0.79 | 0.26 | 36 | 306 | 59.8 | 16.3 |
| | PGR-IV | 3.67 | 0.17 | 3.17 | 2.47 | 0.77 | 0.24 | 21 | 322 | 79.5 | 24.1 |
| | $LSD_{0.05}$ | NS | 0.01 | NS | NS | NS | NS | NS | NS | 15.1 | 5.3 |

[a] NS = non significant.

*Root growth during early vegetative development*

In Experiment 2, the in-furrow applied PGR-IV significantly increased the number of lateral roots and total root length during the first five weeks of cotton seedling growth, compared to the untreated control (Table 2). Taproot length was largely unaffected. One week after planting, PGR-IV had increased the number of lateral roots by 75.3%, total root length by 47.6%, and root dry weight by 29.4%. These differences decreased substantially after two weeks with PGR-IV treated plants exhibiting increases of 14.7%, 31.6% and 21.1% for number of lateral roots, total root length, and root dry weight, respectively. Although the positive root response from the PGR-IV in-furrow application at planting decreased as the plants aged, the number of lateral roots and total root length of PGR-IV treated plants were still greater than those of control plants 5 weeks after planting by 14.9% and 11.1%, respectively. Later measurements were not made.

Five weeks after planting corresponds to the time of first pinhead square appearance (early floral bud development) (Oosterhuis, 1990) and, as such, represents an important stage in cotton development when an adequate supply of water, nutrients and carbohydrate is crucial for adequate square retention (Stewart et al., 1993). PGR-IV may therefore provide an important means of helping to insure adequate assimilate supply to the developing fruit load.

*Dry weight of plant components*

The application of PGR-IV to the soil at planting increased the total dry weight of leaves, stems and roots of cotton seedlings during early growth stages up to 4 weeks (Table 3). The effect was greatest on the roots and smallest on the leaves. In addition, the ratio of root to shoot dry weight of PGR-IV treated plants was higher than that of control plants. These advantages could have an impact on later plant development and boll retention, and thus have possible yield advantages. This was confirmed in related field studies (Oosterhuis and Zhao, 1993).

*Plant nutrient content*

The application of PGR-IV enhanced the concentration in the plant of certain essential mineral nutrients three and five weeks after planting although the results were variable (Table 4). The increased uptake of nutrients as calculated from the nutrient concentration and plant dry matter over time (data not shown) could be expected to have advantages in seedling growth vigour (Holman et al., 1992). There was some evidence that the uptake of P, Mn, Fe, Zn and Cu were significantly increased. There have been previous unsubstantiated reports of increased uptake of P, K, Ca, and Fe following foliar application of PGR-IV (Clarke et al., 1992). The increased uptake of nutrients could be associated with the larger absorption surface of the root system of PGR-IV treated plants or possibly because of a biochemical change in the plant tissues. Further research is needed to explain the mechanism. The immediate effect, however, is an increased nutrient supply to the developing shoot and especially to the developing flower buds (squares) which are prone to abscission from nutrient or water deficiencies. Oosterhuis and Zhao (1993) have reported increased photosynthesis following PGR-IV application and suggested this was related to improved partitioning of assimilates to the developing bolls.

*Cotton yield*

The field trial revealed that fiber yield was significantly increased ($p = 0.05$)from 1049 kg $ha^{-1}$ in the untreated control to 1117 kg $ha^{-1}$ in the foliar-applied PGR-IV treatment, and to 1238 kg $h^{-1}$ in the PGR-IV in-furrow plus foliar treatment, an increase of 6.4% and 18.0% respectively. These results confirm that the positive effects of the in-furrow applied PGR-IV on early seedling establishment (reported above from the growth chamber studies) were manifested as increases in final harvested fiber yield. Boll number was significantly increased ($p = 0.05$) 11.1% from 76 per meter of row in the untreated control to 85 per meter of row in the PGR-IV treated plants. Individual boll weight was not significantly increased, although there have been reports of increases from PGR-IV (Hickey, 1992).

PGR-IV appears to have multiple effects on cotton growth some of which are manifested in increased yield. The yield enhancement from the in-furrow application of PGR-IV at planting was associated with enhanced root growth and nutrient uptake whereas the yield enhancement from the foliar applications has been associated with increased leaf photosynthesis, nutrient partitioning, and boll number (Oosterhuis and Zhao, 1993). Increased root growth could mean increased water and mineral nutrient availability. The immediate effect, however, is an increased supply of water and nutrients to the developing shoot and especially to the developing squares and bolls. Increased boll number following PGR-IV application has been reported to be one of the main positive effects of PGR-IV on fruiting cotton responsible for the increased harvestable yield (Atkins, 1992; Oosterhuis and Zhao, 1993). Guo et al., (1992) reported that a possible mechanism of PGR-IV in soybean was from an increase in polyols associated with increased translocation of assimilate and partitioning to the roots early during growth and later to the developing fruits.

Early seedling growth in the Mississippi river delta occurs during unfavorable conditions, i.e. cool temperature and excessively wet soils. It has been suggested that modifications of hormonal content during stress conditions may enable plants to continue growth despite unfavorable conditions (Lachno, 1983). PGR-IV may function by influencing plant hormone content in favour of enhanced growth under less than ideal conditions.

The beneficial effects of PGR-IV of enhanced root growth and accelerated early season development were reflected in the significant increase in lint yield recorded in the field trial. Additional research is needed to confirm these results and to further explain the underlying mechanism. The positive effect of PGR-IV on root growth and accelerated early-season growth could have very substantial and immediate benefits in cotton production.

**Acknowledgements**

The authors gratefully acknowledge the help of the Director and Staff of the Cotton Branch Station in Marianna, Arkansas, and the assistance of Lynn Janes, Mike Withrow and Changxin Guo.

## References

Atkins R R 1992 Performance of PGR-IV in cotton. Proc. Beltwide Cotton Production Res. Conf., National Cotton Council, Memphis, Tennessee. Vol 1, p 50.

Bonner C M 1992 Cotton production recommendations. Coop. Ext. Ser., Univ. of Arkansas, Miscell. Publ. AG-47-3-92.

Charlton W A 1983 Patterns and control of lateral root initiation. *In* Growth Regulators in Root Development. Eds. M B Jackson and A D Stead. Monograph No. 10. pp 1–13. British Plant Growth Regulator Group, Wantage, U. K.

Clark T H, Stutte C A, Ball R K and Guo C 1992 Root activity in cotton as affected by stress and bioregulants. Proc. Beltwide Cotton Prod. Res. Conf. National Cotton Council, Memphis, Tennessee. Vol. 3, p 1028.

Davies P J 1987 Plant Hormones and Their Role in Plant Growth and Development. Martinus Nijhoff Publishers, Dordrecht.

Guinn G 1982 Causes of square and boll shedding in cotton. USDA-ARS. Technical Bulletin 1672.

Guo C, Stutte C A, Oosterhuis D M and Clark T H 1992 Pinitol concentration in soybean petiole exudates as affected by environmental stress and plant growth regulators. p 83. Proc. Nineteenth Annual Meeting Plant Growth Regulator Soc. Am. San Francisco, California. July 17–20,1992.

Hickey J A 1991 Increasing fruit set, retention, and yield in cotton with PGR-IV. Proc. Beltwide Cotton Prod. Res. Conf. National Cotton Council, Memphis, Tennessee. Vol 1, pp 78–79.

Holman E M, Oosterhuis D M and Hurren R G 1992 Effect of foliar-applied N and K on vegetative cotton. Arkansas Farm Res. 41 (1), 4–5.

Howard B H, Harrison-Murray R S and Fenlon C A 1983 Effective auxin treatment of leafless winter cuttings. *In* Growth Regulators in Root Development. Eds. M B Jackson and A D Stead. Monograph No.10, pp 73–85. British Plant Growth Regulator Group, Wantage, U K.

Jackson M B 1983 Regulation of root growth and morphology by ethylene and other externally applied growth substances. *In* Growth Regulators in Root Development. Eds. M B Jackson and A D Stead. Monograph No.10, pp 103–115. British Plant Growth Regulator Group, Wantage, U K.

Nickell L G 1982 Plant growth regulators: Agricultural Uses. Springer-Verlag, Berlin. 173 p.

Lachno D R 1983 Abscisic acid and indole-3-acetic acid in maize root subject to water, salt and mechanical stress. *In* Growth Regulators in Root Development. Eds. M B Jackson and A D Stead. Monograph No.10, pp 37–54. British Plant Growth Regulator Group, Wantage, U K.

Oosterhuis D M 1990 Growth and development of the cotton plant. *In* Nitrogen Nutrition in cotton: Practical Issues. Eds. W N Miley and D M Oosterhuis. pp 1–24. Proc. Southern Branch Workshop for Practicing Agronomists. Am. Soc. Agron., Madison, Wisconsin.

Oosterhuis D M and Zhao D 1993 Physiological effects of PGR-IV on growth and yield of cotton. *In* Proc. 1993 Cotton Meeting and Summaries of Cotton Research. pp 22–26. Univ. of Arkansas Agric. Exp. Sta. Special Report 162.

Pilet P E 1983 Effective auxin treatment of leafless winter cuttings. *In* Growth Regulators in Root Development. Eds. M B Jackson and A D Stead. Monograph No. 10, pp 15–24. British Plant Growth Regulator Group, Wantage, U.K.

Scott T K 1978 Plant Regulation and World Agriculture. Plenum Press, New York. 575 p.

Stewart M, Hake K, Oosterhuis D, Kerby T, Mauney J and Timpa J 1993 Cotton fruit development - the square. Cotton Physiology Today, National Cotton Council, Memphis, Tennessee 4(1),1–4

Tennant D 1979 A test of a modified line intersect method of estimating root length. J. Ecol. 63,995–1001.

Urwiler M J and Oosterhuis D M 1986 The effect of the growth regulators Pix and IBA on cotton root growth. Arkansas Farm Res. 36(6), 5.

Urwiler M J, Stutte C A, Jourdon S and Clark T H 1987 Bioregulant field evaluations on agronomic crops. Univ. of Arkansas Agric. Exp. Sta. Research Series 385, 28 p.

Urwiler M J, Stutte C A and Clark T H 1988 Field evaluations of bioregulants on agronomic crops in Arkansas. Univ. of Arkansas Agric. Exp. Sta. Research Series 371, 30 p.

*F. Baluška et al. (eds.), Structure and Function of Roots*, 113–119, 1995.


# Modelling of the branching process of the root system of young oil palm seedlings. A Markovian approach

Christophe Jourdan, Hervé Rey and Yann Guedon
*CIRAD / Unité de Modélisation des plantes, BP 5035, 34032 Montpellier Cedex, France*

*Key words: Elaeis guineensis*, Markov chain, modelling, oil palm, root architecture, root branching

## Abstract

The root branching of oil palm, *Elaeis guineensis* Jacq., is modelled by a Markov chain (discrete time, discrete state space stochastic process). This study has been realized on radicles of young oil palm seedlings which are considered as a main axis which ramifies. We define an elementary length unit as the smallest length between two successive lateral roots. The model is based on the analysis of a sequence of events, each event being indexed by the rank of the elementary length unit on the main axis. An event is defined as the state of the length unit, chosen between branched and unbranched. The branching process of the oil palm radicle is modelled by a two-state first order Markov chain. Consequently, the state of the length unit of index n depends only on the state of the previous length unit. The Markov chain is homogeneous i.e. the transition probabilities do not depend on the rank of the elementary length unit.

This study allows us to identify a probabilistic model of root branching which is the first step in the elaboration of an architectural model of the oil palm root system.

## Introduction

In the last few years, some models of root growth and development were investigated (Diggle, 1988; Lungley, 1973; Porter et al., 1986; Rose, 1983). More recently, some models of root system architecture were developed (Atger, 1992; Fitter et al., 1991; Pagès et al., 1989; Tardieu and Pellerin, 1991). In plant modelling, we usually define not a single model but a hierarchy of models: an elementary growth model, an axillary production model and models of the architectural settings (topology and geometry). In this paper, we focalize our interest on the treatment of the axillary root production i.e. the root branching. Several works on the root branching process were initiated in order to characterize it and therefore to understand the theory leading to this process. Concerning Monocotyledons, Riopel (1966, 1969) showed that potential sites for the origin of lateral roots are influenced by the position of the previously established lateral. Moreover, he also found that lateral roots were distributed along the main root in a non-random, dispersed pattern. These results were confirmed by other plant studies (Abadia-Fenoll et al., 1982, 1986; Charlton, 1975; Mallory et al., 1970).

Inter-root distance is one valid measurable criteria for characterizing the root branching process. As the rhizotaxis was not clearly demonstrated, authors tried to find out a quantitative convenient parameter which can explain the lateral root distribution e.g. the "branching index" on maize (Morita et al., 1992) or the "fundamental distance" on tomato roots (Newson et al., 1993).

Mean density of lateral roots is commonly used to characterize the root branching process in many models encountered in literature (Diggle, 1988; Hackett and Rose, 1972; Pagès and Aries, 1988; Pagès et al., 1989; Porter et al., 1986). The root branching process can be estimated with more accuracy by a root segmentation into classes of length where mean branched density is calculated (Pagès et al., 1993; Varney et al., 1991). But approximations still exist owing to the mean calculus. Anyway, the branched density does not express at best the branch distribution along the entire root or in the length classes. Indeed, a same root

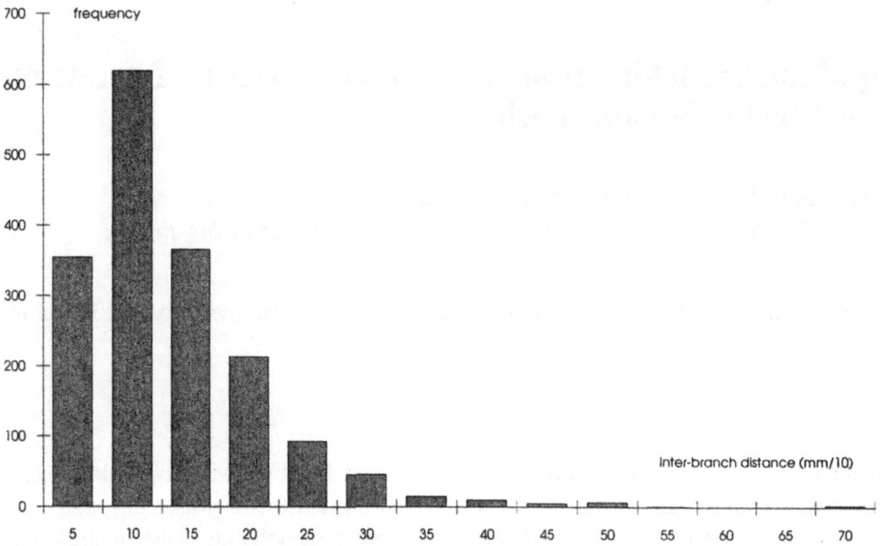


*Fig. 1.* Diagram of inter-branch distances.

may have barely branched zones or highly branched zones, with the same number of roots in both cases, hence with the same average branching probability (branching density), despite a different layout of these ramifications.

In this paper, we present an original treatment of the root branching process which allows us, for a same average density branching, to account for the distribution of inter-lateral root lengths. Therefore, branching of the radicle of young oil palm (*Elaeis guineensis* Jacq.) seedlings is studied through series of successive events which are modelled by a discrete time stochastic process.

## The model: assumptions and derivation

Unlike its aerial counterpart, the root axis has neither nodes nor morphological markers (axillary buds) from which branching occurs. In order to study the root branching process, we observe inter-branch distances. The idea is to determine sequences of events constructed from an index reflecting a biological functionality such as the node rank of aerial axes, so as to study the relationships between successive events. To do this, based on the distribution of inter-branch distances (Fig. 1), we take the minimum distance observed between two branches as the sampling step: somewhat along the lines of the "fundamental distance" separating two lateral roots in tomato (Newson et al., 1993). Thus, the root axis can be sampled and a binary number associated to each elementary length unit: non-branched unit: 0; branched unit: 1 (Fig. 2).

A sequence of events is then defined as being the succession of the different states of the elementary length units (branched, non-branched) observed along the root axis. A direction for running through the sequences of events has to be clearly defined, so as to reflect the causality of the biological studied phenomenon. In our case, the logical direction is that of root growth, i.e. from the base to apex.

A sequence of events can then be characterized by the state probabilities as the function of the rank (Figs. 3 and 4) and by the number of successive length units being in the same state (Figs. 5 and 6).

Based on these characteristics it is possible to put forward hypotheses as to the stochastic process modelling of these sequences. We shall call transition the fact of passing from the state of the rank n length unit to the state of the rank n+1 length unit.

A stochastic process is defined by the number of states, the possible transitions between states (Fig.7) and the transition distributions. If the state probabilities as a function of the rank of the elementary length unit rapidly tend towards constant values, the process is said to be homogeneous i.e. the transition probabilities are independent of the rank (Figs. 3 and 4). The transition distributions can be defined by the dependence between the current event at rank n and the previous event(s). Let us consider a case where the number of successive length units being in the same state are geometrically distributed. The process is then a first order homogeneous Markov chain (Feller, 1969; Karlin and Taylor, 1975). This two-state Markov chain is defined by the following parameters:

– initial probabilities: $\pi_j = P(S_o = j)$
with $j = 0, 1$ where $\pi_0 + \pi_1 = 1$

– transition probabilities: $p_{ij} = P(S_n = j/S_{n-1} = i)$
with $i = 0, 1$ and $j = 0, 1$
where $p_{00} + p_{01} = 1$ and $p_{10} + p_{11} = 1$

$(p_{00}, p_{01})$ and $(p_{10}, p_{11})$ define the two transition distributions of the Markov chain, the first corresponding to the "state 0" memory, the second to the "state 1" memory (Fig. 7). From the definition of the transition distributions, we can deduce the distribution of the number of successive length unit in the same state. In the case where $p_{jj} < 1$, finding n successive length units being in the same state means remaining (n-1) times in this state, then leaving it:

$$d_j(n) = (p_{jj})^{n-1}(1 - p_{jj})$$
$$\text{where } n = 1, 2, ...\infty \text{ and } j = 0, 1$$

This defines a geometric distribution of parameter $(1 - p_{jj})$.

Otherwise, the theory of Markov chains is widely used in signal processing and in automatic control.

## Practical application of the model

### *Experimental material and methods*

The C1001F oil palm seedlings were used for this study. The experiments took place under wet tropical climatic conditions, at the IDEFOR-DPO research station of La Mé (previously IRHO research station), near Abidjan, south of Côte d'Ivoire.

For the analysis, 15 one-month-oil-palm seedlings were observed. The plants grown on Tertiary sandy soil, in tree nursery bags (20 L). The roots were excavated by a slight washing. The radicles are considered as a main axis on which lateral roots are distributed. The relative lengths between two successive branches on the tap root were measured. The following data were analysed with a software developed at the Modelling Unit of CIRAD which runs on a Silicon Graphics workstation.

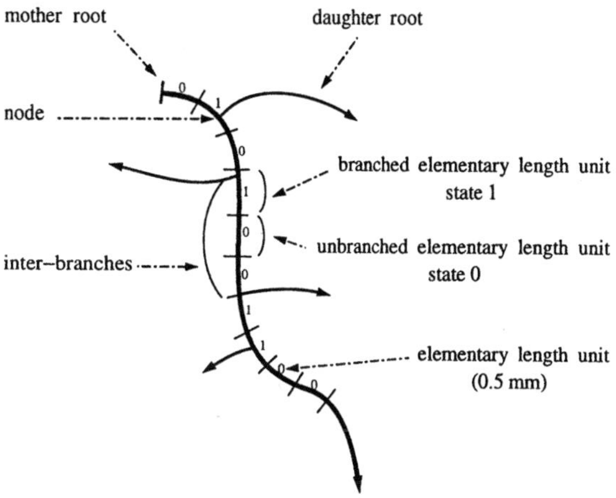


*Fig. 2.* Diagram of an oil palm radicle.

### *Results of using the model*

The results come from the observations on the 15 oil palm seedlings and only individual data are used. On each elementary length unit, the two possible states are the unbranched state (state 0) and the branched state (state 1). The treatment of the data showed that the number of successive unbranched elementary length units (Fig. 5) and the number of successive branched elementary length unit (Fig. 6) were geometrically distributed. These distributions are the state occupancy distributions of a two-state first order Markov chain.

The estimated parameters of the two-state Markov chain are:

– the initial probabilities: $\pi = \binom{\pi_0}{\pi_1} = \binom{0.33}{0.67}$

– the transition probabilities (Fig. 7):

$$P = \begin{pmatrix} p_{00} & p_{01} \\ p_{10} & p_{11} \end{pmatrix} = \begin{pmatrix} 0.53 & 0.47 \\ 0.80 & 0.20 \end{pmatrix}$$

For the unbranched and branched states, the state occupancy distributions are given by:

$$d_0(n) = (0.53)^{n-1}(1 - 0.53)$$
$$d_1(n) = (0.20)^{n-1}(1 - 0.2)$$

The mean of the unbranched state occupancy distribution is 2.14 whereas the mean of the branched state occupancy distribution is 1.25.

## Discussion

Root growth is usually continuous. The apical meristem may slow down but does not stop, unlike its aerial

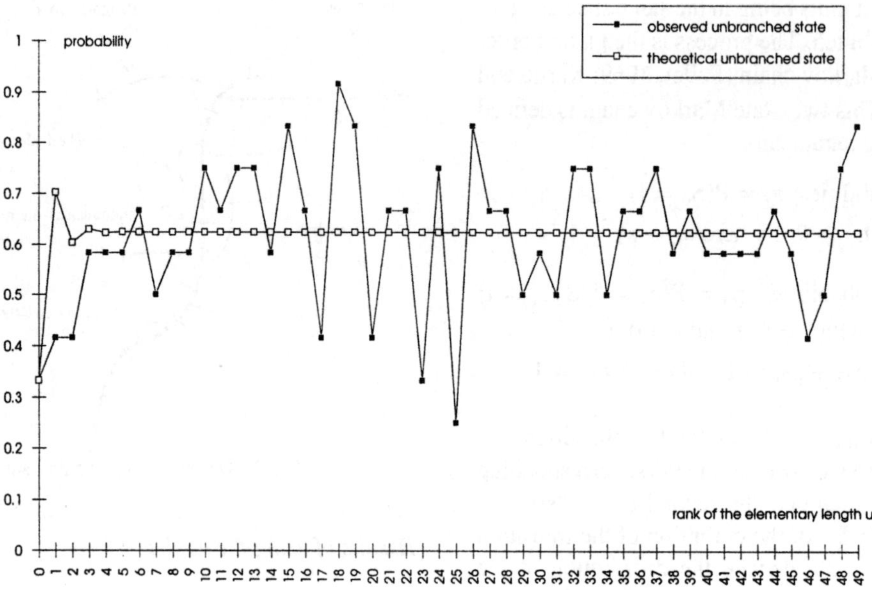


*Fig. 3.* Evolution of the unbranched state probability.

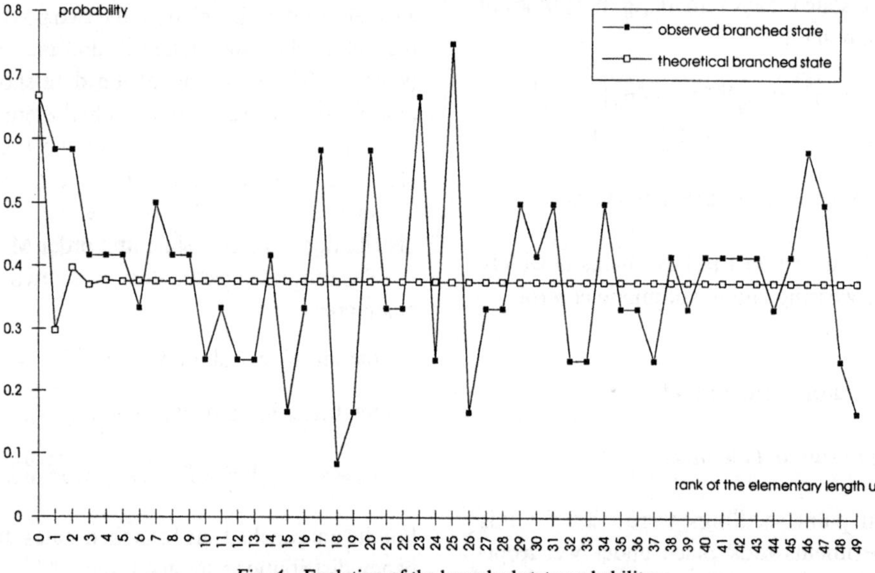


*Fig. 4.* Evolution of the branched state probability.

counterpart. Lateral roots frequently emerge shortly after their initiation and the primordia do not have a dormancy phase. However, cases of abortion or latency period are observed on red pine roots (Wilcox, 1968). The primordia of lateral roots are directly influenced by the apex that generated them and this determines the minimum distance from the apex from which the roots emerge (Barlow, 1989). Moreover, the lateral root growth can be, partially or completely, inhibited by the root apex activity (Dyanat-Nejad and Neville, 1972; Riedacker, 1976). The primordia are established during root differentiation and are expressed when the root reaches a given stage of maturity (Abadia-Fenoll et al., 1982).

The distance between two successive primordia is variable according to species. For each of them, this distance may vary in relation with the root typology and activity (Wilcox, 1968). A spacing modification between two sucessive primordia may be induced by a change of the root speed growth. The slower the root

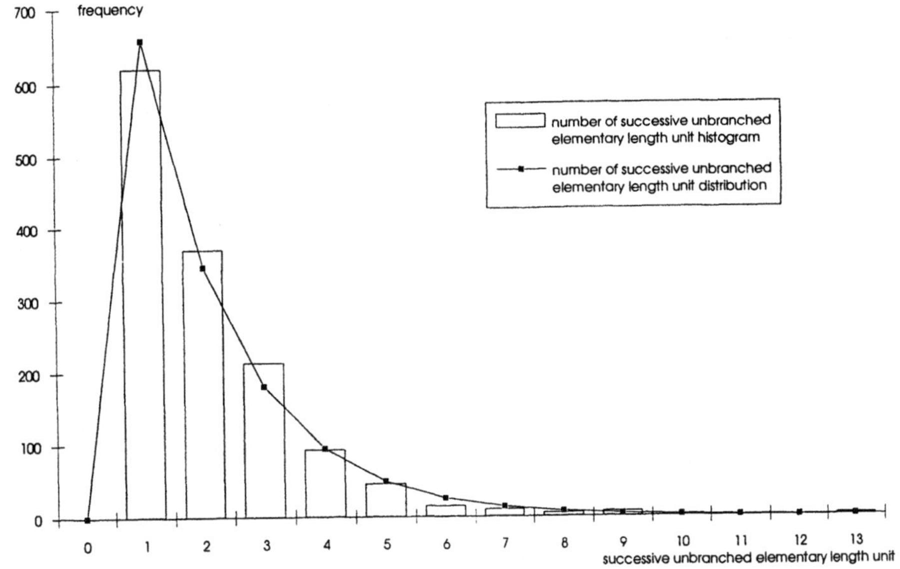


*Fig. 5.* Distribution of the length of the unbranched elementary length unit.

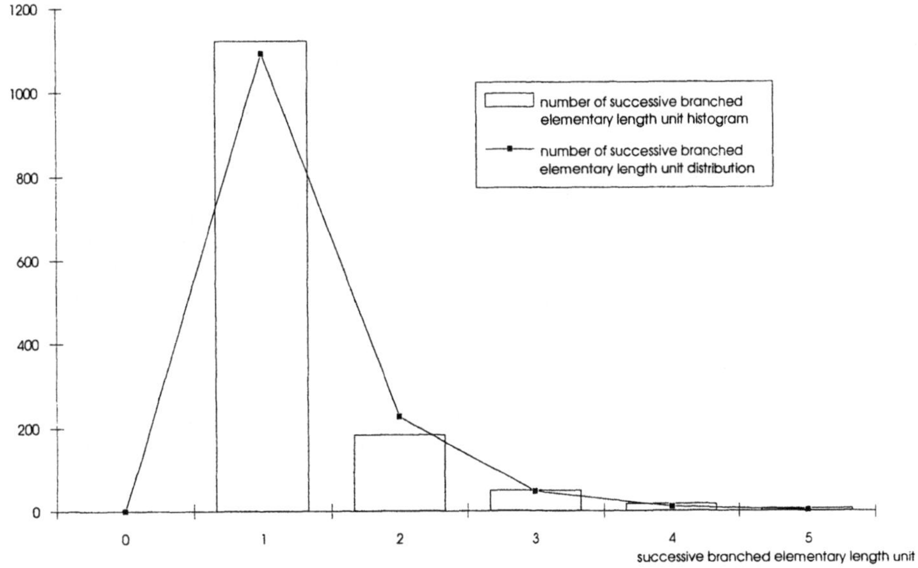


*Fig. 6.* Distribution of the length of the branched elementary length unit.

speed growth will be, the shorter the inter-primordia length will become. The inter-root distance can be reduced but the number of cells between two laterals keeps constant (Wilcox, 1968).

According to our observations (Fig. 1), two successive roots do not occur on the same rank, they are separated by a minimum distance of 0.5 mm. This could be explained by the fact that when a primordium is established, it inhibits the appearance of new primordia in the immediate vicinity (Newson et al., 1993). A

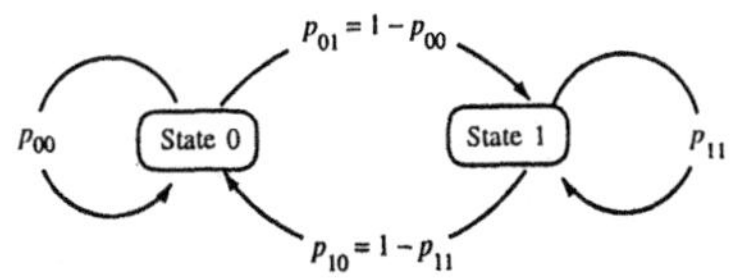


*Fig. 7.* Diagram of the transition probabilities,

zone of suppression therefore seems to be set up near the primordia (Abadia-Fenoll et al., 1982; Charlton, 1975; Mallory et al., 1970; Riopel, 1969).

All these results show that the initiation of primordia is more or less determinated by the root apex itself, but their expression in laterals can differ. The model is able to take into account all these changes even if the causalities are numerous and complex. The pertinence of our stochastic model is to fit the field observations. In a stochastic model, biological hypothesis which define the mathematical approach are not expressed, unlike in any deterministic model. Our model does not explain biological phenomena but it enables to describe them. At this macroscopic description level, we have to check the microscopic scale hypothesis i.e. the cellular and hormonal level. Our dependence hypothesis between successive events does not disagree with the theory of primordia influenced zones and gives a good representation of the observed process. However, in some particular situations (when laterals emerged later behind the acropetal sequence) other parameters are required.

In our study, the branching process of root axes is therefore modelled by a Markov chain. Unlike the usual methods for estimating branching densities, this method accurately indicates the distribution of lateral roots (in respect of each other). In fact, the transition probabilities ($p_{ij}$) reveal the succession of daughter roots along the mother root.

The Markovian approach is therefore more general than a simple branching probability for modelling this biological process with the introduction of dependence between successive events

In case of root verticils i.e. when several laterals are branched on the same level (Mallory et al., 1970), the model is still available. Only one more appropriate variable has to be considered: the number of laterals per verticil.

## Conclusion

This work allowed us to identify a model of root branching characterized by a two-state first order homogeneous Markov chain. It constitutes a first step in the elaboration of an architectural model of the oil palm root system.

It is clear that the stochastic modelling of the branching process may improve predictions of several phenomena. Mineral and water uptake, for example can be mentioned. This will be obtain by the way of simulation on a graphic workstation and by complementary software developed in our laboratory.

## References


Abadia-Fenoll F, Casero P J, Lloret P G and Vidal M R 1986 Development of lateral primordia in decapited adventitious roots of *Allium cepa*. Ann. Bot. 58, 103–107.

Abadia-Fenoll F, Lloret P G, Vidal M R and Casero P J 1982 Pattern distribution of lateral root primordia in *Allium cepa* L. Phyton, Int. J. Exp. Bot. 42, 175–177.

Atger C 1992 Essai sur l'architecture racinaire des arbres. Thèse 3ème cycle, Univ. Sc. Tech. du Languedoc, Montpellier. 287p.

Barlow P W 1989 Meristems, metamers and modules and the development of shoot and root systems. Bot. J. Linnean Soc. 100, 255–279.

Charlton W A 1983 Patterns of distribution of lateral root primordia. Ann. Bot. 51, 417–427.

Diggle A J 1988 ROOTMAP-a model in three-dimensional coordinates of the growth and structure of fibrous root systems. Plant and Soil 105, 169–178.

Dyanat-Nejad H and Neville P 1972 Etude du mode d'action du méristème radical orthotrope dans le contrôle de la plagiotropie des racines chez *Theobroma cacao* L. Rev. Gén. Bot. 79, 319–340.

Feller W 1969 An introduction to the Probability Theory and its Applications. John Wiley and Sons Editors, 3d ed. 509p.

Fitter A H, Stickland T R, Harvey M L and Wilson G W 1991 Architectural analysis of plant root systems. 1- Architectural correlates of exploitation efficiency. New Phytol. 118, 375–382.

Hackett C and Rose D A 1972 A model of the extension and branching of a seminal root of barley, and its use in studying relations between root dimensions. I - The model. Aust. J. Biol. Sci. 25, 669–679.

Karlin S and Taylor H M 1975 A first course in stochastic processes. Academic Press, 2d ed. 557p.

Lungley D R 1973 The growth of root systems - A numerical computer simulation model. Plant and Soil 38, 145–159.

Mallory T E, Chiang S-H, Cutter E G and Gifford E M 1970 Sequence and pattern of lateral root formation in five selected species. Am. J. Bot. 57(7), 800–809.

Morita S, Thongpae S, Abe J, Nakamoto T and Yamazaki K 1992 Root branching in maize. I- "Branching index" and methods for measuring root length. Jpn. J. Crop Sci. 61 (1), 101–106

Newson R B, Parker J S and Barlow P W 1993 Are lateral roots of tomato spaced by multiples of a fundamental distance? Ann. Bot. 71, 549-557.

Pagès L and Aries F 1988 SARAH: modèle de simulation de la croissance, du développement et de l'architecture des systèmes racinaires. Agronomie 8(10), 889–896.

Pagès L, Jordan M O and Picard D 1989 A simulation model of the three-dimensional architecture of maize root system. Plant and Soil 119, 147–154.

Pagès L, Kervella J and Chadoeuf J 1993 Development of the root system of young peach trees (*Prunus persica* L. Batsch): a morphometrical analysis. Ann. Bot 71, 369–375.

Porter J R, Klepper B and Belford R K 1986 A model (WHTROOT) which synchronizes root growth and development with shoot development for winter wheat. Plant and Soil 92, 133–145.

Riedacker A 1976 Etude des possibilités de modifier la morphogénèse juvénile des systèmes racinaires de quelques espèces forestières. Applications pratiques. *In* Proc. XVI IUFRO World Congress, Norway, Div. 1, 376–387.

Riopel J L 1966 The distribution of lateral roots in *Musa acuminata* "Gros Michel". Am. J. Bot. 53(4), 403–407.

Riopel J L 1969 Regulation of lateral root positions. Bot. Gaz. 130 (2), 80–83.

Rose D A 1983 The description of the growth of the root systems. Plant and Soil 75, 405-415.

Tardieu F and Pellerin S 1991 Modélisation de l'architecture du système racinaire du maïs au champ: aspects géométriques. *In* Physiologie et Production du maïs. Colloque 13/15 Nov. 1990, Pau (France). pp 133–141. Ed. INRA Paris.

Varney G T, Canny M J, Wang X L and McCully M E 1991 The branch roots of *Zea*. I - First order branches, their number, sizes and division into classes. Ann. Bot. 67, 357–364.

Wilcox H E 1968 Morphological studies of the root system of red pine, *Pinus resinosa* Ait. I- Growth characteristics and pattern of branching. Am. J. Bot. 55(2), 247–254.

# ABSORPTION, TRANSPORT AND UTILIZATION OF IONS

*F. Baluška et al. (eds.), Structure and Function of Roots,* 123–128, 1995.


# Effects of vanadate on the ATP content, ATPase activity and phosphate absorption capacity of maize roots

J. Sklenar, G.G. Fox, B.C. Loughman, A.D.B. Pannifer and R.G. Ratcliffe
*Department of Plant Sciences, University of Oxford, South Parks Road, Oxford, OX1 3RB, United Kingdom*

*Key words:* ATPase, ATP levels, $^{31}$P NMR, phosphate absorption, vanadate

## Abstract

Although the sensitivity of the plasma membrane $H^+$-ATPase to vanadate is well known, the metabolic response of plant cells to vanadate is less well characterised in vivo and its use as an inhibitor in whole plant experiments has had mixed success. Experiments with maize (*Zea mays*, L.) roots and with purified plasma membrane fractions from the same tissues showed that exposure to vanadate caused: (i) a reduction in the capacity for phosphate uptake; (ii) a reduction in the extractable ATPase activity from the tissue; and (iii) a significant increase in the ATP level. The measurements on the extractable ATPase activity and the ATP level showed that the effect of vanadate developed slowly, apparently reflecting the slow accumulation of intracellular vanadate. The marked effect of vanadate on the ATP level – exposure to 500 $\mu M$ vanadate for 5 h doubled the ATP content of the roots tips – indicates that there is no stringent control over the ATP level in the roots and that the plasma membrane $H^+$-ATPase activity is likely to have a significant role in determining the ATP level under normal conditions.

## Introduction

The relationship between the plasma membrane ATPase activity, the availability of ATP and different aspects of metabolic capacity such as ion absorption has been difficult to establish but it is now generally accepted that specific ion translocating ATPases are present in a wide range of animal and plant species.

Leonard and Hanson (1972) showed that the $K^+$ stimulated ATPase activity of the microsomal fraction of maize roots increases in parallel with the capacity for phosphate absorption during washing of isolated root systems. There appears to be a direct relationship between the capacity for stimulation of ATPase by cations and the specificity of cation transport into oat roots (Sze and Hodges, 1977). ATPase activity and the capacity for $Rb^+$ absorption are severely reduced with similar kinetics when maize roots are allowed to become boron deficient over a period of 18 h and rapid restoration of both activities occurs on the addition of 100 $\mu M$ $H_3BO_3$ to the culture solution in which the plants are growing (Loughman and White, 1984). Results of this type support the view that some portion of the ATPase is directly associated with the transport properties of the plasma membrane.

Sodium vanadate at low concentrations (e.g. 5 $\mu M$) inhibits the transport ATPases of mammalian (Cantley et al., 1978) and fungal cells (Bowman and Slayman, 1979). At higher concentrations (e.g. 50 $\mu M$) ATPase inhibition has also been observed in higher plant cells (Perlin and Spanswick, 1981) and Cocucci et al. (1980) reported that vanadate inhibits ion transport. A correlation between the effects of vanadate on the plasma membrane ATPase and $K^+$ absorption in maize roots was shown by Gallagher and Leonard (1982).

Vanadium exists in several oxidation states and whereas vanadate is a powerful inhibitor of the $Na^+$ $K^+$ ATPase, the reduced form, vanadyl has a negligible effect (Beauge, 1988; Grantham and Glynn, 1979). The effects of vanadate in plant tissues are likely to be influenced by the oxidation state of the system and by the availability of ATP. In this paper assessment of the ATPase activity in the plasma membrane fraction of maize roots was combined with in vivo $^{31}$P NMR experiments and measurement of phosphate absorption with the aim of correlating the effects of vanadate on the ATPase in vivo and in vitro with the metabolic activity, ATP level and transport capacity of the roots.

## Materials and methods

### *Plant material*

Maize seeds (LG11) were germinated in the dark for two or three days at 25°C in rolls of filter paper soaked in 0.1 m*M* $CaSO_4$. Whole primary roots (6–8 cm) were cut from the seedlings into highly purified water at 4°C and used for the isolation of the plasma membrane fraction. Root tips (0.5 cm) were cut into MES buffer (10 m*M* MES/0.1 m*M* $CaSO_4$ pH 6.0) for the NMR experiments. Steles were separated from the cortical tissue of root segments (1–6 cm from the apex) by careful dissection.

### *Plasma membrane fraction*

Primary roots were homogenised at 4°C in 0.25 *M* sucrose, 3 m*M* EDTA, 2.5 m*M* DTT and 25 m*M* Tris-MES (pH 7.7). After filtration through four layers of muslin the homogenate was centrifuged at 55,000 *g* for 30 min, the pellet suspended in 1 mL of buffer solution (0.25 *M* sucrose, 1 m*M* DTT and 1 m*M* Tris-MES pH 7.2) and layered on the top of a 9 mL gradient (7 mL 45% and 2 mL 34% sucrose in 1 m*M* DTT and 1 m*M* Tris-MES pH 7.2). The gradient was centrifuged at 75,000 *g* for 2 h. The membrane band at the 34/45 interface (1.15–1.20 g.mL) was removed, diluted with suspension buffer and centrifuged at 83,000 *g* for 1 h; the final pellet was resuspended and stored at −70°C.

### *Enzyme assay*

ATPase activity was measured at 38°C in a final volume of 1 mL containing enzyme, 3 m*M* ATP (Na salt), 30 m*M* Tris-MES (pH 6.5), 3 m*M* $MgSO_4$ and 25 m*M* $K_2SO_4$. However for some of the measurements summarised in Table 1, the $K_2SO_4$ was omitted from the assay mixture. The reaction was stopped by the addition of 0.64 mL 10% SDS and phosphate released measured by a modified Peterson (1978) procedure. Protein was measured by the method described by the same author (Peterson, 1977). Determinations were carried out in triplicate.

### *Phosphate absorption*

Samples of six 1 cm root tips were weighed and incubated in 10 mL 10 $\mu M$ $KH_2{}^{32}PO_4$ containing 0.1 $\mu$C $^{32}P$ for 1 h in a shaking water bath at 30°C. After rapid filtration on a small Buchner funnel the root tips were washed in successive 10 mL lots of water for 2 min, blotted and transferred to vials for counting by the Čerenkov method (Ross, 1976).

*Table 1.* The effect of $Na_3VO_4$ on the ATPase activity of the plasma membrane fraction from 6 cm maize roots in the presence and absence of 50 m*M* $K^+$

| $Na_3VO_4$ ($\mu M$) | ATPase activity ($\mu$ moles $P_i$ $mg^{-1}h^{-1}$) | |
|---|---|---|
| | 0 m*M* $K^+$ | 50 m*M* $K^+$ |
| 0 | 22.5 ± 0.09[a] | 33.5 ± 0.04 |
| 50 | 22.5 ± 0.04 | 24.9 ± 0.04 |
| 100 | 20.0 ± 0.25 | 21.0 ± 0.25 |
| 200 | 19.8 ± 0.16 | 17.2 ± 0.09 |
| 300 | 19.6 ± 0.10 | 14.5 ± 0.16 |
| 400 | 19.5 ± 0.01 | 14.0 ± 0.64 |
| 500 | 19.3 ± 0.04 | 13.8 ± 0.04 |

[a]Values are means ± SEM of 3 replicates.

### *NMR spectroscopy*

$^{31}P$ NMR spectra were recorded on a Bruker CXP300 spectrometer at 121.49 MHz using a double-tuned $^{13}C/^{31}P$ probehead. Approximately 160, 5 mm root tips were placed in a 10 mm diameter NMR tube and the tissue was supplied with oxygen by circulating an oxygenated medium through the tube at a rate of 8 mL $min^{-1}$ (Lee and Ratcliffe, 1983). The tissue was allowed to equilibrate for 30–40 min before starting the NMR experiment and the temperature of the sample was maintained at 21–22°C. Spectra were recorded in 30 min blocks, corresponding to 3600 scans, using a 45° pulse angle, a 0.5 s recycle time and a spectral window of 10416 Hz. In a typical experiment, two spectra were recorded with the tissue bathed in 10 m*M* MES/0.1 m*M* $CaSO_4$, pH 6.0 and then the circulating medium was switched to a solution containing 500 $\mu M$ sodium vanadate in the same buffer and spectra were recorded for a further 11 h. Chemical shifts were measured relative to the $^1H$-decoupled signal from a capillary containing a 2% v/v solution of the tetraethylester of methylene diphosphonic acid (TEMDP) in water (Kime et al., 1982) and are quoted on the scale that puts the resonance from 85% orthophosphoric acid at 0 ppm.

## Results

### *ATPase activity*

The ATPase activity of the plasma membrane fraction from 6 cm maize roots was significantly inhibited by 100 $\mu M$ vanadate and this inhibition was complete at 300 $\mu M$ (Fig. 1). As expected (Macara et al., 1980),

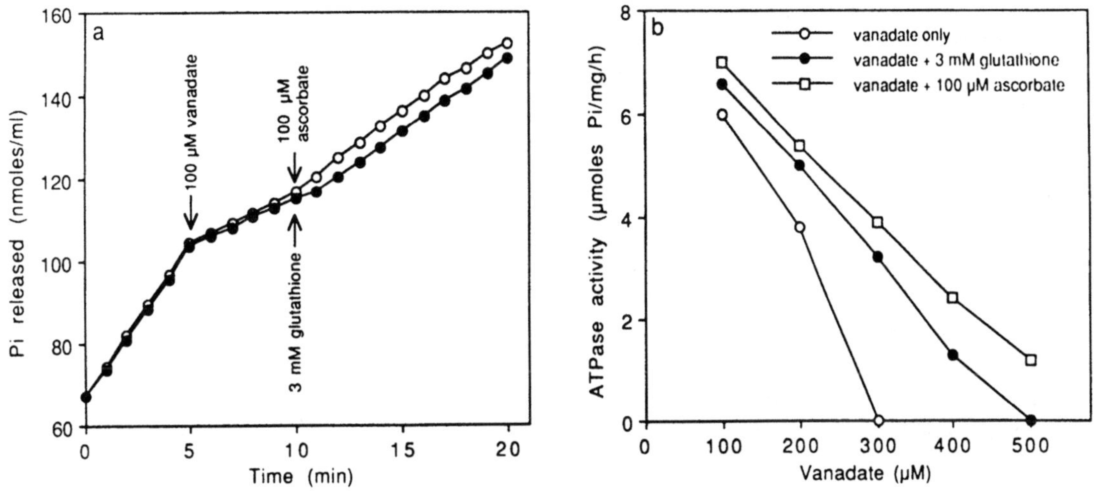


*Fig. 1.* Vanadate inhibition of the plasma membrane ATPase activity isolated from 6 cm maize roots. (**a**) Phosphate release timecourses showing the partial reversal of the inhibition of the ATPase activity by 100 $\mu M$ vanadate in the presence of either 100 $\mu M$ ascorbate or 3 m$M$ glutathione. The timecourses were constructed by removing 1 mL samples for phosphate analysis from a 30 mL assay mixture at 1 min intervals. (**b**) The effect of 100 $\mu M$ ascorbate or 3 m$M$ glutathione on the inhibition of the ATPase activity by vanadate concentrations in the range 100 – 500 $\mu M$. The points in both (a) and (b) are mean values obtained from three experiments.

reducing vanadate to vanadyl by adding 100 $\mu M$ ascorbate or 3 m$M$ glutathione partially reversed the inhibition of the ATPase activity (Fig. 1) and the potential for the intracellular reduction of vanadate, e.g. by sulphydryl groups, needs to be considered when interpreting experiments conducted in vivo. Adding potassium to the plasma membrane fraction increased the ATPase activity and its sensitivity to vanadate (Table 1). The $K^+$ stimulated enzyme was 50% inhibited by 200 $\mu M$ vanadate whereas in the absence of $K^+$ the inhibition was only 12% (Table 1).

Exposure of 6 cm roots to 300 $\mu M$ vanadate affected the activity of the ATPase subsequently extracted from them. The apical 1 cm was examined in addition to the separated cortical and stelar tissue of the remainder of the root. Table 2 shows that the apex was the most active tissue and that after 3 h exposure to vanadate the ATPase activity fell to half that of the control. The cortex also showed a 20% fall but the ATPase activity of the stele was unaffected.

### *Phosphate absorption*

The effect of vanadate at pH 6.0 on the absorption of phosphate over a period of 1 h by 1 cm root tips from 10 $\mu M$ $KH_2{}^{32}PO_4$ is shown in Table 3. Relatively low inhibition occurred up to 300 $\mu M$ vanadate, at which concentration the ATPase was at least 60% inhibited in vitro, and 50% inhibition of P absorption was obtained at 1 m$M$ vanadate. The relatively slow accumulation

*Table 2.* The effect of exposure of maize roots to 300 $\mu M$ $Na_3VO_4$ on ATPase activity of membrane fractions subsequently isolated from the apex, cortex and stele

| Treatment | ATPase activity ($\mu$ moles $P_i$ $mg^{-1}h^{-1}$) | | |
|---|---|---|---|
| | Apex (0–1 cm) | Cortex (1–6 cm) | Stele (1–6 cm) |
| Control | 5.2 ± 0.04[a] | 4.4 ± 0.16 | 2.6 ± 0.01 |
| 300 $\mu M$ $Na_3VO_4$ for 1 h | 5.0 ± 0.04 | 4.2 ± 0.07 | 2..6 ± 0.01 |
| 300 $\mu M$ $Na_3VO_4$ for 3 h | 2.6 ± 0.06 | 3.5 ± 0.02 | 2.6 ± 0.03 |

[a]Values are means ± SEM of 3 replicates.

*Table 3.* The effect of $Na_3VO_4$ on absorption of phosphate by six 1 cm root tips of maize[a]

| $Na_3VO_4$ ($\mu M$) | Phosphate uptake capacity ($\mu$ moles P $g^{-1}h^{-1}$) |
|---|---|
| 0 | 14.85 ± 0.74[b] |
| 100 | 12.34 ± 1.05 |
| 300 | 12.91 ± 0.63 |
| 1000 | 8.80 ± 0.24 |

[a]Absorption from 10 $\mu M$ $KH_2{}^{32}PO_4$ at pH 6.0 and 25°C for 1 h.
[b]Values are means ± SEM of 5 replicates.

of intracellular vanadate into the roots suggested by the NMR experiments (see below) is likely to be a contributory factor and complicates comparison of the two effects.

As with all inhibitor studies, care is needed in interpretation of the results and due regard given to differences in rates of penetration of tissue and possible compartmentation. The different experimental approaches involve different degrees of complexity and comparisons are difficult. The laborious procedure involved in production of sufficient purified ATPase preparations for a number of treatments from different tissues in a single batch of roots makes it difficult to provide facilities for a desirable degree of replication.

*In vivo* $^{31}P$ *NMR*

Exposure to vanadate caused a substantial increase in the NTP pool of the excised root tips (Fig. 2). Control experiments (data not shown, but see Loughman et al., 1989) showed that the NTP signals remained approximately constant, drifting down to approximately 90% of their initial intensity over a period of 12 h, whereas exposure to 500 $\mu M$ vanadate for 5 h caused the signals to double in intensity (Fig. 2 a and d). A comparison of Figure 2a and 2b shows that vanadate had no immediate effect on the spectrum and that it was only after about 3 h (Fig. 2c) that the increase in the NTP signals became detectable. Given the rapid turnover of the cellular NTP pools, this observation indicates that the build-up of intracellular vanadate was slow. The NTP peaks reached their maximum intensity after 5–6 h exposure to vanadate and then remained approximately constant; but if the vanadate was removed from the circulating medium, the signals returned to their initial intensities over a period of about 6 h (data not shown). Vanadate caused no other spectroscopic effects and in particular it had no effect on the position of the pH-sensitive cytoplasmic $P_i$ signal indicating that the cytoplasmic pH was unaffected by exposure to vanadate at pH 6.

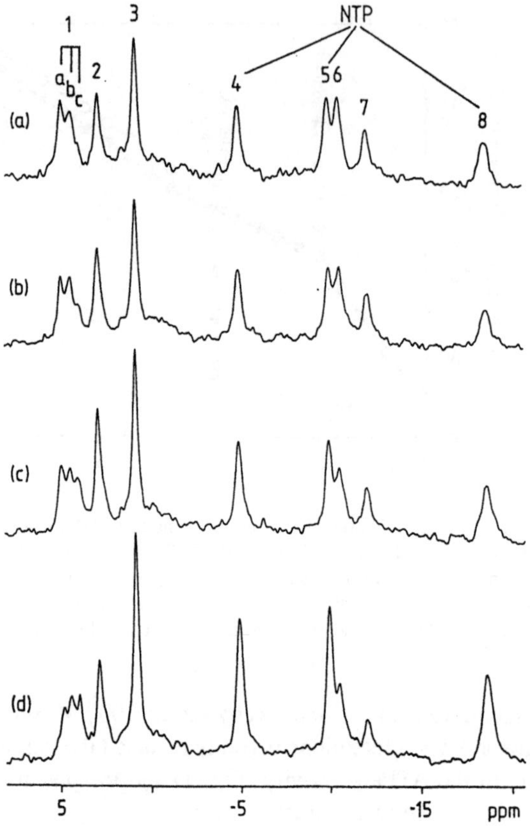


*Fig. 2.* $^{31}P$ NMR spectra of maize root tips during exposure to vanadate. After recording a spectrum (**a**) with the tissue bathed in the standard MES buffer at pH 6, the circulating medium was switched to a solution containing 500 $\mu M$ vanadate in the same buffer and spectra were recorded after (**b**) 30 min, (**c**) 2.5 h and (**d**) 5 h exposure to vanadate. The numbered peaks are assigned to: 1, several phosphomonoesters, including glucose-6-phosphate (1a) and phosphocholine (1c); 2, cytoplasmic inorganic phosphate ($P_i$); 3, vacuolar $P_i$; 4, 5, and 8, the $\gamma$, $\alpha$- and $\beta$-phosphates respectively of nucleoside triphosphate (NTP); 6, UDP-glucose and NAD(P)(H); and 7, UDP-glucose. ATP is the principal contributor to the NTP pool. Vanadate had no immediate effect on the $^{31}P$ NMR spectrum but after 5 h the intensity of the NTP signals had doubled, apparently reflecting the inhibition of the plasma membrane $H^+$- ATPase.

## Discussion

An attempt has been made to correlate the ATPase activity in maize roots with the capacity for ion absorption and with the metabolic state of the tissue. Where such correlations are attempted it is useful to have a suitable inhibitor or metabolic stimulus such as a nutrient deficiency or environmental change that has comparable effects on all the systems and ideally these systems should recover on removal of the constraint. The vanadate ion has some of the requisite properties and we have examined its effect on the $K^+$ $Mg^{2+}$ stimulated ATPase of the plasma membrane fractions and on the capacity for phosphate absorption from concentrations similar to those found in the soil solution. In addition, the response to exposure of root tissue to vanadate has been continuously monitored in vivo for some hours by $^{31}P$ NMR where the metabolic activity of the tissue is reflected in changes in specific intermediates such as ATP and other phosphorylated compounds.

Table 1 shows that the component of the ATPase activated by $K^+$ is sensitive to vanadate whereas in the absence of $K^+$ the inhibitor had little effect. The data showed that $K^+$ not only increased the ATPase activity of the preparation but also that it increased the sensitivity of the non $K^+$-dependent ATPase to vanadate inhibition. The effect of a vanadate pretreatment on the extractable ATPase activity was greatest in the apical region of the root and the vanadate pretreatment had a greater effect on the preparation from the cortical tissue than the stele (Table 2). The lowered activity of the ATPase isolated from the roots after 3 h exposure to 300 $\mu M$ vanadate suggests that the isolated ATPase contained bound vanadate that was not removed during the relatively stringent method of isolation. This could be tested either by using chelating agents, such as EDTA (Kustin and Toppen, 1973), to attempt to displace the bound vanadate or by extracting the ATPase activity from vanadate-treated roots after a period in a vanadate free solution. However it is interesting to note that the NMR data indicated only a slow recovery of the ATP level after the removal of vanadate, suggesting that the restoration of the ATPase activity could well be slow.

In contrast to the data in Table 2, Cowan et al. (1993) have recently shown that the ATPase activity of the plasma membrane fractions isolated from stelar and cortical tissue of 6 cm roots of LG11 maize is inhibited by vanadate to a similar extent. In our experiments the greater inhibition by vanadate with increasing exposure suggests that the accumulation of intracellular vanadate occurred relatively slowly and that while inhibition of the cortical enzyme occurred after 3 h, the stelar enzyme was unaffected because the inhibitor had not yet reached a critical concentration in the tissue. Further evidence for the slow accumulation of vanadate in the tissue is provided by the NMR data (Fig. 2) which showed that the effect of vanadate on the ATP levels was not apparent until more than 3 h exposure of the root tips to the inhibitor. The effects of vanadate in vivo are always going to be mediated by the kinetic processes leading to the accumulation of vanadate in the tissue and as another example of this phenomenon, Brummel et al. (1986) showed that the inhibition of IAA-induced growth by vanadate occurred rapidly in cucumber hypocotyls but at a much slower rate in pea stem segments. Factors limiting the attainment of the critical vanadate concentration include both the rate of vanadate uptake and the possibility of intracellular reduction to vanadyl. The potential importance of the latter is demonstrated by the data in Figure 1 for the isolated ATPase.

ATP is the principal component of the NTP pool in maize root tips (Hooks et al., 1989) and the increase in the NTP signals observed by NMR when the root tips were incubated with vanadate is consistent with an inhibition of the plasma membrane ATPase in vivo. The increase in the NTP pool took place over several hours, suggesting that the accumulation of intracellular vanadate was relatively slow, and this observation is consistent with the data in Tables 2 and 3. The large increase in the NTP pool suggests that the ATP level is not tightly regulated and that the steady state ATP level reflects a simple balance between the rates of synthesis and consumption. Analogous observations have been made with the protein synthesis inhibitor cycloheximide, which increased the steady state ATP level in maize root tips by 50% (Loughman et al., 1989), and it appears that the rate of ATP synthesis does not respond sufficiently, if at all, to compensate for the reduced demand for ATP. If this reasoning is correct, then the fact that the NTP level doubles in response to vanadate indicates that the activity of the plasmalemma ATPase plays a significant role in determining the cellular ATP level under normal conditions.

**Acknowledgements**

JS is grateful for the financial support of an FCO/Soros Fellowship; GGF, BCL and RGR are supported by the Agricultural and Food Research Council, and GGF acknowledges the receipt of a Research Fellowship from the Royal Commission for the Exhibition of 1851.

**References**

Beauge L 1988 Inhibition of translocation reactions by vanadate. Methods Enzymol. 156, 251–267.

Bowman B J and Slayman C W 1979 The effects of vanadate on the plasma membrane ATPase of *Neurospora crassa*. J. Biol. Chem. 254, 2928–2934.

Brummell D A, Hall J L and Armstrong C F 1986 Effect of vanadate on microsomal ATPase activity, acidification and auxin stimulated growth in pea and cucumber. Ann. Bot. 57, 727–735.

Cantley L C, Cantley L G and Josephson L 1978 A characterisation of vanadate interactions with the $Na^+$ $K^+$ ATPase. Mechanistic and regulatory implications. J. Biol. Chem. 253, 7361–7368.

Cocucci M, Ballarin-Dente A and Marré M T 1980 Effects of orthovanadate on $H^+$ secretion, $K^+$ uptake, electric potential difference and membrane ATPase activities of higher plant tissues. Plant Sci. Lett. 17, 391–400.

Cowan D S C, Clarkson D T and Hall J L 1993 A comparison between the ATPase and proton pumping activities of plasma membranes isolated from the stele and cortex of *Zea mays* roots. J. Exp. Bot. 44, 983–989.

Gallagher S R and Leonard R T 1982 Effect of vanadate, molybdate and azide on membrane-associated ATPase and soluble phosphatase activities of corn root. Plant Physiol. 70, 1335–1340.

Grantham J J and Glynn I M 1979 Renal $Na^+$, $K^+$ ATPase: determinants of inhibition by vanadium. Am. J. Physiol. 236, 530–535.

Hooks M A, Clark R A, Nieman R H and Roberts J K M 1989 Compartmentation of nucleotides in corn root tips studied by $^{31}P$ NMR and HPLC. Plant Physiol. 89, 963–969.

Kime M J, Loughman B C, Ratcliffe R G and Williams R J P 1982 The application of $^{31}P$ nuclear magnetic resonance to higher plant tissue. I. Detection of spectra. J. Exp. Bot. 33, 656–669.

Kustin K and Toppen D L 1973 Kinetics of complexation of vanadate anions by ethylenediaminetetra acetic acid and 1,2-dihydroxyanthraquinone. J. Am. Chem. Soc. 95, 3564–3568.

Lee R B and Ratcliffe R G 1983 Development of an aeration system for use in plant tissue NMR experiments. J. Exp. Bot. 34, 1213–1221.

Leonard R T and Hanson J B 1972 Increased membrane-bound ATPase activity accompanying development of enhanced solute uptake in washed corn root tissue. Plant Physiol. 49, 436–440.

Loughman B C, Ratcliffe R G and Schwabe J W R 1989 Galactose metabolism in *Zea mays* root tissues observed by $^{31}P$ NMR spectroscopy. Plant Sci. 59, 11–23.

Loughman B C and White P J 1984 The role of minor nutrients in the control of ion movement across membranes. *In* Membrane Transport in Plants. Eds. A Walker and J Cram. pp 501–502. Academia, Prague.

Macara I G, Kustin K and Cantley L C 1980 Glutathione reduces cytoplasmic vanadate. Mechanism and physiological implications. Biochim. Biophys. Acta 629, 95–106.

Perlin D S and Spanswick R M 1981 Characteristics of ATPase activity associated with corn leaf plasma membranes. Plant Physiol. 68, 521–526.

Peterson G L 1977 A simplification of the protein assay method of Lowry et al. which is more generally applicable. Anal. Biochem. 83, 346–356.

Peterson G L 1978 A simplified method for analysis of inorganic phosphate in the presence of interfering substances. Anal. Biochem. 84, 164–172.

Ross H H 1976 Theory and application of Ĉerenkov counting. *In* Liquid Scintillation Science and Technology. Eds. A A Noujaim, C Ediss and L I Wiebe. pp 79–91. Academic Press, New York.

Sze H and Hodges T K 1977 Selectivity of alkali cation influx across the plasma membrane of oat roots. Cation specificity of the plasma membrane ATPase. Plant Physiol. 59, 641–646.

*F. Baluška et al. (eds.), Structure and Function of Roots*, 129–132, 1995.


# Ion permeability of maize root membrane vesicles: Studies with light scattering

J. Sklenar and B.C. Loughman
*Institute of Botany, Slovak Academy of Sciences, Dubravska cesta 14, 842 23 Bratislava, Slovakia and Department of Plant Sciences, University of Oxford, South Parks Road, Oxford OX1 3RB, UK*

*Key words:* light scattering technique, membrane marker, membrane vesicles, permeability, phase partition

## Abstract

A possible modulation of permeabilities of membrane vesicles to anions and cations was explored by light scattering techniques, evaluated by measuring the capacity of the vesicles to shrink and swell in response to changes of the osmolarity of the incubation medium. Membrane fractions were obtained by phase partition. Purity was evaluated by detection and quantification of membrane enzyme markers: vanadate-sensitive ATPase for the plasma membrane, nitrate-sensitive ATPase for the tonoplast and azide-sensitive ATPase for mitochondria. Membrane vesicles (250 $\mu$g protein) were exposed to hypertonic solutions of salts (0.6 osmolar). Kinetics of the changes in apparent absorbance at 546 nm were observed by the addition of potassium, nitrate and chloride salts. The diffusion of ions into vesicles was induced by an osmotic gradient across the membrane and brought about volume changes of vesicles. Upon addition of vesicles to the different solutions the following ion permselectivity sequences were observed : P$NO_3^-$ > P$Cl^-$ > P$SO_4^{2-}$ and P$K^+$ $\geq$ P$Na^+$ > P$NH_4^+$.

*Abbreviations:* ATP – adenosine 5′-triphosphate, EDTA – ethylene diaminetetraacetic acid, Tris-Mes – (Tris[hydroxymethyl]aminomethane, Mes – (2-[N-Morpholino]ethanesulfonic acid), PEG – polyethylene glycol

## Introduction

Preparations of isolated vesicles represent an ideal system for the study of transport processes across native membranes. It becomes possible to characterize single transport systems without the complications of transport in the intact cell. Solute transport can be studied in the absence of metabolic transformation or compartmentation (Briskin, 1990). A number of investigators have examined transport processes at the plasma membrane (Giannini et al., 1987a, b, 1988 b; De Michelis and Spanswick, 1986); tonoplast (Blumwald and Poole, 1985; Briskin et al., 1985; Poole et al., 1984; Schumaker and Sze 1985), and endoplasmic reticulum (Buckhout 1983, 1984; Bush and Sze, 1986; Giannini et al., 1988a) vesicles. Major progress in plasma membrane-associated transport processes by using isolated vesicles has only been achieved within the last few years.

The purpose of the present study was to investigate the relative permeabilities of the plasma membrane vesicles to various ions by spectrophotometry of salt-induced osmotic swelling. For this reason, plasma membrane vesicles were purified by partitioning in aqueous two- phase systems. It has been shown that the plasma membrane has a high partition in phase systems where chloroplast, mitochondria and endoplasmic reticulum have a low partition (Albertsson et al., 1981 ). In aqueous two-phase systems the mixed population of membrane vesicles will partition between the two phases, partly according to their surface charge and partly according to hydrophilic/hydrophobic properties of their outer surface (Albertsson et al., 1981; Larsson and Andersson, 1979).

## Materials and methods

*Plant material.* Maize seeds (LG 11 ) were germinated in rolls of filter paper soaked in 0.1 m*M* $CaSO_4$ solution in pots at 25°C in the dark. Root material was taken for experiments after three days; whole primary roots were cut from the seedlings and collected in cold distilled water. All the following operations were performed at 4°C.

*Preparation of plasma membrane vesicle.* The primary roots were homogenized with a chilled mortar and pestle in 0.25 *M* sucrose, 3 m*M* EDTA, 2.5 m*M* DTT, and 25 m*M* Tris-Mes (pH 7.7). The homogenate was filtered through four layers of cheesecloth and centrifuged at 10,000 *g* for 15 min in the large swinging rotor TH-641 of a Sorvall centrifuge (OTD B). The supernatant was centrifuged at 55,000 *g* for 30 min and the pellet was suspended in suspension buffer (0.3 *M* sucrose, 1 m*M* DTT, 1m*M* Tris-Mes, pH 7.2), frozen in liquid N2 and stored at -70°C. Before use the suspension was diluted in 5 m*M* potassium phosphate buffer (pH 7.8) containing 0.25 *M* sucrose to a final weight of 6g and subjected to phase partition in an aqueous polymer two-phase system. The 36g phase mixture contained a final concentration of 6.3% (w/w) Dextran T500, 6.3% (w/w) PEG 4000, 0.25 *M* sucrose, 5 m*M* K-phosphate (pH 7.8) and the pellet. It was equally divided to six tubes, mixed by 20–30 inversions of the tube and centrifuged in a swinging bucket rotor at 1,000 *g* for 5 min. 90% of the upper phase was removed with a Pasteur pipette, diluted with an equal volume of the suspension buffer and centrifuged at 100,000 *g* for 1 h to produce the plasma membrane vesicles. Similarly, 90% of the lower phase was removed, diluted with the suspension buffer and centrifuged at 100,000 *g* for 1h.

*Shrinking and swelling.* At zero time the fraction of upper phase (250$\mu$g protein) was diluted with 1.0 mL of the specified solutions and the light scattering recorded every 10 sec. at 21°C as the absorbance at 546 nm ($A_{546}$) with a Beckman DU-70 spectrophotometer. A decrease in $A_{546}$ reflects swelling of the membrane vesicles.

*Enzyme assay.* In order to identify the plasma membrane and assess contamination by other membrane components during the membrane isolation procedure, the following markers were evaluated: Vanadate-sensitive plasmalemma ATPase was assayed in 30 m*M* Tris-Mes, pH 6.5, containing 3 m*M* ATP, 3 m*M* $MgSO_4$, 25 m*M* $K_2$ $SO_4$ and in the presence and absence of 5 m*M* $Na_3VO_4$. 0.02% (w/v) Triton X-100 was added to disrupt right-side-out membrane vesicles such that the vesicles became permeable to ATP without additional stimulatory or inhibitory effect on the ATPase activity. Nitrate-sensitive tonoplast ATPase was assayed in the incubation medium (3 m*M* ATP, 3m*M* $MgSO_4$, 30 m*M* Tris-Mes, pH 8.0, and 0.02% (w/v) Triton X-100) in the presence of 50 m*M* KC1 or $KNO_3$. Azide-sensitive mitochondrial ATPase was measured in 30 m*M* Tris-Mes, pH 8.5, containing 3 m*M* ATP, 3 m*M* $MgSO_4$, 50 m*M* KCl, and 0.02% (w/v) Triton X-100, in the presence or absence of 1 m*M* $NaNO_3$. ATPase activity was determined at 38°C for 30 min. in a reaction volume of 1.0 mL by a modified Peterson (1978) procedure. Proteins were determined by the method of Bradford (1976).

## Results and discussion

*Characterization of the membrane vesicles*

Plasmalemma vesicles were purified by aqueous polymer two-phase partitioning. Table 1 shows the specific activities of the marker enzymes in the microsomal fraction and in the upper phase (U) and the lower phase (L) after phase partition. Low activities of the mitochondrial and tonoplast markers in the U phase were found, resulting in a higher purity of the plasma membrane fraction. These results are in agreement with Briskin et al. (1987) who have shown that plasma membrane markers are distributed almost exclusively in the upper phase while the markers for intracellular membranes are found mainly in the lower phase. Hallberg and Larsson (1981) have shown that intact protoplasts have a high affinity for the polyethylene glycol-rich upper phase. Our plasma membrane vesicles which were partitioned in the upper phase are expected to resemble in properties these protoplasts and represent therefore the right- side-out population, if true, the activity found in the lower phase might represent the inside-out vesicles which had to be excluded from upper phase because of their different surface properties. The presence of the inside-out vesicles population in the lower phase might be the result of thawing the vesicles after their storage at -70°C (Palmgren et al., 1990).

*Permeability properties of the membrane vesicles*

In this study we have examined the movement of cations and anions across the plasma membrane vesicles of maize roots by the use of light scattering techniques. The first requirement of a membrane fraction to be used for the study of transport is a high proportion of osmotically intact vesicles. This can be evaluated

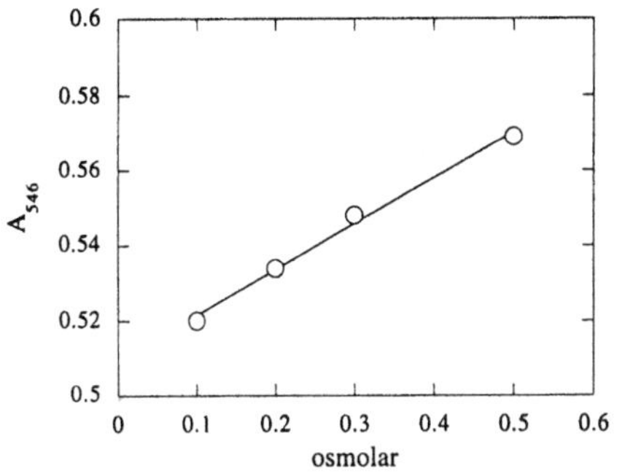


*Fig. 1.* Effect of the osmolarity of the incubation medium on the volume of plasma membrane vesicles. The vesicles (250 $\mu$g protein) were diluted with 1.0 mL sorbitol solutions to give the specified concentrations.

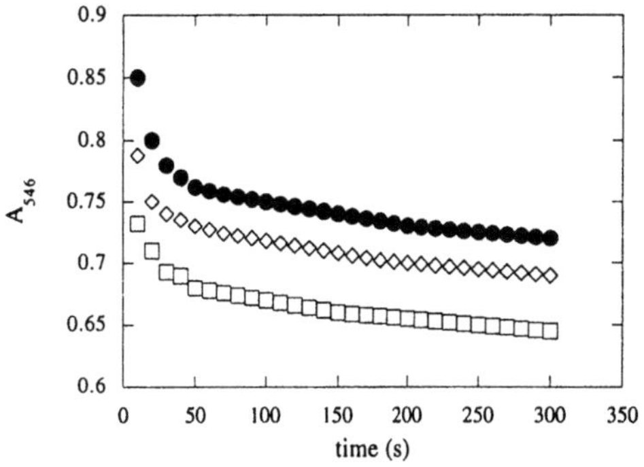


*Fig. 3.* Kinetics of the changes in apparent absorbance at 546 nm induced by the addition of nitrate salts (1.0 mL of 0.6 osmolar) to the suspension of plasma membrane vesicles (250 $\mu$g protein): □$KNO_3$, ○ $NH_4NO_3$, ◇ $NaNO_3$.

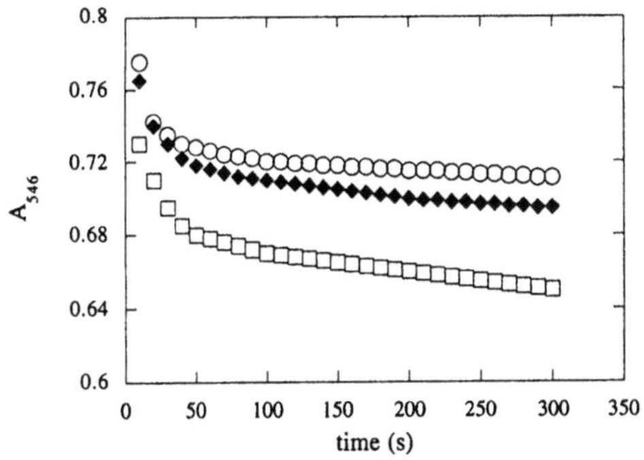


*Fig. 2.* Kinetics of the changes in apparent absorbance at 546 nm introduced by the addition of potassium salts (1.0 mL of 0.6 osmolar) to the suspension of plasma membrane vesicles (250 $\mu$g protein): ◆ KCl, □ $KNO_3$, ○ $K_2SO_4$.

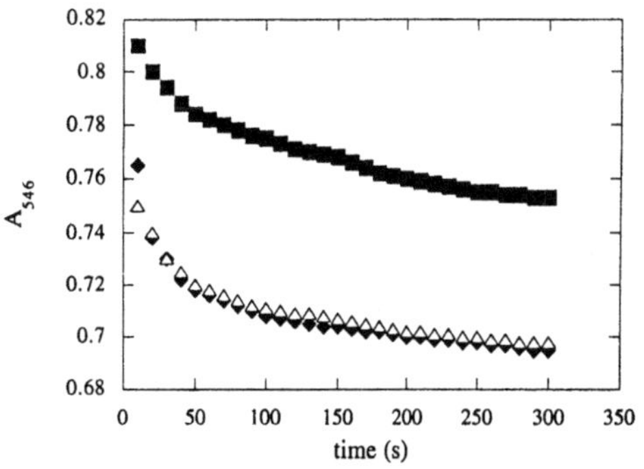


*Fig. 4.* Kinetics of the changes in apparent absorbance at 546 nm induced by the addition of chloride salts (1.0 mL of 0.6 osmolar) to suspension of plasma membrane vesicles (250 $\mu$g protein): ◆ KCl, ■ $NH_4Cl$, △ NaCl.

by measuring the capacity of the vesicles to shrink and swell in response to changes of the osmolarity of the incubation medium (Kaback, 1974). The effect of the osmolarity of the incubation medium on vesicle volume is shown in Figure 1. In a hypertonic solution of sorbitol, which is thought to be nonpenetrating, water rapidly leaves the intravesicular compartment and the vesicles shrink, which results in increased absorbance. In a hypotonic solution, they swell, so that the absorbance falls. Such results indicate that intact vesicles behave in general as osmometers. Use of the osmotic qualities of vesicles is a very useful means of measuring permeability of membranes to solutes (De Michelis et al., 1981; Lehninger, 1964; Stadelman, 1977; Vaandrager et al., 1986). Membrane vesicles were exposed to hypertonic solutions of specified salts (Figs. 2,3,4). When a concentrated solution of salt was added to the vesicle suspension, absorbance decreased due to the dilution of the suspension medium (an instantaneous process), but the vesicles began to lose water to adapt their internal osmotic pressure, resulting in increased absorbance and then, if and when the added solute penetrated the vesicles, their volume increased and light absorbance fell. The vesicles swelled very quickly (within a few seconds) and subsequently shrank as solute molecules entered the vesicles and water, in turn, entered until the volume returned to normal. This slower reswelling phase is dependent on solute influx into the intravesicular space. The rate of the reswelling phase in response to a salt gradient is a function of the total permeability of the vesicle membrane to the least permeant cation or anion (Vaandrager,1986). The recording traces in Figures 2,3,4 showed that the permeability of vesicles to the various ions is different. The results of the osmotic experiments, presented in Figure 2 suggest that $NO_3^-$ and $Cl^-$ are more easily permeating than $SO_4^{2-}$ i.e. the following anion permselectivity sequence was found:

*Table 1.* Comparison of the specific activities of the marker enzymes($\mu$moles Pi/mg protein/h) in the microsomal fraction and in the upper phase (U) and the lower phase (L) after phase partition

| | Microsomal fraction | Phase partition U | Phase partition L |
|---|---|---|---|
| $VO_4$-ATPase | 30.1 ± 1.4 | 31.2 ± 0.2 | 21.7 ± 2.4 |
| $NO_3$-ATPase | 2.5 ± 0.2 | 0.4 ± 0.01 | 2.0 ± 0.01 |
| Azide-ATPase | 2.5 ± 0.04 | 0.2 ± 0.02 | 1.1 ± 0.04 |

$PNO_3^- > PCl^- > PSO_4^{2-}$. In further experiments, salts of the more permeant anions were used with different cations (Figs. 3,4). Comparison of the reswelling curves in Figure 3 with curves in Figure 4 showed that the cation permselectivity sequence was similar i.e. $PK^+ > PNa^+ > PNH_4^+$ for nitrate form and $PK^+ = PNa^+ > PNH_4^+$ for chloride form. There are similarities in influx of $K^+$ and $Na^+$ so that the final sequence can be: $PK^+ \geq PNa^+ > PNH_4^+$.

## References


Albertsson P A, Andersson B, Akerlund H E and Larsson C 1981 Phase partition *In* Methods of Biochemical Analysis, Vol. 28, Ed. D Glick. Wiley-Interscience, New York.

Blumwald E and Poole R J 1985 Na/H-antiport in isolated tonoplast vesicles from storage tissues of *Beta vulgaris*. Plant Physiol. 78, 727–731.

Bradford M M 1976 A rapid and sensitive method for the quantitation of microgram quantities of protein utilizing the principle of protein-dye binding. Anal. Biochem. 72, 248–254.

Briskin D P, Thornley W R and Wyse R E 1985 Membrane transport in isolated vesicles from sugarbeet taproot, II. Evidence for a sucrose/H-antiport. Plant Physiol. 78, 871–875.

Briskin D P, Leonard R T and Hodges T K 1987 Isolation of the plasma membrane. Methods Enzymol. 148, 542–558.

Briskin D P 1990 Transport in plasma membrane vesicles – approaches and perspectives. *In* The Plant Plasma Membrane Structure, Function and Molecular Biology. Eds. C Larsson and I M Moller. pp 154–181. Springer Verlag, Berlin.

Buckhout T J 1983 ATP - dependent $Ca^{2+}$- transport in endoplasmatic reticulum isolated from roots of *Lepidium sativum* L. Planta 159, 84–90.

Buckhout T J 1984 Characterization of $Ca^{2+}$-transport in purified endoplasmatic reticulum membranes from *Lepidium sativum* L. roots. Plant Physiol. 76, 962–967.

Bush D R and Sze H 1986 Calcium transport in tonoplast and endoplasmatic reticulum vesicles isolated from cultured carrot cells. Plant Physiol. 80, 549–555.

De Michelis M I, Pugliarello M C and Rasi-Caldogno F 1981 Osmotic behaviour and permeability properties of vesicles in microsomal preparations from *Pea* internodes. J. Exp. Bot. 32, 293–302.

De Michelis M I and Spanswick R M 1986 $H^+$-pumping driven by the vanadate-ATPase in membrane vesicles from corn roots. Plant Physiol. 81, 542–547.

Giannini J L, Gildensoph L H and Briskin D P 1987a Selective production of sealed plasma membrane vesicles from red beet storage tissue. Arch. Biochem. Biophys. 254, 621–630.

Giannini J L, Gildensoph L H, Ruiz-Cristin J and Briskin D P 1987 Isolation and characterization of sealed plasma membrane vesicles from red beet storage tissue. Plant Physiol. Suppl. 83, 55.

Giannini J L, Gildensoph L H, Reynolds-Niesman I and Briskin D P 1988a Calcium transport in sealed vesicles from red beet storage tissue. I. Characterization of a $Ca^{2+}$-pumping ATPase associated with the endoplasmatic reticulum. Plant Physiol. 85, 1129–1136.

Giannini J L, Ruiz-Cristin J and Briskin D P 1988b Calcium transport in sealed vesicles from red beet storage tissue. II. Characterization of $^{45}Ca^{2+}$ -uptake into plasma membrane vesicles. Plant Physiol. 85, 1137–1142.

Hallberg M and Larsson C 1981 Compartmentation and export of $^{14}CO_2$ fixation products in mesophyll protoplast from the $C_4$ -plant *Digitaria sanguinalis*. Arch. Biochem. Biophys. 208, 121–130.

Kaback H R 1974 Transport studies in bacterial membrane vesicles. Science 186, 882–892.

Larsson C and Andersson B 1979 Two phase methods for chloroplasts chloroplast elements and mitochondria. *In* Plant Organelles, Methodological Surveys (B), Vol. 9. Ed. E Reid. pp 37–46. Ellis Horwood Ltd, Chichester.

Lehninger A L 1964 The mitochondrion - molecular basis of structure and function. pp. 84–204. W A Benjamin Press, New York.

Palmgren M G, Askerlund P, Fredrikson K, Widell S, Sommarin M and Larsson C 1990 Sealed inside-out and right-side-out plasma membrane vesicles. Plant Physiol. 92, 871–880.

Peterson G L 1978 A simplified method for analysis of inorganic phosphate in the presence of interfering substances. Anal. Biochem. 84, 164–172.

Poole R J, Briskin D P, Kratky Z and Johnstone R M 1984 Density gradient localization of plasma membrane and tonoplast from storage tissue of growing and dormant red beet: characterization of proton transport and ATPase in tonoplast vesicles. Plant Physiol. 74, 549–556.

Schumaker K S and Sze H 1985 A $Ca^{2+}/H^+$ antiport system driven by the proton electrochemical gradient of a tonoplast $H^+$-ATPase from oat roots. Plant Physiol. 79, 1111-1117.

Stadelmann E J 1977 Passive transport parameters of plant cell membranes. *In* Regulation of Cell Membrane Activities in Plants, Eds. E Marre and O Ciferri. Elsevier / North Holland. pp. 3–18.

Vaandrager A B, Ploemacher M C and de Jonge H R 1986 Modulation of salt permeabilities of intestinal brush-border membrane vesicles by micromolar levels of internal calcium. Biochim. Biophys. Acta 856, 325–336.

*F. Baluška et al. (eds.), Structure and Function of Roots*, 133–140, 1995.


# The role of plasma membrane-bound activities in nitrate transport into sealed plasma membrane vesicles from *Cucumis sativus* L. roots

G. Kłobus
*Institute of Botany, Wroclaw University, Kanonia 6/8, 50–328 Wroclaw, Poland*

*Key words:* $H^+$-ATPase, nitrate transport, oxidoreductase, plasma membrane

## Abstract

Nitrate uptake was examined on sealed plasma membrane vesicles isolated from cucumber (*Cucumis sativus* L.) roots by two phase system. Latency and lack of trypsin inactivation of $H^+$-ATPase proved the right-side out orientation of plasma membrane vesicles. NADH-loaded plasma membrane vesicles reduced the external ferricyanide. Triton X–100 stimulation of the ferricyanide reduction by vesicles loaded with NADH indicated that oxidoreductase, which catalyzes this reduction is a transmembrane protein, with reactive sites for NADH and ferricyanide on inner and outer sides of membrane, respectively. Nitrate was transported into sealed, right-side out plasma membrane vesicles with imposed proton gradient. The nitrate uptake increased when plasmalemma vesicles were preloaded with ATP. In the presence of vanadate or dicyclohexylcarbodiimide $NO_3^-$ transport into ATP-loaded vesicles was inhibited. Stimulation of nitrate transport was also observed in experiments with NADH-ATP-loaded vesicles but only when ferricyanide was added externally. Addition of quinacrine, an effective inhibitor of plasmalemma oxidoreductase in cucumber roots, to the external solution decreased nitrate uptake by vesicles with NADH and ATP inside. Results presented here are consistent with the hypothesis that proton-nitrate symport is involved in active nitrate transport in plant cells. Moreover, the uptake of nitrate by plasma membrane vesicles is related not only to the plasma membrane ATPase but also to the oxidoreductase activity.

*Abbreviations:* BSA – bovine albumin serum, CCCP – carbonyl cyanide m-chlorophenylhydrazone, DCCD – N,N′dicyclohexylcarbodiimide, DTT-dithiotreitol, EDTA – ethylenediamine tetraacetic acid, PM – plasma membrane, PMSF – phenylmethanesulfonyl fluoride, PVP – polyvinylpyrrolidine

## Introduction

Most nitrate transport investigations have been done with intact plants or suspension cultured cells (Aslam and Oaks, 1975; Dean-Drummond and Glass, 1983; Glass et al., 1985; Ingemarson et al., 1987; Kłobus, 1990; Ward et al., 1989). Results from these studies have lead to a general conclusion that uptake of nitrate is mediated by specific carrier proteins (Jackson et al., 1973; Kłobus et al., 1988) induced by external nitrate (Jackson et al., 1973; Chantarotwong et al., 1976; Goyal and Huffaker, 1986; MacKown and McClure, 1988) requires a significant input of energy (Clarkson, 1986; Glass, 1988). Several lines of evidence have linked nitrate active transport to the proton electrochemical difference generated by the plasma membrane proton pumping ATP-ase (Novacky et al., 1978; Ullrich and Novacky, 1981; Ullrich, 1987; McClure et al., 1990). The alkalinization of the external medium and pH-dependent depolorization of the plasma membrane observed during the absorption of $NO_3^-$ by plants were interpreted in terms of $H^+/NO_3^-$ symport (McClure et al., 1990; Ullrich and Novacky, 1981). On the other hand, results from studies of nitrate uptake determined with $H^+$-ATPase specific inhibitors (diethylstilbestrol, dicyclohexylcarbodiimide and vanadate) indicated that nitrate absorption by plants was not directly coupled to a ATP-dependent proton pump (Deane-Drummond, 1984, 1986; Kłobus, 1990; Thibaud and Grignon, 1981). Our recent investigations with quinacrine, an inhibitor of plasma membrane NADH-ferricyanide oxidoreductase in cucumber roots, have suggested the dependency of nitrate uptake also on the plasma membrane redox activity (Kłobus and Buczek, 1992). A similar suggestion for the diatom *Thalassiosira* was made earlier by Jones and Morel (1988). All above investigations were conducted with intact tissues. In this experimental system the inhibitor effects on nitrate transport are complicated because of the potential for multiple sites of inhibitions. Thus, a simpler system would be more suitable to examine nitrate transport at the membrane level. In the work presented here, we used highly purified plasma membrane vesicles from cucumber roots to study the nitrate transport. Membrane vesicles not only eliminated the experimental problems associated with intact tissues but also allow us to manipulate the composition of both intra- and extravesicular solutions.

## Material and methods

### *Plant material*

Cucumber (*Cucumis sativus* L. var. Wisconsin) seeds were germinated 48 h in darkness at 25°C and than grown hydroponically as described previously (Kłobus, 1990). After germination seedlings were transferred to a nitrate free medium for three days and than to a nitrate containing nutrient solution for one day. The seedlings were grown under 16 h photoperiod (10.3 $Wm^{-2}$) at 25°C by a day and 22°C by a night.

### *Microsome isolation*

Lots of 120 g of cucumber roots were homogenized with 240 $cm^3$ of cold 50 m*M* Hepes – KOH (pH 8.0), 330 m*M* sorbitol, 0.1% BSA, 0.1 m*M* EDTA – $Na_2$, 1 m*M* DTT, 0.5 m*M* PMSF and 0.05% (w/v) PVP in a chilled mortar and pestle. During preparation of plasma membrane vesicles loaded with ATP, NADH or both ATP and NADH, the homogenization medium contained also 2 m*M* ATP or 2 m*M* NADH, or both chemicals. The homogenate was filtered through Miracloth, centrifuged at 15.000 *g* for 10 min and than at 100.000 *g* for 30 min.

### *Preparation of unloaded and loaded right-side out plasma membrane vesicles*

Plasma membrane (PM) vesicles were purified from 15.000–100.000 g microsomal fraction by partitioning in an aqueous polymer two-phase system as described by Larsson (1985). The microsomes were diluted with 5 m*M* K-phosphate (pH 8.0), 5 m*M* KCl, 330 m*M* sorbitol, 0.1 m*M* EDTA and 1 m*M* DTT (unloaded vesicles), 2 m*M* ATP (ATP-loaded vesicles) or 2 m*M* NADH (NADH-loaded vesicles), or both ATP and NADH (ATP, NADH-loaded vesicles). A 24 g phase system containing 6.2% (w/w) Dextran T500, 6.2 (w/w) polyethylene glycol 3350, 330 m*M* sorbitol, 5 m*M* KCl, 5 m*M* K-phosphate (pH 8.0) were used. The contents of tubes were mixed vigorously and centrifuged in a swinging-bucket rotor at 500 *g* for 3 min. The upper phase of system was collected and partitioned again on a fresh lower phase. The final upper phase containing the plasma membranes were diluted several times with 5 m*M* K-phosphate (pH 8.0), 1 m*M* DTT, 0.1 m*M* EDTA and 330 m*M* sorbitol and centrifuged at 120.000 *g* for 30 min. The sedimented plasma membrane pellet was resuspended in 5 m*M* K-phosphate (pH 8.0), 330 m*M* sorbitol, 1 m*M* DTT and 0.1 m*M* EDTA to a final protein concentration of about 1 mg $cm^3$.

### *Enzyme assay*

Specific ATPase activities were determined at 37°C for 10 min according to Gallagher and Leonard (1982) as modified by Sze (1985). The reaction mixture for nitrate-sensitive ATPase contained 20 $\mu$g protein (PM), 33 m*M* Tris-Mes (pH 7.5), 3 m*M* ATP, 2.5 m*M* $MgSO_4$, 50 m*M* KCl, 0.1 m*M* $NH_4$-molybdate, 1 m*M* $NaN_3$, 200 $\mu M$ $Na_3VO_4$, 0.012% Triton X–100 and ± 50 m*M* $KNO_3$. Activity of vanadate-sensitive ATPase was measured in 33 m*M* Tris-Mes (pH 7.5), 3 m*M* ATP, 2.5 m*M* $MgSO_4$, 50 m*M* KCl, 0.1 m*M* $NH_4$-molybdate, 1 m*M* $NaN_3$, 50 m*M* $KNO_3$, 0.012% Triton X–100, 20 $\mu$g protein (PM) and ± 200 $\mu M$ $Na_3VO_4$. In experiments on trypsin digestion of vanadate-sensitive ATPase, plasma membrane vesicles were preincubated 5 min at 38°C in 33 m*M* Tris-Mes (pH 7.5), 330 m*M* sorbitol without or with 5 $\mu M$ trypsin. $P_i$ released during reaction was determined according to Ames (1966) with 0.1% (w/v) sodium dodecyl sulphate included to prevent precipitation (Dulley, 1975). Latency of vanadate-sensitive ATPase was calculated from a difference between the activities determined in the presence and absence of 0.012% Triton X–100.

Antimycine A-insensitive NADH-dependent cyt c reductase was assayed according to Lord et al. (1973). Cyt c oxidase was measured by the method of Hodges and Leonard (1974) and nitrate reductase was assayed according to Hucklesby et al. (1972). Phosphoenolopyruvate carboxylase was determined as described by Spalding and Edwards (1978).

Reduction of ferricyanide by NADH-loaded plasma membrane vesicles was measured spectrophotometrically as the change in absorbance at 420 ($A_{420}$). The assay was run at 25°C in 10 m*M* Tris-Mes (pH 7.5), 330 m*M* sorbitol, 0.8 m*M* $K_3Fe(CN)_6$, 20 $\mu$g protein (plasma membranes) in the absence or presence 0.012% Triton X–100. Correction for nonenzymatic reduction of ferricyanide was made. The rate of ferricyanide reduction was calculated using an excitation coefficient of 1.02 mM $^{-1}$cm$^{-1}$.

Protein was measured according to Bradford (1976) as modified by Ward et al. (1988).

*Table 1.* Marker enzymes activities in microsomes and plasma membrane fractions isolated from *Cucumis sativus* roots. The fractions were isolated as described in Material and methods. The values presented are averages with SE of three replications

| Enzyme activity | Microsomes | Plasma membranes |
|---|---|---|
| Cyt c oxidase[a] | 25.0 ± 0.1 | nd |
| Nitrite reductase[b] | 12.8 ± 0.005 | nd |
| NADH-cyt c Reductase antimycine Insensitives[a] | 1.8 ± 0.03 | nd |
| Latent UDPase[c] | 86.4 ± 0.5 | nd |
| $VO_4$-ATPase[c] | 38.4 ± 0.9 | 104.1 ± 3.2 |
| $NO_3$-ATPase[c] | 2.6 ± 0.03 | 0.08 ± 0.01 |
| PEP-carboxylase[d] | 0.2 ± 0.005 | nd |

[a] nmol cyt c × mg$^{-1}$ protein × min$^{-1}$.
[b] $\mu$mol $NO_2^-$ × mg$^{-1}$ protein × h$^{-1}$.
[c] $\mu$mol $P_i$ × mg$^{-1}$ protein × h$^{-1}$.
[d] nmol PEP × mg$^{-1}$ protein × h$^{-1}$.
nd – not detectable.

*Nitrate uptake by plasma membrane vesicles*

Nitrate uptake was measured at 22°C in 1.5 cm$^3$ total reaction volume. The uptake solution contained 5 m*M* K-phosphate (pH 6.0), 330 m*M* sorbitol, 50 $\mu$*M* $KNO_3$ and 100 $\mu$L resuspended PM vesicles (about 100 $\mu$g protein). At predetermined time points samples were filtrated rapidly under vacuum using 0.45 $\mu$m Millipore filters. Nitrate concentration in an uptake solution was determined by high-performance liquid chromatography and spectrophotometrical detection, according to Thayer and Huffaker (1980). Nitrate uptake by plasma membrane vesicles was calculated from nitrate depletion in the uptake solution at the time specified in the legends of figures and tables.

## Results

The purity of plasma membrane vesicles separated from cucumber roots by an aqueous two-phase partitioning method was estimated on the base of marker enzyme activities (Table 1). The activity of a tonoplast marker (nitrate-sensitive ATPase) was significantly lowered in the plasma membrane fraction in comparison with the microsomes. Cyt c oxidase activity (marker for mitochondria), nitrite reductase (marker for plastid), antimycine A-insensitive NADH-cyt c reductase (endoplasmatic reticulum marker) and latent UDPase activity (Golgi apparatus marker) were below the level of detection. Also the activity of phosphoenol pyruvate carboxylase (cytoplasma marker) was not detectable in the plasmalemma fraction. On the other hand, the PM fraction was three-fold enriched in the vanadate-sensitive ATPase, a specific marker for plasma membranes. The results from Table 1 indicate, that the membrane vesicles used in this experiments were highly enriched in plasma membranes.

The sidedness of the plasma membrane vesicles obtained with the two-phase system was determined by two approaches: the latency of vanadate-sensitive ATPase and trypsin digestion of the enzyme (Larsson et al., 1984; Widell and Larsson, 1990). The highest activity of vanadate-sensitive ATPase was observed in the presence of 0.012% Triton X–100 (Table 2). The enzyme activity measured without detergent was only 3% of ATPase activity determined in the presence of Triton. Trypsin preincubation of the plasma membranes for 5 min. at 38°C in the presence of Triton X–100 abolished almost completely the activity of vanadate-sensitive ATPase. Although, when the plasma membrane vesicles were preincubated with trypsin and then assayed in the presence of Triton X–100 the ATPase activity was recovered (Table 2). Thus, our plasma membrane vesicles were predominantly right-side out orientated.

Ferricyanide reduction by unloaded right-side out plasma membrane vesicles determined in the pres-

*Table 2.* Effect of Triton X-100 and trypsin digestion on vanadate – sensitive ATPase activity of plasma membrane vesicles isolated by partitioning of cucumber root microsomes in an aqueous polymer two-phase system

| Treatment of plasma membrane vesicles | | $VO_4$-sensitive ATPase activity |
|---|---|---|
| Preincubation medium | Incubation medium | $\mu$mol $P_i$ × $mg^{-1}$ protein × $h^{-1}$ |
| none | none | 2.3 ± 0.003 |
| none | Triton X-100 | 75.5 ± 1.3 |
| trypsin, Triton X-100 | none | 6.04 ± 0.03 |
| trypsin | Triton X-100 | 61.9 ± 1 |

The plasmalemma vesicles were preincubated in 33 m$M$ Tris-Mes (pH 7.5) and 330 m$M$ sorbitol without or with 5 $\mu M$ trypsin at 38°C in the absence or presence Triton X-100. Activity of $VO_4$-sensitive ATPase was determined in the reaction mixture without or with Triton X-100, as described in Material and methods. The values presented are averages with SE of three replications.

ence of both NADH and $K_3Fe(CN)_6$ in a reaction medium was traced (Table 3). Triton X–100 included to the reaction medium stimulated significantly the reduction of ferricyanide by plasma membranes. Ferricyanide reduction was also observed during incubation of NADH-loaded vesicles. However, the rate of reduction was much lower as the reduction of ferricyanide by unloaded vesicles observed in the presence of NADH and Triton X–100 in the reaction medium. The results confirm our earlier suggestion (Kłobus et al., 1992) that oxidoreductase is a transplasmamembrane enzyme with reactive sites for NADH and ferricyanide on the inside and outside of the plasmalemma, respectively. Ferricyanide reduction by NADH-loaded vesicles was decreased by quinacrine.

Nitrate was transported into the plasma membrane vesicles when a transmembrane proton concentration gradient was imposed (Fig. 1). In the absence of a pH gradient or in the presence of CCCP (a protonophore) nitrate uptake by PM vesicles was significantly diminished.

Uptake of nitrate by plasma membrane vesicles was dependent on the presence of ATP inside the vesicles (Table 4). ATP-loaded PM vesicles without imposed pH gradient, rapidly absorbed $NO_3$ rapidly whereas unloaded vesicles did not. The most intensive nitrate uptake was observed in experiments with NADH-ATP-loaded vesicles, in the presence of ferricyanide in the uptake solution (Table 4). There was no absorption of nitrate by plasma membrane vesicles loaded only with NADH (data not shown).

Vanadate and DCCD included in the uptake solution diminished the uptake of nitrate by ATP-loaded plasma membrane vesicles (Fig. 2A, B). The inhibition of $NO_3$ uptake by vanadate and DCCD was similar to the effect of both inhibitors on the plasma membrane-bound ATPase (Fig. 2A, B). Also quinacrine, an effective inhibitor of plasmalemma oxidoreductase in cucumber roots (Kłobus et al., 1992) present in the uptake solution decreased the uptake of nitrate by plasma membrane vesicles loaded with NADH and ATP (Fig. 3).

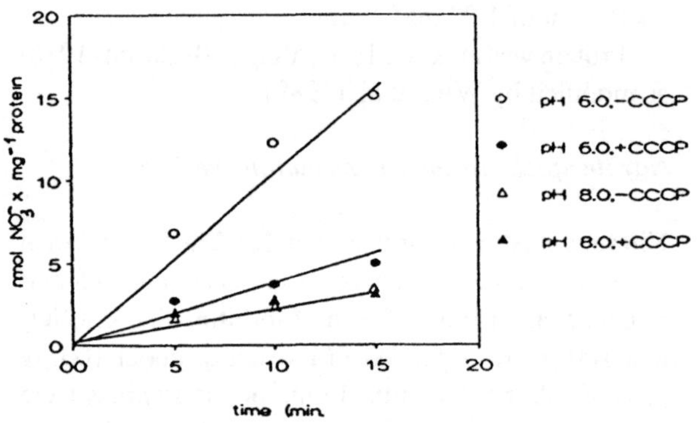


*Fig. 1.* Proton-gradient dependent $NO_3^-$ uptake by right-side out plasma membrane vesicles. Right-side out, ATP-loaded PM vesicles (pH 8.0) were incubated for 30 min at 22°C in uptake solution. The uptake solution included 50 $\mu M$ $KNO_3$, 330 m$M$ sorbitol, 5 $\mu M$ CCCP (●, ▲) and 5 m$M$ K-phosphate pH 6.0 (○, ●) or 8.0 (△, ▲). After 0, 5, 10 and 15 min samples were filtrated under vacuum using 0.45 $\mu$m Millipore filters and $NO_3^-$ concentrations were determined.

*Table 3.* Ferricyanide reduction by unloaded and NADH-loaded plasma membrane vesicles isolated by partitioning of cucumber root microsomes in an aqueous polymer two-phase system

| Plasma membrane vesicles | Reaction solution | Ferricyanide reduction ($\mu$mol $\times$ mg$^{-1}$ protein $\times$ min$^{-1}$) |
|---|---|---|
| unloaded | NADH, $K_3Fe(CN)_6$ | 1.24 ± 0.01 |
| unloaded | NADH, $K_3Fe(CN)_6$, Triton | 22.40 ± 0.5 |
| NADH-loaded | $K_3Fe(CN)_6$ | 8.00 ± 0.5 |
| NADH-loaded | $K_3Fe(CN)_6$, quinacrine | 4.83 ± 0.5 |

Reductions of ferricyanide by NADH-loaded plasma membrane vesicles was measured spectrophotometrically as a change of absorbance at 420 ($A_{420}$) as described in Material and methods. The values presented are averages with SE of five replications.

*Table 4.* Nitrate uptake by unloaded, ATP-loaded and NADH-ATP-loaded plasma membrane vesicles isolated by partitioning of cucumber root microsomes in an aqueous polymer two-phase system

| Plasma membrane vesicles | Uptake solution | $NO_3^-$ uptake (nmol $\times$ mg$^{-1}$ protein) |
|---|---|---|
| unloaded | – | nd |
| unloaded | $K_3Fe(CN)_6$ | nd |
| ATP-loaded | – | 6.01 ± 0.01 |
| ATP-loaded | $K_3Fe(CN)_6$ | 6.19 ± 0.01 |
| NADH, ATP-loaded | – | 6.62 ± 0.01 |
| NADH, ATP-loaded | $K_3Fe(CN)_6$ | 8.62 ± 0.01 |

nd – not detectable. Unloaded, ATP-loaded and NADH-ATP-loaded plasma membrane vesicles (about 100 $\mu$g protein, pH 6.0) were incubated at 22°C in 50 $\mu M$ $KNO_3$, 5 m$M$ K-phosphate (pH 6.0) and 330 m$M$ sorbitol in the absence or presence of $K_3Fe(CN)_6$. After 0 and 15 min of incubation samples were filtrated under vacuum using 0.45 $\mu$m Millipore filters and nitrate concentration in filtrates was determined by a HPLC system. Nitrate uptake was calculated from a difference of nitrate concentration at time 0 and after 15 min of incubation. Each of the assay was repeated at least three times using three different preparations providing the various PM vesicles.

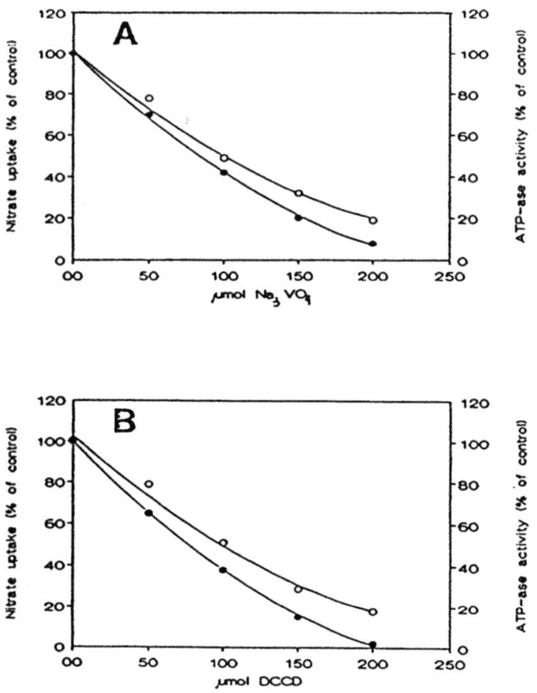


*Fig. 2.* Effect of $Na_3VO_4$ (**A**) and DCCD (**B**) present in the uptake solution of the $NO_3^-$ absorption by ATP-loaded vesicles (○) and activity on plasmalemma-associated ATPase (●). Right-side out ATP-loaded PM vesicles were included for 15 min at 22°C in the uptake solution, containing 50 $\mu M$ $KNO_3$, 330 m$M$ sorbitol, 5 m$M$ K-phosphate (pH 6.0) and 25, 50, 75, or 100 $\mu$moles $Na_3VO_4$ (A) or DCCD (B). Nitrate uptake by vesicles was determined as described in Figure 1. ATPase activity was determined as described in Material and methods without (100 per cent) or with inhibitors.

## Discussion

Aqueous two-phase partitioning was used to isolate plasma membrane preparations of high purity from different plant tissues (reviews, Larsson, 1985; Larsson et al., 1987). The technique of Larsson (1985) was optimized for cucumber roots. Highly enriched plasma membranes devoid of detectable mitochondria, plastids, Golgi membranes, endoplasmatic reticulum and cytoplasma and with traces of tonoplast were obtained (Table 1). It was shown by two independent approaches: latency and trypsin digestion of vanadate-sensitive ATPase, that about 97% of the plasma membrane vesicles was sealed as well as oriented right-side out (Table 2).

Sealed, right-side out PM vesicles were used to investigate the transport of nitrate. When these plasma membrane vesicles equilibrated at pH 8.0 were diluted

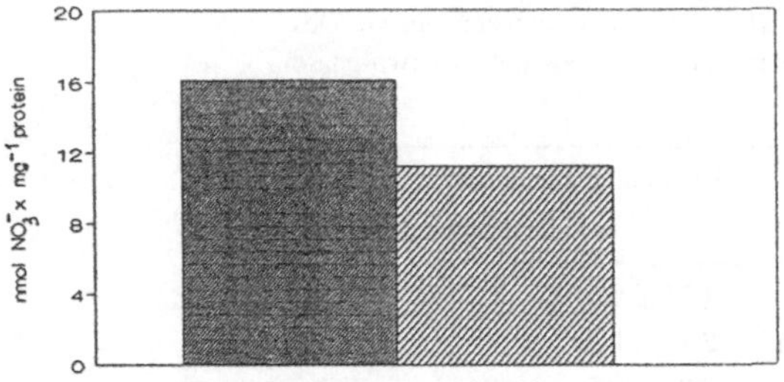


*Fig. 3.* Effect of quinacrine on the $NO_3^-$ absorption by NADH-ATP-loaded vesicles. NADH-ATP-loaded vesicles were incubated for 15 min at 22°C in the uptake solution containing 50 $\mu M$ $KNO_3$, 330 m$M$ sorbitol, 5 m$M$ K-phosphate (pH 6.0) without (■) or with (▨) 200 $\mu M$ quinacrine. Further details as in Figure 2.

into acid uptake solution (pH 6.0), a trans-membrane proton gradient was established, which was used for nitrate transport. In the absence of a proton concentration gradient (the lack of pH jump or in the presence of CCCP – a protonophore) nitrate uptake by plasma membrane vesicles decreased drastically. Thus, nitrate transport across the plasma membrane seems to be coupled to the transmembrane proton gradient. Similar results were obtained by Ruiz-Cristin and Briskin (1991) for transport of $^{36}ClO_3$, an analog of the nitrate ions, in plasma membrane vesicles isolated from maize roots.

In plant cells the proton gradient across the plasmalemma is generated by a proton-pumping ATPase. Since the hydrolysis site of the $H^+$-ATPase is on the cytoplasmic side of the membrane we preloaded a vesicles with ATP to generate the proton gradient instead of an imposed pH gradient. The uptake of nitrate by ATP-loaded vesicles was observed (Table 4). Moreover, ATP-dependent nitrate uptake was inhibited by DCCD and $Na_3VO_4$ in the same manner as the plasma membrane ATPase (Fig. 2A, B). These results provide unequivocal evidence that the proton-pumping ATPase generates a proton motive force used for the symport of $H^+$ and nitrate. It is consistent with the model of $H^+/NO_3^-$ symport proposed by Ullrich and Novacky (1981) and McClure et al. (1990) on the basis of experiments with intact plants.

The most intensive nitrate uptake was observed in NADH, ATP-loaded vesicles but only in the presence of ferricyanide in the uptake solution (Table 4). Activation of $NO_3^-$ uptake by intravesicular NADH and extravesicular ferricyanide was depressed by quinacrine. On the other hand, the plasma membrane vesicles loaded with NADH reduced external ferricyanide and this reduction was inhibited by quinacrine. Thus, we assume that the activation of $NO_3^-$ uptake by NADH and ferricyanide was linked to the activity of a plasma membrane oxidoreductase. During ferricyanide reduction by plant tissues an acidification of cytoplasma and plasma membrane depolarization was observed (Böttger, 1989; Döring et al., 1992; Kłobus et al., 1993; Rubinstein et al., 1992; Ullrich and Guern, 1990). Results of Romani et al. (1985), Hager and Moser (1985) Guern et al. (1988) and Rubinstein and Stern (1986), proved that $H^+$-ATPase was activated by both, the cytoplasma acidification and membrane depolarization. Another type of interaction between oxidoreductase activity and $H^+$-ATPase was suggested by Elzenga et al. (1989). The regulatory regions of $H^+$-ATPase contains SH groups (Serrano, 1989), so it is possible that the reduced pyridine nucleotide may activate this enzyme by affecting its redox state (Bienfait and Lüttge, 1988). These suggestion was supported by the fact that $H^+$-ATPase in plasma membranes from *Elodea* leaves was inhibited by $NAD^+$, NEM and glutathione and stimulated by NADH. Therefore, activity of plasma membrane ATPase is affected by plasmamembrane NADH-ferricyanide oxidoreductase. Taken together, these results proved that, although the ATP-fueled electrogenic pump is an exclusive mechanism for generation a proton motive force used to the symport of nitrate, also the plasma membrane oxidoreductase modulates the $NO_3^-$ transport.

**Acknowledgements**

This study was supported by grant from Polish Committee for Scientific Investigation (KBN, 4.43109102).

**References**

Ames B 1966 Assay of inorganic phosphate, total phosphate and phosphatases. Methods Enzymol. 8, 115–118.

Aslam M and Oaks A 1975 Effect of glucose on the induction of reductase in corn roots. Plant Physiol. 56, 634–639.

Bienfait F and Lüttge U 1988 Mechanisms in Fe-efficiency reactions of higher plants. J. Plant Nutr. 11, 605–629.

Böttger M 1989 Transmembrane electron transfer of NADH loaded right side out vesicles. *In* Plant Membrane Transport: The current position. Eds. J Dainty, M E De Michelis, E Marre and F Rasi-Caldogno. pp 55–60. Elsevier, Amsterdam.

Bradford M M 1976 A rapid sensitive method for quantitation of microgram quantities of protein utilizing the principles of protein dye binding. Ann. Biochem. 72, 248–254.

Chantarotwong W, Huffaker R C, Miller B L and Gransted R C 1976 *In vitro* nitrate reduction in relation to nitrate uptake, nitrate content and *in vivo* nitrate reductase activity in intact barley seedlings. Plant Physiol. 57, 519–522.

Clarkson D T 1986 Regulation of the absorption and release of nitrate by plant cells: A review of current ideas and methodology. *In* Fundamental, Ecological and Agricultural Aspects of Nitrogen Metabolism in Higher Plants. Eds. H Lambers, J J Neeteson and I Stulen. pp 3–27. Martinus Nijhoff, Boston.

Dean-Drummond C E and Glass A D M 1983 Short term studies of nitrate uptake into barley plants using ion-specific electrodes and $ClO_3^-$. I. Control of net uptake by $NO_3^-$ efflux. Plant Physiol. 73, 100–104.

Deane-Drummond C E 1984 Mechanism of nitrate uptake into *Chara corallina* cells: lack of evidence for obligatory coupling to proton pump and a new $NO_3^-/NO_3^-$ exchange model. Plant Cell. Environ. 7, 317–323.

Deane-Drummond C E 1986 Characterisation of $^{36}ClO_4^-$ influx into nitrate deficient $E_1$ *Pisum sativum* seedlings. Evidence for restricted "induction" by nitrate compared with wild type. Plant Sci. 46, 169–173.

Döring L, Lüthje S and Böttger M 1992 Modification of the activity of the plasma membrane redox system of *Zea mays* L. roots by vitamin $K_3$ and dicumarol. J. Exp. Bot. 43 (247), 175–181.

Dulley J R 1975 Determination of inorganic phosphate in the presence of detergents or protein. Anal. Biochem. 67, 91–96.

Elzenga J T, Staal M and Prins H B A 1989 ATPase activity of isolated plasmalemma vesicles of leaves of *Elodea* as affected by thiol reagents and NADH/NAD ratio. Physiol. Plant. 76, 379–385.

Gallagher S R and Leonard R T 1982 Effect of vanadate, molybdate and azide on membrane associated and soluble phosphatase activities of corn roots. Plant Physiol. 70, 1335–1340.

Glass A D M, Thompson R G and Bordeleau L 1985 Regulation of $NO_3^-$ influx in barley. Studies using $NO_3^-$. Plant Physiol. 77, 379–381

Glass A D M 1988 Nitrogen uptake by plant roots. ISI Atlas Science. Anim. Plant Sci. 1, 151–156.

Goyal S S and Huffaker R C 1986 A novel approach and fully automated, microcomputer based system to study kinetics of $NO_3^-$ and $NH_4^+$ transport simultaneously by intact wheat seedlings. Plant Cell Environ. 9, 209–221.

Guern J, Mathieu Y, Ephritikhine G, Ullrich-Eberius C I, Lüttge V, Marre M T and Marre E 1988 Intracellular pH modification linked to the activity of the ferricyanide driven activity of the plasmalemma redox system in *Elodea densa* leaves, *Acer pseudoplatanus*, and *Catharanthus roseus* cells. *In* Plasma Membrane Oxidoreductase in Control of Animal and Plant Growth. Eds. Crane F L, Morre D J and Low H. pp 412. Plenum Press, New York.

Hager A and Moser I 1985 Acetic acid esters and permeable weak acids induce active proton extrusion and extension growth of coleoptile segments by lowering the cytoplasmic pH. Planta 163, 391–400.

Hodges T K and Leonard R T 1974 Purification of a plasma membrane-bound adenosine triphosphatase from plant roots. Methods Enzymol. 32, 392–406.

Hucklesby D P, Dalling M J and Hageman R H 1972 Some properties of two forms of nitrate reductase from corn scutellum. Planta 104, 220–233.

Ingemarson B, Oscarson P, Ugglas M and Larsson C M 1987 Nitrogen utilization in *Lemna*. II. Studies of nitrate uptake using $^{15}NO_3^-$ Plant Physiol. 85, 856–859.

Jackson W A, Flesher D and Hageman R H 1973 Nitrate uptake by dark grown corn seedlings. Plant Physiol. 52, 120–127.

Jones G J and Morel F M M 1988 Plasmalemma redox activity in diatoma *Thalassiosira*. A possible role for nitrate reductase. Plant Physiol. 87, 143–147.

Kłobus G, Ward M R and Huffaker R C 1988 Characteristic of injury and recovery of net $NO_3^-$ transport of barley seedlings from treatments of NaCl. Plant Physiol. 87, 878–882.

Kłobus G 1990 Nitrate uptake and activity of plasmalemma associated ATPase activity in *Cucumis sativus* L. roots. Acta Physiol. Plant. 12, 225–231.

Kłobus G and Buczek J 1992 A possible role of plasmalemma redox activity in nitrate uptake by *Cucumis sativus* L. seedlings. Acta Physiol. Plant. 14, 41–47.

Kłobus G, Olszańska B and Buczek J 1993 Relationship of transplasma membrane redox activity to $H^+$ transport by sealed, in-side out plasma membrane vesicles. Acta Physiol. Plant (*In press*).

Larsson C 1985 Plasma membranes. *In* Modern Methods of Plant Analysis, N.S. Vol. 1, Cell Components. Eds. J F Jackson and H F Linskens. pp 85–104. Springer Verlag, Berlin, Heidelberg, New York.

Larsson C, Kjellbom P, Widell S and Lundborg T 1984 Sidedness of plant plasma membrane vesicles purified by partitioning in aqueous two-phase systems. FEBS Lett. 171, 271–276.

Larsson C, Widell S and Kjellbom P 1987 Preparation of high purity plasma membranes. Methods Enzymol. 148, 558–568.

Lord J M, Kagawa T, Morre T S and Beevers H 1973 Endoplasmic reticulum as the site of lectin formation in castor bean endosperm. J. Cell Biol. 57, 659–667.

McClure P R, Kochian L V, Spanswick R M and Shaff J 1990 Evidence for cotransport of nitrate and protons in maize roots. I. Effects of nitrate on membrane potential. Plant Physiol. 93, 281–289.

Novacky A, Fisher E, Ullrich-Eberius C I, Lüttge U and Ullrich W R 1978 Membrane potential changes during transport of glycine as a neutral amino acid and nitrate in *Lemna gibba* G1. FEBS Lett. 88, 264–267.

Romani G, Marré M T, Bellando M, Alloatti G and Marré E 1985 $H^+$ extrusion and potassuim uptake associated with potential hyperpolarization in maize roots treated with permeant weak acid. Plant Physiol. 79, 734–739.

Rubinstein B and Stern A I 1986 Relationship of transplasmamembrane redox activity to proton and solute transport by roots of *Zea mays*. Plant Physiol. 80, 805–811.

Rubinstein B, Stern A I and Chalmers J D C 1992 Relationship between electron transport across the plasmalemma and pH decrease in the bulk medium. Plant Physiol. 98, 988–994.

Serrano R 1989 Structure and function of plasma membrance ATPase. Ann. Rev. Plant Physiol. Plant Mol. Biol. 40, 61–94.

Spalding M H and Edwards G E 1978 Photosynthesis in enzymatically isolated leaf cells from C A M plant *Sedum telephium* L. Planta 141, 59–63.

Sze H 1985 $H^+$ translocating ATPase: Advance using membrane vesicles. Annu. Rev. Plant Physiol. 36, 175–208.

Thayer J R and Huffaker R C 1980 Determination of nitrate and nitrite by high-pressure liquid chromatography: Comparison with other methods for nitrate determination. Anal. Biochem. 102, 110–119.

Thibaud J B and Grignon C 1981 Electrogenic $NO_3^- - OH^-$ cotransport in corn roots. *In* Plasmalemma and Tonoplast: Their Function in Plant Cells. Eds. D Marre, E Marre, and R Hertel. Proceedings of International Workshop on Plasmalemma and Tonoplast of Plant Cell hold in Strassburg. pp 341–347. Elsevier Biomedical Press, Amsterdam, New York, Strassburg.

Ullrich W R and Novacky A 1981 Nitrate dependent membrane potential changes and their induction in *Lemna gibba*. Plant Sci. Lett. 22, 211–217.

Ullrich C I 1987 Nitrate and ammonium uptake in green algae and higher plants. Mechanism and relationship with nitrate metabolism. *In* Inorganic Nitrogen Metabolism. Eds. W R Ullrich, P J Paricio, P J Syrret and F Castillo. pp 32–38. Springer-Verlag, Berlin.

Ullrich C I and Guern J 1990 Ion fluxes and pH changes induced by transplasmalemma electron transfer and fusicoccin in *Lemna gibba*. Planta 180, 390.

Widell S and Larsson C 1990 A critical evaluation of markers used in plasma membrane purification. *In* The Plant Plasma Membrane. Structure, Function and Molecular Biology. Eds. C Larsson and I M Moller. pp 17–40. Springer Verlag, Berlin, Heidelberg, New York.

Ward M R, Grimes H D and Huffaker R C 1989 Latent nitrate reductase activity associated with the plasma membrane of corn roots. Planta 177, 470–475.

*F. Baluška et al. (eds.), Structure and Function of Roots*, 141–147, 1995.


# Influence of constant nitrate concentration on N uptake and growth of wheat

F. Devienne and B. Mary
*Unité d'Agronomie de Laon-Péronne, INRA, BP 101, 02004 Laon Cedex, France*

*Key words:* nitrate concentration, nitrate uptake, plant growth, *Triticum aestivum* L., wheat

## Abstract

The long-term effect of constant $NO_3^-$ availability on plant growth and $NO_3^-$ uptake rate was investigated on 4- to 7-week old winter wheat plants. The experiment was done under controlled and hydroponic conditions. The range of $NO_3^-$ concentrations tested was 0.1 to 5 m*M* $NO_3^-$. Whole plant growth was maximum at 0.4 m*M* $NO_3^-$, while shoot growth was greatest at 0.8 m*M* $NO_3^-$ . The plant response to $NO_3^-$ availability was mainly due to changes in relative allocation of dry matter between roots and shoots. The specific $NO_3^-$ uptake rate (by unit of root biomass) varied markedly over short time periods (hours to 2–4 days). It was stable during the 4–7 weeks period and was almost independent of the external $NO_3^-$ concentration. The amount of N accumulated by the plant depended more on root biomass than on the specific $NO_3^-$ uptake rate in our culture conditions.

## Introduction

N uptake is an important factor which can limit plant growth and crop yield in field conditions. Increasing the N supply in root medium enhances both growth and N concentration in dry matter (Ingestad and Agren, 1992; Oscarson et al, 1989). Plants also respond to the external N availability by varying their allocation of assimilates between roots and shoots. In most species, the relative biomass partitioning to the roots decreases when the external N availability increases (Clarkson, 1985; Hilbert, 1990).

In temperate agricultural conditions, $NO_3^-$ seems to be the most common form of N absorbed by the plants growing in the field (Redinbaugh and Campbell, 1991). The regulation of $NO_3^-$ uptake by the whole plant has been the subject of many investigations. Some authors highlight that net $NO_3^-$ uptake rate is regulated by internal factors such as growth rate and C assimilation (Clement et al., 1978; Clarkson, 1986; Gastal and Saugier, 1989), fluxes of carbohydrates from shoots to roots (Macduff and Jackson, 1992; Raper et al., 1991), amino acids and amides (Lee et al., 1992; Muller and Touraine, 1992), or malate (Ben Zioni et al., 1971; Touraine et al., 1992). Others showed that net $NO_3^-$ uptake increases mainly with $NO_3^-$ availability in the medium (Oscarson et al, 1989; Rao and Rains, 1976; Siddiqi et al., 1990). Long-term studies about N uptake in relation to growth have been performed under field conditions. Total N concentration in the plant has been showed to decline during plant growth and relationship between total N and accumulated biomass has been proposed (Greenwood et al., 1991; Lemaire and Salette, 1984). However, wheat plants being fed with plenty of N in controlled or field conditions can accumulate additional N (in form of $NO_3^-$ and reduced N) without any increase in dry matter (Justes, 1993; Zhen and Leigh, 1990), suggesting that N supply can regulate N uptake even when growth rate is maximal.

These conflicting results have led us to study the effect of different nitrate availability (kept constant) on $NO_3^-$ uptake and growth of established winter wheat plants (4 to 7 week-old, from tillering to beginning of stem extension). This paper examines this effect at different time scales varying from four hour to several weeks.

*Table 1.* Mineral composition of the nutrient solutions (m*M*). All the solutions also contained micronutrients (Bo, 0.25 mg $L^{-1}$; Mo, 0.027 mg $L^{-1}$; Mn, 0.5 mg $L^{-1}$; Cu, 0.06 mg $L^{-1}$; Zn, 0.23 mg $L^{-1}$) and iron as FeEDDHA (ethylene diamine di(o-hydroxyphenyl)acetic acid, 1.2 mg $L^{-1}$)

| $NO_3^-$ concentration (m*M*) | $KNO_3$ | $Ca(NO_3)_2$ | $Ca(H_2PO_4)_2$ | $KH_2PO_4$ | $MgSO_4$ | NaCl |
|---|---|---|---|---|---|---|
| 0.1 | | 0.05 | 0.30 | 0.50 | 0.80 | 0.20 |
| 0.25 | | 0.125 | 0.25 | 0.60 | 0.80 | 0.20 |
| 0.4 | | 0.20 | 0.20 | 0.70 | 0.80 | 0.20 |
| 0.5 | | 0.25 | 0.20 | 0.70 | 0.80 | 0.20 |
| 0.8 | | 0.40 | 0.10 | 0.90 | 0.80 | 0.20 |
| 1.5 | 0.10 | 0.70 | | 1.10 | 0.80 | 0.20 |
| 3.0 | 0.70 | 1.15 | | 1.10 | 0.80 | 0.20 |
| 3.0 | 1.60 | 1.70 | | 1.10 | 0.80 | 0.20 |

## Materials and methods

### *Plant culture*

Seeds of wheat (*Triticum aestivum* cv. Courtot) were germinated on filter paper moistened with distilled water at 25°C in the dark. After 3 days, the seedlings were transferred onto the lids of 230 L-continuous flow, hydroponic-culture systems. Each system consisted of an upper compartment in which the plant roots were suspended in 130 L nutrient solution and a lower reservoir containing 100 L nutrient solution. The nutrient solution was continuously circulated between the two compartments and air bubbled in the upper compartment in order to homogenize and aerate the solution around the roots. Each system contained 100 plants on an area of 0.5 $m^2$.

Three identical systems were located in a growth chamber programmed tor day/night temperature of 20/12°C which was maintained at ± 1°C, at 85% relative humidity (± 10% within a 24h period and between days), and with a 16h/8h light/dark cycle. The photosynthetic photon flux density was 250 $\mu$mol $m^{-2}$ $s^{-1}$ at canopy height (± 20% between the different experiments). The composition of the nutrient solution is given in Table 1. Three experiments were conducted successively in the growth chamber, corresponding to various concentrations of $NO_3^-$ in solution: 0.25 and 0.5 m*M* $NO_3^-$ for the first one, 0.1, 0.4 and 0.8 m*M* $NO_3^-$ for the second one and 1.5, 3.0 and 5.0 m*M* $NO_3^-$ for the third experiment. Aliquots of nutrient solution were taken at the end of each light period and analysed for $NO_3^-$ by a rapid spectrophotometric method (Armstrong, 1963) with a precision of 10%. The analysis was used to determine the amount of $Ca(NO_3)_2$ needed each day to restore approximately the nominal concentration. The nutrient solution was also completely renewed twice a week to avoid depletion or accumulation of other nutrients. The actual $NO_3^-$ concentration of the solutions varied between 85 and 105% of the nominal value.

### *Measurements*

Growth and N uptake were studied from tillering (4 week-old plants) to the beginning of stem extension (7 weeks). Six plants were harvested every week, the shoots and roots of each plant being separated. Organs were weighted, dried for 48 hours at 75°C, reweighed, ground and mixed thoroughly. Total N was determined using a combustion nitrogen analyser (Carlo Erba NA 1500, Dumas method).

The hourly $NO_3^-$ uptake rates were determined during a night/day cycle at the 28th day of growth for the 8 $NO_3^-$ concentrations used. Two aliquots of each nutrient solution were sampled every four hours. $NO_3^-$ concentration in the nutrient solution was determined precisely (±2%) by continuous flow colorimetry. Interference due to organic molecules was eliminated by the use of a dialysis membrane. $NO_3^-$ ions were reduced to nitrite on a Cd column and stained by the Griess-Ilosvay method (Henriksen and Selmer-Olsen, 1970). $NO_3^-$ uptake rate was calculated from changes in the external $NO_3^-$ concentration. Six plants were harvested at the middle of the light period to accurately measure the root dry matter.

Daily $NO_3^-$ uptake rate was determined precisely using an accurate measurement of $NO_3^-$ concentration (by continuous flow colorimetry). Root dry matter at each date was estimated from measurements made every week. The net $NO_3^-$ uptake rate was determined from the changes in the $NO_3^-$ concentration in the medium. Weekly $NO_3^-$ uptake rate was determined by accounting for the daily addition of $Ca(NO_3)_2$ (see above).

## Results

Plant growth was severely limited at the lowest concentration (0.1 m*M* $NO_3^-$) in the medium, particularly at the end of the experiment (beginning of stem extension at 7 weeks). Whole plant growth was maximum at 0.4 m*M* $NO_3^-$, while shoot growth reached a maximum at 0.8 m*M* (Fig. 1a). Maximal N and dry matter accumulations in the whole plant were obtained at 0.4 m*M* $NO_3^-$, but the proportion of N allocated to the shoot increased continuously from 0.1 to 1.5 m*M* $NO_3^-$ (Fig. 1b). N concentration in the whole plant increased sharply between 0.1 and 0.4 m*M* $NO_3^-$ and reached a plateau at 1.5 m*M* $NO_3^-$. N concentration in shoot increased slightly from 1.5 to 5 m*M* whereas it declined steadily in roots.

The wheat response to $NO_3^-$ availability was mainly caused by changes in dry matter partitioning between roots and shoots: the root weight ratio RWR (ratio between root and total plant weight) first increased as the $NO_3^-$ concentration increased from 0.1 to 0.4 m*M* (Fig. 2), when the whole plant growth was maximum. It then decreased rapidly as the external $NO_3^-$ concentration went from 0.4 to 1.5 m*M* and was more or less constant between 1.5 and 5 m*M*. The RWR decreased with time at all the concentrations studied. However this drop was greater at 49 days for the concentrations from 1.5 to 5 m*M*, for which shoot growth and N concentration were maximal. Over this range of concentration, this decrease is mainly due to a important slowing down of root growth after 42 days of culture (data not shown).

The net $NO_3^-$ uptake rate varied markedly within a period of 24 hours (Fig. 3): a large increase

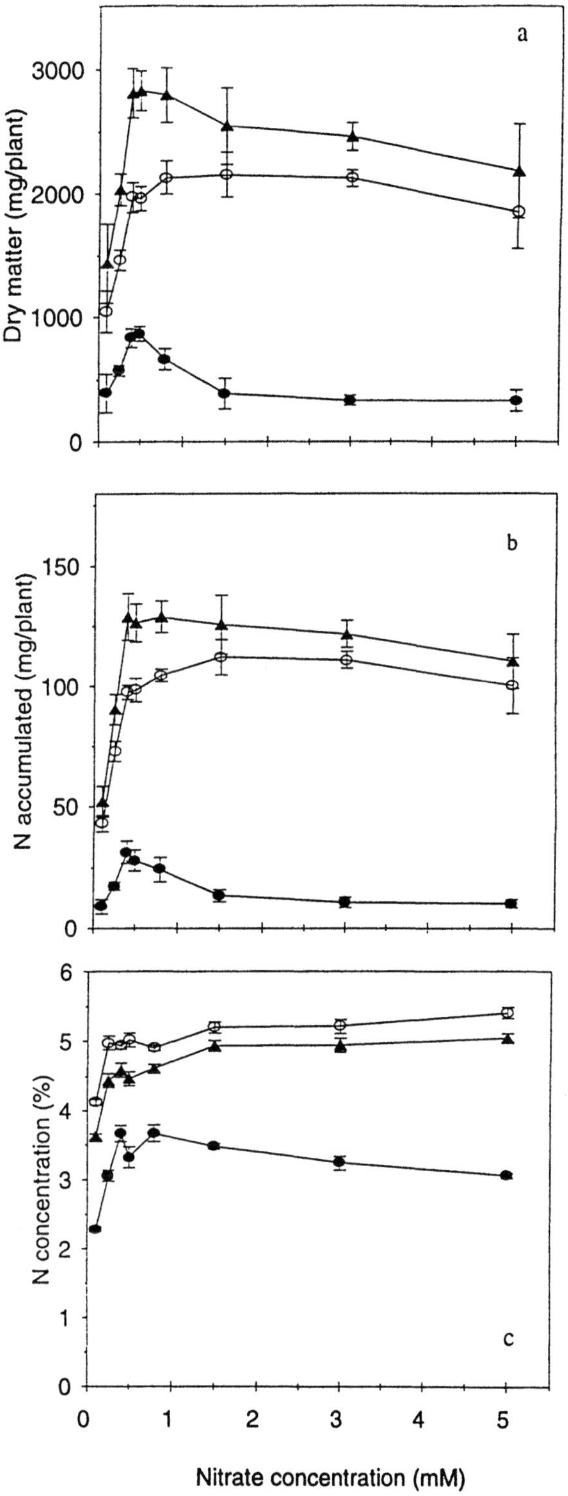


*Fig. 1.* Dry matter (**a**), N accumulated (**b**) and N concentration (**c**) in the shoot (○) the roots (●) and the whole plant (▲) of the wheat plant versus nutrient concentration (7 week-old plants, beginning of the stem extension). The vertical bars indicate standard errors.

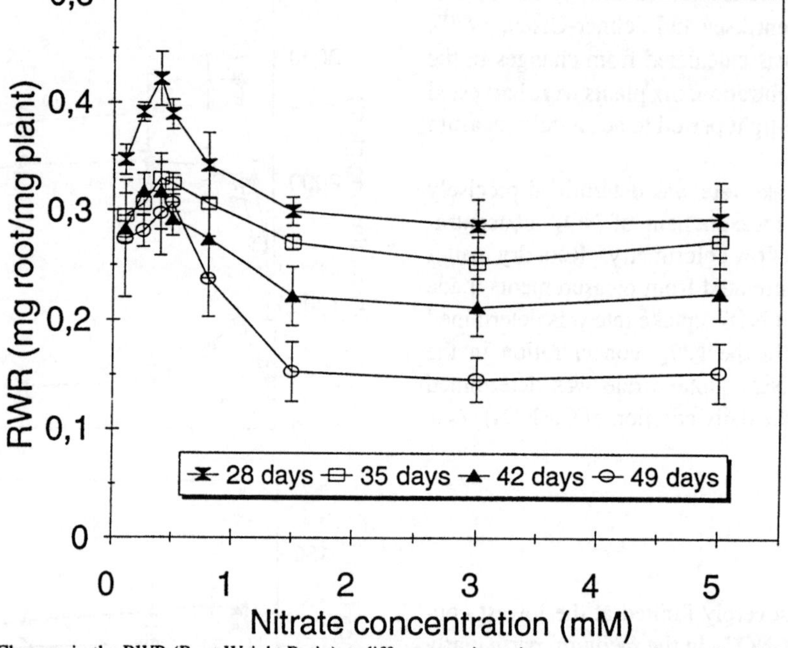


*Fig. 2.* Changes in the RWR (Root Weight Ratio) at different nutrient nitrate concentrations and at different dates

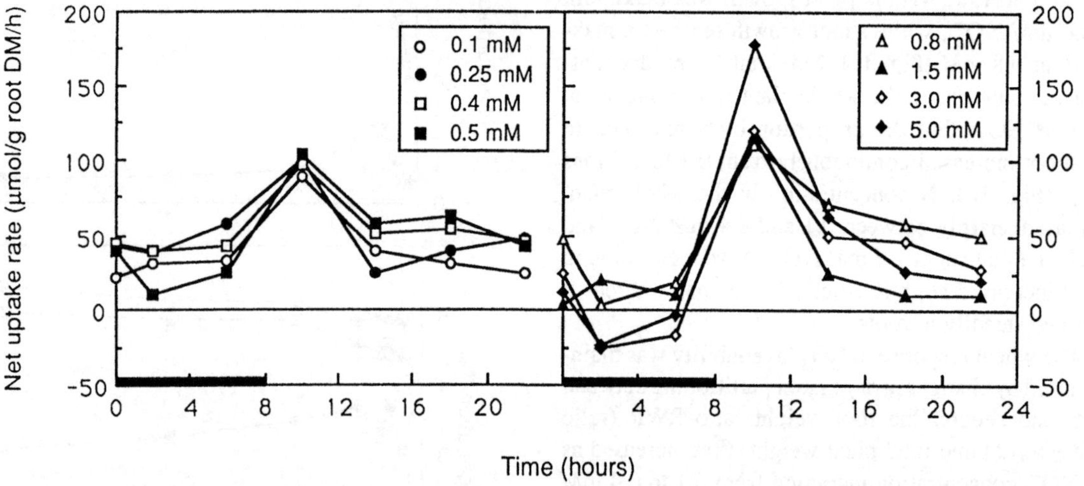


*Fig. 3.* Net nitrate uptake rate over 24 hours at the 8 nitrate concentrations used. Measurements were made on the 28th day of the experiment. The black bars indicate the dark period. Each point represents a single measurement (there is one culture system for each treatment).

occurred at the beginning at the light period followed by a decrease. A nearly constant value of $NO_3^-$ - uptake rate was obtained at the end of the light period. The net uptake rate varied also greatly over periods of 2–4 days (Fig. 4). The higher the concentration in the nutrient solution, the greater was the amplitude of the oscillations in $NO_3^-$ - uptake. However during a longer time-interval of a week, the net specific absorption rate (per unit of root mass) was much more stable (Table 2). The specific uptake rate was weakly dependant on the external $NO_3^-$ concentration, since it varied less than 30% between 0.4 and 5 m*M* $NO_3^-$.

## Discussion

Variations of $NO_3^-$ uptake during a night/day cycle have been already reported (Hansen, 1980; Le Bot and

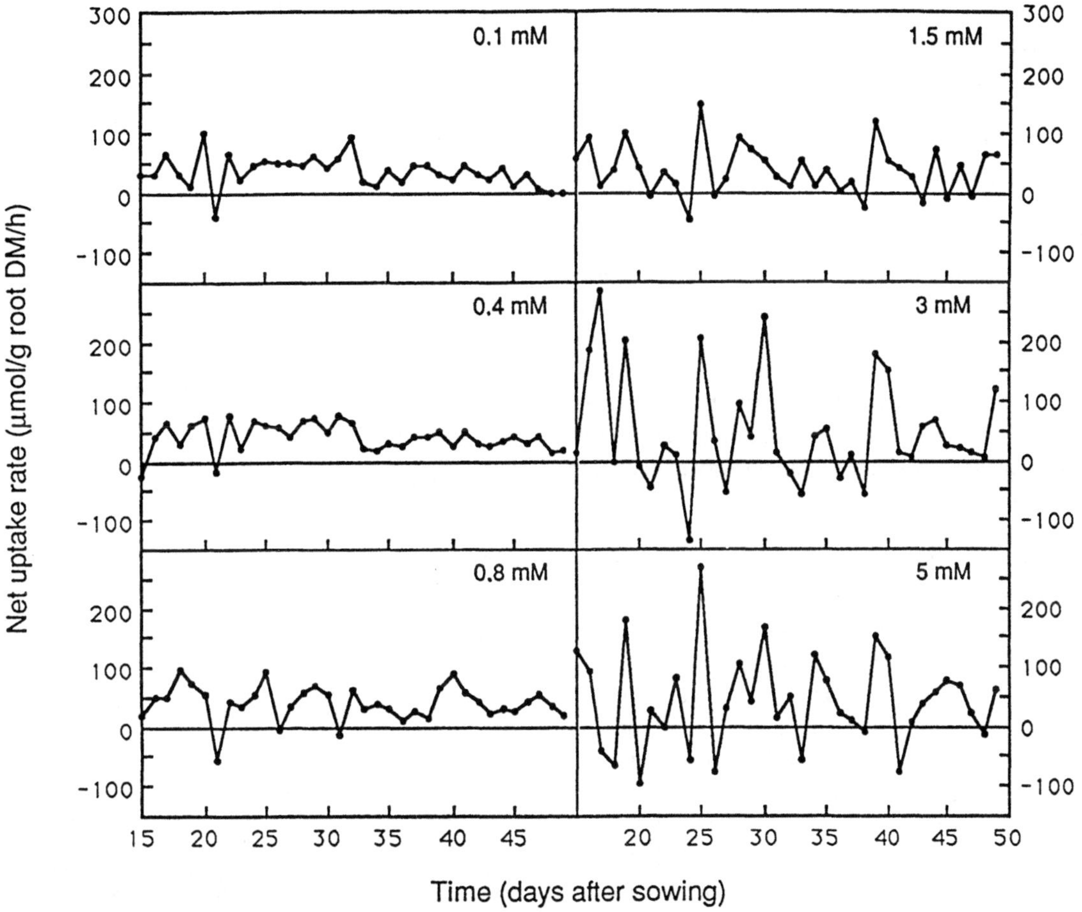


*Fig. 4.* Net specific nitrate uptake rate during plant growth at 6 of the 8 nitrate concentrations used in the study. The treatments 0.25 and 0.5 m*M* gave the same pattern as 0.4 m*M*. Each point represents a single measurement (there is one culture system for each treatment).

*Table 2.* Average specific $NO_3^-$ uptake rate during 3 1-week periods at 8 $NO_3^-$ concentrations. The last line shows the averages for all periods. Parenthesis indicate standard errors calculated from the replicates of $NO_3^-$ concentration for each date

| Time period (days) | 0.1 m*M* | 0.25 m*M* | 0.4 m*M* | 0.5 m*M* | 0.8 m*M* | 1.5 m*M* | 3.0 m*M* | 5.0 m*M* |
|---|---|---|---|---|---|---|---|---|
| | ($\mu$mol g$^{-1}$ root DM h$^{-1}$) | | | | | | | |
| 29–35 | 45.7 (2.6) | 51.8 (5.1) | 47.3 (0.7) | 52.1 (4.8) | 38.2 (4.4) | 37.6 (4.6) | 45.7 (4.1) | 59.8 (2.8) |
| 36–42 | 33.1 (1.8) | 39.6 (4.5) | 38.1 (2.3) | 36.6 (2.1) | 43.8 (3.1) | 33.4 (2.3) | 40.7 (4.9) | 31.4 (3.6) |
| 43–49 | 15.6 (1.3) | 31.1 (0.7) | 29.1 (4.7) | 28.2 (1.4) | 31.7 (3.3) | 31.2 (5.1) | 44.6 (4.2) | 44.9 (3.7) |
| Average | 31.5 (1.7) | 40.8 (3.4) | 38.2 (2.7) | 39.0 (2.7) | 37.9 (3.1) | 34.1 (4.4) | 43.7 (4.3) | 45.4 (3.5) |

Kirkby, 1992). The large increase in the uptake rate observed at the beginning of the light period can be related to the induction of nitrate reductase by light in the leaves and to the subsequent translocation of malate from shoot to roots which stimulates $NO_3^-$ absorption (Ben Zioni et al., 1971; Touraine et al., 1992). The following decrease could result from a retroinhibition of $NO_3^-$ absorption by amino acids translocated from the shoot and accumulated in the roots during this period (Muller and Touraine, 1992). A recent work shows that the regulation of $NO_3^-$ absorption during the dark period would not depend on amino acids or malate translocation, but rather on sucrose translocated to the roots via the phloem sap (Delhon, 1993).

There have been few data showing changes of $NO_3^-$ uptake rates over periods of 2–4 days. Tolley-Henry et al. (1988) obtained such results with soybean and suggested that these variations were correlated with the rhythm of leaf appearance. This cannot be the case for wheat because leaves appear continuously during tillering. Uptake fluctuations might result from variations of the relative humidity in the atmosphere (the only factor significantly varying in the growth chamber) which can affect water consumption of the plants. However nitrate absorption is apparently not related to water uptake (Hansen, 1980; Le Bot and Kirkby, 1991). At present, we have no satisfactory explanation of these fluctuations.

The amplitude of the changes in $NO_3^-$ uptake depended on the $NO_3^-$ concentration in the medium, suggesting that the external concentration influences the regulation of $NO_3^-$ uptake. However, the short-term variations (within a day) could not explain the net $NO_3^-$ absorption by the whole plant over a longer period of time (over a week), under our conditions of constant N nutrition and up to stem extension stage. Net $NO_3^-$ uptake rate depended little on $NO_3^-$ availability in the medium on a weekly time scale. The long-term effect of various external $NO_3^-$ concentrations on N uptake was mainly due to variations in partitioning of C assimilates between roots and shoot. The plant maintained a high RWR at the expense of shoot growth when N availability in the medium was limiting plant growth. In contrast, the RWR decreases with time when N supply was sufficient to support maximal growth. In these conditions, the amount of N accumulated by the whole plant depends more on the extension of the root system (the root biomass decreased by a factor of 3 between 0.4 and 5 m*M*) than on the specific $NO_3^-$ uptake rate which varied from about 30% over the same range of concentrations.

This study shows that the external $NO_3^-$ concentration influences mainly the amplitude of the variations of specific $NO_3^-$ uptake rate (within 24 hours). Its most important effect is on plant growth and dry matter allocation between roots and shoots.

## Acknowledgements

We thank G Alavoine for $NO_3^-$ analysis and O Delfosse for total N determinations.

## References

Armstrong F A J 1963 Determination of nitrate in water by ultraviolet spectrophotometry. Anal. Chem. 35, 1292–1294.

Ben Zioni A, Vaadia Y and Lips S 1971 Nitrate uptake by roots as regulated by nitrate reduction products of the shoot. Physiol. Plant. 24, 288–290.

Clement C R, Hopper M J and Jones L H P 1978 The uptake of nitrate by *Lolium perenne* from flowing nutrient solution. J. Exp. Bot. 29, 453–464.

Clarkson D T 1985 Factors affecting mineral nutrient acquisition by plants. Annu. Rev. Plant Physiol. 36, 77–115.

Clarkson D T 1986 Regulation of the absorption and release of nitrate by plant cells: A review of current ideas and methodology. In Fundamental, Ecological and Agricultural Aspects of Nitrogen Metabolism in higher Plants. Eds. H Lambers, J J Neeteson and I Stulen. pp 3–27. Martinus Nijhoff Publishers, Dordrecht, The Netherlands.

Delhon P 1993 Régulation de l'absorption racinaire du nitrate par les parties aériennes: effet de la lumière. Ph. D. Thesis, E.N.S.A. Montpellier, 116 p.

Gastal F and Saugier B 1989 Relationships between nitrogen uptake and carbon assimilation in whole plants of tall fescue. Plant Cell Environ. 12, 407–418.

Greenwood D J, Gastal F, Lemaire G, Draycott A, Millard P and Neeteson J J 1991 Growth rate and N% of field grown crops: theory and experiments. Ann. Bot. 67, 181–190.

Hansen G K 1980 Diurnal variation of root respiration rates and nitrate uptake as influenced by nitrogen supply. Physiol. Plant. 48, 421–427.

Henriksen A and Semler-Olsen A 1970 Automatic methods for determining nitrate and nitrite in water and soil extract. Analyst 95, 514–518.

Hilbert D W 1990 Optimization of plant root:shoot ratios and internal nitrogen concentrations. Ann. Bot. 66, 91–99.

Ingestad T and Agren G I 1992 Theories and methods on plant nutrition and growth. Physiol. Plant. 84, 177–184.

Justes E 1993 Diagnostic de la nutrition azotée du blé, à partir de la teneur en nitrate de la base de la tige. Application au raisonnement de la fertilisation. Ph. D. Thesis, Institut National Agronomique Paris-Grignon, 227 p.

Le Bot J and Kirkby E A 1992 Diurnal uptake of nitrate and potassium during the vegetative growth of tomato plants. J. Plant Nutr. 95, 247–264.

Lee R B, Purves J V, Ratcliffe R G and Saker L R 1992 Nitrogen assimilation and the control of ammonium and nitrate absorption by maize roots. J. Exp. Bot. 43, 1385–1396.

Lemaire G and Salette J 1984 Relation entre dynamique de croissance et dynamique de prélèvement d'azote pour un peuplement de graminées fourragères. I- Etude de l'effet du milieu. Agronomie 4, 423–430.

Macduff J H and Jackson S B 1992 Influx and efflux of nitrate and ammonium in Italian ryegrass and white clover roots: comparisons between effects of darkness and defoliation. J. Exp. Bot. 43, 525–535.

Muller B and Touraine B 1992 Inhibition of $NO_3^-$ uptake by various phloem-translocated amino acids in soybean seedlings. J. Exp. Bot. 43, 617–623.

Oscarson P, Ingemarsson B and Larsson C M 1989 Growth and nitrate uptake properties of plants grown at different relative rates of nitrogen supply. Plant Cell Environ. 12, 779–785.

Rao K P and Rains D W 1976 Nitrate absorption by barley. I-Kinetics and energetics. Plant Physiol. 57, 55–58.

Raper C D, Vessey J K and Henry L T 1991 Increase in nitrate uptake by soybean plants during interruption of the dark period with low intensity light. Physiol. Plant. 81, 183–189.

Redinbaugh M G and Campbell W H 1991 Higher plant responses to environmental nitrate. Physiol. Plant. 82, 640–650.

Siddiqi M Y, Glass A D M, Ruth T J and Rufty T W 1990 Studies of the uptake of nitrate in barley. I-Kinetics of $NO_3^-$ influx. Plant Physiol. 93, 1426–1432.

Tolley-Henry L C, Raper C D Jr and Granato T C 1988 Cyclic variations in nitrogen uptake rate of soybean plants: effects of external nitrate concentration. J. Exp. Bot. 39, 613–622.

Touraine B, Muller B and Grignon C 1992 Effect of phloem-translocated malate on $NO_3^-$ uptake by roots of intact soybean plants. Plant. Physiol. 99, 1118–1123.

Zhen R G and Leigh R A 1990 Nitrate accumulation by wheat (*Triticum aestivum*) in relation to growth and tissue N concentrations. Plant and Soil 124, 157–160.

F. Baluška et al. (eds.), Structure and Function of Roots, 149–154, 1995.


# Al avoidance and Al tolerance of *Mucuna pruriens* var. *utilis*: Effects of a heterogeneous root environment and the nitrogen form in the root environment

K. Hairiah[1], I. Stulen[2], M. van Noordwijk[3] and P.J.C. Kuiper[2]
[1]*Brawijaya University, Faculty of Agriculture, Malang 64145, Indonesia,* [2]*Department of Plant Biology, P. O. Box 14, 9750 AA Haren (Gn), The Netherlands and* [3]*ICRAF c/o Forest Research and Development Center, Jalan Gunun Batu 5, P.O. Box 161, Bogor, 16001, Indonesia*

*Key words:* Al avoidance, Al tolerance, ammonium, *Mucuna,* nitrate, split-root

## Abstract

In *Mucuna pruriens* var. *utilis*, grown with nitrate-N in a hydroponic split-root system, an Al avoidance reaction of root growth was observed, which was ascribed to local P stress in the Al containing compartment. The Al avoidance reaction was similar to the avoidance of *Mucuna* roots of acid subsoil in the field where roots grew preferentially in the topsoil. In the present paper the effect of different N forms ($NO_3^-$ and $NH_4^+$) on the reactions of *Mucuna* to Al were studied, since in acid soils N is present as a mixture of $NO_3^-$ and $NH_4^+$. No interaction between the N form and Al toxicity was found. A hydroponic split-root experiment with $NH_4NO_3$ nutrition, which is comparable to the situation in the field, showed that under these conditions Al avoidance did not occur. It is concluded that a relation between the Al avoidance reaction of *Mucuna* and P stress is still likely.

*Abbreviations:* $D_r$ – root diameter, $L_{pr}$ – total root length per plant, $L_{rw}$ – specific root length, NRA – nitrate reductase activity, S/R – shoot : root ratio

## Introduction

Physiological studies and screening tests for selection and breeding of Al tolerant crops are generally performed in homogeneous root media, such as nutrient solution and well mixed soil preparations. A heterogeneous soil profile, however, allows the plant to choose, because the root system encounters different conditions in the various soil layers. In a model experiment for a heterogeneous root environment Hairiah et al. (1992) showed for the velvet bean, *Mucuna pruriens*, grown in a hydroponic split-root system, that root systems avoided Al by the preferential development of roots in the non-Al half of the split-root system, despite the fact that *Mucuna* roots tolerated Al in a homogeneous hydroponic system (Hairiah et al., 1990). The Al avoidance reaction in the hydroponic split-root system is similar to the avoidance response to acid subsoil under natural conditions: *Mucuna* roots hardly grew in the acid subsoil when the topsoil is present, but when the topsoil was removed, and there was no alternative, they grew unimpeded in the subsoil (Hairiah et al., 1991a). The observed Al avoidance reaction at the root system level is basically different from the Al exclusion by individual roots which was described as Al avoidance, by e.g. Marschner (1991) and Miyasaka et al. (1989). Further experiments with a split-root system, in which extra P was supplied, showed that the Al avoidance reaction was related to a local response of the plants to differences in P supply. An increase in P supply resulted in the disappearance of the Al avoidance reaction (Hairiah et al., 1992).

In previous hydroponic split-root experiments (Hairiah et al., 1992, 1993), N was supplied as $NO_3^-$. In acid soils, however, N is present as a mixture of $NH_4^+$ and $NO_3^-$. The N form may influence Al tolerance of the plants, either directly through competition between $NH_4^+$ and Al for cell wall binding sites (Klotz and Horst, 1988), or indirectly through effects on the balance of cation-anion uptake (McCain and Davies,

1983) and hence on rhizosphere pH and the concentration of monomeric Al and Ca in soil solution (Blamey et al., 1983; Jaillard, 1987).

The objectives of the present paper were: 1) to compare the effect of different N forms ($NO_3^-$ and $NH_4^+$) on Al tolerance, and 2) to study the Al avoidance reaction of *Mucuna* with $NH_4NO_3$ nutrition, in order to find out whether the Al avoidance reaction is also found with an $NH_4^+ : NO_3^-$ ratio of 1, within the range of 0.3 - 1.2 found for soil-solution from Lampung, Indonesia where field experiments were performed (Handayanto et al., 1992).

## Material and methods

*Plant material*
Seeds of *Mucuna pruriens* var. *utilis* (L) DC var. *utilis* (Wall. ex. Wright) Baker ex Burckt were germinated as before (Hairiah et al., 1990), and grown in a glasshouse (Hairiah et al., 1993).

*Comparison between $NH_4^+$ and $NO_3^-$ nutrition*
Mucuna seedlings were grown in 30-L containers with N as $NO_3^-$ or $NH_4^+$ (replacing $KNO_3$ and $Ca(NO_3)$ by Cl salts) in modified 1/4 strength Hoagland solutions. Al (110 $\mu M$) was supplied as $AlCl_3$. Each container contained 14 plants, and each treatment was repeated 3 times in a randomized block design. Since previous experiments showed that the increase of the solution pH as a result of $NO_3^-$ uptake drastically reduced the monomeric Al concentrations, the pH of the solutions was adjusted daily, and the monomeric Al concentration was also determined (Hairiah et al., 1990, 1992). Root diameter, $D_r$, and total root length per plant, $L_{rp}$, were measured as described before (Hairiah et al., 1990). Nitrate reductase activity (NRA) of the shoot was assayed in vivo after Jaworski (1971) as described by Stulen et al. (1981).

*Split-root experiments in $NH_4NO_3$*
A modified 1/4 strength Hoagland solution with the $NH_4^+ : NO_3^-$ ratio of 1 was circulated in the system (250 L). The pH of the solution was adjusted and the water uptake compensated daily (Hairiah et al., 1992, 1993). Al (185 $\mu M$) was added as $AlCl_3$. Daily P additions were started one week after transfer of the plants to the system; other nutrients were regularly added on the basis of expected uptake rates (Hairiah et al., 1992). Plants with two halves of the root system exposed to the same solution, either with (+/+) or without (0/0) Al, were compared with plants with two halves in different solutions (0/+), in a randomized block design. Solution samples were taken every other day for determination of inorganic monomeric Al (Hairiah et al., 1991b, 1992; Kerven et al., 1989), P (colorimetric with molybdate blue), $NH_4^+$ (colorimetric, with Nessler reagent) and $NO_3^-$ (Orion 93–07 nitrate electrode). Plants were harvested after 4 weeks in solution. Chemical analysis of macro elements and statistical analysis of the data was as described by Hairiah et al. (1992).

## Results

*Comparison between $NH_4^+$ and $NO_3^-$ nutrition*
With $NH_4^+$ nutrition, shoot and root dry weight were reduced significantly ($p < 0.05$) and shoot:root ratio (S/R) was decreased compared to $NO_3^-$ nutrition (Table 1). The presence of 110 $\mu M$ reduced shoot and root dry weight: by 14 and 11%, respectively, and had no significant effect on S/R. The presence of Al significantly increased average root diameter, $D_r$, and reduced total root length per plant, $L_{rp}$; $NH_4^+$ -nutrition had no effect on average root diameter, but it reduced $L_{rp}$. A significant ($p < 0.05$) interaction between the effects of N and Al was found for the specific root length, $L_{rw}$, and also for $L_{rp}$, with high values for the $NO_3^-$ -fed plants without Al.

The total P content of the shoot was higher with $NH_4^+$ than with $NO_3^-$ nutrition (Table 2). The presence of Al decreased total P significantly in the shoot of the $NO_3^-$ -fed plants, but for the $H_2PO_4^-$ concentration the decrease was not statistically significant. In roots of $NO_3^-$ fed plants a signifcant effect of Al was found on the $H_2PO_4^-$ concentration, but not on the total P concentration. With $NH_4^+$ nutrition no effect of Al on the $H_2PO_4^-$ or total P concentration in root or shoot was found.

*Split-root experiment with $NH_4NO_3$ as N source*
Figure 1 shows the concentration of monomeric Al and pH of the solution during the experiment. Al-monomeric concentration was about 60–80% of the added Al concentration of 185 $\mu M$. The pH of the solution was approximately constant during the experiment; only after 20 days a slight acidification occurred. The $NO_3^-$ concentration in the solution decreased slightly faster than the $NH_4^+$ concentration (Fig. 2). *Mucuna* thus showed a slight preference for $NO_3^-$. The

*Table 1.* Effect of Al and N form on shoot yield and various root parameters (numbers followed by different letters indicate significant differences ($p < 0.05$): ns = no significant differences ($p > 0.05$). Plants were harvested after 1 week. The interaction between Al × N treatment was not significant

| Treatment | | S dw | R dw | S/R | S | R | $D_r$ | | $L_{rw}$ | $L_{rp}$ |
|---|---|---|---|---|---|---|---|---|---|---|
| N | Al | (g) | | dw | (% dm)[a] | | | (mm) | ($m\ g^{-1}$) | ($m\ pl^{-1}$) |
| $NO_3^-$ | 0 | $0.84^a$ | $0.29^a$ | $2.9^a$ | $16^{ns}$ | $4.9^{ns}$ | $0.58^b$ | | $107^a$ | $31^a$ |
| | + | $0.73^a$ | $0.26^{ab}$ | $2.8^{ab}$ | 18 | 5.4 | $0.66^a$ | | $70^b$ | $18^b$ |
| $NH_4^+$ | 0 | $0.64^b$ | $0.26^{ab}$ | $2.5^{ab}$ | 19 | 5.2 | $0.57^b$ | | $74^b$ | $19^b$ |
| | + | $0.53^b$ | $0.22^b$ | $2.4^b$ | 19 | 5.2 | $0.67^a$ | | $90^{ab}$ | $20^b$ |

[a]% dm; percentage dry matter, $D_r$ - root diameter (mm), $L_{rw}$ - specific root length ($m\ g^{-1}$), $L_{rp}$ - root length (m) per plant, S- shoot, R - root.

*Table 2.* Effect of Al and N form on N, P and Al concentration (m*M*) expressed on tissue water (= fresh minus dry weight) basis; NRA = nitrate reductase activity ($\mu mol\ g\ fw^{-1}\ h^{-1}$; numbers followed by different letters indicate significant differences ($p < 0.05$); ns = no significant differences ($p>0.05$)

| Treatment | | $NO_3^-$ | | $N_{tot}$ | | NRA | $H_2PO_4^-$ | | $P_{tot}$ | | Al |
|---|---|---|---|---|---|---|---|---|---|---|---|
| N | Al | R[a] | S | R | S | S | R | S | R | S | P |
| $NO_3^-$ | 0 | $51^a$ | $65^a$ | $212^a$ | $526^{ns}$ | $1.63^a$ | $3.5^a$ | $13^{ns}$ | $11^{ns}$ | $36^b$ | $1.7^c$ |
| | + | $47^a$ | $39^a$ | $189^b$ | 589 | $1.27^a$ | $1.5^b$ | 5 | 17 | $26^c$ | $25^a$ |
| $NH_4^+$ | 0 | $8^b$ | $11^b$ | $163^d$ | 500 | $2.47^b$ | $2.7^b$ | 14 | 13 | $43^a$ | $3.0^c$ |
| | + | $23^b$ | $11^b$ | $175^c$ | 502 | $2.67^b$ | $2.7^a$ | 14 | 12 | $43^a$ | $16^b$ |

[a]R = root; S = shoot; P = whole plant.

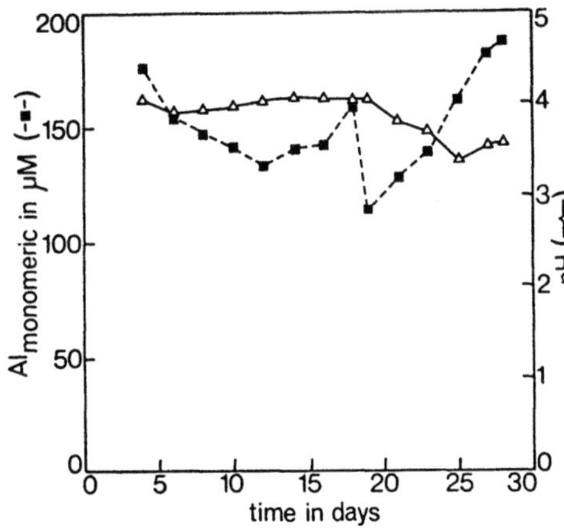


*Fig. 1.* Inorganic monomeric Al concentration and pH of the $NH_4NO_3$ nutrient solution during the experiment.

presence of Al had no effect on the N preference of *Mucuna* (Fig. 2), nor on shoot and root dry weight (Table 3). Plants growing on $NH_4NO_3$ had on average a 23 and 31% lower shoot and root dry weight than plants growing simultaneously in the same glasshouse in nutrient solution with $NO_3^-$ in as the sole N source. A pH stat was used to maintain the pH of the solution (Hairiah et al., 1992).

The *Mucuna* plants with a choice for Al (0/+) produced significantly ($p < 0.05$) more roots with Al (+) than without (0); 64% of root weight was found in the +Al side (Table 3), in contrast to previous experiments with $NO_3^-$ (Hairiah et al., 1992, 1993). Table 3 shows the cation concentrations in shoot and roots. In the roots Al reduced the $Ca^{2+}$ concentration and increased Al and total cation concentration significantly ($p < 0.05$), but had no significant effect on $K^+$ and $Mg^{2+}$ concentrations. In the shoot, Al increased $K^+$, Al, and total cation concentrations significantly ($p < 0.05$) and had no effect on $Ca^{2+}$ and $Mg^{2+}$ concentrations.

## Discussion

Previous experiments showed that *Mucuna* was moderately tolerant to Al with $NO_3^-$ as the N source (Hairiah et al., 1990). In contrast to results with soybean, in

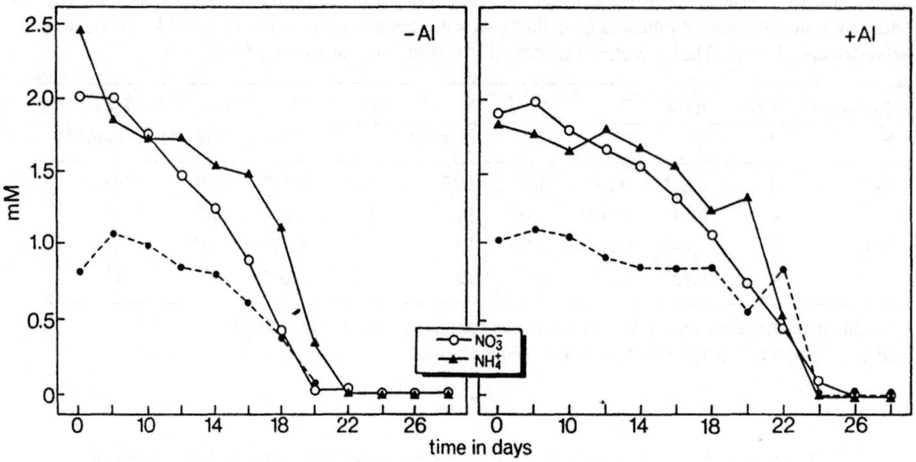


*Fig. 2.* Concentration of $NO_3^-$ and $NH_4^+$ in the $NH_4NO_3$ nutrient solution during the experiment. $=NO_3^-$, $=NH_4^+$, = ratio between $NO_3^-$ and $NH_4^+$.

*Table 3.* Cation concentration (mmol$_c$ $L^{-1}$, expressed on tissue water (fresh-dry weight) basis) of *Mucuna* after 4 weeks on $NH_4NO_3$. Plants were grown with two halves of the root system exposed to the same solution, either with (+/+) or without Al (0/0), or with two halves in different solutions (0/+)

| Treatment | dw (g) | Rww ($g\ g^{-1}$) | $K^+$ | $Ca^{2+}$ | $Mg^{2+}$ | $Al^{3+}$ | TC |
|---|---|---|---|---|---|---|---|
| **Root** | | | | | | | |
| 0/0 | 7.89$^{ns}$ | 21.2$^{ns}$ | 81$^{ns}$ | 4.35$^{ns}$ | 9$^{ns}$ | 2.32$^{b}$ | 96$^{b}$ |
| 0/+ 0 side | 2.93* | 21.4 | 81 | 4.00$^{ab}$ | 10 | 2.86$^{b}$ | 98$^{b}$ |
| + + side | 5.21* | 23.6 | 91 | 3.05$^{bc}$ | 11 | 11.2$^{a}$ | 116$^{a}$ |
| +/+ | 7.47 | 22.3 | 101 | 2.92$^{c}$ | 11 | 9.9$^{a}$ | 125$^{a}$ |
| **Shoot** | | | | | | | |
| 0/0 | 43$^{ns}$ | 4.84$^{ns}$ | 120$^{c}$ | 102$^{ns}$ | 18$^{ns}$ | 0.336$^{b}$ | 241$^{b}$ |
| 0/+ | 51 | 4.91 | 136$^{b}$ | 101 | 16 | 0.50$^{b}$ | 254$^{b}$ |
| +/+ | 48 | 4.86 | 180$^{a}$ | 92 | 15 | 1.546$^{a}$ | 289$^{a}$ |

Rww = relative water weight = (fresh - dry) /dry weight; TC = $K^+ + Ca^{2+} + Mg^2 + Al^{3+}$; numbers followed by different letters indicate significant differences ($p < 0.05$); ns = no significant differences ($p > 0.05$).
*Total root dw was not significantly different; root weight on the + side of the 0/+ treatment was significantly higher than on the 0 side.

which $NH_4^+$ -N increased Al tolerance of soybean compared to $NO_3^-$ -N (Klotz and Horst, 1988), the present experiments showed no statistically significant interaction between N form (100% $NH_4^+$ -N or 100% $NO_3^-$ -N) and Al toxicity in *Muncuna*. No growth reduction was found when *Mucuna* was grown on $NH_4NO_3$ at an Al concentration which reduced growth with $NO_3^-$ -N. It should be realized that $NH_4^+$ or $NH_4NO_3$ nutrition by itself reduced growth of *Mucuna* compared to growth on $NO_3^-$. $NH_4^+$ nutrition led to a reduced $Ca^{2+}$ and $Mg^{2+}$ concentration in the plant, similar to the effect of Al. Whether this reduction in divalent cation concentrations was the cause of reduced plant growth remains questionable.

No indications were obtained that the presence of Al disturbed N assimilation or that it influenced the

$NH_4^+/NO_3^-$ uptake ratio of *Mucuna*, in contrast to results for sorghum (Keltjens and Van Ulden, 1987). The presence of Al had hardly any effect on the uptake balance of cations and anions and hence on the external pH effect of the plant (Hairiah, 1992)

Foy et al. (1965) formulated the hypothesis that differences in Al tolerance between wheat cultivars were based on a difference in $NH_4/NO_3$ uptake ratio. Subsequent research, however, has shown that differences in rhizosphere pH between wheat cultivars may be the result, rather than the cause of differences in Al tolerance (Blamey et al., 1990; Miyasaka et al., 1989; Taylor, 1991).

When grown on $NH_4NO_3$ *Mucuna* roots showed a preferential root growth on the Al side of a split-root (0/+) treatment after 4 weeks. This response is opposite to the Al avoidance reaction found in earlier experiments on $NO_3^-$ -N (Hairiah et al., 1992, 1993). In this case Al avoidance was interpreted as a response to Al-induced P shortage in the roots, due to Al-$PO_4$ precipitation (Hairiah et al., 1992, 1993). The absence of Al avoidance for plants growing in an $NH_4NO_3$ solution might be based on an improved P supply, due to less Al-$PO_4$ precipitation in the root apoplast because of an acidifying rather than alkalinizing effect on rhizosphere pH, especially around the root tip (Römheld and Marschner, 1986). The lower Al concentration in the roots with $NH_4^+$ -N (Table 2) could be due to competition for binding sites in the root apoplast between $NH_4^+$ and Al (Klotz and Horst, 1988). An improved P status with $NH_4^+$-N could, in part, be explained by the fact that overall plant growth was slower and the same amount of P was supplied in both studies. Alternatively, the absence of the Al avoidance reaction may indicate a reduced Al toxicity in an $NH_4^+$ environment. The preferential root development on the Al side could also be interpreted as Al protection against $NH_4^+$ toxicity, but such protection is still speculative.

In conclusion, Al avoidance did not occur when $NH_4NO_3$ was the N source; on the contrary, a preferential root development was found on the +Al side. The similarity between subsoil avoidance by *Mucuna* roots in the field and Al avoidance in split-root solution culture studies, is less convincing with a more realistic $NH_4^+ : NO_3^-$ ratio of 1. A relation between Al avoidance and P stress, however, is still likely.

## References


Blamey F P C, Edwards D G and Asher C J 1983 Effects of aluminium, OH : Al and P : Al molar ratios, and ionic strength on soybean root elongation in solution culture. Soil Sci. 136, 197–207.

Blamey F P C, Wheeler D M, Edmeades D C and Christie R A 1990 Independence of differential aluminium tolerance in *Lotus* on changes in rhizosphere pH or excretion of organic ligands. J. Plant Nutr. 13, 713–728.

Foy C D, Burns G R, Brown J C and Fleming A L 1965 Differential aluminum tolerance of two wheat varieties associated with plant-induced pH changes around their roots. Soil Sci. Soc. Am. Proc. 29, 64–67.

Hairiah K 1992 Aluminium tolerance of *Mucuna*, a tropical leguminous cover crop. Thesis, Groningen, The Netherlands.

Hairiah K, Stulen I and Kuiper P J C 1990 Aluminium tolerance of the velvet beans *Mucuna pruriens* var. *utilis* and *Mucuna deeringiana*. I. Effects of aluminium on growth and mineral composition. *In* Plant Nutrition - Physiology and Applications. Ed. M L Van Beusichem. pp 365–374. Kluwer Academic Publishers, Dordrecht, The Netherlands.

Hairiah K, Van Noordwijk M and Setijono S 1991a Tolerance to acid soil conditions of the velvet beans *Mucuna pruriens* var. *utilis* and *M. deeringiana*. I Root development. Plant and Soil 134, 95–105.

Hairiah K, Van Noordwijk M and Stulen I 1991b Determination of inorganic monomeric aluminium with the 60 s pyrocatechol violet technique. Institute for Soil Fertility Research, Nota 244.

Hairiah K, Van Noordwijk M, Stulen I and Kuiper P J C 1992 Aluminium avoidance by *Mucuna pruriens* var. *utilis*. Physiol. Plant. 86, 17–24.

Hairiah K, Van Noordwijk M, Stulen I, Meijboom F W and Kuiper P J C 1993 Phosphate nutrition effects on aluminium avoidance of *Mucuna pruriens* var. *utilis*. Environ. Exp. Bot. 33, 75–83.

Handayanto E, Nuraini Y, Purnomosidi P, Hanegraaf M, Agterberg G, Hassink J and Van Noordwijk M 1992 Decomposition rates of legume residues and N-mineralization in an ultisol in Lampung. Agrivita 15, 75–86.

Jaillard B 1987 Techniques for studying the ionic environment at the soil/root interface. *In* Methodology in Soil-K Research. pp 247–261. Intern. Potash. Inst., Baden bei Wien, Austria.

Jaworski E G 1971 Nitrate reductase assay in intact plant tissues. Biochem. Biophys. Res. Comm. 43, 1274–1279.

Keltjens W G and Van Ulden P S R 1987 Effects of Al on nitrogen ($NH_4^+$ and $NO_3^+$) uptake, nitrate reductase activity and proton release in two sorghum cultivars differing in Al tolerance. Plant and Soil 104, 227–234.

Kerven G L, Edwards D G, Asher C J, Hallman P S and Kokot S 1989 Aluminium determination in soil solution. II. Short-term colorimetric procedures for the measurement of inorganic monomeric aluminium in the presence of organic acid ligands. Austr. J. Soil Res. 27, 91–102.

Klotz F and Horst W J 1988 Effect of ammonium and nitrate-nitrogen nutrition of aluminium tolerance of soybean (*Glycine max* L.). Plant and Soil 111, 59–65.

Marschner H 1991 Mechanisms of adaptation of plants to acid soils. Plant and Soil 134, 1–20.

McCain S and Davies M S 1983 The influence of background solution on root responses to aluminium in *Holcus lanatus* L. Plant and Soil 73, 425–430.

Miyasaka S C, Kochian L V, Shaff J E and Foy C D 1989 Mechanisms of aluminum tolerance in wheat. An investigation of genotypic differences in rhizosphere pH, $K^+$, and $H^+$ transport and root-cell membrane potentials. Plant Physiol. 91, 1188–1196.

Römheld V and Marschner H 1986 A simple method for non-destructive measurements of pH and root exudates at the root-soil interface (rhizosphere). Transaction of the XII Congress Int. Soc. Soil Sci., Hamburg. pp 937–938.

Stulen I, Lanting L, Lambers H, Posthumus F, Van de Dijk S and Hofstra R 1981 Nitrogen metabolism of *Plantago lanceolata* as dependent on the supply of mineral nutrients. Physiol. Plant. 51, 93–98.

Taylor G J 1991 Current views of the aluminum stress response; the physiological basis of tolerance. Curr. Top. Biochem. Physiol. 10, 57–93.

*F. Baluška et al. (eds.), Structure and Function of Roots*, 155–161, 1995.


# Phosphate transfer in VA mycorrhizas. Special mechanisms or not?

F.A. Smith[1], S. Dickson[2], C. Morris[2], R.J. Reid[1], M. Tester[1] and S.E. Smith[2]
[1]*Department of Botany, The University of Adelaide, S.A. 5005, Australia and* [2]*Department of Soil Science, The University of Adelaide, S.A. 5005, Australia*

*Key words:* ion channels, ion transport, mycorrhizas, phosphate, roots


## Abstract

The widespread symbiosis between plants and vesicular-arbuscular (VA) mycorrhizal fungi involves uptake of phosphate by fungal hyphae, translocation into the root, and transfer to the plant across the fungus-root interface. We have suggested that this transfer may involve membrane transport mechanisms that are only present (or are activated) once the symbiosis is established. We describe here two studies designed to clarify this issue. Firstly, we have measured phosphate inflow into non-mycorrhizal and mycorrhizal roots of *Allium porrum* (leek) and have used image analysis to calculate phosphate fluxes across the fungus – root interface. These fluxes fall in the range 4–13 nmol $m^{-2}s^{-1}$, as in other plant tissue. Secondly, we have compared effluxes of chloride and phosphate from the macro-alga *Chara*, cereal roots, and the ericoid (endo-)mycorrhizal fungus *Hymenoscyphus ericae* in culture. The aim was to ascertain if phosphate can pass through the anion channels that, when open, allow efflux of chloride. Experimental treatments involved cytoplasmic acidification and, for *Chara*, increased cytoplasmic calcium produced by electrical stimulations. We conclude that the control of phosphate efflux differs from that of chloride efflux. If 'phosphate channels' exist, the nature of the signals for opening and closing remains unknown. With all the material tested, phosphate efflux was much lower than the calculated values for fluxes at the VA mycorrhizal interface. This supports our view that phosphate transfer at the interface indeed involves 'special mechanisms' such as channels or carriers that are essentially inoperative in free-living material.


## Introduction

The symbiosis between plants and mycorrhizal fungi is extremely widespread, and different types of mycorrhizas are classified on the basis of types of fungi, types of host plant and structural modifications of the infected plant roots. With the exception of orchid mycorrhizas, there is a more-or-less common pattern of nutrient exchanges: the mycorrhizal fungus essentially acts as a physiological extension of the root, absorbing nutrients from the soil and transferring them to the host in exchange for organic carbon. In general terms there is a bidirectional source-sink relationship that involves regulated membrane transport across a fungus-root interface. The nutrient that has received most attention is phosphate, and positive mycorrhizal growth responses associated with improved phosphate nutrition are well established.

In ectomycorrhizas, this interface is entirely intercellular (extracellular), occurring at the contact between the fungal sheath and underlying root cells, and also where the fungal hyphae traverse the cortex (the 'Hartig net'). In endomycorrhizas there is again an intercellular phase – the fungal hyphae that traverse epidermis and cortex – and also an intracellular phase, where the fungus penetrates the walls of these cells. In the most widespread endomycorrhizal symbiosis, vesicular-arbuscular (VA) mycorrhizas, the surface area of this intracellular phase is relatively large, due to the development of the highly branched fungal arbuscules (Cox and Tinker, 1976).

We are interested in the mechanisms of nutrient transport across mycorrhizal interfaces and in the factors that control transport. We have previously suggested that transfer of phosphate may involve membrane transport mechanisms that are "special" in the sense that they are only present (or are only activated) once the mycorrhizal symbiosis is established (Smith and Smith, 1989, 1990; Tester et al., 1992). In thinking along these lines, we have been influenced by evidence that phosphate is not readily lost from plant or fungal cells, unlike ions such as potassium, chloride, etc. (Beever and Burns, 1980; Cairney and Smith, 1992).

We discuss here two studies designed to clarify these issues. Firstly, we have measured uptake of phosphate by non-mycorrhizal and VA mycorrhizal *Allium porrum* (leek) and have used image-analysis to calculate phosphate fluxes from fungus to host, on the basis of the surface area of the interface. Secondly, we have tested a range of plant material for the presence of anion channels that would allow efflux of phosphate in response to appropriate metabolic signals. We have measured and compared efflux of chloride and phosphate from the macroalga *Chara corallina* (widely used in studies of ion transport), roots of cereals and the ericoid (endo-)mycorrhizal fungus *Hymenoscyphus (Pezizella) ericae.* This fungus was used as a model for a VA mycorrhizal fungus since the latter cannot be grown in culture. Otherwise, the aim of using a range of species was to see if it was possible to generalise about the existence of anion channels in plants and fungi without embarking on a detailed study such as 'patch-clamping' (Tyerman, 1992).

The values for phosphate fluxes obtained in the two studies have been compared with those in other plant material, the aim being to see if it is in fact necessary to postulate 'special' transport mechanisms that operate in VA mycorrhizas.

## Materials and methods

### *Phosphate uptake and transfer in VA mycorrhizal symbiosis*

Leeks (*Allium porrum* cv. Musselburgh) were grown from seed in pots containing 600 g of a soil/sand mix containing low levels of available phosphate. Plants were either not inoculated or inoculated with pot culture material of *Glomus sp.* City Beach, grown previously on *Trifolium subterraneum.* Plants were grown in the glasshouse, watered to 12.5% soil moisture three times a week, and received basic nutrients minus P. Methods for harvest and determination of weight and phosphate contents have been described previously (Son and Smith, 1988) and methods for estimation of mycorrhizal infection for this experiment have been described by Smith and Dickson (1991). Inflow of phosphate per metre of root was calculated for three successive growth periods, with plants harvested at 21, 42 and 63 days, using the method of Brewster and Tinker (1972). Inflow via the fungal hyphae was calculated by subtracting inflows into non-mycorrhizal plants from those for the corresponding mycorrhizal plants for each growth period. Numbers and uninvaginated perimeters of active arbuscules and intercellular hyphae in sectioned roots were obtained with the Chromatic Colour Image Analysis system (L.R. Jarvis – Leading Edge), as described by Smith and Dickson (1991). The uninvaginated perimeters were converted to invaginated perimeters using a factor of 2.48. This combined the factor of 2.07 obtained by Cox and Tinker (1976) for the ratio of the invaginated arbuscular perimeter to the perimeter of the host cell, for *Allium cepa* (onion) infected by *Glomus mosseae*, and a ratio of 1.2 for the perimeter of the infected host cell to uninvaginated arbuscular perimeter in our experiments. The values for invaginated arbuscular perimeter per cross-section were then converted to arbuscular surface area per metre using the standard stereological formula (see Cox and Tinker, 1976). Surface areas of intercellular hyphae per metre of root were calculated assuming that the hyphae were cylindrical or, in the relatively few cases where there were two or more hyphae per intercellular space, that the bundle represented a single cylinder. The error here was small. The calculated areas of interface take into account that not all infected sections contained arbuscules. Values for percentage infection include regions with either hyphae alone, or hyphae plus arbuscules.

At 21 and 42 days, 50 sections from a single plant were used. At 63 days, values are means ± standard errors of means for values from 50 sections of each of three replicate plants. Direct measurements of the perimeters of the arbuscules and intercellular hyphae were not available for the first harvest but were estimated using the numbers at 21 days and mean values for the perimeters at 42 days (Smith and Dickson, 1991). Again, the error would be small.

### *Effluxes of chloride and phosphate*

Isolated giant internodal cells of *Chara corallina* were bathed for 7–14 days in an artificial pondwater containing 0.1 m*M* $K_2SO_4$, 0.5 m*M* $CaSO_4$, and either 0.1 m*M* or 1.0 m*M* NaCl, labelled with high-specific activity $^{36}Cl$, after which efflux of $^{36}Cl$ into non-radioactive solutions was measured as described by Smith and Reid (1990). Efflux of $^{32}P$ was measured similarly, following loading in the basic solutions plus 0.1 m*M* $NaH_2PO_4$ labelled with $^{32}P$. Effects of cytoplasmic acidification were tested by measuring efflux at an external pH of 5.0, and transferring cells to solutions containing 0.5 m*M* butyric acid (Smith and Reid, 1990). To test the effects of action potentials on efflux, cells were electrically stimulated by passing small depolarizing electric currents through individual cells six times over a period of 10 min.

Effluxes of chloride and phosphate from roots of *Triticum aestivum* (wheat) and *Hordeum vulgare* (barley) respectively were measured with roots from plants grown for 4 or 5 days in 0.5 m*M* $CaSO_4$ and then transferred for 2 days to solutions containing $^{36}Cl$ or $^{32}P$. The loading solution for experiments with $^{36}Cl$ contained 0.1 m*M* NaCl and 0.1 m*M* $KH_2PO_4$, and that for experiments with $^{32}P$ contained 0.1 m*M* $KH_2PO_4$, plus 0.5 m*M* $CaSO_4$ in both cases. Roots were excised and rinsed before measurement of efflux into non-radioactive solutions over successive 10 min or 15 min periods. For experiments with $^{36}Cl$, the efflux solution was the same as that for uptake, without added $^{36}Cl$. For experiments with $^{32}P$, the solution contained 0.5 m*M* $Na_2SO_4$, 0.5 m*M* $CaSO_4$ and 2 m*M* Na/MES buffer at pH 5.0. In both cases, 1 m*M* butyrate at pH 5.0 was added at known times, and effects on efflux were tested. It is assumed that any artefacts arising from the use of excised roots applied to both control and test conditions.

Efflux from *Hymenoscyphus ericae* was measured using small mats of fungus that were cultured in a solution containing 0.2 m*M* $NH_4Cl$, 0.3 m*M* $MgSO_4$, 0.2 m*M* NaCl, 0.2 m*M* $CaCl_2$, 5.4 $\mu M$ FeEDTA, 0.3 $\mu M$ thiamine-HCl, 1 m*M* glucose, 0.1 m*M* $KH_2PO_4$ plus 10 m*M* MES buffer (pH 5.0). Small pieces of fungus, about 1 $cm^2$, were cut from the mats and placed in the culture medium, labelled with high specific activity $^{36}Cl$ or $^{32}P$, for 24 hours. Efflux was measured into non-radioactive solutions of the same composition as above, minus phosphate. The aim of omitting phosphate from efflux solutions was to produce a 'source-sink' effect to maximise the chances of stimulating phosphate efflux. Fortuitously, the experimental pH of 5.0, chosen to allow rapid influx of butyric acid, would be expected to approximate to that in the apoplast at the mycorrhizal interface, given the operation of $H^+$-extruding ATPases (Smith and Smith, 1990).

Radioactivity in efflux solutions was measured by scintillation counting, and was converted to amounts of ions effluxed either using the known internal concentrations for $Cl^-$ efflux from *Chara corallina* or in all other cases by assuming that internal specific activities equalled the specific activities of the radioactive loading solutions. With the latter method, the effluxes are minimum estimates since internal specific activities could have been lower than external specific activities in some cases.

## Results

### *Phosphate fluxes across the VA mycorrhizal interface*

Table 1 shows inflow of phosphate into *Allium porrum*, over three growth periods. Declining inflow with time, into both non-mycorrhizal and mycorrhizal plants, could have several causes, such as ageing of roots, i.e. a decreasing proportion of root length being involved in phosphate uptake, or progressive depletion of phosphate in the pots. Mycorrhizal plants had higher inflows over the the three periods, allowing calculation of phosphate transferred by the mycorrhizal fungus (Table 1 ). Table 2 shows the percentage infection (which was high), at the three harvest times, and the surface area of interface : arbuscules, intercellular hyphae, and the total per metre of root. Table 3 shows the phosphate fluxes across the interface over the three growth periods, with values ranging from 3.7 to 12.8 nmol $m^{-2}s^{-1}$. Inclusion of the intercellular hyphae increasingly affects the calculated fluxes over time.

A similar experiment was also done with *Glomus mosseae* (Smith and Dickson, 1991). In this case the calculated P inflows via the fungus were lower than those shown in Table l. This was associated with slightly higher percentage infections (73–84% over the three growth periods). Although diameters of arbuscules in the two fungi were similar, numbers of arbuscules and intercellular hyphae per m of root, and hence areas of interfaces, were greater with *G. mosseae*, resulting in calculated phosphate fluxes that fall in a lower range: 1.7 to 3.2 nmol $m^{-2}s^{-1}$.

### *A search for anion channels that allow efflux of phosphate*

Exposure of cells of *Chara corallina* to low concentrations of butyric acid causes cytoplasmic acidification by up to 0.6 units, and massive efflux of chloride (Smith and Reid, 1990). Figure 1a shows this effect and Figure 2a shows that chloride efflux is also greatly increased during electric action potentials. These results demonstrate the opening of anion channels by changes in intercellular ionic concentrations: cytoplasmic pH in the former treatment, possibly associated

*Table 1.* Inflows of phosphate into nonmycorrhizal (-M) and mycorrhizal (+M) *Allium porrum* (leek); with inflows via fungus calculated by subtraction

| Growth period (days) | Inflow: pmol $m^{-1}$ (root) $s^{-1}$ | | |
|---|---|---|---|
| | -M | +M | via fungus |
| 0–21 | 18.0 | 33.0 | 15.0 |
| 21–42 | 2.7 | 25.0 | 22.3 |
| 42–63 | 2.9 | 12.0 | 9.1 |

*Table 2.* Percentages of the root length infected, and areas of arbuscular and hyphal interface infected, at three harvest times

| Harvest time | Infection | Interface area : $m^2$ $m^{-1}$ (root) $\times 10^{-3}$ | | |
|---|---|---|---|---|
| (days) | (%) | Arbs. | Hyphae | Total |
| 21 | 63.5 ± 4.1 | 2.33[1] | 0.31[1] | 2.64[a] |
| 42 | 68.8 ± 1.9 | 2.65 | 0.99 | 3.64 |
| 63 | 82.1 ± 3.4 | 0.44 | 0.89 | 1.33 |

[a]estimates; for derivation see Materials and methods.

with increased cytoplasmic calcium – though the latter effect has not been demonstrated – and increased cytoplasmic calcium (but no change in cytoplasmic pH) in the second treatment (see Tyerman, 1992). Figure lb and 2b demonstrate that there was little response of the low phosphate efflux to internal acidification, at least initially, and no change during action potentials. In individual cells $^{32}P$ efflux initially increased by not more than about 100% following internal acidification, with a later more dramatic increase. The large error bars in Figures 1b and 2b reflect differences in efflux between individual cells even before the experimental treatments were applied.

Much of the $^{32}P$ absorbed by charophytes under these conditions remains as inorganic phosphate (Smith, 1966; also unpublished results with *Chara*). Assuming, therefore, that the phosphate species that effluxes during acidification is $H_2PO_4^-$, the increase could be caused by conversion of cytoplasmic $HPO_4^{2-}$ to $H_2PO_4^-$. With a pKa of about 7.2 for this reaction, a change in cytoplasmic pH from 7.6 to 7.0 would be more than enough to produce this effect. In other words, it is not necessary to postulate increased opening of channels if efflux is limited by the cytoplasmic concentration of $H_2PO_4^-$. Our confidence in this interpretation is increased by the lack of effect of electrical stimulation on phosphate efflux.

*Table 3.* Fluxes of phosphate from fungus to *Allium porrum* (leek). Values are obtained by dividing inflow (Table 1) by average areas of interface between two harvests (from Table 2)

| Growth Period | Phosphate flux (nmol $m^{-2}s^{-1}$) | |
|---|---|---|
| (days) | Arbs. only | Arbs. + hyphae |
| 0–21 | 12.8 | 11.3 |
| 21–42 | 9.0 | 7.1 |
| 42–63 | 5.9 | 3.7 |

Figure 3a shows that intracellular acidification greatly increased chloride efflux from excised wheat roots. Efflux of phosphate from barley roots was not affected (Fig. 3b). During the loading period, net uptakes of chloride ($^{36}Cl$) and phosphate ($^{32}P$) were both about 10 $\mu$mol $g^{-1}$ over 24 h, but subsequent efflux of phosphate was much smaller than that of chloride (Fig. 3).

Figures 4a and 4b compare efflux of chloride and phosphate from *Hymenoscyphus ericae*. Intracellular acidification caused significant increases in efflux of both anions. Net uptake of chloride ($^{36}Cl$) during the loading period was smaller than that into the cereal roots (less than 2 $\mu$moles $g^{-1}$). Net uptake of phosphate (P) was about the same as into the cereal roots (about 10 $\mu$moles $g^{-1}$). The increases in phosphate

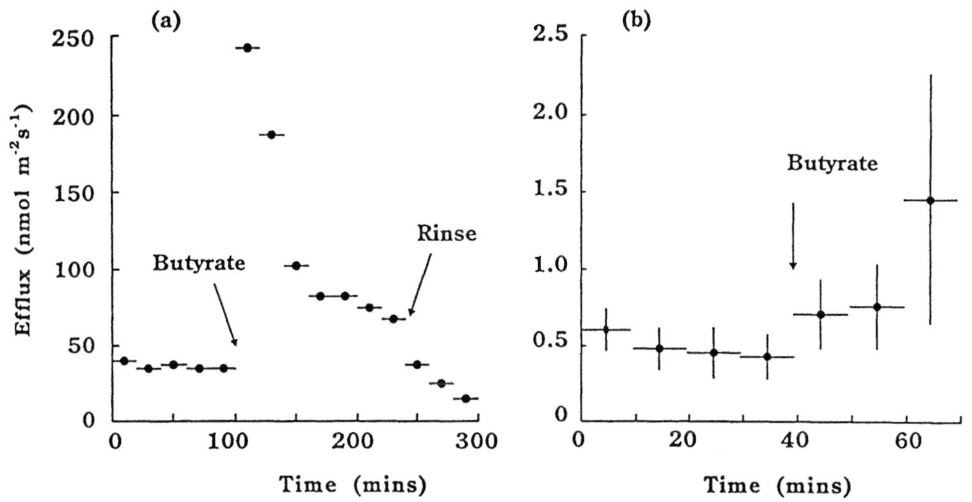


*Fig. 1.* Effects of 0.5 m*M* butyric acid (pH 5.0) on (**a**) efflux of chloride from a single cell of *Chara corallina*, after rinsing for 10 min; and (**b**) mean efflux of phosphate from eight cells of *Chara corallina* after rinsing for 90 min. Error bars are standard errors of the mean (s.e.m.).

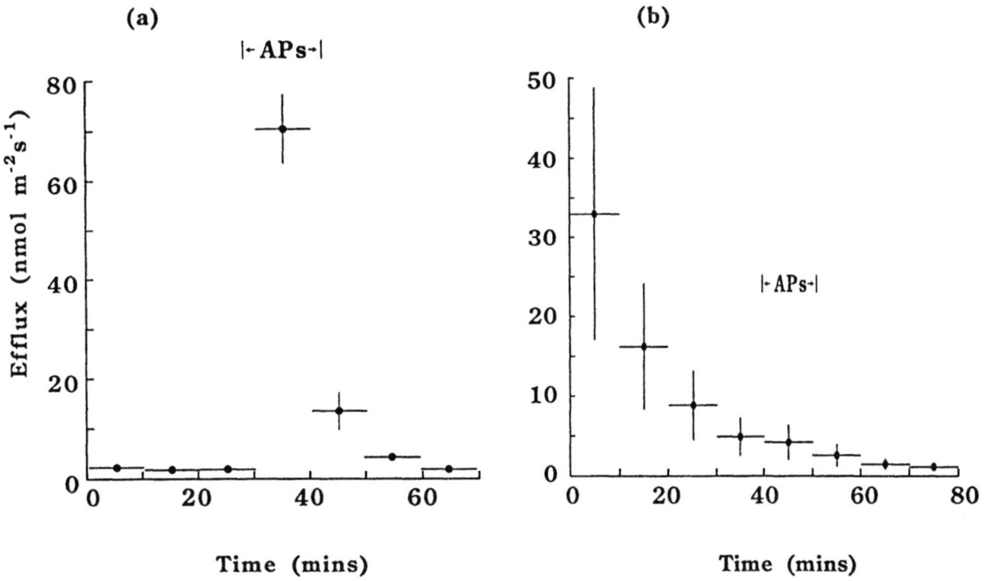


*Fig. 2.* Effects of action potentials (AP) on efflux of (**a**) chloride and (**b**) phosphate from cells of *Chara corallina*. Cells were first rinsed for 30 min (a) or 2 min (b). Values are means ± s.em's for six cells (a) or four cells (b).

efflux from the fungus could have been caused by an increased cytoplasmic concentration of $H_2PO_4$ following acidification, as in *Chara*, rather than by increased opening of channels.

## Discussion

These two studies support our previous suggestions (Smith and Smith, 1989, 1990; Tester et al., 1992) that mycorrhizal infection involves novel mechanisms for phosphate efflux at fungus/host interfaces, or abnormally enhanced efflux compared with free-living material. Our calculations of phosphate fluxes across the interface give values similar to the flux (13 nmol $m^{-2}$ $s^{-1}$) calculated by Cox and Tinker (1976) for *G. mosseae*/*Allium cepa*, using essentially the same methods. We are aware that the factors used to calculate surface areas of the interface still involve some assumptions and sources of error, possibly arising for example, from different fungus/host combinations. However, our values for diameters of leek cells and arbuscules of

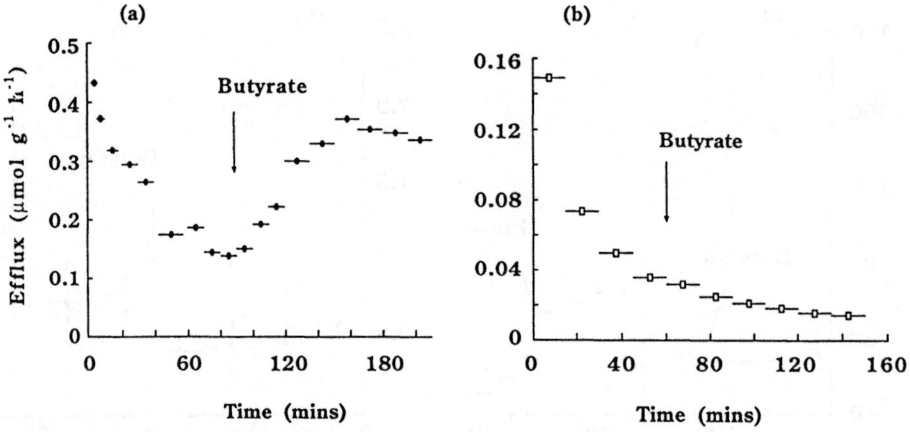


*Fig. 3.* Effects of 1.0 m*M* butyric acid (pH 5.0) on (**a**) efflux of chloride from a batch of excised wheat roots and (**b**) efflux of phosphate from a batch of excised barley roots, after rinsing for 20 min (a), or 40 min (b)

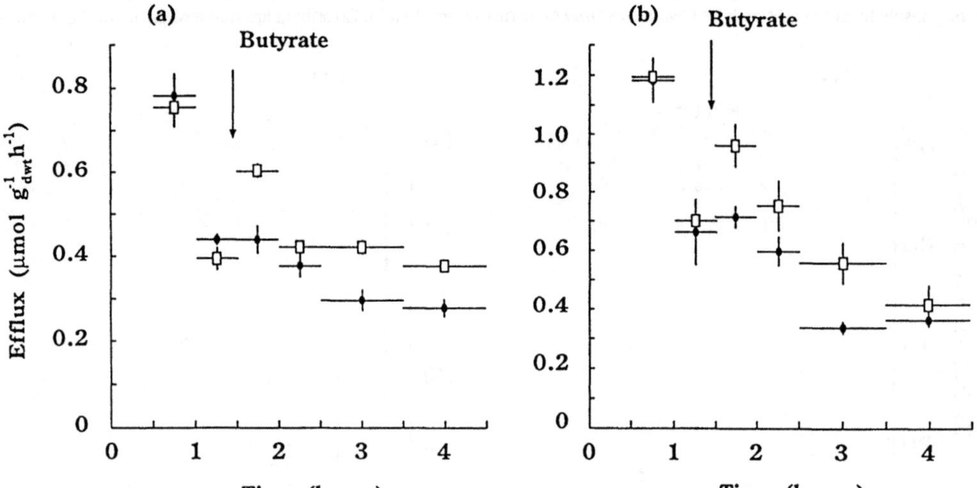


*Fig. 4.* Effects of 2.0 m*M* butyric acid (pH 5.0) on efflux of (**a**) chloride and (**b**) phosphate from *Hymenoscyphus ericae* after rinsing for 20 min. Closed symbols : controls (no butyrate added); open symbols : butyrate added at the times shown

*Glomus mosseae* are very similar to those obtained by Alezxander et al. (1989) for the *G fasciculatus/Allium cepa* symbiosis.

Steady-state phosphate efflux from *Chara corallina* was lower than chloride efflux, and also lower than phosphate influx measured under similar conditions, i.e. about 10 nmol $m^{-2}s^{-1}$. Low efflux cannot be explained by irreversible conversion of inorganic phosphate to organic phosphate, as much of the phosphate that is absorbed by charophytes is accumulated as inorganic phosphate in the cytoplasm and vacuole (Smith, 1966). Phosphate efflux from cereal roots was also lower than chloride efflux. Effluxes of chloride and phosphate from *Hymenoscyphus ericae* were quite similar but also low. They can be converted to a surface area basis by using the measured radius of the hyphae (1.17 μm), and assuming the hyphae to be long cylinders with a density of 1 g $cm^{-3}$ and a fresh weight/dry weight ratio of 10. The average efflux of phosphate obtained from 70 experiments was 0.79 μmoles $g^{-1}$ (dry weight) $h^{-1}$. This equates to only 13 pmol $m^{-2}$ $s^{-1}$, 100–1000 times lower than the calculated fluxes across the VA mycorrhizal interface. Thus if *H. ericae* is indeed a model for a VA mycorrhizal fungus, then novel mechanisms (or greatly increased rates) for phosphate efflux in symbiosis must be postulated. Effluxes of phosphate from a free-living ectomycorrhizal fungus (*Pisolithus tinctorius*) have been measured by Cairney and Smith (1992); however, fluxes on a surface area basis have not been calculated. Efflux from this fungus was stimulated by increasing the external $K^+$ concentration to 50 m*M*, an effect that we have not found in *Hymenoscyphus ericae*. As far as uptake across the interface into the host plant is concerned, there is no need to postulate novel mechanisms, since influxes of 1–10 nmol $m^{-2}$ $s^{-1}$ are not high for plant cells in general (Bieleski and Ferguson, 1983).

An important question relates to differences between different VA mycorrhizal fungi. As noted above, *Glomus* sp. City Beach transferred more phosphate per $m^2$ of interface than did *G. mosseae*. Hence, the former fungus seems more efficient in transferring phosphate. Whether this relates to the transport mechanisms at the interface or to other factors such as differences in concentration at the interface resulting from different rates of delivery from the soil, remains to be seen.

Finally, our brief search for a phosphate channel in a variety of plant material revealed that, if phosphate efflux occurs through channels, they do not respond to treatments that open anion channels that allow chloride efflux from similar material (Tyerman, 1992). Hence the molecular mechanism for efflux of phosphate – whether slow or rapid – from plants and fungi is still obscure. Further progress in determining the control of phosphate transfer at the interface will depend on the formidable task of measuring the physiological conditions there (Smith and Smith, 1990). Otherwise it will remain extremely hazardous to use free-living material as model systems.

## Acknowledgement

This work was supported by the Australian Research Council.

## References

Beever R E and Burns D J W 1980 Phosphorus uptake, storage and utilization by fungi. Adv. Bot. Res. 8, 128–219.

Bieleski R L and Ferguson I B 1983 Physiology and metabolism of phosphate and its compounds. *In* Encyclopedia of Plant Physiology, Vol 15A. Inorganic Plant Nutrition. Eds. A Lauchli and R L Bieleski. pp 422–449. Springer-Verlag, Berlin

Brewster J L and Tinker P B 1972 Nutrient flow rates into roots. Soils Fert. 35, 355–359.

Cairney J G W and Smith S E 1992 Effect of monovalent cations on efflux of phosphate from the ectomycorrhizal fungus *Pisolithus tinctorius*. *In* Mycorrhizas in Ecosystems. Eds. D J Read, D H Lewis and A H Fitter. pp 352–355. CAB International, Wallingford, Oxon, UK.

Cox G and Tinker P B 1976 Translocation and transfer of nutrients in vesicular-arbuscular mycorrhizas. I. The arbuscule and phosphorus transfer: a quantitative ultrastructural study. New Phytol. 77, 371–378.

Smith F A 1966 Active phosphate uptake by *Nitella translucens*. Biochim. Biophys. Acta 126, 94–99.

Smith F A and Reid R J 1991 Biophysical and biochemical regulation of cytoplasmic pH in *Chara corallina* during acid loads. J. Exp. Bot. 42, 173–182.

Smith F A and Smith S E 1989 Membrane transport at the biotrophic interface: an overview. Aust. J. Plant Physiol. 16, 33–43.

Smith S E and Dickson S 1991 Quantification of active vesicular-arbuscular mycorrhizal infection using image analysis and other techniques. Aust. J. Plant Physiol. 18, 637–648.

Smith S E and Smith F A 1990 Structure and function of the interfaces in biotrophic symbioses as they relate to nutrient transport. New Phytol. 114, 1–38.

Son C L and Smith S E 1988 Mycorrhizal growth responses: interaction between photon irradiance and phosphorus nutrition. New Phytol. 108, 305–314.

Tester M, Smith F A and Smith S E 1992 The role of ion channels in controlling solute exchange in mycorrhizal associations. *In* Mycorrhizas in Ecosystems. Eds. D J Read, D H Lewis and A H Fitter. pp 348–351. CAB International, Wallingford, Oxon, UK.

Tyerman S D 1992 Anion channels in plants. Annu. Rev. Plant Physiol. Mol. Biol. 43, 351–373.

*F. Baluška et al. (eds.), Structure and Function of Roots*, 163–167, 1995.


# Phosphatase induction under stress conditions in wheat

A. Szabó-Nagy and L. Erdei
*Institute of Biophysics, Biological Research Center, Hungarian Academy of Sciences, P.O.Box 521, Szeged H-6701, Hungary*

*Key words:* acid and alkaline phosphatases, osmotic stress, phosphorus deficiency, pyrophosphatase, salinity stress, wheat (*Triticum aestivum*) shoot

## Abstract

The effects of phosphorus deficiency and osmotic (NaCl or polyethylene glycol, PEG) stresses on phosphatase activity were compared in the soluble fraction of wheat shoot (*Triticum aestivum* L.cv. Kharchia, a known salt tolerant cultivar). Phosphatase activity increased under P deficiency and water stress conditions, as well. The phosphatase activities in the supernatant were assayed using ATP, Na-pyrophosphate and p-nitro-phenylphosphate as substrates at different pH-s. Changes in phosphatase isoenzyme patterns under different stress conditions were followed using phosphocellulose column chromatography. In control plants two main phosphatase peaks were found. Both were present under different stress conditions, but in different proportions. There were also changes in the relative activities towards different substrates. Phosphatase isoenzyme pattern differences found under P deficiency, osmotic and salt stresses, gave a possible explanation for the differences in substrate specificities and in specific activities of the supernatants. The $Mg^{2+}$- activated alkaline phosphatase activity, specific for the shoot, decreased under all stress conditions, and it did not bind to P-11 column providing an easy separation from the acid phosphatases.

*Abbreviations:* AcPase – acid phosphatase, ATP – adenosine 5′-triphosphate, MES – 2-(N-morpholino)ethanesulfonic acid, NaPP – sodium pyrophosphate, PEG – polyethylene glycol, $P_i$ – inorganic phosphate, pNPP – p-nitrophenylphosphate, PPase – pyrophosphatase

## Introduction

Acid phosphatases (AcPase) (orthophosphoric-monoester phosphohydrolase with an acid pH optimum, EC. 3.1.3.2) are widely distributed in plants and they are thought to play a role in phosphorus metabolism. AcPase activities were reported from cytoplasmic and cell wall fractions of wheat (Hasegawa et al., 1976) and barley roots (Panara et al., 1990), from cell wall, mitochondrial and microsomal fractions of bean hypocotyl (Mizuta and Suda, 1980), aleurone particles of rice grains (Yamagata et al., 1980) and cytoplasmic and membrane fractions of yam tubers (Kamenan and Diopoh, 1983).

Alkaline inorganic pyrophosphatase (PPase) (pyrophosphatehydrolase EC 3.6.1.1.) catalyzes the hydrolysis of inorganic pyrophosphate to orthophosphate (Cooperman, 1982). All known PPases require divalent metal ions for their maximum activities, $Mg^{2+}$ being the preferred cation. The inorganic pyrophosphatase activities were observed in leaves and other plant parts (Gama Brandao and Aoyama, 1992; Krishnan and Gnanam, 1988). In the cell, a part of the PPase activity is associated with the vacuolar membrane (Chanson, 1990) and with the chloroplast (Krishnan and Gnanam, 1988).

The induction of soluble phosphatases in phosphorus deficient plants is an ubiquitous phenomenon. Under P deficient conditions, tissue extracts from tomato (Besford, 1979; Boutin et al., 1981) and wheat (Sedenko and Zaitseva, 1984; Szabó-Nagy et al., 1987, 1992) showed elevated phosphatase activity. The increased level of phosphatase activity correlated well with a low level of $P_i$ (Barrett-Lennard and Greenway, 1982; Besford, 1979; Boutin et al., 1981; Dracup et al., 1984; O'Connel and Grove, 1985; Sedenko and Zaitseva, 1984; Szabó-Nagy et al., 1987; Ueki and Sato, 1977). Similar increases of phosphatase activities were

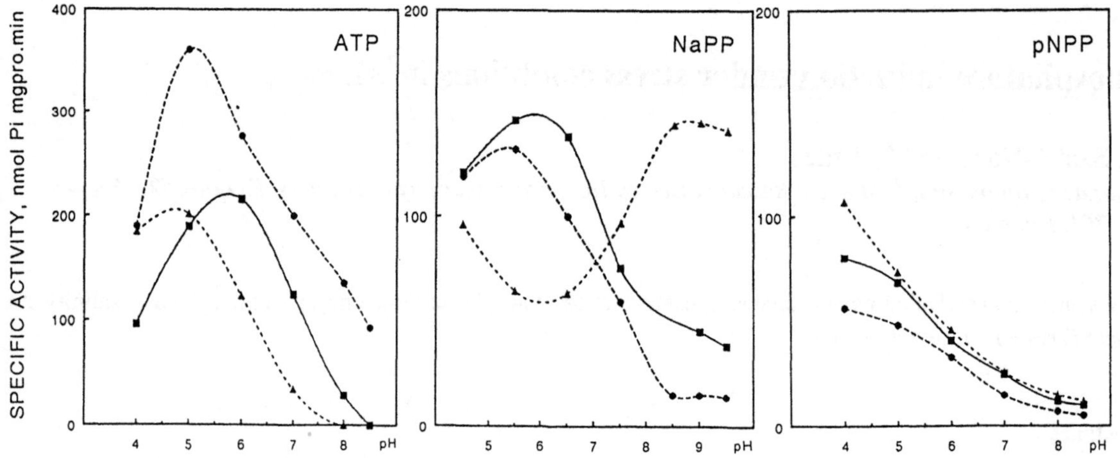


*Fig. 1.* Substrate and pH dependence of the phosphatase activities in 30,000 g supernatant from control Kharchia shoot. Activities were measured using 2 m*M* ATP, NaPP or pNPP as substrates in the absence (■) or presence of 10 m*M* $Mg^{2+}$ (▲) or 2 m*M* $Ca^{2+}$ (●). 25 m*M* Tris-Mes buffer system was used to generate the pH range.

observed under water deficient conditions (Barrett-Lennard and Greenway, 1982; Barrett-Lennard et al., 1982; Jonas et al., 1990), however, without considerable changes in total P levels of the tissues.

Growth under salt and osmotic stresses of different varieties of wheat seedlings was reduced at higher than 50 m*M* NaCl concentrations (Erdei and Trivedi, 1989). The effects of different stress conditions as P deficiency, ionic (NaCl) and non-ionic osmotic (mannitol or PEG) stresses were compared in wheat varieties differing in water stress tolerance (Szabó-Nagy et al., 1992). Water content in stressed plants decreased in each case, especially in shoots. As expected, P content decreased under P deficiency in both roots and shoots while under non-ionic osmotic stress conditions it was unchanged. Kharchia was consequent in maintaining high levels of $K^+$ under NaCl stress as well, both in calli and whole plants. This character appears to be genotype-specific and may be of adaptive value under stress (Sharma, 1989; Trivedi et al., 1991).

The purpose of the present work was to investigate changes in phosphatase activities that occur in wheat under P deficiency and water stress, caused by ionic (NaCl) and non-ionic (PEG) osmotic stresses. In particular, the changes in the phosphatase isoenzyme patterns of the cytoplasmic fraction from shoot were investigated.

## Materials and methods

Wheat (*Triticum aestivum* L. cv. Kharchia) seedlings were grown hydroponically under controlled conditions in a growth chamber (23/18°C, 11/13 h day/night period, day-time illumination at 250 $\mu$mol $m^{-2}s^{-1}$ light intensity, 85% relative humidity of the air). Plants grown in a complete nutrient solution were used as controls. For stress conditions, nutrient solutions were supplemented with 200 mOsm (checked by osmometer) NaCl or PEG 6000 at the 12th day for one week. In case of P deficiency, phosphorus was omitted from the beginning.

Soluble fraction of shoot was prepared using 50 g fresh weight, by homogenization with an Ultra Turrax (IKA-Werk, Staufen, FRG) for 3 × 2 min in 150 mL homogenization buffer consisting of 5 m*M* EDTA, 25 m*M* Tris-Mes (pH 7.2) and 250 m*M* sucrose. The homogenate was centrifuged at 10,000 *g* for 15 min and the supernatant at 30,000 *g* for 60 min to remove cell debris.

Purification of phosphatases was carried out on a fibrous phosphocellulose (Whatman, Type P11) column eluting with an increasing (0-1 *M*) concentration gradient of NaCl (Szabó-Nagy et al., 1987). One hundred mL of the 30,000 *g* supernatant obtained from 3-week-old shoots was applied onto a phosphocellulose column (2.5 × 20 cm) equilibrated with 25 m*M* Tris-Mes buffer (pH 7.2). The same buffer (100 mL) was used for the washing of the unbound proteins. Fractions (6.5 mL) were collected at 60 mL $hr^{-1}$ and assayed for AcPase and PPase activities.

One hundred or 25 μL supernatant were used for phosphatase activity determination. (The fractions after chromatography were not dialysed, the reaction mixture contained variable concentration of NaCl, but 100 m*M* NaCl in the assay caused only slight effects on the phosphatase activity.) The acidic phosphatase activity was measured in a reaction mixture (0.5 mL) containing 25 m*M* Tris-Mes (pH 6.5 or 5.0 as stated) and either 2 m*M* ATP as Na-salts, Na- pyrophosphate (Reanal, Budapest, Hungary) or p-nitro-phenylphosphate (Calbiochem, San Diego, Calif., USA) as substrates. Alkaline phosphatase activity was measured in the same buffer system (pH 8.5) in the presence of 10 m*M* $MgCl_2$ and 2 m*M* Na-pyrophosphate. After 30 min incubation at 30°C, reactions were terminated by the addition of 0.1 mL 20% (w/v) trichloroacetic acid.

Phosphatase activity was determined from the amount of $P_i$ liberated from the substrates (Lindeman, 1958) and followed as changes in the absorbance at 730 nm. Protein content was determined according to Markwell et al. (1978) with bovine serum albumin as a standard.

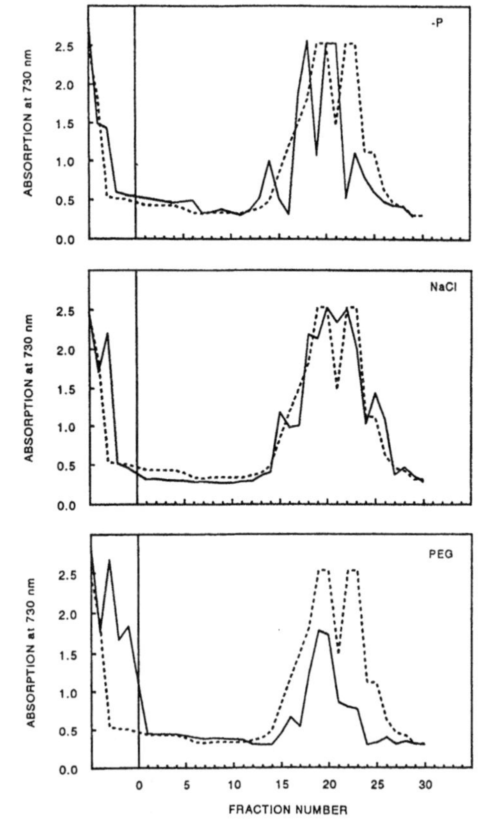


*Fig. 2.* Phosphocellulose (P 11) chromatography of the 30,000 *g* supernatant (100 mL) from Kharchia wheat shoot homogenates. For elution 0 *M* (100 mL)-1 *M* (100mL) NaCl gradient was used. Plants were grown in P-deficient growth solution or in the presence of 200 m Osmol NaCl or PEG. Control results are presented as dotted lines. One representative of the independent series (3- in case of P deficient and Control plants, and 2- under NaCl or PEG treatment) can be seen on the figures. The fraction volume was 6.5 mL, from each of them 25μL was used for phosphatase activity measured in the presence of 2 m*M* ATP at pH 6 (25 m*M* Tris-Mes).

## Results and discussion

Earlier we have shown (Szabó-Nagy et al., 1992) that phosphatase levels in wheat roots increased under P deficiency but were nearly unchanged in salt- and PEG-stressed plants. In shoots, in addition to P deficiency, salt and PEG treatments were also inductive factors. The $Mg^{2+}$-activated alkaline pyrophosphatase activity in shoots of control plants was very pronounced but it was decreased by all stress conditions while in the roots it was practically absent.

Kharchia, a known salt tolerant wheat cultivar, is a suitable object to investigate the effects of extreme environmental conditions. In the present study this cultivar was used, and the enzymatical changes in the shoot supernatant was followed. First the pH dependence (in 25 m*M* Tris-Mes buffer system) and ion-requirement of the phosphatase activities in the supernatant were determined in the presence of different substrates (2 m*M* ATP, NaPP or pNPP) (Fig. 1). In general, there were higher phosphatase activities with ATP than with NaPP and pNPP for the supernatants. The pH optimum for ATP without any added ion was at pH 6, which is similar to the pH optimum of the plasma membrane ATPase (Szabó-Nagy et al., 1989). The optimum was shifted to a more acid pH in the presence of $Mg^{2+}$ and $Ca^{2+}$. While $Mg^{2+}$ did not increase the activity, $Ca^{2+}$ caused a significant increase of it. The effect of these cations is opposite compared to those on the plasma membrane ATPase. This result suggests that the activity measured in the supernatant does not originate from the plasma membrane.

In the presence of NaPP, the pH optimum was similar to that found for ATP in the absence of divalent cations: the activity was decreased by $Ca^{2+}$ in the whole pH range, while $Mg^{2+}$ was inhibitory only at the acid pH range. Above pH 8, $Mg^{2+}$ caused a very significant activation similarly to that found for alkaline pyrophosphatase in maize germ (Gama Brandao and Aoyama, 1992).

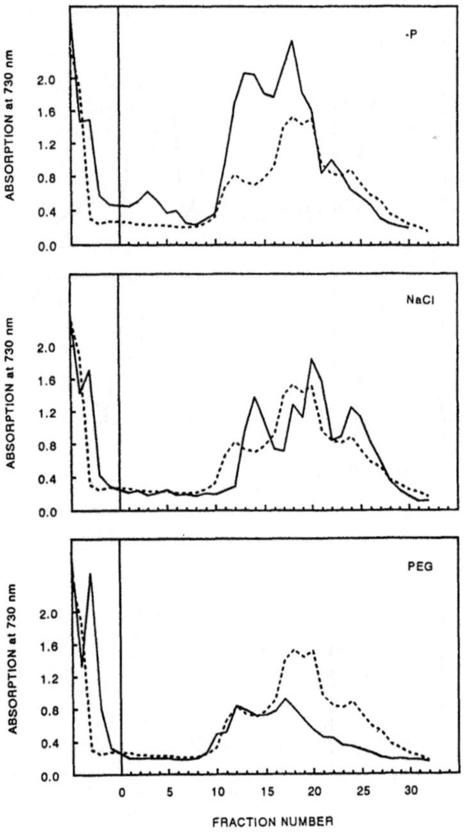


*Fig. 3.* Phosphocellulose chromatography as in Figure 2. Phosphatase activity was measured in the presence of 2m*M* pNPP as substrate at pH 5 (25 m*M* Tris-Mes) using 100$\mu$L eluate.

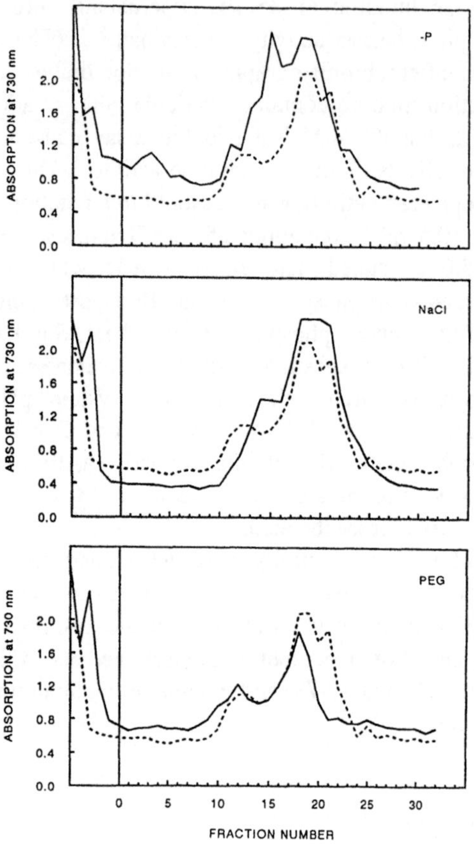


*Fig. 4.* Phosphocellulose chromatography as in Figure 2. Phosphatase activity was measured in the presence of 2m*M* NaPP as substrate at pH 6 using 100$\mu$L eluate.

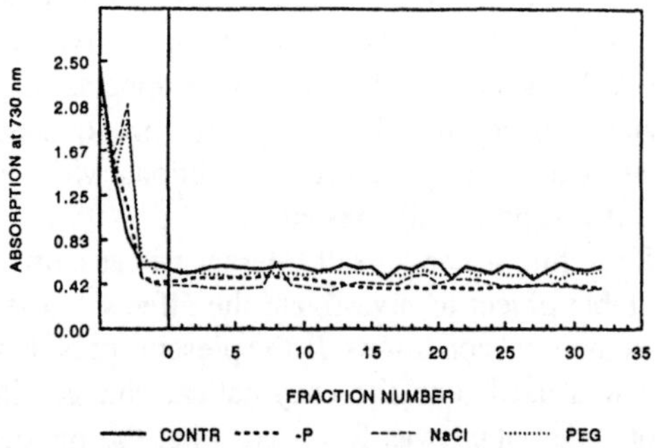


*Fig. 5.* Phosphocellulose chromatography as in Figure 2. The alkaline pyrophosphatase activity was measured in the presence of 2 m*M* NaPP as substrate, 10 m*M* $Mg^{2+}$ and 100$\mu$L eluate at pH 8.5 (25 m*M* Tris-Mes).

The phosphatase activity measured in the presence of pNPP showed the lowest pH optimum and this activity was affected only slightly by cations.

On the basis of their substrate-, pH- and ion-requirements, the following conditions were chosen for the determination of the different phosphatase activities: AcPase in the presence of pNPP at pH 5.0, with ATP and NaPP at pH 6.0; alkaline PPase in the presence of NaPP and 10 m*M* $Mg^{2+}$ at pH 8.5.

For the separation of the phosphatase isoenzymes, induced or decreased by different stress conditions, cellulose phosphate column chromatography was used. The fractionation of the different AcPases present in the soluble fractions were assayed with the above mentioned substrates (Figs. 2-5). In case of control plants, two main phosphatase peaks were detected on the basis of their ATP hydrolysing activities (Fig. 2) similarly to those found in root supernatant from MV-8 wheat cultivar (Szabó-Nagy et al., 1987) and from barley (Panara et al., 1990). In the present experiments these two forms appeared under P deficiency and salt stress but in the PEG-treated plants the second peak decreased significantly. Under P deficiency and salt stress, additional new peaks appeared.

The activities detected in the presence of pNPP (Fig. 3) showed similar phosphatase isoenzyme patterns under the investigated stress conditions. The new peak present in P deficient shoot had much higher pNPP hydrolysing activity than in control, and the opposite were detected after PEG treatment. The phosphatase activities in the presence of pNPP were totally inhibited by 0.1 m*M* molybdate, in agreement with data for a cytoplasmic AcPase from barley roots (Panara et al., 1990). In the presence of NaPP (Fig. 4) as substrate, similar effects were detected as in case of ATP and pNPP. The alkaline PPase activity was determined in the presence of $Mg^{2+}$ at pH 8.5 with NaPP. It is seen from Figure 5 that the alkaline PPase did not bind to the phosphocellulose column, therefore, it did not disturb the measurements of AcPase activities.

These results suggest that cation exchanger fibrous cellulose phosphate chromatography provides a suitable method for the separation of AcPases from the alkaline PPase. The results also indicate that the NaPP hydrolysing activity measured at pH 6 is not due to the activity of the alkaline PPase which is a separate enzyme.

## Acknowledgements

We would like to thank Mónika Ördög, Erika Kónya and Etelka Mitrov for their technical assistance. This project was financially supported by the Hungarian Scientific Research Foundation (OTKA I/4. T 5074).

## References

Barrett-Lennard E G and Greenway H 1982 Partial separation and characterization of soluble phosphatases from leaves of wheat grown under phosphorus deficiency and water deficit. J. Exp. Bot. 33, 694–704.

Barrett-Lennard E G, Robson A D and Greenway H 1982 Effect of phosphorus deficiency and water deficit on phosphatase activities from wheat leaves. J. Exp. Bot. 33, 682–693.

Besford R T 1979 Quantitative aspects of leaf acid phosphatase activity and the phosphorus status of tomato plants. Ann. Bot. 44, 153–161.

Boutin J P, Provot M and Roux L 1981 Effect of cycloheximide and renewal of phosphorus supply on surface acid phosphatase activity of phosphorus deficient tomato roots. Physiol. Plant. 51, 353–360.

Chanson A 1990 Use of the pyrophosphatase activity as a reliable tonoplast marker in maize roots. Plant. Sci. 71, 199–207.

Cooperman B S 1982 The mechanisms of action of yeast inorganic pyrophosphatase. Methods Enzymol. 87, 526–548.

Dracup M N H, Barrett-Lennard E G, Greenway H and Robson A D 1984 Effects of phosphorus deficiency on phosphatase activity of cell walls from roots of subterranean clover. J. Exp. Bot. 35, 466–480.

Erdei L and Trivedi S 1989 Responses to salinity of wheat varieties differing in drought tolerance. *In* Plant Water Relations and Growth under Stress. Eds. M Tazawa et al. pp 201–208. MYU K K, Tokyo.

Gama Brandao M E and Aoyama H 1992 Kinetic properties of maize germ cytoplasmic inorganic pyrophosphatase. Plant Physiol. Biochem. 30, 753–760.

Hasegawa Y, Lynn K R and Brockbank W J 1976 Isolation and partial characterization of cytoplasmic and wall bound acid phosphatases from wheat roots. Can. J. Bot. 54, 1163–1169.

Jonas O A, Pereyra M C, Cabeza C, Golberg A D and Ledent J F 1990 Activity of nitrate reductase and acid phosphatase in leaves of wheat, after a period of cessation of watering. Cereal. Res. Commun. 18, 299–305.

Kamenan A and Diopoh J 1983 Properties of two membrane-bound acid phosphatases compared with those of a cytoplasmic acid phosphatase from *Dioscera cayenensis rotundata*. Plant Sci. Lett. 32, 305–312.

Krishnan V A and Gnanam A 1988 Properties and regulation of $Mg^{2+}$-dependent chloroplast inorganic pyrophosphatase from *Sorghum vulgare* leaves. Arch. Biochem. Biophys. 260, 277–284.

Lindeman W 1958 Observations on the behavior of phosphate compounds in *Chlorella* at the transition from dark to light. *In* Proceedings of the II. International Conference of UN on the Peaceful Uses of Atomic Energy. 24, 8–15.

Markwell M A K, Hass S M, Bieber L L and Tolbert N E 1978 A modification of the Lowry procedure to simplify protein determination in membrane and lipoprotein samples. Anal. Biochem. 87, 206–210.

Mizuta S and Suda S 1980 A comparative study of multiple acid phosphatases in cellular fractions of bean hypocotyl. Ann. Bot. 45, 369–382.

O'Connel A M and Grove T S 1985 Acid phosphatase activity in karri (*Eucalyptus diversicolor* F. Muell. in relation to soil phosphate and nitrogen supply. J. Exp. Bot. 36, 1359–1372.

Panara F, Pasqualini S and Antonielli M 1990 Multiple forms of barley root acid phosphatase: purification and some characteristics of the major cytoplasmic isoenzyme. Biochim. Biophys. Acta 1037, 73–80.

Sharma S K 1989 Effect of salinity on growth, ionic and water relations of three wheat genotypes differing in salt tolerance. Indian J. Plant Physiol. 32, 200–205.

Sedenko D M and Zaitseva M G 1984 The activity of phosphatases in wheat roots in relation to plant supply with phosphorus. Fiziol. Rast. 31, 146–152.

Szabó-Nagy A, Oláh Z and Erdei L 1987 Phosphatase induction in roots of winter wheat during adaptation to phosphorus deficiency. Physiol. Plant. 70, 544–552.

Szabó-Nagy A, Erdei L and Galiba G 1992 Induction of soluble phosphatases under ionic and non-ionic osmotic stresses in wheat. J. Plant Physiol. 140, 629–633.

Szabó-Nagy A, Bérczi A and Erdei L 1989 Plasma membrane purification from roots of sunflower by phase partitioning. Acta Biochim. Biophys. Hung. 24, 203–211.

Trivedi S, Galiba G, Sankhla N and Erdei L 1991 Responses to osmotic and NaCl stress of wheat varieties differing in drought and salt tolerance in callus cultures. Plant Sci. 73, 227–232.

Ueki K and Sato S 1977 Regulation of phosphatase synthesis by orthophosphate in cultured tobacco cells. Plant Cell Physiol. 18, 1253–1263.

Yamagata H, Tanaka K and Kasai Z 1980 Purification and characterization of acid phosphatase in aleurone particles of rice grains. Plant Cell Physiol. 21, 1449–1460.

*F. Baluška et al. (eds.), Structure and Function of Roots*, 169–173, 1995.


# Changes in respiration and ultrastructure of radicle mitochondria during early germination in isolated wheat embryos

O. Gašparíková, T. Pšenáková, M. Čiamporová, I. Haviarová, M. Luxová and S. Kašiarová
*Institute of Botany, Slovak Academy of Sciences, Dubravska cesta 14, 842 23 Bratislava, Slovak Republic*

*Key words:* cytochrome and alternative pathways, germination, mitochondrial structure, respiration, wheat embryo, *Triticum durum*

## Abstract

The contributions of the cytochrome and alternative pathways to total respiration were assessed together with mitochondrial structure during the early stages of germination of isolated wheat embryos (*Triticum durum*, cv. Appulo). Sharp acceleration in oxygen uptake occurred within minutes after the embryos had been exposed to water. The initial increase in total dark respiration was due to an increase in the activity of the cytochrome pathway. The presence of the alternative pathway was observed from 0.5 h to 3 h and later. The capacity of alternative pathway progresively increased during germination, amounting to 70 percent of the control rate of respiration after 24 h. However, its involvement was rather low at 24 h after imbibition. The development of mitochondrial structure during this period indicated that the initial acceleration of $O_2$ uptake resulted from the activation of pre-existing mitochondria and by an increase in the number of cristae per mitochondrion. An increase in the relative number of mitochondria and in the number of cristae per mitochondrial cross section was responsible for the next increase of $O_2$ uptake between 6 and 24 h. There was not a continuous increase in the formation of mitochondrial cristae or in the relative number and volume of mitochondria in the cortical cells of radicle during the first 24 h of germination of the isolated wheat embryos. Similarly, the total rate of respiration as well as the activities of the cytochrome and alternative pathways increased more rapidly in the early stages than after 6 h of imbibition.

*Abbreviations:* AP-alternative pathway, SHAM-salicylhydroxamic acid

## Introduction

During seed germination there is a marked increase in respiratory activity (Bewley and Black, 1978). This increase may be due to the repair and activation of pre-existing mitochondria, or to mitochondrial biogenesis (Attucci et al., 1991; Ehrenshaft and Brambl, 1990; Morohashi and Bewley, 1980; Morohashi et al., 1981; Morohashi, 1986). Mitochondria with very few cristae occurred in the cells of dry embryo radicles of *Lactuca sativa* (Paulson and Srivastava, 1968), *Pisum sativum* (Yoo, 1970), *Zea mays* (Deltour and Bronchart, 1971), *Medicago sativa* (Singh, 1977), and *Hordeum vulgare* (Buvat and Robert, 1992). Increase in the number of cristae was observed after 6 to 72 h of imbibition and germination in different plant species (Deltour and Bronchart, 1971; Hallam et al., 1972; Klein and Ben Sahul 1966; Srivastava and Paulson, 1968; Yoo, 1970). Water uptake, cell enlargement and respiratory activity of plant embryos seem to be mobilized very early and, appear to be also related to quantitative changes in the mitochondrial population. We therefore, decided to investigate ultrastructural and quantitative changes in mitochondria of the radicle cortical cells together with the pattern of development of respiratory pathways in wheat embryo during the first 24 h of imbibition and germination.

## Material and methods

### *Plant material*

Embryos of *Triticum durum*, cv. Appulo were isolated manually from dry mature seeds and incubated in 2% sucrose containing 10 $\mu g.mL^{-1}$ chloramphenicol (SERVA Heidelberg) at 25°C in the dark. Measure-

*Table 1.* Fresh weight and dry weight of embryo together with respiration rates during germination

| Hours after imbibition | Fresh wieght [a] $mg^{-1}$ embryo$^{-1}$ | Dry weight[b] $mg^{-1}$embryo$^{-1}$ | Total $O_2$ uptake nmol $O_2$ (g.FW)$^{-1}$ | nmol $O_2$(g.DW)$^{-1}$.s$^{-1}$ |
|---|---|---|---|---|
| 0 | 0.47±0.016 | 0.33±0.011 | - | - |
| 0.5 | 0.95±0.0017 | 0.33±0.014 | 6.55±1.4 | 18.78±3.6 |
| 3 | 0.96±0.035 | 0.33±0.036 | 9.10±0.5 | 30.76±5.2 |
| 6 | 0.98±0.28 | 0.3 1±0.07 | 10.95±1.1 | 37.42±5.4 |
| 24 | 2.03±0.06 | 0.36±0 | 10.70±0.9 | 61.11±5.2 |

[a]n = 4–5 batches of 10 embryos each.
[b]n = 3–4 single measurements.
Mean values ± SD.

ments of respiration rates and microscopical observations were made at designated intervals (0, 0.5, 3, 6, 18 and 24 hours from the beginning of incubation).

*Respiration of embryos*

Respiration of 10 – 15 embryos germinated from 0 to 24 hrs after imbibition was measured polarographically at 25°C in 3 mL of 10 m*M* phosphate buffer (pH 6.5) using Clarktype oxygen electrode (YSI, model 5331). Rates of $O_2$ uptake were determined 10 min after transfer and are expressed on the basis of embryo fresh weight. To check for possible side effects from the use of SHAM and to determine the optimal concentration to use embryo respiration was titrated with different concentrations of SHAM in both the absence and presence of 1 m*M* KCN (Møller et al., 1988).

*Electron microscopy*

Embryos were fixed with 3% glutaraldehyde and 1% osmium tetroxide (both dissolved in 0.1 *M* Na-cacodylate buffer, pH 7.2). The embryos were dehydrated in ethanol and propylene oxide and embedded in Epon-Araldite resin mixture. Ultrathin sections were cut at 0.5 mm from the radicle apex and stained with uranyl acetate and lead citrate. Cortical cells at this distance were studied with a Tesla BS 500 electron microscope. The negatives were printed at a standard magnification (21,600 ×) and a stereological analysis conducted using universal point lattice (Weibel, 1963; c.f. Mráz and Polónyi, 1988), in two embryos for each time interval of germination. Numerical and volume densities, i.e. the ratios of number and volume of mitochondria, to the unit cell volume were evaluated. The mean volume of mitochondrial particles and the number of cristae per mitochondrial cross section were determined in the same electron micrographs.

*Table 2.* Sensitivity to SHAM together with activities of cytochrome ($v_{cyt}$) and alternative pathway ($v_{alt}$) during wheat embryo germination. All the values are mean of three separate experiments ± SD

| Hours after imbibition | Appropriate SHAM-conc. | nmol $O_2$ (g.FW)$^{-1}$ $v_{cvt}$ | $v_{alt}$ |
|---|---|---|---|
| 0.5 | 5 m*M* | 6.00±0.7 | 0 |
| 3 | 10 m*M* | 9.10±0.5 | 1.85±0.35 |
| 6 | 10 m*M* | 10.9 ±1.1 | 1.90±0.9 |
| 24 | 15 m*M* | 10.7 ±0.9 | 1.20±0.3 |

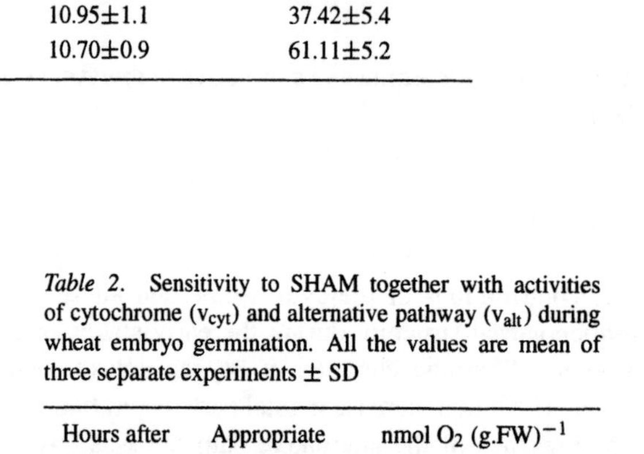


*Fig. 1.* The changes in components of embryo respiration during the first stages of germination. Percentages of respiration (expressed as percentage of total $O_2$ consumption) resistant to SHAM ($v_{cyt}$), to 1 m*M* KCN ($v_{alt}$) and to SHAM + KCN ($v_{res}$). n=3–4, *bars*±SD.

## Results and discussion

*Respiration*

Table 1 summarizes the $O_2$ uptake by the germinating embryos, together with the changes in their fresh and dry weights. The rate of total respiration rapidly increased during the first 3 hrs of imbibition when enormous enlargement of the embryo was observed in particular due to the high flux of water. The dry weight of the embryo was almost the same during the first 24 hrs of the germination and this could explain the higher increase of respiration when expressed on dry weight basis. In Table 2 we report the appropriate SHAM - concentrations at various stage of imbibition, that fully inhibited the alternative pathway without any side effects. Increase in SHAM concentration alone had little effect on respiration of 30 min imbibed embryos, indicating that the activity of the alternative pathway ($v_{alt}$) was virtually absent in dry wheat embryos. During the next period, differences were observed and higher concentrations of SHAM were used (Table 2). We evaluated the maximum flow of electrons when one of the pathways was inhibited. Figure 1 shows the percentages of SHAM resistance (equivalent to the activity of the cytochrome pathway, $v_{cyt}$) together with the percentages of KCN resistance (equivalent to the capacity of the alternatlve pathway, $v_{alt}$). A residual component of respiration ($v_{res}$) was absent in embryos imbibed for the first 6 h, since a combination of 5 or 10 m*M* SHAM and 1 m*M* KCN completely inhibited $O_2$ uptake. Therefore only the percentages of 24 hrs embryos were corrected for residual respiration shown in Figure 1.

Changes in the activities of the cytochrome and the alternative pathway ($v_{cyt}$ and $v_{alt}$, respectively), together with the values of capacity of the alternative pathway are shown in the Table 2 and Figure 1, and indicate that no alternative pathway capacity was present following the initial exposure to water. Inhibition by KCN became less complete only later (after 3 hours from the beginning of imbibition) and progressive development of the capacity for the alternative pathway activity started.

Several lines of evidence already suggest that both pathways pre-exist in mitochondria of dry and early imbibed seeds. Ehrenshaft and Brambl (1990) showed, that in maize radicles, the increase in respiration depends on a cytochrome pathway conserved in an active form in quiescent desiccated embryos. In the case of the SHAM - sensitive pathway, Leprince et al. (1992) found that, as for the cytochrome pathway, a pre-existing alternative respiratory system is also conserved in dry maize tissue. In the light of these data and our own results we supposed that during the very early stages of germination of wheat embryos respiration predominantly proceeds via the cytochrome pathway. A pre-existing alternative oxidase can also be conserved in the dry embryo, but it needs a longer time for its activation.

*Electron microscopy*

The differentiation of mitochondrial structure in cortical cells of the wheat-embryo radicle was accompanied by an increase in the number of cristae during imbibition (Figs. 2A - D). Quantitative evaluation of the number of cristae per cross section of mitochondrial particle has shown that this increase was uneven, with an initial increase up to 6 hrs, very small changes from 6 to 18 hrs and again an increase up to 24 hrs (Table 3). Similarly, the relative number and the relative volume of the mitochondria did not increase continually during the first 24 hrs of germination (Table 3).

Changes in mitochondrial number and volume obviously depend on the interrelationships between expansion and division of both cells and mitochondria during imbibition. The numerical and volume densities of mitochondria decreased during the first 3 hrs of imbibition. During the next three hours, the relative volume as well as the volume of individual particles increased, while the relative number decreased. These quantitative changes are in accord with the growth pattern of isolated wheat embryos observed by Bewley and Black (1978). The authors also observed an enormous enlargement of the embryos during the first hour of imbibition, while the embryonal growth slowed down during the next 15 hours. It is known that only cell enlargement occurs in the first hours of embryonal growth. Therefore, the decrease in the relative number and volume of mitochondria could be due to the initial enormous cell enlargement. During the next hours of germination, slowing down of cell expansion can be supposed. Thus, no increase in the relative number occured, while the enlargement of mitochondrial particles could be responsible for the relative volume increase. The increase in numerical density as well as the decrease in particle volume between 6 and 18 hrs could be due to division of mitochondria during this period. Cell division has been reported to start 18 (Marciniak et al., 1987) and 24 to 32 hrs (Baluška and Kubica, 1984) after the beginning of wheat embryo hydration. Similarly, in our isolated embryos, cell division might have occurred by 18 hrs. Thus, the relative volume of mitochondria could decrease while the numerical density could increase, probably in relation to the smaller volume of daughter cells.

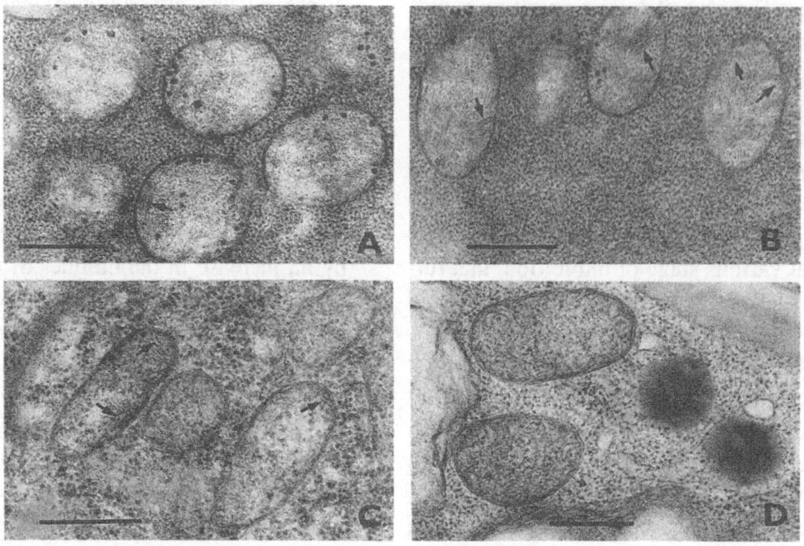


*Fig. 2.* **Mitochondria in the cells of radicle cortex during the first hours of germination of isolated wheat embryos. The *bar* in the figures represents 0.5 $\mu$m. A. Mitochondria with light stroma and very few cristae (arrow) in dry embryo. B. Development of cristae (arrows) after 3 hrs of imbibition. C. Granular appearance of stroma with ribosomes and cristae (arrows) after 6 hrs of germination. D. Dense stroma, ribosomes and numerous cristae are characteristic of mitochondria after 24 hrs of germination.**

*Table 3.* Quantitative evaluation of mitochondria in cortical cells of radicle during the first 24 hours of germination

| Hours of germination | Numerical density [mitoch./$\mu m^3$] | Volume density [$\mu m^3/\mu m^3$] | Volume particle [$\mu m^3$] | Number of particle per section |
|---|---|---|---|---|
| 0 | 3.22 | 0.156 | 0.11886 | 1.29±0.65 |
| 3 | 1.81 | 0.092 | 0.07180 | 1.45±0.65 |
| 6 | 1.56 | 0.111 | 0.10740 | 2.29±0.77 |
| 12 | 3.14 | 0.132 | 0.08330 | 2.32±0.66 |
| 18 | 3.67 | 0.097 | 0.06880 | 2.28±0.63 |
| 24 | 1.99 | 0.087 | 0.05600 | 3.62±0.77 |

Mean values ± SD.

The decrease in relative values of volume and number of mitochondria as well as the decrease in the volume of mitochondrial particles recorded by 24 hrs of germination could indicate an increased volume of cortical cells and division of mitochondria. After 24 hrs of germination, the mitochondria reveal a well developed system of cristae (Fig. 2D). Previous studies have shown that if longer periods of imbibition and germination are investigated, the differentiation of the mitochondrial compartment appears as a process of continuous increase in cristae (Buvat and Robert, 1992;Deltour and Bronchart, 1971; Srivastava and Paulson, 1968; Yoo, 1970). However, the study of the isolated wheat embryo have indicated decreases and increases in the development of cristae and, particularly of the quantitative characteristics of the mitochondrial compartment during the first 24 hours of imbibition. The pattern of the structural changes could contribute to the dynamics in the development of respiratory activity of germinating embryos.

## References

Attucci S, Saint-Ges V, Carde J P, Raymond P and Pradet A 1991 Oxidative phosphorylation by mitochondria extracted from dry sunflower seeds. Plant Physiol. 95, 390–398.

Baluška F and Kubica Š 1984 Changes in DNA content and chromatin condensation in relation to growth processes during germination of wheat. Biológia 39, 1059–1066.

Bewley J D and Black M 1978 Physiology and biochemistry of seeds in relation to germination. 1. Development, germination, and growth. Springer Verlag, Berlin, Heidelberg New York.

Buvat R and Robert G 1992 Cytologie ultrastructurale de l' hydratation germinative des primordiums de racines des semences de l' Orge (*Hordeum vulgare* L.). I. L' embryon sec, l' état de maturité. Ann. Sc. Nat. Bot. Biol. Vég., 13 Sér., 4, 147–159.

Deltour R and Bronchart R 1971 Changements de l'ultrastructure des cellules de *Zea mays* au debout de la germination. Planta 97, 197–207.

Ehrenshaft M and Brambl R 1990 Respiration and mitochondrial biogenesis in germinating embryos of maize. Plant Physiol. 93, 295–304.

Hallam N D, Roberts B E and Osborne D J 1972 Embryogenesis and germination in rye *Secale cereale* L. II. Biochemical and fine structural changes. Planta 105, 293–301.

Klein S and Ben-Sahul Y 1966 Changes in cell fine structure of lima bean axes during early germination. Can. J. Bot. 44, 331–339.

Leprince 0, van der Werf A, Deltour R and Lambers H 1992 Respiratory pathways in germinating maize radicles correlated with desiccation tolerance and soluble sugars. Physiol. Plant. 85, 581–588.

Marciniak B, Bucholc M and Buchowicz J 1987 Early DNA synthesis during the germination of wheat embryos. Phytochemistry 26, 331–334.

Møller J M, Berczi A, van der Plas H W and Lambers H 1988 Measurement of the activity and capacity of the alternative pathway in intact plant tissues: Identification of problems and possible solutions. Physiol. Plant. 72, 642–649.

Morohashi Y 1986 Patterns of mitochondrial development in reserve tissues of germinated seeds: a survey. Physiol. Plant. 66, 653–658.

Morohashi Y and Bewley J D 1980 Development of mitochondrial activities in pea cotyledons. Plant Physiol. 66, 637–640.

Morohashi Y, Bewley J D and Yeung E C 1981 Biogenesis of mitochondria in imbibed peanut cotyledons: Influence of the axis. J. Exp. Bot. 32, 605–613.

Mráz P and Polónyi J 1988 Metódy elektrónovej mikroskopie živočišnych tkaniv (Methods in electron microscopy of animal tissues).Veda, Vyd. SAV, Bratislava. 315p.

Paulson P E and Srivastava L M 1968 The fine structure of the embryo of *Lactuca sativa* I. Dry embryo. Can. J. Bot. 46, 1437–1445.

Singh A P 1977 Fine structure of the dormant embryo of *Medicago sativa*. Am. J. Bot. 64, 1008–1022.

Srivastava L M and Paulson R E 1968 The fine structure of the embryo of *Lactuca sativa* II. Changes during germination. Can. J. Bot. 46, 1447–1455.

Yoo B Y 1970 Ultrastructural changes in cells of pea embryo radicles during germination. J. Cell Biol. 45, 158–171.

*F. Baluška et al. (eds.), Structure and Function of Roots,* 175–180, 1995.


# Comparison of phytate in radicle, plumule, scutellum and endosperm of *Zea mays* and *Zea diploperennis* seeds

M. Mikuš[1,2], A. Lux[1,3], D.C. Crans[2], P.K. Shin[2] and J. Krištín[1]
[1]*Department of Plant Physiology, Comenius University, Mlynská dolina B2, 842 15 Bratislava, Slovakia and* [2]*Cell and Molecular Biology Program and Department of Chemistry, Colorado State University, Fort Collins, CO 805 23, USA.* [3]*Corresponding author*

*Key words:* phytate, radicle, scutellum, X-ray microanalysis, *Zea diploperennis, Zea mays*


## Abstract

Phytin element composition and phytate distribution and content were determined in endosperm, scutellum, radicle and plumule of mature *Zea diploperennis* and *Zea mays* grains. The embryo of *Z. diploperennis* and *Z. mays* contained 90% and 86% of total grain phytate, respectively, with most of the phytate localized in the scutellum and radicle. All examined parts of *Z. diploperennis* had a higher phytate concentration than *Z. mays*. A strong similarity in phytate distribution was found in both species despite variability in grain size and year of production. Phytate was stored as phytin in globoid crystals with variable diameters. The size of the globoid crystals reflected the phytate content in each tissue. The largest globoid crystals as well as the highest phytate content were found in the scutellum of *Z. diploperennis*. Energy-dispersive X-ray microanalysis showed that globoid crystals of both species were dominated by P, K, and Mg. Phytate content in different embryonic parts correlated with the amount of K and Mg, but not with the less abundant elements (Zn, Ca and Fe) found in the tissues. Even though some differences in the elemental composition of phytin and the phytate content were found, a similarity in phytate distribution patterns in grains of these species may reflect their close taxonomic relationship.


*Abbreviations:* AAS — atomic absorption spectroscopy, DW — dry weight, EDX — energy-dispersive X-ray, DW — dry weight, GC(s) — globoid crystal(s), ICP — ion-coupled plasma, NMR — nuclear magnetic resonance.

## Introduction

*Zea diploperennis* is a new teosinte that was first reported from Mexico in 1977. Its discovery evoked great interest owing to its potential importance in the evolution of modern maize (*Zea mays* L.). To assess the relationship with modern maize, the morphology, cytology and isozymic variation of *Z. diploperennis* have been studied (Doebley and Iltis, 1981; Doebley, 1983; Doebley et al., 1984; Mastenbroek et al., 1982; Pasupuleti and Galinat, 1982). Recent reports support the view that *Z. diploperennis* is a primitive offshoot of the evolutionary line leading to maize (Doebley, 1983). Moreover, *Z. diploperennis* is used in maize improvement programs as a valuable source of germplasm and it appears to be an ideal breeding source for the transfer of different genetic traits including regeneration capacity (Swedlung and Locy, 1988), high tillering ability, and resistance to various virus and virus-like diseases of maize (Nautl, 1980; Nautl et al., 1982).

Phytate (*myo*-inositol 1, 2, 3, 4, 5, 6, -hexa*kis*phosphate) has long been known as the storage form of phosphorus (P) in seeds (Maga, 1982), representing from 60% to 80% of total seed P (Raboy, 1990). Phytate is predominately stored in the form of the globoid crystals (GCs) visible as distinct inclusions inside of protein bodies (Lott, 1984). The correct location of phytate, the amount and form of storage determine the pattern of phytate metabolism during seed development and germination. The understanding of these processes will be advantageous for the breeding processes or improving programs, which are often performed through tissue cultures using immature or mature embryos as starting material.

In the present study the phytate distribution, form of storage, and elemental composition of globoid crystals were examined in grains of *Z. mays* and *Z. diploperennis* using various methods including electron microscopy, energy-dispersive X-ray (EDX) microanalysis, UV spectrophotometry and $^{31}P$ NMR spectroscopy.

## Materials and methods

*Plant material*
The samples were obtained from mature grain of *Zea mays* (cv. 265-1, 356-1, CE 268, TR 256) and *Zea diploperennis* Iltis, Doebley and Guzman (collected in Southern Jalisto, Mexico in 1986 and 1988). Grains were divided into endosperm, scutellum, radicle and plumule with a razor blade and analyzed.

*Phytate extraction*
Grain parts were frozen in liquid nitrogen, ground in a pre-chilled mortar and pestle, and homogenized with 10 mL of 0.2 *N* HC1 per g of the sample dry weight (DW). The mixture was stirred at room temperature for 20 minutes and then centrifuged at 16000 *g* for 10 min. Phytate was determined in these extracts colorimetrically and by $^{31}P$ NMR spectroscopy. The NMR spectra were recorded at pH 10.0 ($\pm$ 0.5) after addition of EDTA (0.1 *M* final concentration) to the supernatant obtained from HC1-extracted material.

*Phytate determination*
Phytate was determined colorimetrically in HC1 extracts of grain parts (Huang and Lantzsch, 1983). Residual iron was measured by a bipyridine reagent after formation of ferric phytate.

*Phosphorus determination*
Inorganic phosphate ($P_i$) was measured in all extracts as described previously (Ames, 1966).

*$^{31}P$ NMR spectroscopy*
The $^{31}P$ NMR spectra of HC1 extracts were recorded at 202 MHz (11.7 T) using a Bruker AM-500 spectrometer. The spectra were recorded using parameters described previously (Crans et al., 1993). The chemical shifts are all reported in reference to 85% $H_3PO_4$ at 0 ppm (external reference).

*Electron microscopy and EDX analyses*
The basal part of primary roots were cut from mature grains of both species with a razor blade, and fixed with glutaraldehyde in 60% ethanol according to the procedure of Lott et al., (1984). After fixation, the samples were dehydrated with ethanol and propyleneoxide, and embedded in Durcupan ACM (FLUKA). Blue sections (approximately 200-230 nm thick) were picked up from redistilled water immediately after sectioning and placed on formvar-carbon-coated copper grids. The sections were viewed without post-staining in a transmission electron microscope (JEOL 2000 FX) operated at 80 kV in STEM mode. The elemental composition of globoid crystal (GC) regions was determined with EDX analysis system LINK AN 10/85 S using identical parameters for each analysis as described previously (Mikuš et al., 1992).

*Elemental content determination by ion coupled plasma (ICP)*
The metal analyses were carried out at the Soil Testing Laboratory, Colorado State University using an ICP unit (Model 75 Plasma Atomcomp, manufactured by Thermo-Jarrell Ash).

## Results

*Phytate distribution and content*

The distribution of phytate was measured by UV spectrophotometry and $^{31}P$ NMR spectroscopy in radicles, plumules, scutella, and endosperms of grains from 4 varieties and cultivars of *Z. mays* and the results were compared to *Z. diploperennis* (Table 1). Phytate present in various grain parts was shown as a percentage of the phytate distribution (total grain phytate) and as the phytate concentration (phytate content per unit of DW). Most of the phytate found in the grains of both species was present in the scutellum (Table 1). The lowest content of phytate was found in the samples of endosperm which included aleurone tissue. Very similar patterns of phytate distribution within the embryo were found in all varieties and cultivars of *Z. mays* and *Z. diploperennis* with the radicle containing approximately twice the amount of phytate found in the plumule (Table 1). The embryo parts containing the highest proportion of phytate (scutellum and radicle) also showed the highest phytate concentration (g $g^{-1}$ tissue DW). Phytate concentrations were higher in *Z. diploperennis* than in *Z. mays* for all grain parts (Table 1).

*Table 1.* Phytate distribution and concentration in grains of *Z. mays* and *Z. diploperennis*. Radicle (RAD) plumule (PLUM), scutellum (SCUT) and endosperm (END) were analyzed separately. The values are the averages of three repetitions of duplicate analyses with confidence limit higher than 95%

| Species (Line and cultivar) | Distribution of phytate P (% of total grain phytate P) | | | | Concentration of phytate P ($mg\ g^{-1}$ DW) | | | |
|---|---|---|---|---|---|---|---|---|
| | RAD | PLUM | SCUT | END | RAD | PLUM | SCUT | END |
| *Z. mays* | | | | | | | | |
| Tva 265-1 | 4.9 | 1.6 | 90.3 | 3.2 | 5.1 | 2.1 | 18.3 | 0.06 |
| Tva 356-1 | 3.4 | 2.0 | 91.7 | 2.9 | 3.2 | 1.5 | 16.5 | 0.06 |
| CE 268 | 4.9 | 2.0 | 91.6 | 1.5 | 8.8 | 4.1 | 22.9 | 0.04 |
| TR 256 | 2.8 | 1.4 | 91.4 | 4.4 | 3.9 | 2.0 | 15.4 | 0.12 |
| *Z. diploperennis* | | | | | | | | |
| 1986 | 5.9 | 3.1 | 86.9 | 4.1 | 10.9 | 3.7 | 29.9 | 0.23 |
| 1988 | 5.0 | 2.8 | 88.4 | 3.8 | 9.8 | 5.0 | 31.3 | 0.22 |

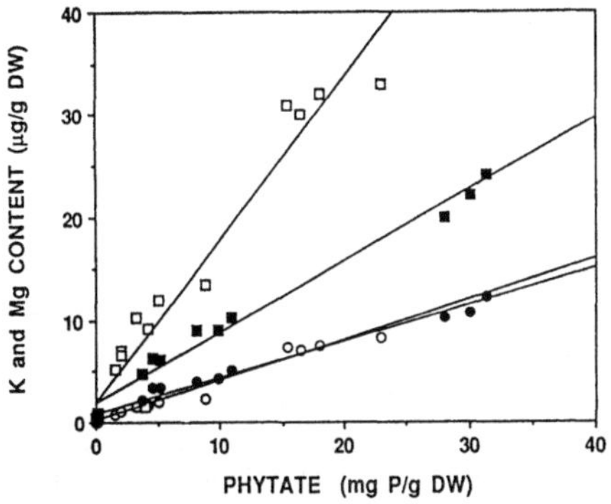


*Fig. 1.* Relationship between the phytate content in different grain parts of various varieties of *Z. mays* (open symbols) and *Z. diploperennis* (solid symbols) and the levels of tissue K (□,■) and Mg (○,●) as determined by ICP.

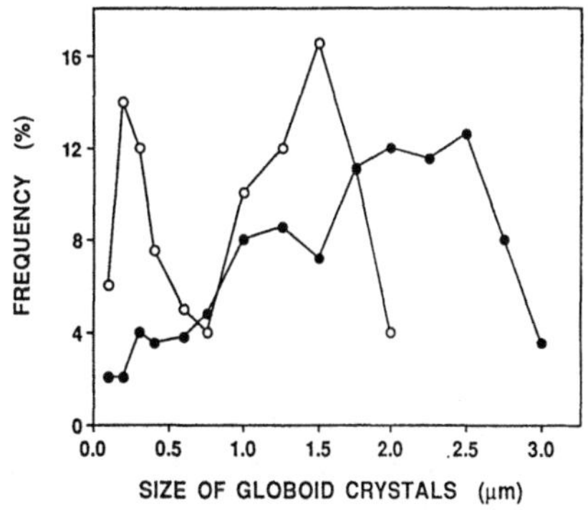


*Fig. 2.* Comparison of the frequency of GCs with different sizes in scutellum of *Z. mays* (○) and *Z. diploperennis* (●).

*Table 2.* Percentage of GCs from *Z. mays* (115 analyses) and *Z. diploperennis* (98 analyses) scutellum that contained measurable Na, Mg, K, Ca, Fe and Zn

| Species | Element | | | | | |
|---|---|---|---|---|---|---|
| | Na | Mg | K | Ca | Fe | Zn |
| *Z. mays* | 28 | 85 | 97 | 7 | 40 | 47 |
| *Z. diploperennis* | 15 | 95 | 100 | 0 | 40 | 17 |

### *Distribution and content of cations*

The elemental composition of each grain part was determined using ICP spectroscopy. We found significantly higher K levels in the samples of *Z. mays* then in those of *Z. diploperennis*, while other cations showed similar content per g DW for both species. Total levels of cations were compared to the phytate content. A positive correlation between the levels of two major cations (K and Mg) and phytate levels was found in each examined grain part (Fig. 1) The K/phytate ratio was significantly higher for *Z. mays* samples whereas the Mg/phytate ratio was similar for both species (Fig. 1). The content of the other analyzed elements (Ca, Zn and Fe) showed no association with phytate levels. The relationship between K, Mg and phytate was studied further by EDX analysis yielding information on the composition of phytate salts (phytin) in the two species.

Electron-microscopical analyses showed that phytate was stored as a phytin and existed as numerous electron-dense GCs with a regular roundish shape (Mikuš et al., 1992). Even though the size of GCs

*Table 3.* EDX analyses of GCs from scutellum of *Z. mays* (115 analyses) and *Z. diploperennis* (98 analyses). P, K, and Mg peak-to-background or element-to-element values are averages of X-ray counts either after background subtraction ([a]) or after peak-to-background calculation ([b])

| Ratio Elements | *Z. mays* | *Z. diploperennis* | t -test $p = 0.01^c$ |
|---|---|---|---|
| K:P[a] | 0.54 ± 0.14 | 0.79 ± 0.16 | S |
| K:P[b] | 0.97 ± 0.40 | 1.47 ± 0.56 | S |
| Mg:P[a] | 0 09 ± 0.03 | 0.08 ± 0.03 | NS |
| Mg:P[b] | 0.26 ± 0.15 | 0.22 ± 0.14 | NS |
| K[b] | 3.15 ± 1.00 | 4.24 ± 1.44 | S |
| P[b] | 3.39 ± 0.66 | 3.12 ± 0.95 | NS |
| Mg[b] | 0.78 ± 0.35 | 0.66 ± 0.31 | NS |

[c] S= significant, NS = not significant.

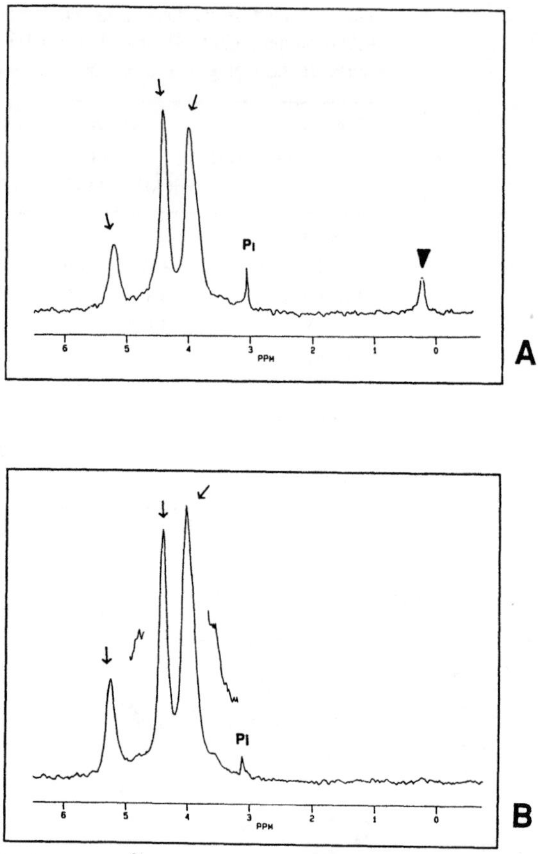


*Fig. 3.* Typical $^{31}$P NMR spectra of HC1 extracts prepared from *Z. mays* (**A**) and *Z. diploperennis* (**B**) radicles showed the presence of the resonances for inorganic phosphate ($P_i$), phosphodiesters (solid arrowhead), phytate (arrows) and other phosphomonoesters (detailed regions in B).

varied in samples prepared from equivalent areas of both species, specific size patterns of the GCs were observed when specimens of the two species were compared (Fig. 2). The size of GCs varied for *Z. mays* and *Z. diploperennis* from 0.2 $\mu$m to 2.0 $\mu$m and from 0.2 $\mu$m to 3.0 $\mu$m, respectively. The tissue with the highest phytate content (scutellum of *Z. diploperennis*) contained the largest GCs and a comparison of the GC size distribution with that of *Z. mays* scutellum is showed in Figure 2. EDX analyses of GCs in sectioned resin-embedded material revealed the presence of considerable P, K, and Mg, but K was the most abundant cation. Less than half of the GCs contained detectable amounts of other elements (Na, Ca, Fe and Zn) (Table 2.). Calcium was found exclusively in small GCs of *Z. mays*, often together with a relatively high amount of Zn (for further detail see Mikus et al., 1992). All samples of *Z. diploperennis* lacked Ca and some also lacked Zn. Mg:P ratios, as well as P and Mg peak-to-background values of GCs, did not differ significantly between species (Table 3) even when groups of GCs with the same sizes were compared (not documented). However K:P ratios and K peak-to-background values were significantly higher in GCs of *Z. mays*, and was particularly evident in the scutella (Table 3). All these EDX results are in agreement with the presence of significantly higher K content found by ICP analyses in *Z. mays* samples.

*Phytate and other P compounds in the radicles*

Samples of the radicles analyzed colorimetrically were also studied by $^{31}$P NMR spectroscopy (Fig. 3). Phytate was the major P compound in the HC1 extracts containing approximately 85% and 83% of total P in *Z. mays* and *Z. diploperennis*, respectively. The amount of $P_i$ (8 ± 3% of total P) was similar for radicles of both species. Quantitation using NMR confirmed colorimetrical results on phytate content and showed some species-specific differences in phosphorus compounds (Fig. 3). We found higher content of the phosphodiesters (identified as phospholipids) in the extracts of *Z. mays* radicles in comparison to *Z. diploperennis*. The radicle of the latter contained a higher content of non-phytate phosphomonoesters for example small signals for glucose 6-phosphate and 3-phosphocholine were observed.

## Discussion

Seeds of *Z. mays* and *Z. diploperennis* have almost all the phytate stored in the embryo, and thus they differ from all other cereals which contain most of their phytate in cells of the aleurone layer (Maga, 1980). The phytate distribution previously described for maize grains by O' Dell et al. (1972) is confirmed in our study and furthermore we determine the phytate distribution and content in various parts of the maize embryo. We found that the scutellum and radicle stored most of the embryo phytate in both species. Seeds of *Z. diploperennis* from two harvests and four varieties of *Z. mays* seeds grown under different conditions in two locations (Slovakia and Colorado, USA) were studied to minimize the possibility that environmental factors (Raboy and Dickinson, 1984) and/or genetic diversity (Miller et al., 1980) modified the phytate levels in the analyzed grains. All analyzed parts of the *Z. diploperennis* embryo were, however, higher in phytate content in comparison to *Z. mays*. We conclude that the phytate content is a species-specific feature.

Using electron microscopy, EDX and ICP spectroscopy we investigated the relationship between phytate levels in analyzed grain parts, their elemental composition and the form of phytate storage and elemental composition of phytate salts (phytin). Phytate content in each sample was found to be correlated with the size of the GCs as well as tissue K and Mg levels determined by ICP. The presence of surprisingly similar tissue Mg:phytate ratios for both species was in contrast to the significantly higher tissue K:phytate ratios, as well as K:P and K peak-to-background values, for *Z. mays* determined by EDX. The presence of higher amounts of K phytate in *Z. mays* might be a result of evolutionary selection for this feature, since this salt is highly water soluble (Cheryan, 1980). The high solubility of K phytate would facilitate the access to phytate upon hydration. The presence of free phytate is presumably advantageous, since only free phytate is a suitable substrate for phytases (Gibson and Ullah, 1990).

An additional species-specific difference in the P compounds of the radicles was observed by means of $^{31}$P NMR spectroscopy. We observed higher levels of phosphomonoesters in the extract of *Z. diploperennis* radicles which could reflect different metabolic activities during dormancy. The metabolism of dormant seeds is, however, very restricted due to low water content and little is known about its nature (Bewley and Black, 1986). It is difficult to judge the significance of the observed findings, since the proper content of non-phytate phosphomonoesters could not be determined due to the presence of phytate signals in the same region of the $^{31}$P NMR spectra.

We can conclude that the phytate distribution and elemental composition of GCs are as similar as can be expected for closely related species, and that the process of development to modern maize from an ancestor similar to *Z. diploperennis* was not accompanied by any major changes in phytate distribution or form of phytate storage. Some species-specific features in embryos of *Z. mays* and *Z. diploperennis* were found, however. These differences must be investigated further to substantiate a possible physiological significance during germination and post-germinative growth.

## Acknowledgements

We thank the Colorado State Experiment Station and Slovak Ministry of Education for funding this work. We thank Dr Eric E Roos at the National Seed Storage Laboratory, Colorado State University, Fort Collins, USA and Dr M Nesticky, The Institute of Corn, Slovakia and Facultad de Ciencias Biologicas UANL, Monterrey, Mexico, for providing us with the seed material.

## References

Ames B N 1966 Assay of inorganic phosphate, total phosphate and phosphatases. Methods Enzymol. 8, 115-118.

Bewley J D and Black M 1985 Seeds. Physiology of Development and Germination. Plenum Press, New York, London. 367 p.

Cheryan M 1980 Phytic acid interaction in food systems. CRC Crit. Rev. Food Sci. Nutr. 13, 297-335.

Crans D C, Mikuš M and Marshman R W 1993 $^{31}$P NMR examination of phosphorus metabolites in the aqueous acidic, and organic extracts of *Phaseolus vulgaris* seeds. Anal. Biochem. 209, 85-94.

Doebley J F 1983 The maize and teosinte male inflorescence: A numerical taxonomic study. Ann. Missouri Bot. Gard. 70, 32-70.

Doebley J F and Iltis H H 1981 Taxonomy of *Zea* (Gramineae). 1. A subgenic classification with key to taxa. Am. J. Bot. 67, 982-993.

Doebley J F, Goodman M M and Stuber C W 1984 Iso-enzymatic variation in *Zea* (Gramineae). Syst. Bot. 9, 203-218.

Gibson D M and Ullah A B J 1990 Phytases and their action on phytic acid. *In* Inositol Metabolism in Plants. Eds. D J Morre, W Boss and F Loewus. pp 77-92. Alan R. Liss, New York.

Huang W and Lantzsch H J 1983 Sensitive method for the rapid determination of phytate in cereals and cereal products. J. Sci. Food Agric. 32, 1423-1426.

Lott J N A 1984 Accumulation of seed reserves of phosphorus and other minerals. *In* Seed Physiology. I. Development. Ed. D R Murray. pp 139-166. Academic Press, Sydney.

Lott J N A, Goodchild D J and Craig S 1984 Studies of mineral reserves in pea (*Pisum sativum*) cotyledons using low-water-content procedures. Can. J. Bot. 11, 459-469.

Maga J A 1982 Phytate: its chemistry, occurrence, food interactions, nutritional significance and methods of analysis. J. Agric. Food Chem. 30, 1-9.

Mastenbroek I, Cohen C E and Dewet J M J 1982 Seed protein and seedling isoenzyme patterns of *Z. mays* and its closest relatives. Biochem. System. Ecol. 9, 179-183.

Mikuš M, Bobák M and Lux A 1992 Structure of protein bodies and elemental composition of phytin from dry germ of maize (*Zea mays* L.). Bot. Acta 105, 26-33.

Miller G A, Youngs V L and Oplinger E S 1980 Environmental and cultivar effects on oat phytic acid concentration. Cereal Chem. 57, 189-191.

Nautl L R 1980 Maize bushy stunt and corn stunt: A comparison of disease symptoms pathogen host ranges and vectors. Phytopathol. 70, 659-662.

Nautl L R, Gordon D T, Domsteegt U D and Iltis H H 1982 Response of annual and perennial teosintes (*Zea*) to six maize viruses. Plant Disease 66, 61-62.

O' Dell B L, De Boland A R and Koirtyohann S R 1972 Distribution of phytate and nutritionally important elements among the morphological components of cereal grains J. Agric. Food Chem. 207, 18-721.

Pasupuleti C V and Galinat W C 1982 *Zea diploperennis*. 1. Its chromosomes and comparative cytology. J. Hered. 73, 168-170.

Raboy V 1990 Biochemistry and genetics of phytic acid synthesis. *In* Inositol Metabolism in Plants. Eds. D J Morre, W Boss and F Loewus. pp 52-73. Alan R. Liss, New York.

Raboy V and Dickinson D B 1984 Effect of phosphorus and zinc nutrition on soybean seed phytic acid and zinc. Plant Physiol. 75, 1094-1098.

Swedlung B and Locy R D 1988 Somatic embryogenesis and plant regeneration in 2-year old cultures of *Zea diploperennis*. Plant Cell Rep. 7, 144-147.

*F. Baluška et al. (eds.), Structure and Function of Roots,* 181–187, 1995.


# The effect of vanadate on growth and phospholipid levels in the root and hypocotyl of bean seedlings (*Phaseolus vulgaris* L.)

Miloš Mikuš[1,2], Paul K. Shin[2] and Debbie C. Crans[2*]
[1]*Department of Plant Physiology, Comenius University, Mlynská dolina B2, 842 15 Bratislava, Slovakia and*
[2]*Cell and Molecular Biology Program and Department of Chemistry, Colorado State University, Fort Collins, CO 80523, USA*

*Key words:* hypocotyl, *Phaseolus vulgaris*, phospholipids, post-germinative growth, root, vanadate

## Abstract

Vanadate treatment accelerated seed germination rate and increased fresh weight accumulation by snap bean seedlings during early post-germination. This stimulatory effect of $V_i$ was followed by inhibition of root growth after 56 h. After 100 h, vanadium accumulated in the root of $V_i$ treated seedlings, whereas insignificant accumulation occurred in the hypocotyl until 144 h. Vanadium accumulation in roots was associated with a decrease in their total phospholipid content. This was in contrast to increasing levels of total phospholipids in the hypocotyl of $V_i$-treated seedlings compared to control. $^{31}P$ NMR analyses of the phospholipid profiles of control and $V_i$-treated seedlings showed organ-specific differences in levels of phosphatidylinositol and an unidentified phospholipid in both root and hypocotyl. The inhibition of root growth was specifically associated with changes in an unidentified phospholipid, whose level decreased with decreasing root growth. We conclude that $V_i$-induced growth effects were associated with the organ-specific vanadium accumulation, levels of total phospholipids and changes in phospholipid profiles.

*Abbreviations:* FW – fresh weight, NMR – nuclear magnetic resonance, PA – phosphatidic acid; PC – phosphatidylcholine, PE – phosphatidylethanolamine, PI – phosphatidylinositol, PL(s) – phospholipid(s), PS – phosphatidylserine, TLC – thin layer chromatography, V – vanadium, $V_i$ – vanadate

## Introduction

The high vanadium (V) content in soils such as the Rocky Mountain Plateau (Parker et al., 1978) and the increased levels of V emissions (Lee, 1983) combined with the sensitivity of various plant species to V compounds (Aller et al., 1990; Byczkowski et al., 1979; Morrell, 1986) may affect future cultivation and crop yields. It is well known that V affects a multitude of biological processes in animals (Chasteen, 1990). Recent interest in V derivatives as compounds which may replace insulin (McNeill et al., 1992), which enhance yield (Wendt et al., 1993) and as a growth factor (Gresser et al., 1987; Nechay et al., 1986) have suggested the importance of exploring the detailed mechanism of action for V in intact biological systems. V in the form of vanadate ($V_i$) is generally perceived as a potent enzyme inhibitor (Chasteen, 1990), although several cases of enzyme stimulation have recently been reported (Crans et al., 1992; Drueckhammer et al., 1989). The interaction of various ATPases from different sources including plants (Perlin and Spanswick, 1981) with $V_i$ has been studied extensively since $V_i$ is a potent inhibitor of this enzyme. In recent study (Sklenár et al., 1994) the inhibitory effect of $V_i$ on maize root ATPases was associated with increased ATP levels in roots of $V_i$ treated seedlings. Given the number of enzymes and their respective metabolic processes that can be potently affected by $V_i$, it would be premature to attribute all the $V_i$ induced effects to the interaction with ATPases (Nechay, 1986).

*Phaseolus vulgaris* was used for our studies since it can tolerate moderate levels of V (Kaplan et al., 1990; Kohno, 1986) with significant accumulation of V in the root (Kaplan et al., 1990). Our previous study of phosphorus metabolism in bean embryonic axes showed, that $V_i$ has no influence on inorganic phosphate and phytate levels (Crans et al., 1994). Little is known concerning the effects of $V_i$ on phospholipid (PL) and phosphoinositide metabolism, although a few reports suggested that $V_i$ may affect these processes (Khan and Malhotra, 1987; Morita et al., 1992; Paris et al., 1987). In the present paper, we describe an association between $V_i$ induced growth effects (or V accumulation) with $V_i$ induced changes in PL levels and profiles in seedlings of *Phaseolus vulgaris*.

## Materials and methods

### *Plant material*

Snap bean seeds (*Phaseolus vulgaris* L., cv. Shamrock), donated by Rogers Brothers Seed Company to the National Seed Storage Laboratory at Colorado State University, were used for all studies. Seeds were germinated between layers of wet paper towels at 25°C. Sample material was collected from germinating seeds at 6, 12, 18, 24, 36 and 48 hrs as well as from seedlings grown for 56, 96, 107, 120, 144, and 156 h in water or 1 m*M* ammonium vanadate ($V_i$). The cotyledons were removed from the seedlings, the root and hypocotyl were separated and frozen in liquid nitrogen. All samples were stored at - 80°C before being analyzed.

### PL extraction

PLs of bean embryonic axes were extracted using a chloroform-methanol-HCl extraction (Murthy et al., 1989). The frozen material was ground, then ice-cold methanol (2.0 mL), chloroform-methanol (1/2, v/v, 1.7 mL), chloroform (1.1 mL), and 2.4 *M* HCl (0.5 mL) were added sequentially. The suspension was centrifuged at 9,000 *g* for 15 min and the resulting pellet was reextracted with 2 mL of chloroform-methanol mixture (1/2, v/v). The aqueous phase was reextracted with 1.1 mL of chloroform. The combined organic extract was mixed with 2.5 mL of 50% aqueous methanol and 0.3 mL of 2.4 *M* HCl and was centrifuged. The phases were separated, and the organic phase was evaporated to dryness by rotary vacuum evaporation.

### PL characterization and quantitation

The extracted PLs were analyzed by $^{31}$P NMR spectroscopy, TLC, and colorimetrical analysis of the total phosphorus content. The PL samples were prepared for $^{31}$P NMR analysis using a procedure described previously (Meneses and Glonek, 1988). An EDTA-methanol reagent was prepared just before used by mixing 1 mL of aqueous 0.1 *M* EDTA ($2Na^+$ or $2K^+$) with 3 mL of dry methanol. The EDTA-methanol reagent (1 mL) was added to 2 mL of the PL extract. The $^{31}$P NMR spectrum of the chloroform phase was then recorded.

TLC was used to separate PLs and to confirm the results obtained by $^{31}$P NMR. PLs were separated using potassium oxalate (1%) impregnated silica gel TLC plates using chloroform/acetone/methanol/glacial acetic acid/water (40/15/13/12/8, by vol.) (Jolles et al., 1979) or chloroform/methanol/6.5 *N* $NH_4OH$ (130/30/4.7, by vol.) (Miyazawa and Fujino, 1978) as an eluent systems. PLs extracts were simultaneously analyzed with authentic compounds (purchased from Sigma). All the lipid spots were visualized after spraying with cupric acetate or by charring. PLs were detected by immersion into the phosphomolybdate reagent (Lendrath et al., 1990). Sugars were visualized with $\alpha$-naphtol, and amines were revealed by spraying with ninhydrin, Dragendorff reagent or chloroplatinic acid. Quantification of the total PL content and the content of each TLC-separated PL species was carried out by total phosphorus determination (Ames, 1966). All values presented here are averages of a minimum of duplicate experiments.

### $^{31}$P NMR spectroscopy

The $^{31}$P NMR spectra were recorded at 202 MHz (11.7 T). The spectra were recorded using previously described method (Crans et al., 1993). The spectra were recorded using an external lock with spectral windows of 40–50 ppm, a 90° pulse angle, and an acquisition time of approximately 0.3 sec and routinely a relaxation delay of 3 sec was used. The number of scans acquired was 750–1500. The chemical shifts were referenced to an external standard, 85% $H_3PO_4$ at 0 ppm.

## Results and discussion

### *Snap bean germination and growth in the presence of $V_i$*

Cultivation of snap bean seeds with 1 m*M* $V_i$ during their germination and early seedling development stimulated both the germination and growth during the first 48 h (Fig. 1). These stimulatory effects were followed by a period of main root growth inhibition, which was accompanied by changes in the morphology of the seedlings: the seedlings turned brown and hypocotyl diameter increased. In addition, $V_i$ treatment accelerated the development and growth of lateral roots. Growth inhibition in the presence of $V_i$ was reported previously for bean seedlings (Kaplan et al., 1990) and other species (Hidalgo et al., 1988; Lux et al., 1986). To identify the metabolic processes responding to $V_i$ treatment, we examined the phosphorus metabolism in bean seedlings during early stages of seedling growth. We found no differences in inorganic phosphate and phytate levels of bean seedlings during post-germinative growth in water or 1 m*M* $V_i$ (Crans et al., 1994). This findings would be unexpected, if $V_i$ simply acted as a phosphate analog, however, as observed in mammals, the biological responses to V and/or $V_i$ are more complex (Chasteen, 1990). In this study we examined the possibility that the PL metabolism may be associated with the $V_i$ induced growth effects. The level of total PLs in bean embryonic axes (measured per g FW) decreased from beginning of germination in both control and $V_i$ treated seeds (Fig. 1). During the first 56 h from imbibition the $V_i$ treatment slightly accelerated the decrease in total PLs (Fig. 1). This was paralleled by the radicle emergence which, although emerging earlier than control, occurred when similar PL levels were reached.

### *$V_i$ uptake and FW accumulation in root and hypocotyl during post-germinative growth*

$V_i$ uptake and fresh weight accumulation in the bean seedling root and hypocotyl were determined to examine their correlation with the changes in growth and PL levels. The content of endogenous V in the hypocotyl and root was measured by the Ion Coupled Plasma spectroscopy at different times after 56 h of growth in water or 1 m*M* $V_i$ (Fig. 2). After approximately 100 hrs, the V accumulated most in the root (Fig. 2A); we specifically measured the V content in root, hypocotyl (Fig. 2B) and cotyledons (not documented). The ability of roots to accumulate much of the absorbed V was reported for several plant species (Kaplan et al., 1990; Morrell et al., 1986). Even though the uptake of $V_i$ by roots of mature plants does not involve an active transport (Welch, 1973), the lag in V accumulation we observed in roots of bean seedlings suggests that some factors interfere with $V_i$ uptake. This lag may be associated with intracellular reduction of vanadate to vanadyl cation (Morrell et al., 1986; Willsky, 1984). The ability of young bean roots to restrict transport to the shoot was very effective, since a significant decrease in V content per gram of hypocotyl FW was found when V accumulation in root started (Figs. 2A, B). However, after 144 h V levels in the hypocotyl started to increase (Fig. 2B). This suggests that the ability of roots to prevent V transport to aerial parts of seedling is limited and that eventually the V levels increase in tissues other than the root.

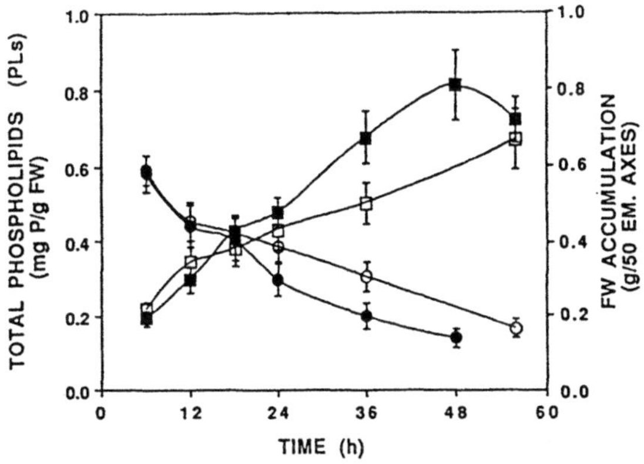


*Fig. 1.* Changes in the growth (□, ■) and PL content (○, ●) were determined in the embryonic axes of germinating *Phaseolus vulgaris* (L.) seeds treated with water (open symbols) or 1 mM $V_i$ (solid symbols). Values are averages of two repetitions of triplicate experiments.

The effects of $V_i$ treatment on the total PL levels in hypocotyl and root during post-germinative growth was also determined. The total PL levels revealed organ-specific changes in the presence of $V_i$ (Fig. 2). An increase in the total PL content was observed in both hypocotyl (Fig. 2A) and root (Fig. 2B) of $V_i$ treated seedlings in comparison to control. In roots, this increase was followed by a gradual decrease, observed when the endogenous V accumulation began (Fig. 2B). Inhibitory effect of $V_i$ on PL levels was observed also in needles of jack pine (Khan and Malhotra, 1987). The difference in the total PL levels in roots and hypocotyls of vanadate treated seedlings may reflect a higher concentration of active forms of V in roots compared to

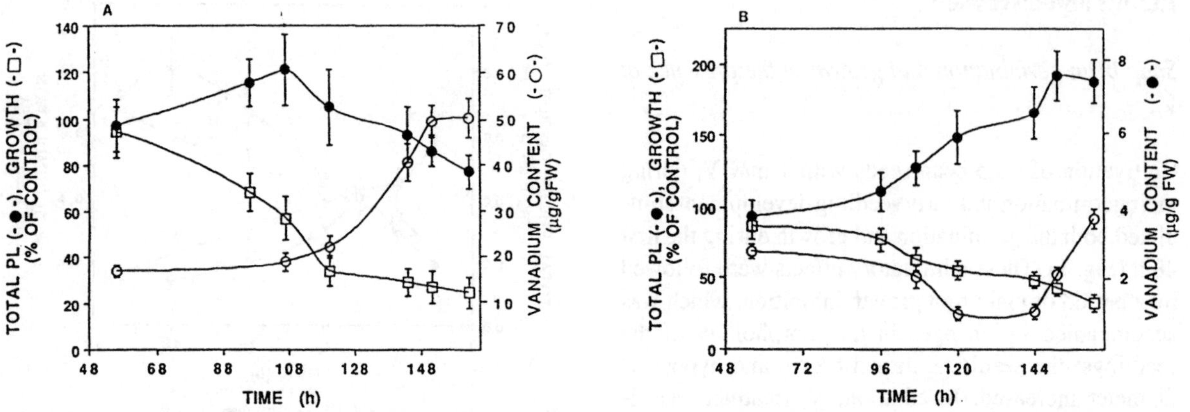


*Fig. 2.* Changes in $V_i$ treated seeds during post-germinative growth in FW accumulation (□), and total PL content (●). The endogenous V levels (○) were determined in root (**A**) and hypocotyl (**B**) tissues of the $V_i$ treated seedlings during post-germinative growth. Values are averages of two repetitions of triplicate experiments.

hypocotyl. Alternatively, the enzymes related to PL metabolism in root may respond differently to V than such enzymes in hypocotyl. The difference in the total PL levels in roots and hypocotyls of vanadate treated bean seedlings was explored further by analysis of PL patterns using $^{31}P$ NMR spectroscopy and TLC.

*PL profiles in hypocotyl and root of bean seedlings*

In the control, phosphatidylcholine (PC), phosphatidylethanolamine (PE), phosphatidyl-inositol (PI), phosphatidic acid (PA) and an unidentified PL (with a $^{31}P$ NMR resonance at 1.95 ppm) accounted for more than 85% of PLs in both hypocotyl and root at 56 hrs (Fig. 3). Some differences in the PL profiles between these two parts observed at 56 hrs (Fig. 3) remained present during the following seedling growth (Table 1, Figs. 4, 5). All major PLs except PA showed higher concentration in hypocotyl than in root (Table 1). Similarly relative levels (% of total PLs) of PC, PI and unidentified PL were higher in hypocotyl than in root (Figs. 4, 5). Organ-specific PL profiles were reported previously (Nishihara and Kito, 1978) and may be traced to tissue specific PL profiles as documented for stele and cortex of maize root (Cowan et al., 1993). Growth of bean seedlings was associated with continuing decrease in content of PLs per g of FW (Table 1). Regardles the decrease in the PLs, a relative level of PI was found to be almost constant in both hypocotyl and root from 96 hrs (Fig. 4). Similarly, in seedlings of other species the relative levels of PI were found to undergo less changes in comparison to other membrane PLs during post-germinative growth (Rakhimov et al., 1986; Shewry et al., 1973). These findings may reflect a functional difference between PI and the structural membranous PLs (PC, PE) and the existance of PL-specific mechanisms controlling PL composition of cellular membranes early after germination.

*Effects of $V_i$ on PL profiles of bean seedlings*

During first 56 h $V_i$ induced no observable changes in the PL profile of the hypocotyl tissue (Fig. 3A, B), however, it significantly altered the PL profile of the root tissue (Fig. 3C, D). Although PA was still the major PL in the root, the contributions by PE, PI, and the unidentified PL were higher in comparison to control (Fig. 3C, D). During further growth of $V_i$ treated seedlings the major differences in PL profiles in comparison to the control were specifically found for PI and for the unidentified PL in both hypocotyl and root tissues.

In root tissues, the PI levels in the $V_i$ treated seedlings were twice as high as in the control when examining the changes in the levels of these PLs as a function of time (Fig. 4). This pattern was not affected by the V accumulation observed in the root after 100 hrs (Fig. 2A). On the other hand, the PI level in the hypocotyl decreased and continued to fall significantly below that of the control as the V levels in the hypocotyl started to increase (Fig. 4). It is possible that the changes in the PI levels can produce altered levels of inositolphosphatides and inositolphosphates. Both these goups of metabolites have been implicated in the regulation of cell proliferation (Heim and Wagner, 1987) and related processes of tumourigenesis in

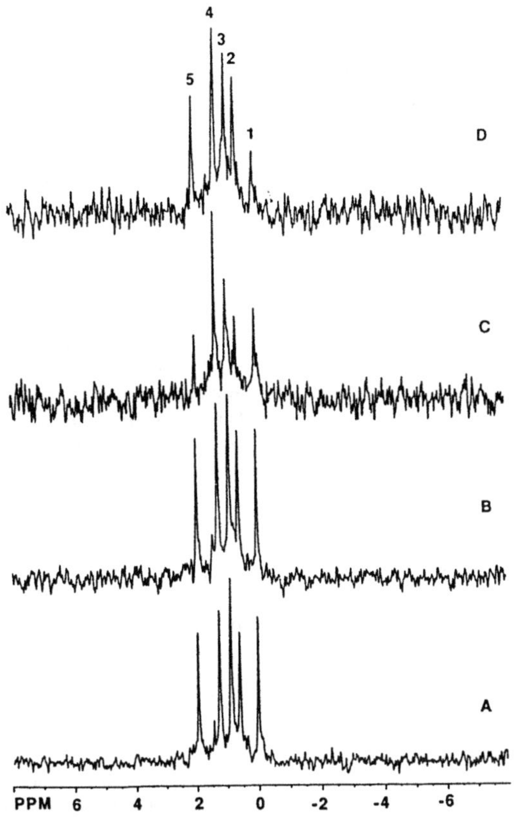


*Fig. 3.* $^{31}P$ NMR spectra were acquired from the PL extracts of hypocotyls (**A, B**) and roots (**C, D**) of seedlings grown in water (A, C) and 1 mM $V_i$ (B,D). Peaks represent: 1, PC; 2, PI; 3, PE+ traces of PS; 4, PA; 5, an unidentified PL. Relative content of PS determined by TLC represented less than 5% of total PLs in all samples.

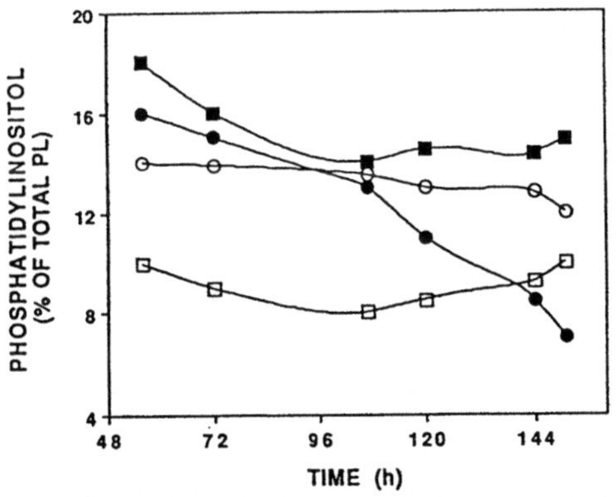


*Fig. 4.* The levels of PI were determined from $^{31}P$ NMR spectra of PLs extracted from hypocotyl (solid symbols) and root (open symbols) of seedlings grown in water (□, ■) or 1 mM $V_i$ (○, ●). Values are averages of two repetitions of duplicate experiments and are accurate within 6% of the values.

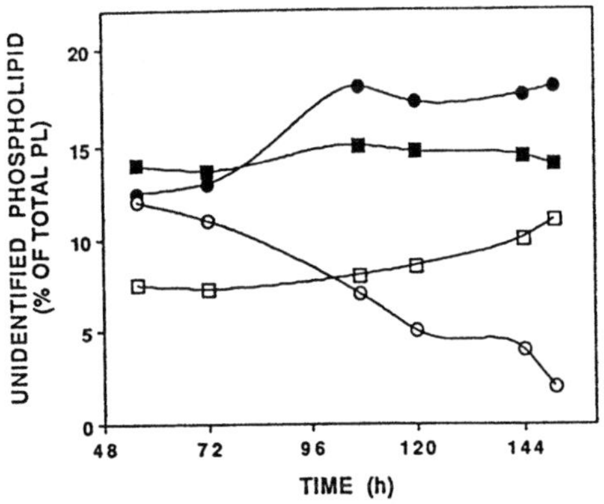


*Fig. 5.* The levels of an unidentified PL (peak 5 in Fig. 3) were determined from $^{31}P$ NMR spectra of PLs extracted from hypocotyl (solid symbols) and root (open symbols) of seedlings grown in water (□, ■) or 1 mM $V_i$ (○, ●). Values are averages of two repetitions of duplicate experiments and are accurate within 6% of the values.

plants (Srivastava and Guha-Mukherjee, 1992). The elevated PI levels in the root of V treated seedlings could therefore be associated with lowered rates of root growth. The possibility that inositolphosphates are associated with the $V_i$ induced growth inhibition should be investigated further since the changes in the levels of inositolphosphates in the presence of V have been described in mammalian tissues (Morita et al., 1992; Paris et al., 1987).

The changes in levels of the unidentified PL was associated with the FW accumulation in root of $V_i$ treated seedlings. The unidentified PL level decreased by a factor of four in the root, whereas in the hypocotyl, a minor increase is observed during 156 hrs of $V_i$ treatment in comparison to control (Fig. 5). The structure of this PL is under investigation, since its levels were associated with the growth of root. While the $^{31}P$ NMR chemical shift of this unidentified PL supports the possibility that it is a lysophosphatidic acid, analysis by TLC showed that this is not a plausible candidate (data not shown). Further spectroscopic characterization by $^{1}H$ and $^{13}C$ NMR, and mass spectroscopy of the isolated unidentified PL will be necessary to elucidate its structure. The structural information will further assists the studies concerning function and metabolism of this PL.

## Conclusions

The stimulation and growth inhibition of bean seedlings by $V_i$ may be associated with the tissue specific changes in PL metabolism. $V_i$ appears to effect

*Table 1.* The effect of 1 mM $V_i$ treatment on the levels ($\mu$g P $g^{-1}$FW) of phosphatidylcholine (PC), phosphatidylinositol (PI), phosphatidylethanolamine (PE) plus phosphatidylserine (PS), phosphatidic acid (PA) and unidentified phospholipid (UNPL) in root and hypocotyl of bean seedlings. Table shows representative data of one experiment. Since the PL levels were associated with developmental stage of the seedling, PL levels varied as growth varied and this was observed in repeated experiments

| Sample | Phospholipid ($\mu$g P $g^{-1}$ FW) | | | | |
|---|---|---|---|---|---|
| | PC | PI | PE+PS | PA | UNPL |
| Dry embryonic axes | 46.5 | 11.5 | 32.8 | <3.0 | 0 |
| Root-control | | | | | |
| 56 | $23.1^a$ | $11.2^b$ | $23.6^a$ | $24.4^a$ | $7.9^b$ |
| 107 | $2.2^c$ | $3.1^c$ | $7.8^c$ | $10.3^b$ | $3.5^c$ |
| 151 | $1.3^c$ | $2.0^c$ | $4.2^c$ | $4.2^c$ | $2.0^c$ |
| Root - $V_i$ | | | | | |
| 56 | $6.8^c$ | $17.6^a$ | $27.3^a$ | $22.7^a$ | $11.6^a$ |
| 107 | $5.7^c$ | $6.5^c$ | $10.6^b$ | $6.7^c$ | $3.1^c$ |
| 151 | $2.8^c$ | $5.2^c$ | $7.7^c$ | $5.8^c$ | $0.7^c$ |
| Hypocotyl-control | | | | | |
| 56 | $13.7^b$ | $16.4^a$ | $22.9^a$ | $18.3^a$ | $16.4^a$ |
| 107 | $10.3^b$ | $8.4^c$ | $14.9^b$ | $8.1^c$ | $9.3^b$ |
| 151 | $7.1^c$ | $6.2^c$ | $12.4^b$ | $3.4^c$ | $6.9^b$ |
| Hypocotyl - $V_i$ | | | | | |
| 56 | $13.1^b$ | $17.1^a$ | $19.9^a$ | $20.3^a$ | $15.4^a$ |
| 107 | $7.8^c$ | $11.3^b$ | $20.4^a$ | $11.8^b$ | $15.7^a$ |
| 151 | $3.3^c$ | $2.3^c$ | $11.7^b$ | $4.5^c$ | $6.6^b$ |

$^{a,b,c}$ represent statistically different values.

the PI levels and levels of the unidentified PL differently in the root and hypocotyl of the seedlings. The levels of the unidentified PL was associated with the V induced changes in FW accumulation in both root and hypocotyl tissues. These observations suggest that such a compound can play an important role in the growth.

## Acknowledgment

We thank the Colorado State Experimental Station for funding this work by grant to D C Crans. We thank Dr R W Marshman for technical assistance and Drs E Roos and S Sowa, National Seed Storage Lab, Fort Collins, for supplying seed material.

## References

Aller A J, Bernal J L, del Nozal M J and Deban L 1990 Effects of selected trace elements on plant growth. J. Sci. Food Agric. 51, 447–479.

Ames B N 1966 Assay of inorganic phosphate, total phosphate and phosphatases. *In* Methods Enzymology. Eds. E F Neufeld and V Ginsburg. Vol. 8. pp 115–118. Academic Press, New York.

Byczkowski J Z, Lychlinski L and Tiuchkiewiez J 1979 Interaction of vanadate with respiratory chain of rat liver and wheat seedling mitochondria. Int. J. Biochem. 10, 1007–1014.

Chasteen N D 1990 Vanadium in Biological Systems: Physiology and Biochemistry. Kluwer Academic Publishers, Boston. 225 p.

Cowan D S C, Cooke D T, Clarkson D T and Hall J L 1993 Lipid and sterol composition of plasma membranes isolated from stele and cortex of maize roots. J. Exp. Bot. 44, 991–994.

Crans D C, Simone C M and Blanchard J S 1992 Chemically induced modifications of cofactor specificity of glucose-6-phosphate dehydrogenase. J. Am. Chem. Soc. 114, 4926–4928.

Crans D C, Mikuš M and Mashman R W 1993 Comparison of phosphorus metabolites in various extraction procedures of *Phaseolus vulgaris* seeds using $^{31}$P NMR spectroscopy. Anal. Biochem. 209, 85–94.

Crans D C, Mikuš M and Friehauf B R 1994 Phytate metabolism in bean seedlings during post-germinative growth. J. Plant Physiol. (*In press*).

Drueckhammer D G, Durrwachter J R, Pederson R L, Crans D C, Daniels L and Wong C H 1989 Reversible and in situ formation of organic arsenates and vanadates as organic phosphate mimics in enzymatic reactions: Mechanistic investigation of aldol reactions and synthetic applications. J. Org. Chem. 54, 70–77.

Gilkes N R, Herman E M and Chrispeels M J 1979 Rapid degradation and limited synthesis of phospholipids in the cotyledons of mung bean seedlings. Plant Physiol. 64, 38–42.

Gresser M J, Tracey A S and Stankiewicz P J 1987 The interaction of vanadate with tyrosine kinases and phosphatases. Adv. Protein Phosph. 4, 35–57.

Heim S and Wagner K G 1987 Enzymatic activities of the phosphatidylinositol cycle during growth of suspension cultured plant cells. Plant Sci. 48, 167–173.

Hidalgo A, Navas P and Garcia-Herdugo G 1988 Growth inhibition induced by vanadate in onion roots. Environ. Exp. Bot. 28, 131–136.

Jolles J, Wirtz K W A, Schotman P and Gipsen W H 1979 Pituitary hormones influence polyphosphoinositide metabolism in rat brain. FEBS Lett. 105, 110–114.

Kaplan D I, Adriano D C, Carlson C L and Sajwan K S 1990 Vanadium: Toxicity and accumulation by beans. Water Air Soil Pollut. 49, 81–91.

Khan A A and Malhotra S S 1987 Effect of vanadium, nickel and sulphur dioxide on polar lipid biosynthesis in jack pine. Phytochemistry 26, 1627–1630.

Kohno Y 1986 Vanadium induced manganese toxicity in bush bean plants grown in solution culture. J. Plant Nutr. 9, 1261–1272.

Lee K 1983 Vanadium in the aquatic ecosystem. *In* Aquatic Toxicology. Ed. J O Nriagu. pp 155–187. John Wiley & Sons, New York.

Lendrath G, Nasner A and Kraus L 1990 Behaviour of vegetable phospholipids in thin-layer chromatography. Optimization of mobile phase, detection and direct evaluation. J. Chromatogr. 502, 385–392.

Lux A, Mikuš M and Koval Š 1986 Effect of vanadium on root growth and regeneration of the root cap in maize (*Zea mays* L.). Act. Fac. Rerum. Nat. Univ. Comenianae, Physiol. Plant. 22, 29–36.

McNeill J H, Yuen V G, Hoveyda H R and Orvig C 1992 Bis(maltolato)oxovanadium(IV) is a potent insulin mimic. J. Medic. Chem. 35, 1489–1491.

Meneses P and Glonek T 1988 High resolution $^{31}$P NMR of extracted phospholipids. J. Lipid Res. 29, 679–689.

Miyazawa T and Fujino Y 1978 Chemical constitution of N-acylethanolaminephosphatides in pea seed. J. Food Sci. 43, 98–101.

Morita T, Motoyashiki T, Tsuruzono Y, Kanagawa A, Tominaga N and Ueki H 1992 Rapid increase of inositol 1,4,5-trisphosphate content in isolated rat adipose tissue by vanadate. Chem. Pharm. Bull. 40, 2242–2244.

Morrell B G, Lepp N W and Phipps D A 1986 Vanadium uptake by higher plants: Some recent developments. Environ. Geochem. Health 8, 14–18.

Murthy P P N, Renders J M and Keranen L M 1989 Phosphoinositides in barley aleurone layers and gibberellic acid-induced changes in metabolism. Plant Physiol. 91, 1266–1269.

Nechay B R, Nanninga L B, Nechay P S E, Post R L, Grantham J J, Macara I G, Kubena L F, Phillips T D and Nielsen F H 1985 Role of vanadium in biology. FASEB (Fed. Am. Soc. Exp. Biol.) 45, 123–132.

Nishihara M and Kito M 1978 Changes in the phospholipid molecular species composition of soybean hypocotyl and cotyledons after dedifferentiation. Biochim. Biophys. Acta 531, 25–31.

Paris S, Chambard J C and Pouysségur J 1987 Coupling between phosphoinositide breakdown and early mitogenic events in fibroblasts. Studies with fluoroaluminate, vanadate and pertussis toxin. J. Biol. Chem. 262, 1977–1983.

Parker R D R, Sharma R P and Miller G W 1978 Vanadium in plants, soil and water in the Rocky Mountain region and its relationship to industrial operations. Trace Subs. Environ. Health 12, 340–350.

Perlin D S and Spanswick R M 1981 Characterization of ATPase activity associated with corn leaf plasma membranes. Plant Physiol. 68, 521–526.

Rakhimov M M, Abdullaeva M M, Balikhanov M N and Babaev M U 1986 Role of phytase and phospholipase in changes of phospholipids during germination of cotton seeds. Sov. Plant Physiol. 33, 858–865.

Shewry P R, Pinfield N J and Stobart A K 1973 Phospholipids and the phospholipid fatty acids of germinating hazel seeds (*Corylus avellana* L.). J. Exp. Bot. 24, 1100–1105.

Sklenár J, Fox G G, Loughman B C, Pannifer A D B and Ratcliffe R G 1994 Metabolic effects of vanadate in roots of *Zea mays*. Plant and Soil (This Volume).

Srivastava V and Guha-Mukherjee S 1992 Phosphatidylinositol turnover in crown gall tumour. Phytochemistry 31, 773–774.

Welch R M 1973 Vanadium uptake by plants. Absorption kinetics and the effects of pH, metabolic inhibitors, and other anions and cations. Plant Physiol. 51, 828–832.

Wendt C W, Hicks S K and Baker R B 1993 Preparation of vanadyl carboxylates as crop yield promoters. Patent U.S US 5,186,738, February 16 1993.

Willsky G R, White D A and McCabe B C 1984 Metabolism of added orthovanadate to vanadyl and high-molecular-weight vanadates by *Saccharomyces cerevisiae*. J. Biol. Chem. 259, 13273–13281.

*F. Baluška et al. (eds.), Structure and Function of Roots,* 189–193, 1995.


# Distribution of soil invertase in relation to the root systems of *Picea sitchensis* (Bong.) Carr. and *Acer pseudoplatanus* L. during development of young plants

D. Vaughan[1], B.G. Ord[1], S.T. Buckland[2], E.I. Duff[2] and C.D. Campbell[1]
[1]*Macaulay Land Use Research Institute, Craigiebuckler, Aberdeen, AB9 2QJ, UK and* [2]*Scottish Agricultural Statistics Service, Macaulay Land Use Research Institute, Craigiebuckler, Aberdeen AB9 2QJ, UK*

*Key words: Acer*, invertase, *Picea*, rhizosphere, roots, statistics, trees

**Abstract**

A method is described for sampling rhizosphere soil under newly established *Picea sitchensis* and *Acer pseudoplatanus*. The technique involves taking soil samples to a depth of 150 mm at 100 mm intervals along transects, each 45° from its neighbour, radiating from the base of the stem. Invertase activities were measured in the soil samples and compared to their activities in fallow and rhizosphere soils. When the field soil was dry, the tree root systems were carefully excavated to retain as many fine roots as possible. The distribution of the soil invertase was matched to the spatial distribution of the roots showing the precise position of the rhizosphere relative to the initial 'blind' soil sampling. Statistics were applied to derive equations for calculating the percentage enzyme activity relative to that found in rhizosphere soil at various locations radiating from the base of the stem. This information was subsequently applied to soil sampled under trees of the same age as those excavated to give a non-destructive method for sampling rhizosphere soil routinely from under a large number of trees.

## Introduction

Plant roots exude water soluble organic substances which influence microbial activity and other soil processes in the rhizosphere (Marschner, 1991). The simplest definition of the rhizosphere is that volume of soil influenced by the root and is usually assumed to extend from the root surface (rhizoplane) out into the soil for a few millimetres, or possibly a few centimetres in extreme cases (Campbell and Greaves, 1990). About 20–30% of assimilated carbon may be lost from roots annually into the rhizosphere (Cheshire and Mundie, 1990; Whipps and Lynch, 1986). Of this amount 25% is lost directly through root exudation, the remaining 75% comprising mucigel, sloughed off cells and dead rootlets (Lambers, 1987). Microbial activity is, therefore, highest in the rhizosphere and least in the background soil. Most of the information on microbial activity in the rhizosphere comes from studies on herbaceous plants. The impact which tree root exudates have on soil processes in the rhizosphere has been a neglected area of work.

One major problem in the way of studying roots in the field is that it is impossible to observe a root growing in the soil and the main difficulties encountered in such studies have been highlighted by Harper et al. (1990). Sampling rhizosphere soil from under trees may pose a problem because the spatial distribution of the roots, particularly fine roots, has seldom been established. The spatial distribution of roots will vary between different plants of the same species, while remaining characteristic for that species. Hence it is not possible to determine with certainty the presence of rhizosphere soil in the vicinity of trees during routine soil sampling. In this report we assess the distribution of the activities of the enzyme invertase, indicative of rhizosphere activity, in soil sampled under newly established trees in a farm forest and match these activities to the distribution of the root system, and hence rhizosphere, of trees carefully excavated from soil. The data are used to derive a formula for predicting the percentage of rhizosphere soil at various distances from the base of young trees of the same age during frequent routine sampling.

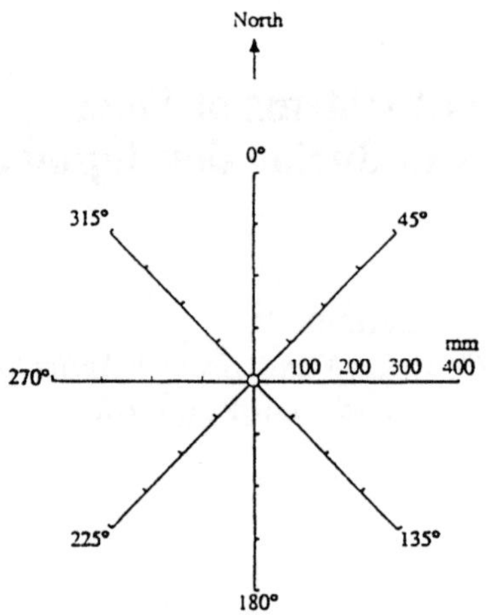


*Fig. 1.* Plan of sampling technique showing transects radiating from trunk of tree. The zero degree transect always faced north. Sampling positions at 100, 200, 300 and 400 mm are also shown.

## Materials and methods

### *Location of trees*

A farm forestry experiment was set up on 3.5 ha of agricultural land at Lower Affleck farm 15 km north-northwest of Aberdeen (National grid reference NJ 864239) in May 1990. The soil, of the Thistlyhill series derived from basic igneous rocks, comprises brown forest soils with gleying and is imperfectly drained (Lilly, 1990). The trees, *Picea sitchensis* (Bong.) Carr. and *Acer pseudoplatanus* L. were planted at a conventional UK forest density of 2000 stems per hectare (2.24 m apart).

### *Soil sampling*

Soil was sampled in August 1992 along straight line transects, at 100 mm intervals, to a depth of 150 mm. Each transect, which radiated from the base of the stem, was at 45° to its neighbour giving eight transects per tree (Fig. 1). The zero degree transect always faced north. Soil was also sampled 10 m away from the trees to give a background soil and shaken from the immediate vicinity of excavated tree roots < 2 mm diameter to give a rhizosphere soil. All soil was passed through a 2 mm sieve and used immediately for enzyme analysis. Moisture contents were measured after drying 5 g soil at 100°C for 48 h.

### *Invertase assays*

Invertase activity was determined by incubating 1g soil in 10 $cm^{-3}$ unbuffered 150 mol $m^{-3}$ sucrose for 1 h at 35°C (Vaughan and Malcolm, 1984).

### *Root systems*

The root systems of three trees of both species were carefully excavated from dry soil, washed with water, suspended and photographed. Tracings were made of the root systems directly from the photographs and the positions of the transects accurately marked. In Figures 2a and 4a, only the position of the zero transect is shown.

### *Statistical treatment*

Models for the decline in invertase activity in the soil with distance from the tree stem were fitted to the data recorded at intervals of 100, 200, 300 and 400 mm along the transects from the base of the stem. Each of the eight samples from a fixed distance (x in the equation below) was estimated in duplicate giving 16 determinations for each of the three trees per species. This gave 48 samples at each distance for each species which were averaged. A logistic curve with the lower asymptote constrained to the background level of invertase was found to model the data adequately. The fitted curves were rescaled so that the predicted values could be expressed as percentage of rhizosphere soil activity by distance from the tree. Comparison of the upper asymptote with invertase activities in the rhizosphere soil, shaken from roots, allowed estimation of the percentage of rhizosphere soil in the top 150 mm of soil at the base of the tree.

The model for this curve may be expressed as:

$$Y = c/[1 + \exp\{-b(x - m)\}]$$

where

x = distance from stem of tree
Y = proportion of rhizosphere soil activity at distance x
m = distance of the point of inflection from stem of tree
b = slope parameter
c = proportion of rhizosphere soil activity at tree base

The above model but with c = 1 was fitted to the data for each of the two tree species. The constant c was then estimated by dividing the predicted maximum invertase activity by the observed invertase activity for rhizosphere soil shaken from roots.

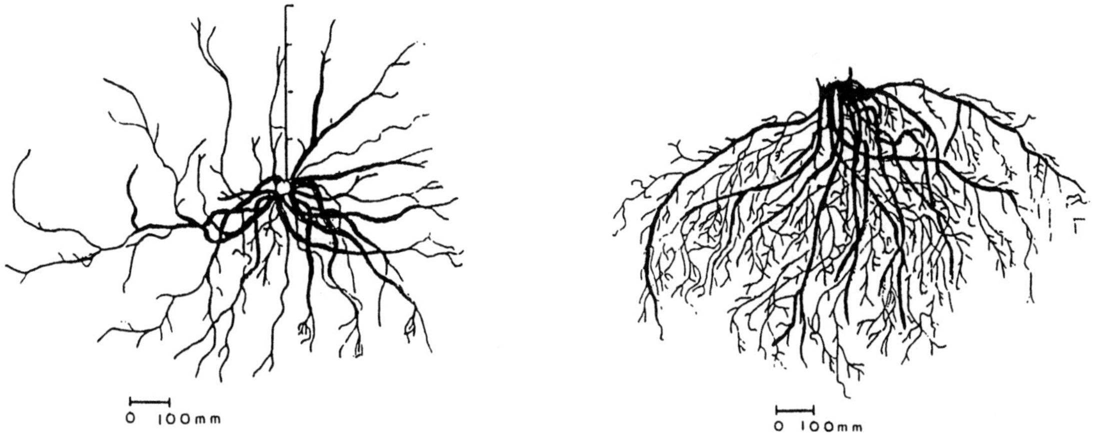


*Fig. 2.* Morphology of root system of 2-year old *Picea sitchensis* showing (**a**) horizontal and (**b**) vertical views. The position and sampling distances are shown in the horizontal view for the 0° transect. Soil invertase activities under this tree are shown in Table 1.

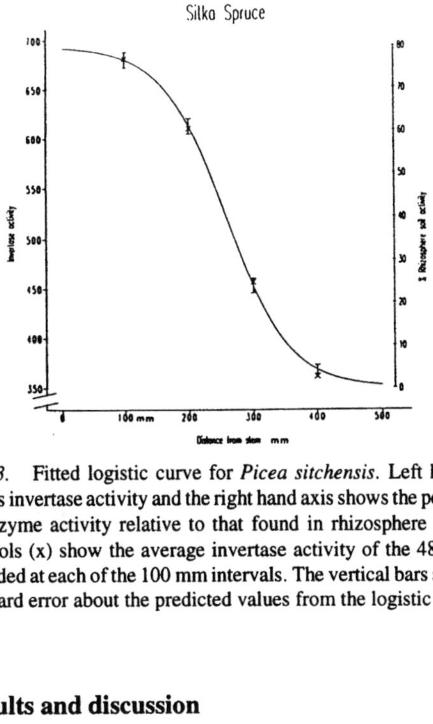


*Fig. 3.* Fitted logistic curve for *Picea sitchensis*. Left hand axis shows invertase activity and the right hand axis shows the percentage of enzyme activity relative to that found in rhizosphere soil. The symbols (x) show the average invertase activity of the 48 samples recorded at each of the 100 mm intervals. The vertical bars show ± 1 standard error about the predicted values from the logistic curve.

*Table 1.* Invertase activities for soils sampled at 100 mm intervals along eight transects radiating from the centre of 2-year old *Picea sitchensis*

| Transect (°) | Distance (mm) | | | |
|---|---|---|---|---|
| | 100 | 200 | 300 | 400 |
| | ($\mu$g glucose formed $g^{-1}$ soil $h^{-1}$) | | | |
| 0 | 616 573 | 327 351 | 482 531 | 336 349 |
| 45 | 697 671 | 487 523 | 341 347 | 342 357 |
| 90 | 735 705 | 635 672 | 491 472 | 362 351 |
| 135 | 779 747 | 621 583 | 341 337 | 339 341 |
| 180 | 792 756 | 541 562 | 392 417 | 435 472 |
| 225 | 791 772 | 682 725 | 422 391 | 362 341 |
| 270 | 732 741 | 731 782 | 521 479 | 339 357 |
| 315 | 637 694 | 543 569 | 342 341 | 329 351 |
| Mean | 715 | 583 | 415 | 360 |

Duplicate samples at each location were analysed, corresponding to the two observations per cell in the table. Three determinations of background activity gave a mean of 350 $\mu$g glucose formed $g^{-1}$ soil $h^{-1}$ with a standard error of 3, and five determinations for rhizosphere soil gave a mean of 781 and a standard error of 5.

## Results and discussion

### *Root distribution of* Picea sitchensis

The morphology, horizontal and vertical views, of one of the excavated roots of *P. sitchensis* is shown in Figure 2. The fibrous root system comprises, in addition to the primary roots, a substantial amount of fine roots. In this specimen, the roots were not evenly distributed with most of the mass being towards the south. Hence the distribution of rhizosphere soil was not evenly spread along the transects. This variation was reflected in the distribution of invertase activity in the soil sampled under the tree at fixed distances from the stem (Table 1), although the replicate determinations for each sample point are in close agreement. There is, however, a continuous decrease in the soil invertase activity with increasing distance from the stem. The substantial variation in enzyme activity prevented reliable fitting of a logistic curve to data for each tree. Thus the invertase determinations at each distance were averaged across trees of a single species and a curve fitted to these means of 48 determinations

(Fig. 3). For *P. sitchensis* the fitted logic curve for the percentage of rhizosphere soil activity (Y) at a given distance from the tree fitted the equation:

$$Y = 1/[1 + \exp\{0.020 (x - 258.0)\}]$$

The derived curve gave lower and upper asymptotes for invertase activity of 350 and 694 $\mu$g glucose formed $g^{-1}$ soil $h^{-1}$ respectively. The lower value was constrained to equal that of the background soil. The invertase activity in the rhizosphere soil was 781 mg glucose formed $g^{-1}$ soil $h^{-1}$ so that the upper asymptote value was 79.8% that of the rhizosphere soil. The axis showing percentage of rhizosphere soil activity (Fig. 3) has been rescaled so that Y = 1 in the above equation corresponds to 79.8%. Expressed in this way the data show the percentage activity relative to the rhizosphere (activity 100%) which can be obtained by sampling at known distances.

*Root distribution of* Acer pseudoplatanus

The morphology of one of the excavated root systems of *A. pseudoplatanus* is shown in Figure 4. The primary roots were thicker and more branched than those of *P. sitchensis*, forming a tap root system, but there were fewer fine roots. The roots were not distributed evenly throughout the soil and this was shown in the soil invertase activity along the transects (Table 2). Thus there was virtually no activity comparable to that found in rhizosphere soils along transects 0° and 315°. In comparison with *P. sitchensis* there was a continuous decrease in soil invertase activity, increasing with distance from the stem, and good replication between the duplicate determinations at fixed distances along the transects. The invertase activity in the rhizosphere soil of *A. pseudoplatanus* was considerbly higher than that in the corresponding soil of *P. sitchensis* being 976 $\pm$ 6.23 (Table 2) and 781 $\pm$ 5.03 (Table 1) $\mu$g glucose formed $g^{-1}$ soil $h^{-1}$ respectively.

The fitted logic curve for *A. pseudoplatanus* was:

$$Y = 1/[1 + \exp\{0.011 (x - 252.5)\}]$$

This curve gave lower and upper asymptotes for invertase activity of 346.3 and 845.3 $\mu$g glucose formed $g^{-1}$ soil $h^{-1}$ respectively. The 845.3 value was 79.2% of that measured in rhizosphere soil (Table 2). The axis showing percentage of the rhizosphere soil activity in Figure 5 has been rescaled so that Y = 1 in the above equation corresponds to 79.2%.

*Table 2.* Invertase activities in soils sampled at 100 mm intervals along eight transects radiating from the centre of 2-year old *Acer pseudoplatanus*

| Transect (°) | Distance (mm) | | | | | | | |
|---|---|---|---|---|---|---|---|---|
| | 100 | | 200 | | 300 | | 400 | |
| | ($\mu$g glucose formed $g^{-1}$ soil $h^{-1}$) | | | | | | | |
| 0 | 413 | 383 | 363 | 332 | 381 | 387 | 342 | 363 |
| 45 | 782 | 726 | 410 | 336 | 347 | 379 | 372 | 341 |
| 90 | 972 | 1021 | 798 | 872 | 366 | 347 | 362 | 352 |
| 135 | 997 | 971 | 382 | 427 | 526 | 471 | 375 | 364 |
| 180 | 1012 | 954 | 591 | 655 | 732 | 784 | 473 | 421 |
| 225 | 994 | 938 | 721 | 852 | 357 | 341 | 372 | 363 |
| 270 | 375 | 427 | 688 | 725 | 585 | 541 | 329 | 341 |
| 315 | 497 | 526 | 352 | 341 | 362 | 357 | 339 | 359 |
| Mean | 749 | | 553 | | 454 | | 367 | |

Duplicate samples per location correspond to the two observations per cell in the table. Three determinations of background activity gave a mean of 346 $\mu$g glucose formed $g^{-1}$ soil $h^{-1}$ with a standard error of 3, and five determinations for rhizosphere soil gave a mean of 976 and a standard error of 6.

*Application of technique*

The distribution of roots provides information on the extent and density of plant root systems, yet methods used in such studies are usually tedious (Escamilla et al., 1991). Our technique is simple and can be applied to newly planted forest systems where the distribution of roots can be evaluated in relation to soil microbial activity. Because the maximum activity is found in rhizosphere soil, and the least in background soil (Sparling, 1985), the data can be assessed in relation to changes in percentage rhizosphere activity with increasing distance from the tree trunk so that the region of maximum activity can be identified. Once the region of maximum activity has been identified, the procedure can be applied to large numbers of samplings of soils from under trees of the same age without having to excavate the root systems. By reference to the fitted logistic curves, a correction factor can then be applied to the data to express the results on a 100% rhizosphere soil basis. However it should be borne in mind that for each tree, 8 soil samples should be taken along transect lines at a fixed distance from the stem and pooled.

Changes in the position of the rhizosphere in relation to young, growing trees must be measured yearly to maintain the accuracy of the method. In the 2-year

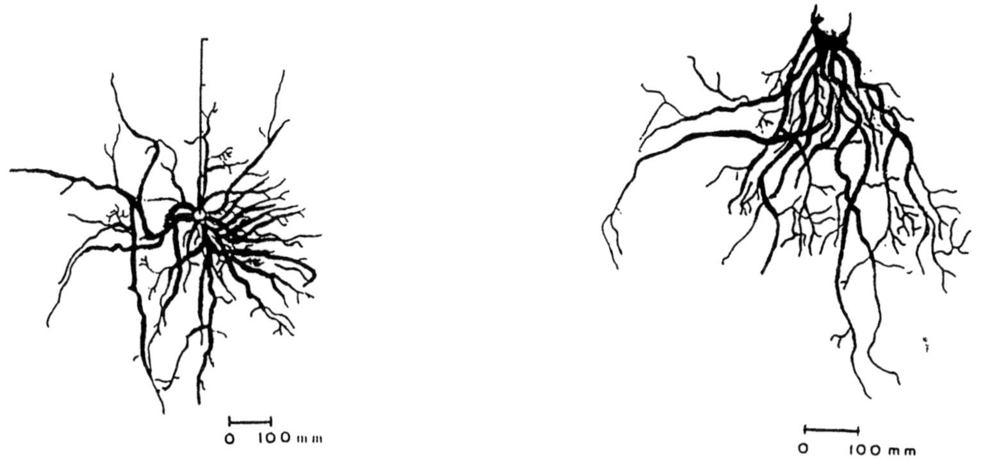


*Fig. 4.* Morphology of root system of *Acer pseudoplatanus* showing (**a**) horizontal and (**b**) vertical views. The position and sampling distances are shown in the horizontal view for the 0° transect. Soil invertase activities under this tree are shown in Table 2.

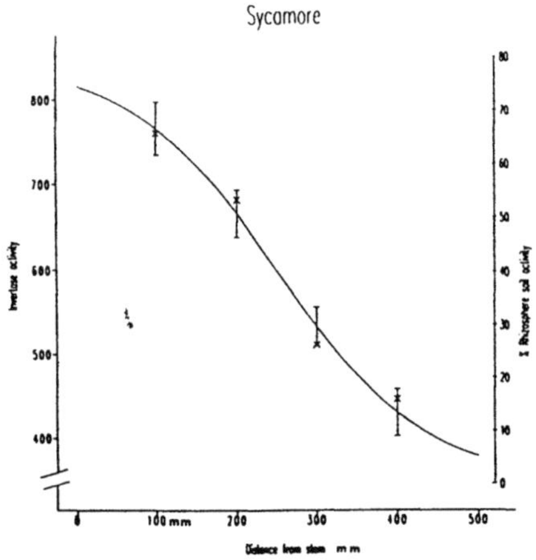


*Fig. 5.* Fitted logistic curve for *Acer pseudoplatanus*. Left hand axis shows invertase activity and the right hand shows the percentage of enzyme activity relative to that found in rhizosphere soil. The symbols (x) show the average invertase activity of the 48 samples recorded at each of the 100 mm intervals. The vertical bars show ± 1 standard error about the predicted values from the logistic curve.

old trees sampled, there were no interactions between the roots of different trees, so that it will be important to assess the valdity of the technique as the forest system matures. Although soil invertase activity under trees changes during the growing season (unpublished data), this has no effect on the validity of the technique described above for any given sampling time during the growing season.

## Acknowledgement

We are grateful to the Scottish Office Agricultural and Fisheries Department for funding this project.

## References

Campbell R and Greaves M P 1990 Anatomy and community structure of the rhizosphere. *In* The Rhizosphere. Ed. J M Lynch. pp 11–34. John Wiley and Sons, Chichester.

Cheshire M V and Mundie C M 1990 Organic matter contributed to soil by plant roots during the growth and decomposition of maize. Plant and Soil 121, 107–114.

Escamilla J A, Comerford N B and Neary D G 1991 Soil-core break method to estimate pine root distribution. Soil Sci. Soc. Am. J. 55, 1722–1726.

Harper J L, Jones M and Sackville Hamilton N R 1990 The evolution of roots and the problems of analysing their behaviour. *In* Plant Root Growth – An Ecological Perspective. Ed. D Atkinson. pp 49–59. Blackwell Scientific Publications, Oxford.

Lambers H 1987 Growth, respiration and symbiotic associations: the fate of carbon translocated to roots. *In* Root Development and Function. Eds. P J Gregory, J V Lake and D A Rose. pp 125–145. Cambridge University Press.

Lilly A 1990 The soils of the Lower Affleck farm forestry experimental site. Macaulay Land Use Research Institute, Aberdeen, April 1990.

Marschner H 1991 Plant-soil relationships: acquisition of mineral nutrients by roots from soils. *In* Plant Growth: Interactions with Nutrition and Environment. Eds. J R Porter and D W Lawlor. pp 125–155. Cambridge University Press.

Sparling G P 1985 The soil biomass. *In* Soil Organic Matter and Biological Activity. Eds. D Vaughan and R E Malcolm. pp 223–262. Martinus Nijhoff, Dordrecht.

Vaughan D and Malcolm R E 1984 Influence of assay conditions on invertase activity in different Scottish soils. Plant and Soil 76, 339–347.

Whipps and Lynch J M 1986 The influence of the rhizosphere on crop productivity. Adv. Microb. Ecol. 9, 187–244.

# ABSORPTION AND TRANSPORT OF WATER

*F. Baluška et al. (eds.), Structure and Function of Roots,* 197–208, 1995.


# Water transport across roots

Ernst Steudle
*Lehrstuhl Pflanzenökologie, Universität Bayreuth, Universitätstrasse 30, D-95440 Bayreuth, Germany*

*Key words:* composite transport model, hydraulic conductivity, osmometer model, reflection coefficient, root water transport, solute permeability


## Abstract

Usually, roots are looked at as rather perfect osmometers with the endodermis being the 'root membrane' which is equivalent to the plasma membrane of cells. However, this 'single-equivalent-membrane model' of the root does not explain the findings of a variable hydraulic resistance of roots as well as of differences between hydraulic and osmotic water flow and of low reflection coefficients of roots. Recent work with the root pressure probe is reviewed and discussed which indicates that the simple osmometer model of the root has to be extended by incorporating its composite structure, i.e. the fact that there are different parallel pathways for water in the root, namely, the cell-to-cell and apoplasmic path. The new 'composite transport model of the root' readily explains the experimental findings mentioned above. Pressure probe work with roots in which the endodermis was punctured to create an additional parallel path as well as anatomical studies support the model.


## Introduction

Roots are very efficient in taking up water from the soil and meet the requirements of the transpiring shoot. This means that the hydraulic conductivity of roots and its components (cell membranes, cell walls, etc.) has to be rather high. On the other hand, roots have to be sufficiently 'tight' to prevent a substantial leakage of nutrients once accumulated in the xylem. Thus, the 'passive' permeability of roots to solutes has to be low compared with that of water. The 'optimization' of the problem has led to the 'invention' of the root endodermis. This structure is thought to have rather perfect semipermeable properties in that water passes the membranes much faster than nutrient ions (usually the permeability of cell membranes to water is larger by 3 to 8 orders of magnitude than that for solutes). In addition, the Casparian band nearly completely interrupts the passage of water and hydrophilic solutes across the walls. In principle, this picture may be true. However, there are some problems with it because, in some instances, water and solute relations of roots deviate from those predicted from the simple model. The deviations include:

(i) Water flow is not linearly related to the driving force (water potential difference across root) over the entire range of forces (cf. Fiscus, 1975; Kramer, 1983; Steudle, 1989, 1992, 1993, 1994; Weatherley, 1982).

(ii) There is an offset of pressure (water potential) which has to be overcome before water is taken up into the root (Passioura, 1988).

(iii) There is some variability in the hydraulic conductivity of roots which is related to the nature of the driving force used. Osmotic water flow can be much smaller than hydrostatic. (Hallgren et al., 1994; Steudle et al., 1987; Steudle, 1989, 1992, 1994).

(iv) Root reflection coefficients are substantially lower than unity for solutes present in the xylem sap ($\sigma_{sr}$ = 1 refers to the perfect osmometer and $\sigma_{sr}$ = 0 to an osmotic barrier which does not distinguish between water and solute) (Steudle and Frensch 1989; Steudle and Brinckmann, 1989; Steudle, 1989, 1992, 1994).

These points are important since they either refer directly to the root hydraulics ($Lp_r$) or to the forces driving water across roots ($\sigma_{sr}$). Evidence for deviations comes from work of different laboratories which use quite different techniques such as the pressure-chamber technique (Fiscus, 1975), stop-flow techniques (Miller, 1985; Pitman et al., 1981), or the balancing pressure technique (Passioura, 1988). A lot of evidence concerning items (iii) and (iv) has been collected in the past few years using the root pressure probe (Azaizeh and Steudle, 1991; Birner and Steudle, 1993; Heydt and Steudle, 1991; Melchior and Steudle, 1993; Peterson et al. ,1993; Steudle, 1989, 1992, 1993, 1994; Steudle et al., 1993). In the following, this work is reviewed and discussed in relation to transport models of roots. The 'composite transport model of the root' is presented as a new concept which explains the deviations. The root pressure probe technique has been described in detail in many original contributions (see above and Hallgren et al., 1994; Melchior and Steudle, 1994).

## Transport models of roots

In the traditional view, the endodermis is thought to be the main hydraulic and osmotic barrier in roots. Since root cell protoplasts are considered to represent a hydraulic resistance much larger than that of the root apoplast, the preferred path for water in the cortex is the wall path, i.e. water moves around the protoplasts. At the endodermis, the Casparian band blocks the apoplasmic passage and a transport across two membranes is required. In the stele, transport is again mainly apoplasmic up to the vessels which (when mature) represent a path of high conductance. In this picture, the endodermis with the fully developed Casparian band provides a rather perfect semipermeable barrier. Roots, like cells, have been regarded to be nearly ideal osmometers (Weatherley, 1982). Since the endodermis is considered to be 'the root membrane', the osmometer model may also be called a 'single-equivalent-membrane model' (Dainty, 1985). This implies that 'the root membrane' behaves like a 'homogenous membrane', an assumption which may be questioned.

Although the term may be not used directly, many textbooks refer to the root as a rather perfect (semipermeable) osmometer. This is so because this model fulfills two requirements for a proper functioning of roots: (i) Water uptake (or loss) may be rate-limited or controlled at the endodermis. (ii) The loss (leakage) of nutrients accumulated in the xylem by active processes is effectively prevented. The first requirement has been thought to be important to protect the root cortex against rapid changes in xylem water potential as they occur in the transpiring plant, or to dampen the stelar tissue against changes in soil water potential. At least in young, non-suberized roots, the first requirement is not fulfilled since these roots have high radial hydraulic conductivities and quickly equilibrate with their immediate surroundings. The second requirement (roots impervious to nutrients and other solutes) seems to be fulfilled. Here, most of the evidence comes from observations of the radial movement of apoplasmic dyes or other tracers which do show that the endodermis is an effective barrier for solutes (Clarkson, 1991; Marschner, 1986).

## Single-equivalent-membrane model of root: quantitative formulation of transport

The analogy between the cell and root osmometer has led to the application of basic transport equations of irreversible thermodynamics to the root system which have been successfully used at the cell level (Fiscus, 1975; Steudle et al., 1987). The root has been treated as a two-compartment system with an internal compartment (xylem tracheary elements) and an external (soil solution or root medium) separated by a membrane-like barrier. The transport equations for roots describe the flow of water (or of volume in the more precise notation of the theory) per unit surface area of a root and the forces (hydrostatic or osmotic pressure gradients) driving it, as well as the solute (nutrient) flow (again per unit root surface area). In the description, interactions between flows, i.e. between the water (volume) ($J_{Vr}$) and the solute ($J_{sr}$) flow are taken into account: The latter play an important role in roots and are well-known from guttation, exudation and other root pressure phenomena. If we accept the two-compartment, root membrane model, and, for the sake of simplicity, consider the flow of only one solute (subscript's') besides the water flow, we get (Steudle, 1989, 1992, 1993, 1994):

$$J_{Vr} = -\frac{1}{A_r}\frac{dV_x}{dt} = Lp_r \cdot P_r - Lp_r \cdot \sigma_{sr} \cdot RT \cdot (C_s^x - C_s^o), \quad (1)$$

and

$$J_{sr} = -\frac{1}{A_r}\frac{dn_s^x}{dt} = P_{sr} \cdot (C_s^x - C_s^o) + (1-\sigma_{sr}) \cdot \overline{C}_s \cdot J_{Vr} + J_{sr}^*. \quad (2)$$

Here, $A_r$ and $V_x$ are the root surface area and the volume of mature xylem, respectively. $P_r$ = root pressure or tension in root xylem (atmospheric pressure is taken as the reference); $\sigma_{sr}$ = root reflection coefficient; $C_s^{o,x}$ = concentration of solute 's' in the medium (superscript 'o') or in the xylem (superscript 'x'); $n_s^x$ = amount of solute 's' in the xylem; $P_{sr}$ = permeability coefficient of root; $\overline{C}_s$ = mean concentration of 's' in the root ($\approx (C_s^o + C_s^x)/2$). $J_{sr}^*$ = active solute flow.

It should be noted that, according to Eq. (1), water flow has a hydrostatic and osmotic component. Osmotic and hydrostatic pressure gradients are related to water flow by different coefficients ($Lp_r$ and $\sigma_{sr} \cdot Lp_r$, respectively). This expresses the fact that there are interactions between water and solutes as they pass the membrane which cause the differences. It is seen from Eq.(1) that hydrostatic and osmotic gradients yield the same flow (i.e. they show the same apparent conductivity) only for a semipermeable membrane or barrier ($\sigma_{sr} = 1$; $P_{sr} = 0$). If the selectivity of the membrane for the given solute 's' is reduced ($\sigma_{sr} < 1$), the osmotic flow is smaller by the factor of $\sigma_{sr}$ for the same gradient applied. At an extreme, $\sigma_{sr} = 0$. This means that there is no selectivity and osmotic gradients cause no water flow. Conditions for the latter situation are found in the apoplast (such as in the root; see below).

According to Eq. (2), the solute flow ($J_{sr}$) has three different components. The diffusional component ($= P_{sr} \cdot (C_s^x - C_s^o)$) represents the passive permeation of solute across the root cylinder according to Fick's first law. The second term on the right side is the solvent drag which quantifies the effect of water flow on solute movement. In roots, this component will usually be negligible. The third component ($J_{sr}^*$) represents the active transport of 's', i.e., it relates solute flow to metabolism, e.g. to the splitting of ATP by an ATPase. $J_{sr}^*$ provides a possibility of a regulation of solutes in roots. However, since $J_{sr}^*$ affects concentrations ($C_s^x$), it will also affect water flow. The relative contribution of solute flow should be most important at low or zero transpiration. It is easily verified from Eqs. (1) and (2) that, at zero water and solute flow ($J_{sr}$, $J_{Vr} = 0$), the steady-state root pressure ($P_{ro}$) will be (Birner and Steudle, 1993; Steudle, 1994):

$$P_{ro} = -\sigma_{sr} \cdot RT \frac{J_{sr}^*}{P_{sr}}. \qquad (3)$$

It is plausible that $P_{ro}$ increases with increasing solute pumping ($J_{sr}^*$ being more negative) and decreases with increasing $P_{sr}$ (increasing leak of solutes across the root cylinder). As a borderline case, Eqs. (1) and (2) incorporate the perfect osmometer ($\sigma_{sr} = 1$; $P_{sr} = 0$). However, they extend the simple model in that they allow for more realistic properties of roots, namely, some leakage of solutes ($P_{sr} > 0$) and some deviation from properties of the ideal osmometer ($\sigma_{sr} < 1$). Despite this extension, the idea of a homogeneous 'root membrane' and the simple two-compartment model still forms the basis of the concept given by Eqs. (1) and (2)

The model has been successfully used to explain experimental data. Most importantly, the linear relationships between flows and forces have been basically verified in experiments. However, the two-compartment, single-equivalent-membrane model has been criticized and more complex models have been proposed such as models consisting of three or more compartments in series (Newman, 1976). In fact, one could think of multi-compartment models in which each cell layer represents a compartment. In other models, the compartmentation of the root into apoplast and symplast has been considered to be important: solutes are transferred between compartments and water follows changes in osmotic pressure ('canal model of root', Katou and Taura, 1989).

Multi-compartment models require detailed information about the transport across borders of different compartments, their sizes, internal concentrations etc. These data are usually not available. In addition, all complex series compartment models are not easily handled. Therefore, it has been claimed that the root-membrane model should persist for practical reasons and deviations would have to be interpreted in terms of variable transport coefficients ($Lp_r$, $\sigma_{sr}$, $P_{sr}$) or active components ($J_{sr}^*$) (Dainty, 1985). However, this point of view does not seem to be convincing. An alternative would be to use the simple approach as a starting point and to extend it, if necessary, which in turn requires techniques to measure relevant transport parameters. Such concepts are briefly reviewed in the following. First, some evidence is presented in favor of the simple model. Then, extensions are given to correct for non-linear force/flow relations. This is followed by the presentation of a new root model which also accounts for deviations, but, for the first time considers different parallel transport pathways in the root. The concept of an osmotic barrier exhibiting transport properties of a homogenous membrane is given up and a 'composite barrier model' is used to describe water and solute transport across roots still using the basic concepts of irreversible thermodynamics.

## Evidence for the single-equivalent-membrane model of the root

Most of the evidence for the model comes from experiments in which excised roots freely exude xylem sap.

In early studies, the differences between the osmotic concentrations of xylem sap and root medium was found to be proportional to the (osmotic) water flow in these experiments (for a review, see Anderson, 1976). Also in the absence of a concentration difference, there was a 'non-osmotic' water flow which is understandable if there is an active uptake of solutes which drags some water along with it so that, at the steady state, $J^*_{sr}/J_{Vr} = C^x_s$. In other experiments, the rate of exudation was increased by pressurizing the root system or by applying a vacuum to the cut root surface (Cruz et al., 1992; Fiscus, 1975; Mees and Weatherley, 1957). It was found that, in general, pressure-volume curves were curvilinear and the apparent $Lp_r$ increased with increasing flow. Rather than rejecting the simple model, this was interpreted as being due to a dilution effect in the root xylem, i.e. part of the force driving the flow did vary with $J_{Vr}$ (Fiscus, 1975).

In other exudation experiments, 'stop flow techniques' were used. A stationary root exudation (constant $J_{Vr}$) was first established which was then stopped (i.e. $J_{Vr}$ set zero) by either increasing the osmotic pressure of the medium or by applying a hydrostatic pressure pulse to the cut surface of the root (Miller, 1985; Pitman et al., 1981). These experiments yielded values similar to those obtained from the earlier measurements of steady-state flows. There were also indications of a $\sigma_{sr} < 1$ (Miller, 1985). Values of $Lp_r$ obtained by different techniques were similar, although the comparison of the data obtained by using either osmotic or hydrostatic pressure gradients yielded a fairly broad spectrum of $Lp_r$ values. There were clear indications of a lower $Lp_r$ in the presence of osmotic forces than in the presence of hydrostatic (Cruz et al., 1992; Hallgren et al., 1994; Steudle, 1989; Steudle et al., 1987).

**Deviations from the simple model: root pressure probe measurements**

The trend of $Lp_r$(osmotic) < $Lp_r$(hydrostatic) continued when the root pressure probe was introduced. The technique has been explained in detail in earlier publications (see, for example, Hallgren et al., 1994; Melchior and Steudle, 1994). It is based on the measurement of root pressure in excised roots using a little manometer adapted from the conventional cell pressure probe. Root pressure is measured by tightly fixing the excised root to the equipment without interrupting xylem tracheary elements. This yields the steady-state root pressure ($P_{ro}$ in Eq.(3)) which is a measure of the active pumping of a root. Root pressures and transport properties of roots can be measured either with root tips, segments, or whole root systems. In a single root tip, measurements are usually possible for one to two days. In root systems, measurements have been performed for up to 10 days (see below).

Hydraulic data (hydraulic conductivity of root, half time of water exchange between root and medium) are obtained by either changing the root pressure with the aid of the equipment or by changing the osmotic pressure of the medium to induce an osmotic water flow. In a variation of the technique, $J_{Vr}$ and $Lp_r$ are obtained by pressure clamp (Steudle and Frensch, 1989). Thus, there are different techniques to evaluate $Lp_r$. As already mentioned, the $Lp_r$ obtained with the technique using hydrostatic gradients was larger and in some cases much larger (one to three orders of magnitude) than that obtained by osmotic gradients, although this did not hold for all species investigated. Thus, the trend was similar to that known from the literature (see above), but could be worked out much more precisely.

In contrast to the simple exudation technique, root pressure probe techniques also provided data of solute parameters ($\sigma_{sr}$ and $P_{sr}$). Reflection coefficients were determined by changing the osmotic concentration of the root medium and recording the maximum change in root pressure at $J_{Vr} = 0$. For roots of herbaceous species, reflection coefficients were substantially smaller than unity and ranged between 0.5 and 0.8 even for solutes for which cell membranes exhibit a $\sigma_s \approx 1$ (salts, sugars, PEG, mannitol, urea, etc.). Careful investigation showed that the result could not be explained by unstirred layers (Steudle and Frensch, 1989). Hence, the simple single-equivalent-membrane model of the root in which the root membrane represented a nearly semipermeable barrier (see above) was questioned (Steudle et al., 1987; Steudle, 1989, 1992, 1993, 1994). Despite the low reflection coefficient, permeability coefficients of roots were low which demonstrated that the roots used were not unusually leaky. Absolute values of $P_{sr}$ for salts, sugars and mannitol were of an order of magnitude similar to those of root cells. In studies with fluorescent dyes it was determined that solutes could not flow across the cortical apoplast in the area where the root was sealed to the probe (Peterson and Steudle, 1993). If there were such a flow, this would have artifactually increased the permeability of the system.

## Longitudinal vs. radial transport

Roots were also checked for a substantial contribution of the longitudinal (axial) hydraulic resistance (Frensch and Steudle, 1989; Melchior and Steudle, 1993). It was found that the longitudinal component was usually negligible at distances from the root tip where xylem had already matured (about 20 mm in young corn roots and 30 mm in onion roots grown in hydroponics). Using the root pressure probe, both longitudinal and radial hydraulic resistances have been worked out along the developing root to determine how both components would influence the efficiency of roots for water uptake. The results showed that, except for the tip regions of roots, the simple two-compartment model which neglects the axial component provides a good basis for the evaluation of data. If the axial component were co-limiting transport along with the radial, this would complicate the situation and more sophisticated models would have to be used (Frensch and Steudle, 1989).

Axial resistances obtained with the root pressure probe for maize and onion have been compared with those calculated from Poiseuille's law. In maize, calculated values were smaller (conductances higher) by a factor of 2 to 5 than the measured. Near the base and the tip of onion roots grown in hydroponics, measured and calculated axial resistances were similar. However, at distances between 200 and 300 mm from the apex measured values were smaller by more than an order of magnitude than the calculated, probably, because of remaining cross walls between vessels members. During root development, the axial resistance in onion was reduced by three orders of magnitude. The differences between measured and calculated values of axial hydraulic resistances are known from the literature. They may be referred to the fact that the assumption of ideal cylindrical capillaries does not hold or to difficulties in identifying all vessels in stained cross sections (Frensch and Steudle, 1989; Melchior and Steudle, 1993).

In maize, the radial hydraulic conductivity did not change at distances between 20 and 120 mm from the tip where the Casparian band of the endodermis was already developed but no suberin lamellae were present. In onion roots, the radial hydraulic conductivity decreased at distances larger than 150 mm from the tip which was correlated with suberin lamellae in the endodermis and/or with the maturation of the exodermis (Melchior and Steudle, 1993).

## Negative root pressure

In order to test whether or not the transport parameters measured with the root pressure probe in the range of positive pressures would also apply to the range of negative pressures (tensions in the xylem), the technique was also used in the range of negative pressures using young maize roots as the objects (Heydt and Steudle, 1991). This test is important because it may be argued that parameters ($Lp_r$, $P_{sr}$, $\sigma_{sr}$) could change when switching to tensions in the xylem (the condition usually occuring in the transpiring plant). Tensions (negative pressures) are not easily created in the measuring system because of the danger of cavitations. The system had to be kept free of any air bubbles prior to the experiments. This means that air also had to be removed from the root tissue by vacuum infiltration. Under these conditions, tensions of as large as - 0.2 MPa (- 2 bar) could be created by applying hypertonic solutions to the root. Osmotic experiments yielded similar $Lp_r$, $P_{sr}$, and $\sigma_{sr}$ values in the presence of tensions as were obtained at positive root pressures. It was also established that infiltration did not change the transport properties of roots. The experiments indicated that, at least for maize, there was no change in transport properties when switching from an overpressure to tensions in the root xylem.

## Combination of root and cell pressure probes

In order to test current models of preferred pathways for water in the root cylinder, hydraulic measurements have also been performed at the level of root cells by combining the cell and root pressure probes (Steudle and Jeschke 1983; Steudle et al., 1987; Steudle and Brinckmann, 1989; Zhu and Steudle, 1991). Neglecting the apoplasmic component, the data allowed an estimation of root $Lp_r$ using the measured cell Lp. The calculated root $Lp_r$ was then compared with the measured. Provided that water flow was mainly from cell to cell, the calculated and measured $Lp_r$ should be similar. However, if there were a substantial apoplasmic component, the measured $Lp_r$ should be larger than the calculated. For some species (barley, bean), the data indicated a preference for a cell-to-cell transport, regardless of whether hydrostatic or osmotic gradients were used to drive water flows across the roots. In others (maize, onion), the measured $Lp_r$ was much larger than expected from the cell Lp when hydrostatic gradients were applied. When osmotic gradients were used, transport occurred mainly from cell to cell. These findings indicated that, at least in some species, the transport pathways changed with the driving force used. In hydrostatic measurements, flow was mainly around protoplasts, whereas in osmotic it was from cell to cell. It has been concluded that the differences between species may be related to the developmental state of the endodermis. If the Casparian band effectively interrupts water flow, this would result in a preferred cell-to-cell transport across the root cylinder regardless of the nature of the driving force. If there were differences, this should point to either an apoplasmic water flow in the endodermis or to flow in regions in the endodermis lacking an intact Casparian band (e.g., root primordia; see below).

Simultaneous measurements at the cell and root level have been also performed (Zhu and Steudle, 1991). These experiments did show how changes in the water potential (osmotic pressure) of the root medium propagated across the root cortex and eventually reached the xylem. From the stationary changes in cell and root pressure, reflection coefficients were evaluated for both cortical cells and the root during the experiment. The data demonstrated that reflection coefficients decreased for cells with increasing distance from the root surface which was in agreement with the composite transport model. However, when the cell $\sigma_s$ was measured with the roots not attached to the root pressure probe, the $\sigma_s$ was close to unity as expected from the composite transport model (Azaizeh and Steudle, unpublished). In the presence of an osmotic gradient between xylem and medium, there will be a counterflow of water in the root which should effect not only the overall root reflection coefficient but also that of living tissue cells. Thus, the experiments in which the cell $\sigma_s$ has been measured in the presence and in the absence of osmotic and hydrostatic gradients between xylem and medium are a further indication for the validity of the model.

By measuring the initial water flow, the simultaneous measurement of hydraulic properties at the cell and root levels also allowed an estimation of the hydraulic conductivity of the apoplasmic path ($Lp_{cw}$; Zhu and Steudle, 1991). This was done by comparing the cell-to-cell component with the overall flow. Data of $Lp_{cw}$ are not easy to obtain for technical reasons. However, they are badly needed for modelling of water relations of plant tissue.

**Effects of external factors on root hydraulics: high salinity and anoxia**

The root pressure probe technique may be employed to evaluate the effects of external factors on the ability of roots to take up water and solutes (nutrients). The most important factors would be: low temperature, high salinity, drought, mycorrhiza, and anoxia. To date, effects of salinity and anoxia have been studied in detail. The precise mechanism(s) of the adverse effects of high salinity on the root are still not fully understood. They may include osmotic effects as well as toxicity caused by certain ions (Cheeseman, 1988; Epstein, 1985). Reductions of growth could be caused by an inhibition of water uptake due to a reduced Lp of root cells. This, in turn, may result in a reduced overall hydraulic conductivity of the roots ($Lp_r$). It is known that effects of salinity could, in part, be reversed by calcium (Cramer et al., 1986; Lynch et al., 1987). It is thought that at high salinity, $Na^+$ competes for $Ca^{2+}$ at the cell membranes and this has deleterious effects on the proper function and integrity of the membranes. In order to determine the effects of high salinity on the water relations of roots, Azaizeh and Steudle (1991) performed root pressure probe experiments using young maize roots. The hydraulic conductivity of the cell membranes of cortex cells was measured as well (Azaizeh et al., 1992). Maize seedlings grown in a nutrient solution plus 100 m*M* NaCl showed an $Lp_r$ which was reduced by 30 to 60% compared to the control (1/5 Hoagland solution; 0.5 m*M* Ca). Increased levels of Ca (10 m*M*) had an ameliorative effect on the $Lp_r$ of salinized roots so that the water permeability increased again, thus, increasing the availability of water for the plant. It was also shown that the effects were much larger at the cell than at the root level. High salinity (100 m*M* NaCl) caused a reduction of the cell Lp by a factor of 3 to 6 which was reversed by a factor of 2 to 3 in the presence of calcium. The differences in the effects of NaCl on either the root $Lp_r$ or the cell Lp were due to the fact that in the roots the water flow was mainly around cells under hydrostatic conditions (see above). The findings of a reduced root $Lp_r$ are in line with those of Munns and Passioura (1984) and O'Leary (1969) for bean and lupin. Conversely, salinity had no effect on $Lp_r$ in barley, tomato, and sunflower (Munns and Passioura, 1984; Shalhevet et al., 1976). The findings are at variance with those of Tyerman et al. (1989) who found no change of the root cell Lp of *Nicotiana tabacum* at high salinity, i.e. for a plant which can grow reasonably well at high salinity (Flowers et al., 1986)

Similar to increased salinity, anoxia also has a negative effect on root $Lp_r$ as shown in root pressure probe experiments with maize roots (Birner and Steudle, 1993). It is likely that the reason for the decrease of $Lp_r$ was also due to a decrease in the hydraulic conductivity at the level of cell membranes (Lp). In addition to the changes in water transport reported in the literature (e.g. Everard and Drew, 1987; Zhang and Tyerman, 1991), the results obtained with the root pressure probe allowed quantification of changes in root pressure, the permeability of roots to nutrients ($P_{sr}$), and of the selectivity of roots ($\sigma_{sr}$). The stationary root pressure of maize roots strongly decreased during anoxia to reach a low value after several hours. Since ion leakage from the roots (permeability coefficient, $P_{sr}$) was also reduced at a constant $\sigma_{sr}$, this indicated that the active uptake of solutes (nutrient salts) was inhibited (Eq.(3)). Remarkably, roots treated in anoxia for several hours did not recover as quickly as they changed their transport properties. In the experiments, great care was taken to characterize the physiological status of the roots by performing osmotic experiments and other tests which proved that, despite anoxic conditions, the roots were still properly functioning as osmometers.

The two examples presented above demonstrate that the root pressure probe may be used to determine changes in both water and solute uptake of roots in response to environmental stresses such as pollutants, low pH, heavy metals etc. Combining the approach with measurements at the cell level using the cell pressure probe makes the method even more powerful, since this combination reveals deeper insight into mechanisms.

## Modified roots: root steaming, dissecting, and puncturing

Work with modified roots provides another tool for getting a deeper insight into the mechanisms of water and solute flow in roots. Root properties may be modified by removing part of the cortex by scraping or by microdissection (Peterson et al., 1993). Parts of roots may be also killed by steaming, thus, removing the entire barrier for radial water flow except for that of the lateral walls of mature vessels (Peterson and Steudle, 1993). Another possibility of modifying roots is the puncturing of the endodermis with needles (e.g. with the tip of a cell pressure probe) to create small holes of some ten $\mu$m in diameter (Steudle et al., 1993).

The puncturing of the endodermis offers the possibility to create additional 'apoplasmic by-passes' which, by area, are only a fraction of $10^{-2}$ to $10^{-3}\%$ of the entire surface area of the endodermis. It may be expected that this would affect the hydraulic conductivity, solute permeability, and reflection coefficients of roots and should allow one to extrapolate to the possible size of by-passes in the intact system. The experiments indicated that only small by-passes (leaks) were required to considerably reduce root pressure and to substantially lower $\sigma_{sr}$ at a fairly constant hydraulic conductivity of the root. It should be noted that usually puncturing did not result in a root pressure of zero. This means that, according to the pump/leak model of the root (Eq.(3)), the passive leak of ions could be compensated for by active pumping. Thus, the by-passes present in the intact system such as those across the root primordia, across the root tip, or some leakiness of Casparian bands could be sufficient to cause a $\sigma_{sr} < 1$ as found. On the other hand, the fact that the root hydraulic conductivity (half-time of water exchange) did not increase significantly upon puncturing, would be simply explained by the high overall $Lp_r$ of the intact system. Thus, at least for young maize roots, the radial hydraulic resistance of roots is more evenly distributed across the living tissue. The result contradicts the general view that the endodermis represents the main hydraulic resistance in roots. By contrast, the puncturing experiments did show that the endodermis was the main barrier for solutes, a conclusion which would be in line with the traditional view of the function of the endodermis. The puncturing experiments demonstrated that only small apoplasmic by-passes would substantially reduce $\sigma_{sr}$ and substantially increase $P_{sr}$. In maize, a careful quantitative evaluation of apoplasmic pathways in the root endodermis lacking a Casparian band (root primordia, root tip) showed that these would be sufficient to explain the low reflection coefficients (Steudle et al., 1993). Thus, the puncturing experiments did support the composite transport model.

Results from experiments in which roots were scraped or dissected were in line with the findings in the puncturing experiments. As long as the endodermis was not damaged by the manipulations, root pressure remained high or recovered after healing of the endodermis. The hydraulic conductivity of the modified roots only increased substantially when considerable amounts of cortical tissue were removed.

In the steaming experiments, the manipulation was most severe. Killing of a few millimeters of a young maize root of a total length of about 100 mm resulted in an immediate decrease in root pressure to nearly zero and in an increase of the overall $Lp_r$ from which the hydraulic conductivity of the steamed zone could be evaluated. The $Lp_r$ of the steamed zone was larger than that of the intact root by a factor of 3 to 13. In the steamed zone, the remaining radial hydraulic resistance could be attributed to the walls of mature xylem vessels. This allowed measurement of the lateral hydraulic conductivity of the walls of the vessels ($Lp_x$). It turned out that, although $Lp_x$ was much higher than $Lp_r$, the contribution of the lateral hydraulic resistance to the overall radial resistance was not negligible.

## Tree roots

At a first glance, tree roots or better tree root systems grown in soil may be considered to be objects which, because of their size and heterogeneity, are much more difficult to handle with the root pressure probe than roots of herbaceous species which can be easily grown in hydroponics. Fortunately, this is not the case. In some respects, roots of woody species are even easier to handle. For example, when tree root systems are fixed to the root pressure probe, there is practically no danger of crushing the woody xylem in the sealing area (as sometimes occurs with herbs). In addition, excised root systems of tree seedlings may be used for longer periods of time. For example, with oak seedlings, root pressure could be measured up to ten days.

To date, root systems of saplings of oak (*Quercus robur* and *Q. petraea*), walnut (*Juglans regia*), and Norway spruce (*Picea abies*) have been investigated with the pressure probe (Steudle and Meshcheryakov, in preparation; Hallgren et al., 1994). Trees were grown in sand culture and watered regularly with nutrient solution. This provided a defined osmotic pressure in the soil solution and made possible a rapid exchange of the regular soil solution for solutions to which osmotica were added (osmotic experiments). Prior to the experiments, the trees were well watered and shoots enclosed in plastic bags to prevent transpiration and to fully hydrate the plants so that upon excision no or little air entered the xylem. Root systems were allowed to exude for some time before they were attached to the root pressure probe to build up a steady root pressure which took several hours. Then experiments were performed (for a detailed description, see Hallgren et al., 1994) which involved typical root pressure relaxations (osmotic and hydrostatic), but also pressure clamp experiments and experiments in which the pressure chamber was combined with the root pressure probe. The pressure chamber was also used to perform classical root exudation under pressure (Fiscus, 1975). The different approaches allowed us to cross check the validity of root pressure probe data. In order to express water flows and root hydraulic conductance on the basis of root surface area, the root systems were removed from the sand after the experiments and stained with toluidine blue for better contrast. Surface areas were then determined using an image analyzing system.

Root pressures of the two ring-porous trees were 0.05 to 0.15 MPa (0.5 to 1.5 bar) and much larger than those of the conifer (-0.001 to +0.004 MPa; -10 to + 40 cm of water column) which is in agreement with literature data. Although the exchange rates for water between xylem and soil solution were fast in hydrostatic experiments, the hydraulic conductivity was, on average, smaller by an order of magnitude than that usually observed in herbaceous species. The difference may be explained by differences in the root anatomy and may be mainly due to the larger degree of suberization of the tree roots. For oak, the contribution of fine roots (diam.: < 1.0 mm) to the overall surface area was 80% of the total and thus the absolute value of $Lp_r$ did not change very much assuming water flow occurred only in fine roots. Thus, per unit surface area, tree roots exhibited a hydraulic resistance higher than that of roots of non-woody species. Nevertheless, half-times of water flow equilibration were short because of the large surface area of roots (a few seconds or even smaller). This means that the root systems would tend to have water potentials close to those of the soil. However, this is only true in the presence of hydrostatic gradients. In osmotic experiments, half-times of tree roots were larger than in hydrostatic by one to three (!) orders of magnitude and the $Lp_r$ was smaller by the same factor. Thus, the large differences found in some herbaceous species between osmotic and hydrostatic water flow was even more pronounced in trees.

As already discussed, the large differences between hydrostatic and osmotic hydraulic conductivities suggest differences in the transport pathways. In terms of the composite transport model they are understandable. If the model applies, the large differences in $Lp_r$ would then also suggest smaller reflection coefficients. These were found. Reflection coefficients for salts, sugars, mannitol etc. were as small as $\sigma_{sr} = 0.1$ to 0.5, i.e. they were, on average, smaller by a factor of two than those measured for herbs (see above). Low reflection coefficients were found in trees in spite of very low permeability coefficients which were usually not measurable with the root pressure probe. The apparent contradiction is also solved by the model.

Low reflection coefficients and low osmotic $Lp_r$ should be important during osmotic or water stress (drought) which may require some osmoregulation of roots. We may conclude from the data that, under these conditions, osmotic processes in tree roots may be slow. On the other hand, the data indicate that tree roots represent an efficient hydraulic system under well-watered conditions and in the presence hydrostatic gradients driving water flow. It would be interesting to see if hydraulic properties ($Lp_r$) would change in roots during water stress. This could be checked with the root pressure probe. Other interesting areas of application would be effects of low pH or toxic metals, effects of mycorrhiza, or of low temperature on both hydraulic properties and active and passive solute (nutrient) relations.

## Composite transport model of root

The root pressure probe data briefly reviewed above as well as other evidence strongly indicate deviations from the single-membrane-equivalent model. Structurally, there is a basis for deviations from the simple model because of the existence of different parallel pathways for water and solutes such as the apoplasmic and cell-to-cell path, or root zones of different permeability patterns. This would enhance the composite character of roots due to different tissues arranged in series (see above). There is considerable experimental and theoretical evidence which indicates that the model has to be extended. Mainly, the evidence is based on (i) the finding of low root reflection coefficients ($\sigma_{sr}$) and of apoplasmic by-passes of solutes, (ii) apparent differences between osmotic and hydrostatic root $Lp_r$, and (iii) the effect of water flow on the absolute value of root $Lp_r$.

These findings are directly related to basic functions of roots. They are important for the limitation and control of water and solute flows across them. Changes of root $Lp_r$ and $\sigma_{sr}$ will directly affect the efficiency of roots for taking up water. They may be readily explained in terms of a new root model which has been termed a 'composite transport model of the root'. The model integrates the different experimental findings listed above as well as known anatomical features of roots. It focuses on the fact that there are parallel transport elements in the root. Obvious parallel transport pathways are the apoplasmic and cell-to-cell path, but also root zones of different states of development. The theory of the transport properties of composite membranes have been investigated in detail by Kedem and Katchalsky (1963). Although the root system is a much more complicated structure, basic results of the theory may be applied.

The model is depicted in Figure 1. The root interior (xylem) is separated by the root cylinder from the medium. Due to the active uptake of solutes (nutrients) into the root xylem the internal concentration is increased and a root pressure is built up. Along the cell-to-cell path (superscript 'cc') the uptake of water will be driven by the osmotic pressure difference which is fully effective since the reflection coefficient here is close to unity. Along the parallel apoplasmic path (superscript 'cw'), the reflection coefficient will be close to zero, and there will be very little water flow into the root along this path. However, along the apoplasmic path there will be a backflow of solution caused by the overpressure in the xylem. Hence, the hydrostatic pressure difference developed in the system at zero water flow will be smaller than the osmotic which means that $\sigma_{sr} < 1$ as found experimentally. For a system with two parallel pathways such as in Figure 1, the overall reflection coefficient ($\sigma_{sr}$) would be (Steudle, 1989, 1992, 1993, 1994):

$$\sigma_{sr} = \gamma^{cc} \cdot \frac{Lp^{cc}}{Lp_r} \sigma_s^{cc} + \gamma^{cw} \cdot \frac{Lp^{cw}}{Lp_r} \sigma_s^{cw}. \quad (4)$$

$Lp^{cc,cw}$ = hydraulic conductivity of the cell-to-cell and apoplasmic pathway, respectively. $\gamma^{cc,cw}$ = fractional contributions of cross-sectional areas of pathways to the overall root area. $Lp_r$ = overall hydraulic conductivity ($Lp_r = \gamma^{cc} \cdot Lp^{cc} + \gamma^{cw} \cdot Lp_{cw}$. From Eq. (4) it is evident that, provided that the apoplasmic passage across the root exhibits a fairly low $\sigma_s^{cw}$ (e.g., $\sigma_s^{cw} \approx 0$) and a fairly high $Lp^{cw}$, this component could still contribute to the overall $\sigma_s$ although the cross-sectional area of the apoplast is small. The physical reason for the effect resulting in a low $\sigma_{sr}$ is that, in the presence of two parallel pathways exhibiting different values of $\sigma$ and Lp, opposing water flows would be created in the root which would result in a circulation flow of water. Thus, the model explains the low $\sigma_{sr}$. According to Eq.(4), the overall reflection coefficient is a weighted mean of the reflection coefficients of individual pathways which contribute according to their

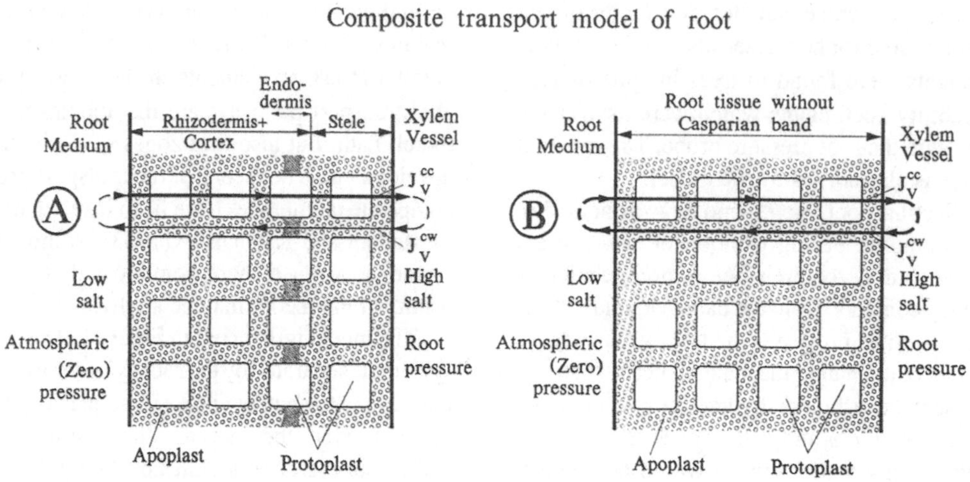


*Fig. 1.* Diagrammatic scheme of composite transport in a plant root. **(A)** denotes transport across the root having an endodermis and **(B)** an area of the root lacking a Casparian band (root primordia or root tip). For the sake of simplicity, rhizodermis and cortex (plus endodermis) are represented by three cell layers and the stelar tissues by a single layer. Different routes of radial flow are denoted, i.e. the flow across the apoplast (superscript 'cw') and the cell-to-cell path (superscript 'cc'). The figure indicates that there will be an osmotic water flow along the cell-to-cell path ($J_V^{cc}$) directed from the root medium into the xylem (stele) as long as the osmotic concentration in the xylem is larger than that in the medium. The reflection coefficient along this path will be high ($\sigma_s^{cc} \approx 1$). However, as the root pressure builds up in the absence of transpiration, there will be a backflow of water (xylem solution) across the non-selective apoplasmic path ($\sigma_s^{cw} \approx 0$) in (B) and, perhaps, also in (A), if the mature Casparian band is somewhat permeable to water. Opposing water flows or a circulation flow of water across the root result. Opposing flows exactly cancel at a zero net water flow across the root. Under these conditions, the steady-state root pressure will be smaller than expected from the difference in osmotic pressures, i.e. $\sigma_{sr}$ will be smaller than unity. For further explanation, see text.

hydraulic conductance. The permeability of pathways is not involved and may be still low so that, despite a low $\sigma_{sr}$, the overall $P_{sr}$ (leak rate for solutes) is still sufficiently low to provide a high retention of nutrients within the xylem.

With respect to the differences between osmotic and hydraulic $Lp_r$, the model predicts that in the apoplasmic path the effective osmotic driving force should be low when osmotic gradients are applied to the root. Consequently, the osmotic flow would be low despite a high potential hydraulic conductance of the apoplasmic path. By contrast, the apoplasmic path could be fully used in the presence of hydrostatic gradients. It is clear from the model that a complete interruption of the apoplasmic path (Casparian band) would cause the osmotic and hydrostatic flows to be similar as found for some species. Thus, the variability in root water flow would largely depend on the existence of discrete apoplasmic by-passes lacking Casparian bands and on the developmental state (tightness) of the Caspariarn band itself.

Similarly, a dependence of the root hydraulic conductivity ($Lp_r$) on the water flow could be understood by the model. At low or zero water flows, the hydraulic resistance of the system would be high because flows would, in part, counterbalance. However, as the hydrostatic gradient across the system is increased by the onset of transpiration, the situation will change and water will flow across the root in the same direction on both pathways.

The model (Eq. (4)) predicts that in roots having a small overall $Lp_r$ (such as the roots of trees; see above), the relative contribution of apoplasmic by-passes to the overall value of $\sigma_{sr}$ should be larger than in roots having a higher $Lp_r$ (such as the roots of herbs). Thus, the model explains the differences in $\sigma_{sr}$ values between herbaceous and woody species.

The composite transport model explains important deviations in root transport from that of a nearly perfect osmometer which, in part, have been known for a long time. The model has a sound physical and anatomical basis and predicts that, despite of the deviations, roots will still exhibit basic functions of an osmometer.

## Acknowledgement

This work was supported by a grant from EUROSILVA (project no. 39473C).

## References


Anderson W P 1976 Transport through roots. *In* Encyclopedia of Plant Physiology, Vol. 2, Part B, Transport in Plants. Eds. U Lüttge and M G Pitman. pp 129–156. Springer-Verlag, Heidelberg.

Azaizeh H and Steudle E 1991 Effects of salinity on water transport of excised maize (*Zea mays* L.) roots. Plant Physiol. 97, 1136–1145.

Azaizeh H, Gunse B and Steudle E 1992 Effects of NaCl and $CaCl_2$ on water transport across root cells of maize (*Zea mays* L.) seedlings. Plant Physiol. 99, 886–894.

Birner T P and Steudle E 1993 Effects of anaerobic conditions on water and solute relations and active transport in roots of maize (*Zea mays* L). Planta 190, 474–483.

Cheeseman J M 1988 Mechanisms of salinity tolerance in plants. Plant Physiol. 87, 547–550.

Clarkson D T 1991 Root structure and sites of ion uptake. *In* Plant Roots, the Hidden Half. Eds. Y Waisel, A Eshel and U Kafkafi. pp 417–453. Marcel Dekker, New York.

Cramer G A, Epstein E and Läuchli A 1986 Effect of NaCl and $CaCl_2$ on ion activities in complex nutrient solutions and root growth of cotton. Plant Physiol. 81, 792–797.

Cruz R T, Jordan W R and Drew M C 1992 Structural changes and associated reduction of hydraulic conductance in roots of *Sorghum bicolor* L. following exposure to water deficit. Plant Physiol. 99, 203–212.

Dainty J 1985 Water transport through the root. Acta Hort. 171, 21-31.

Epstein E 1985 Salt tolerant crops: origins, development, and prospects of the concept. Plant and Soil 89, 183–198.

Everard J D and Drew M C 1987 Mechanisms of inhibition of water movement in anaerobically treated roots of *Zea mays* L. J. Exp Bot. 38, 1154–1165.

Fiscus E L 1975 The interaction between osmotic- and pressure-induced water flow in plant roots. Plant Physiol. 55, 917–922.

Flowers T J, Flowers S A and Greenway H 1986 Effects of sodium chloride on tobacco plants. Plant Cell Environ. 99, 645–615.

Frensch J and Steudle E 1989 Axial and radial hydraulic resistance to roots of maize (*Zea mays* L.). Plant Physiol. 91, 719–726.

Hallgren S W, Rüdinger M and Steudle E 1994 Root hydraulic properties of spruce measured with the pressure probe. Plant and Soil (*In press*)

Heydt H and Steudle E 1991 Measurement of negative root pressure in the xylem of excised roots. Effects on water and solute relations. Planta 184, 389–396.

Katou K and Taura T 1989 Mechanism of pressure-induced water flow across plant roots. Protoplasma 150, 124–130.

Kedem O and Katchalsky A 1963 Permeability of composite membranes. Part 2: Parallel elements. Trans. Far. Soc. 59, 1931–1940.

Kramer P J 1983 Water Relations of Plants. Academic Press, Orlando. 489 p.

Lynch J, Cramer G R and Läuchli A 1987 Salinity reduces membrane-associated calcium in corn root protoplasts. Plant Physiol. 83, 390–394.

Marschner H 1986 Mineral Nutrition in Higher Plants. Academic Press, London. 674 p.

Mees G C and Weatherley P E 1957 The mechanism of water absorption by roots. I. Preliminary studies on the effects of hydrostatic pressure gradients. Proc. Roy. Soc. London, Ser. B 147, 367–380.

Melchior W and Steudle E 1993 Water transport in onion (*Allium cepa* L.) roots. Changes of axial and radial hydraulic conductivity during root development. Plant Physiol. 101, 1305–1315.

Melchior W and Steudle E 1994 Hydrostatic and osmotic hydraulic conductivities of onion (*Allium cepa* L.) roots. Structure and Function of roots. Eds. F Baluska, O Gasparikova and M Ciamporova. Kluwer Academic Publishers, Dordrecht.

Miller D M 1985 Studies of root function in *Zea mays*. III. Xylem sap composition at maximum root pressure provides evidence of active transprt in the xylem and a measurement of the reflection coefficient of the root. Plant Physiol. 77, 162–167.

Munns R and Passioura J B 1984 Hydraulic resistance of plants. III. Effects of NaCl in barley and lupin. Aust. J. Plant Physiol. 11, 351–359.

Newman E I 1976 Interaction between osmotic- and pressure-induced water flow in plant roots. Plant Physiol. 57, 738–739.

O'Leary J W 1969 The effect of salinity on permeability of roots to water. Isr. J. Bot. 18, 1–9.

Passioura J B 1988 Water transport in and to the root. Annu. Rev. Plant Physiol. Plant Mol. Biol. 39, 245–265.

Peterson C A, Murrmann M and Steudle E 1993 Location of the major barrier(s) to the movement of water and ions in young roots of *Zea mays* L. Planta 190, 127–136.

Peterson C A and Steudle E 1993 Lateral hydraulic conductivity of early metaxylem vessels in *Zea mays* L. roots. Planta 189, 288–297.

Pitman M G, Wellfare D and Carter C 1981 Reduction of hydraulic conductivity during inhibition of exudation from excised maize and barley roots. Plant Physiol. 67, 802–808.

Shalhevet J, Maass E V, Hoffmann G J and Ogata G 1976 Salinity and the hydraulic conductance of roots. Physiol. Plant. 38, 224–232.

Steudle E 1989 Water flow in plants and its coupling to other processes: an overview. Methods Enzymol. 174, 183–225.

Steudle E 1992 The biophysics of plant water: compartmentation, coupling with metabolic processes, and water flow in plant roots. *In* Water and Life: Comparative Analysis of Water Relationships at the Organismic, Cellular, and Molecular Levels. Eds. G N Somero, C B Osmond and C L Bolis. pp 173–204. Springer-Verlag, Berlin.

Steudle E 1993 Pressure probe techniques: basic principles and application to studies of water and solute relations at the cell, tissue, and organ level. *In* Water Deficits: Plant Responses from Cell to Community. Eds. J A C Smith and H Griffiths. pp 5–36. Bios Scientific Publishers Ltd., Oxford.

Steudle E 1994 The regulation of plant water at the cell, tissue and organ level: role of active processes and of compartmentation. *In* Flux Control in Biological Systems. Ed. E-D Schulze. pp 237–299. Academic Press, San Diego.

Steudle E and Brinckmann E 1989 The osmometer model of the root: water and solute relations of *Phaseolus coccineus*. Bot. Acta 102, 85–95.

Steudle E and Frensch J 1989 Osmotic responses of maize roots. Water and solute relations. Planta 177, 281–295.

Steudle E and Jeschke W D 1983 Water transport in barley roots. Planta 158, 237–248.

Steudle E, Murrmann M and Peterson C A 1993 Transport of water and solutes across maize roots modified by puncturing the endodermis: further evidence for the composite transport model of the root. Plant Physiol. 103, 335–349.

Steudle E, Oren R and Schulze E D 1987 Water transport in maize roots. Plant Physiol. 84, 1220–1232.

Tyerman S D, Oats P, Gibbs J, Dracup M and Greenway H 1989 Turgor-volume regulation and cellular water relations of *Nicotiana tabacum* roots grown in high salinities. Aust. J. Plant Physiol. 16, 517–531.

Weatherley P E 1982 Water uptake and flow in roots. *In* Encyclopedia of Plant Physiology, Vol. 12B. Eds. O L Lange, P S Nobel, C B Osmond and H Ziegler. pp 79–109. Springer-Verlag, Berlin.

Zhang W H and Tyerman S D 1991 Effect of low $O_2$ concentration and azide on hydraulic conductivity and osmotic volume of the cortical cells of wheat roots. Aust. J. Plant Physiol. 18, 603–613.

Zhu G L and Steudle E 1991 Water transport across maize roots: simultaneous measurement of flows at the cell and root level by double pressure probe technique. Plant Physiol. 95, 305–315.

*F. Baluška et al. (eds.), Structure and Function of Roots*, 209–213, 1995.


# Hydrostatic and osmotic hydraulic conductivities and reflection coefficients of onion (*Allium cepa* L.) roots

Walter Melchior and Ernst Steudle[1]
*Lehrstuhl Pflanzenökologie, Universität Bayreuth, D-95440 Bayreuth, Germany.* [1]*Corresponding author*

*Key words: Allium cepa*, composite transport model, hydraulic conductivity, root development, root pressure probe, water transport

**Abstract**

Axial hydraulic conductivity of segments of onion (*Allium cepa* L.) roots was measured using the root pressure probe and calculated from stained cross sections according to Poiseuille's law. The contribution of the longitudinal path to overall hydraulic resistance of the roots was negligible except for the apical 30 mm. Therefore, total hydraulic conductivity represented a good approximation of the radial hydraulic conductivity ($Lp_r$) when expressed on the basis of the 'effective surface area'. Hydrostatic $Lp_r$ (measured using a gradient in hydrostatic pressure) was fairly constant in the apical root zones (30 to 150 mm from the tip), but was lower in more basal zones (at distances >150 mm from the tip). Thus, basal root parts contributed to a smaller extent to total water uptake of the roots. $Lp_r$ was also measured by applying gradients in osmotic pressure. The osmotic hydraulic conductivity was smaller than the hydrostatic by one to two orders of magnitude and reflection coefficients of the roots for nutrient salts were significantly smaller than unity (0.18–0.88), indicating that roots did not behave like 'ideal osmometers' as commonly assumed. A composite transport model is discussed to account for the non-semipermeability of roots. The essentially nonselective apoplasmic path in parallel with the highly selective cellular path could explain both low reflection coefficients for nutrient salts and the dependency of water flow on the nature of the driving force.

*Abbreviations:* $A_r$ – geometric surface area of the root [$m^2$], $Lp_{rh}$ ($Lp_{ro}$) – root hydraulic conductivity per unit effective surface area in hydrostatic (osmotic) experiments [$m \cdot s^{-1} \cdot MPa^{-1}$], $L_{rh}$ ($L_{ro}$) – hydraulic conductance in hydrostatic (osmotic) experiments [$m^3 \cdot s^{-1} \cdot MPa^{-1}$], $L_x$ – axial hydraulic conductivity [$m^4 \cdot s^{-1} \cdot MPa^{-1}$], $P_r$ – root pressure [MPa], $\sigma_{sr}$ – root reflection coefficient.

## Introduction

During root development, conductivities for radial and axial water flow vary largely. Axial hydraulic conductivity ($L_x$) increases during xylem development. On the other hand, radial hydraulic conductivity ($Lp_r$) may decrease as the result of hydrophobic depositions in the endodermis or exodermis. Therefore, it is to be expected that the relative contributions of $L_x$ and $Lp_r$ to the total hydraulic conductivity of the root would change along the root and that water uptake would mainly occur in the apical root zones (Boyer, 1985). The influence of $L_x$ on total conductivity has been considered in a detailed study of water uptake of maize where it was found that the apical 15 mm were not contributing to water uptake because of immature metaxylem (Frensch and Steudle, 1989). Thus, the 'effective surface area' for water transport was less than the total root surface area.

In a common view, the root is considered as an ideal osmometer, i.e. as a system of an inner (xylem) and an outer (soil solution) compartment separated by a single barrier which allows water to pass but is not permeable for solutes. In this picture, nutrients are only taken up by active transport whereas the semipermeable barrier is necessary for an accumulation of nutrients in the xylem. However, for roots of various species, low reflection coefficients were found despite low permeability coefficients (maize: $\sigma_{sr}$ = 0.3 to 1.0, Miller, 1985; Steudle and Frensch, 1989; oak: $\sigma_{sr}$ = 0.2 to 0.5, Steudle, this volume; Norway spruce: $\sigma_{sr}$ = 0.1 to 0.2, Hallgren et al., this volume). This means that the ratio of changes in root pressure in response to changes in osmotic pressure of the outer medium was less than one, although concentration gradients equilibrated very slowly (indicating a non leaky barrier).

When hydrostatically and osmotically driven pressure relaxations measured with the root pressure probe were compared, some species showed differences of one order of magnitude or even more (Hallgren et al., this volume; Steudle and Frensch, 1989; Steudle), indicating different predominant pathways for water flow in hydrostatic and osmotic experiments. These pathways were assumed to be the apoplasmic (cell wall) path and the cellular path (Steudle, 1992). They should have quite different selective properties and could also differ greatly in their hydraulic conductance.

There are numerous studies on water transport across maize roots which, when grown hydroponically, developed neither Casparian bands nor suberin lamellae in the exodermis before emergence of laterals. However, hydrophobic barriers in the apoplasmic or cellular path could largely influence the relative contribution to the overall water uptake. Therefore, we have chosen onion roots in which exodermal Casparian bands and suberin lamellae started to develop at distances from the root tip of 50 to 60 mm (Perumalla and Peterson, 1986). This was close to the zone where xylem was mature and had a high hydraulic conductance (Melchior and Steudle, 1993). Thus, detailed information could be gained about changes of $Lp_r$ and $L_x$ along a developing root. These data serve as the basis for modelling the water uptake of roots.

## Materials and methods

One-year old onions (*Allium cepa* L.) obtained from Vanderhave Seeds (NL-4410AA Rilland, The Netherlands) were kept for 9 d at 6°C and for 3 d at 20 to 21°C on vermiculite moistened with one-third Miller solution and were then transferred to hydroculture with one-third Miller solution (Melchior and Steudle, 1993). When roots had been in hydroculture for one to two weeks, they were cut and connected to the root pressure probe (Fig. 1). Roots were left on the probe for at least 24 h, by which time they built up root pressures of $P_r$ = 0.24–0.62 MPa. Initially, the elasticity of the system ($dP_r/dV_S$) and the rate constant of pressure relaxations ($k_{wr}$) were determined on closed end segments by applying gradients in hydrostatic pressure by moving a metal rod into (out of) the probe which directly changed root pressure. The hydraulic conductivity

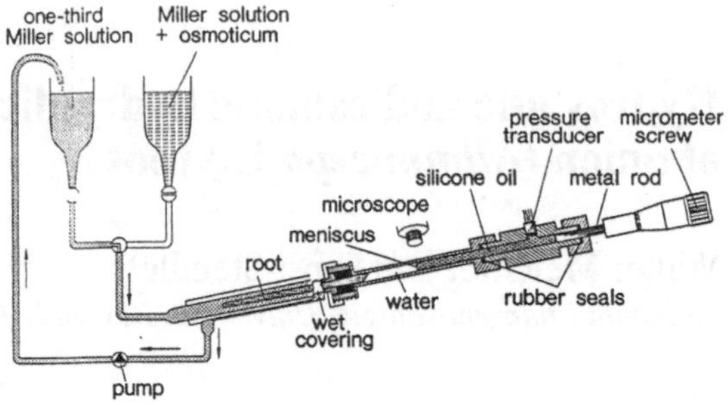


*Fig. 1.* Root pressure probe for measuring root pressure and hydraulic conductivity of roots. Root pressures were monitored continuously by a pressure transducer. Water flows were induced by moving the metal rod (changing root pressure directly) or by exchanging the nutrient solution for a solution with extra osmotic solute or vice versa. The elasticity of the system ($\Delta P_r/\Delta V_s$) was determined by inducing fast volume changes (quantified under the microscope) and measuring the response in root pressure.

was calculated as:

$$Lp_r = \frac{k_{wr}}{A_r \cdot dP_r/dV_s} \quad (1)$$

where $A_r$ = effective geometric surface area of the root (Steudle et al., 1987).

In the hydrostatic experiments, the time course of $P_r$ was not a simple exponential function but should be described as the sum of three exponential functions which could be easily separated by eye because of large differences in the rate constants (half times). The main (second) phase was attributed to the radial water flow and its rate constant was computed after fitting the stationary end pressure to the $P_r(t)$ function (see Melchior and Steudle, 1993).

For the determination of axial hydraulic conductivity ($L_x$), the apical part (> 45 mm) of the root end segment was excised and pressure relaxations were performed on the remaining open segments. $L_x$ was based on the length of the segment (1):

$$L_x = \frac{k_{wr} \cdot l}{dP_r/dV_S} \quad (2)$$

Lignified xylem vessels were stained with berberine sulfate (1 g $L^{-1}$, 1 h) and aniline blue (5 g $L^{-1}$, 1 h) and photographed under a fluorescence microscope (Brundrett et al., 1988). From the radius of the vessels, $L_x$ was calculated according to a modified form of Poiseuille's law which considered the elliptical cross section of the vessels (Melchior and Steudle, 1993; Nonweiler, 1975).

Prior to osmotic experiments, $P_r$ had to be stable within ± 3 kPa $h^{-1}$ (0.03 bar $h^{-1}$). This was rarely fulfilled. Although many roots showed a straight increase in $P_r$ over the first day and evening, $P_r$ usually dropped down in the morning of the second day. With these roots, no experiments could be performed. Therefore, osmotic relaxations could be measured only on a few roots and, for this reason, we tested only one osmotic solute in detail, i.e. potassium nitrate ($KNO_3$), but checked if other nutrient salts (potassium chloride, sodium and ammonium nitrate) and mannitol gave similar results. As shown by Birner and Steudle (1993), $KNO_3$ is not permeating into the stele because of high potassium concentrations already present in the xylem. The addition of the osmotic solute to the nutrient solution resulted in an exponential decrease in root pressure, the rate constant ($k_{wr}$) of which was directly correlated to the hydraulic conductivity (Eq. 1). After changing back to medium without osmoticum (endosmotic experiment), $P_r$ increased and the rate constant gave the endosmotic $Lp_{ro}$.

At the end of osmotic relaxations, $P_r$ did not always reach a stationary value but in some cases continued to fall because of the long duration of the osmotic experiments. Therefore, this linear decrease in root pressure was subtracted from the $P_r(t)$ function before calculating rate constants.

Reflection coefficients of the roots were calculated from:

$$\sigma_{sr} = \frac{\Delta P_r}{\Delta \pi^o} \tag{3}$$

where $\Delta P_r$ = change in stationary root pressure and $\Delta\pi^0$ = change in osmotic pressure of the outer medium. The osmotic concentration was determined by measuring the freezing temperature of the solution (Osmomat 030, Gonotec, D-10823 Berlin).

## Results

Axial hydraulic conductivity ($L_x$) increased with increasing distance from the tip (z) in the apical 200 mm (Fig. 2). $L_x$ values ranged over more than three orders of magnitude, reflecting the development of xylem from small protoxylem vessels to the large elements of the late metaxylem. Axial hydraulic conductivities measured with the root pressure probe and calculated from vessel dimensions were similar at the tip and at the base, but differed by more than one order of magnitude at z = 200–300 mm. This probably indicates that in lignified vessels of the metaxylem, entire or parts of cross walls had remained, which cannot be seen in cross sections and therefore lead to an overestimation of calculated $L_x$.

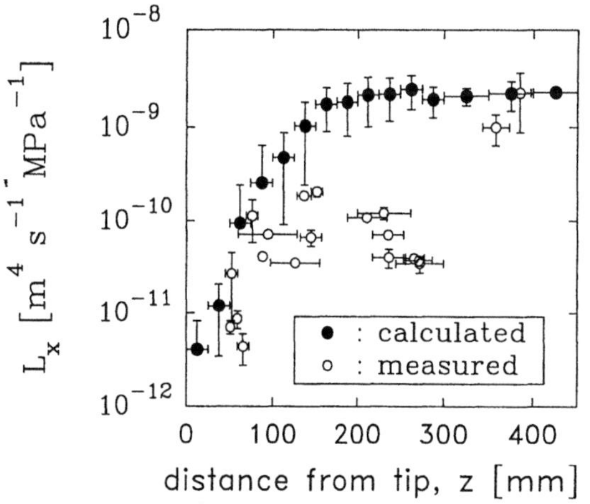


*Fig. 2.* Axial hydraulic conductivity ($L_x$) calculated from Poiseuille's law (filled symbols) and measured with the root pressure probe (open symbols) as a function of the distance from the root tip (z). $L_x$ ranged over three orders of magnitude. Error bars represent the SD (n = 2–14 and n = 3–5 for calculated and measured $L_x$ respectively). Note the logarithmic $L_x$ axis.

In hydrostatic experiments, the total hydraulic conductance ($L_{rh}$) increased with increasing surface area ($A_r$) of the end segment in the apical 150 mm (corresponding to $A_r$ = 550 mm$^2$; Fig. 3A). However, for $A_r$ > 550 mm$^2$, no further change of $L_{rh}$ could be measured. The slope of the regression line in Figure 3A was the radial hydraulic conductivity, $Lp_{rh} = 1.4 \cdot 10^{-7} m \cdot s^{-1} \cdot MPa^{-1}$. On the other hand, the intercept with the $A_r$-axis corresponded to the apical root zone with limited axial water transport. This hydraulically isolated zone had a surface area of 90 mm$^2$ (corresponding to a length of 30 mm) and should be always subtracted from the total surface area of the root when the 'effective surface area' is needed as reference.

When water flow was induced by a gradient in osmotic pressure, radial hydraulic conductivity was smaller by one to two orders of magnitude than in hydrostatic experiments (Fig. 3B). This indicated different transport pathways for water flow in the two types of experiments (see Discussion). Reflection coefficients for potassium nitrate ranged from 0.18 to 0.72 (mean $\pm$ SD = 0.42 $\pm$ 0.17; 17 measurements on 8 roots) and from 0.35 to 0.88 for the other osmotic solutes (KCl, $NaNO_3$, $NH_4NO_3$, mannitol; 5 roots measured). This showed that roots did not behave like 'ideal osmometers' which should have reflection coefficients of unity (perfect semipermeability).

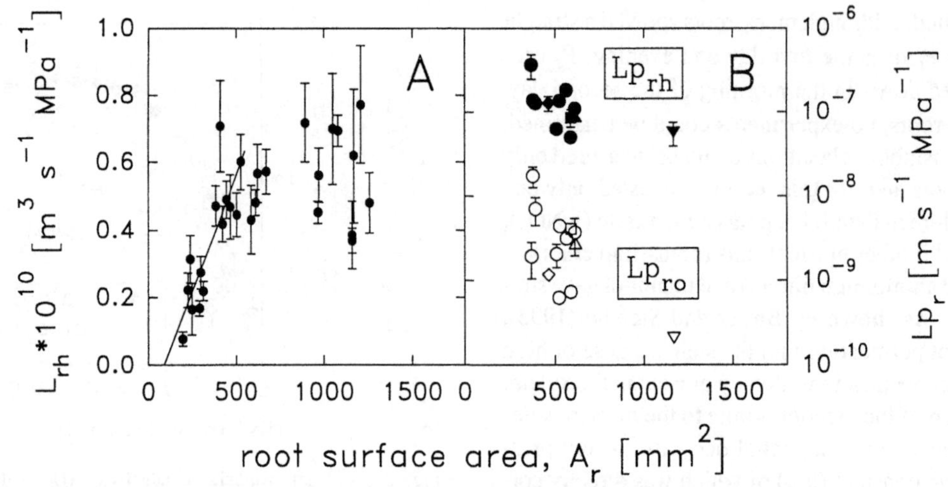


*Fig. 3.* Total hydraulic conductance in hydrostatic experiments ($L_{rh}$, A) and hydraulic conductivity (per effective surface area) in hydrostatic and osmotic experiments ($Lp_{rh}$ and $Lp_{ro}$, B). A: With increasing surface area of the root ($A_r$) up to 550 $mm^2$, $L_{rh}$ shows a proportional increase. For $A_r > 550$ $mm^2$ (root length > 150 mm), no significant change of $L_{rh}$ could be measured. B: $Lp_{rh}$ was larger than $Lp_{ro}$ by a factor of 8 to 270, indicating that predominant water flow was through different pathways during hydrostatic or osmotic relaxations (filled symbols: $Lp_{rh}$, open symbols: $Lp_{ro}$; ●, ○: $KNO_3$; ▼, ▽: $NaNO_3$; ▲, △: $NH_4NO_3$; ■, □: KCl; ◆, ◇: mannitol).

## Discussion

The results show that both axial and radial hydraulic conductivities undergo substantial changes during root development. The data from Figure 3A indicate that basal root zones had low water uptake rates. This could not be directly correlated with early stages in the formation of Casparian bands in the exodermis or suberin lamellae in endodermis or exodermis, which started to develop at 50 to 60 mm from the root tip (data not shown and Perumalla and Peterson, 1986). The ring of Casparian band and suberin lamellae in the exodermis was first discontinuous and probably did not limit water transport.

Axial (longitudinal) resistance could be neglected at distances greater than 30 mm from the tip. Thus, the simple two-compartment model of the root in which the xylem is separated from the outer medium by a single barrier could be used to calculate radial hydraulic conductivities ($Lp_{rh}$ and $Lp_{ro}$) from total hydraulic conductances of the root. However, there are strong indications that this model has to be extended to adequately describe radial water transport in roots, arising from this work on onion as well as from experiments with other species. In fact, the composite nature of pressure-time curves (three overlaying exponential functions) mentioned in the Material and methods is well known from tracer experiments in which different rate constants were attributed to different barriers in a series arrangement. Recently, Peterson et al. (1993) have shown that in maize roots with a state I endodermis (endodermis with Casparian bands, but without suberin lamellae), the water permeability was evenly distributed within the root cortex and did not become markedly reduced in the endodermis. Hydrostatic pressure relaxations following three exponential functions were reported for maize roots by Peterson and Steudle (1993). The authors measured the contribution of the radial hydraulic resistance of xylem vessels to the overall radial hydraulic resistance and found that the former was small but not negligible. These results show that a series arrangement of hydraulic resistances rather than a unique barrier may govern radial water uptake of roots.

Large differences of one to two orders of magnitude between hydraulic conductivities in hydrostatic vs. osmotic experiments and reflection coefficients ($\sigma_{sr}$) of significantly less than unity are other points not explained by the two-compartment model of the root. An extension of this model, based on the theory of Kedem and Katchalsky (1963), was named the 'composite transport model of the root' (Steudle, 1992). It considers the composite structure of the root in terms of parallel pathways for water transport on the cellular level (apoplasmic vs. cell-to-cell water transport) or at the level of different root zones with different water transport coefficients (different selectivity, different hydraulic conductivity etc.) One prediction of the model is that at low gradients in hydrostatic pressure, opposing water flows in the apoplast and from cell to cell occur which should lead to an overall reflection coefficient of less than unity. The application of the theory of Kedem and Katchalsky (1963) yields for water flows driven by gradients in osmotic pressure:

$$\sigma_{sr} \cdot Lp_{ro} = \gamma^{cc} \cdot \sigma_s^{cc} \cdot Lp^{cc} + \gamma^{apo} \cdot \sigma_s^{apo} \cdot Lp^{apo} \quad (4)$$

where $\gamma$ = fractional contribution of pathways to overall cross-sectional area, $\sigma_s$ = reflection coefficient, and Lp = hydraulic conductivity of pathways; superscripts 'apo' and 'cc' denote apoplasmic and cell-to-cell transport. Equation (4) shows that even with a highly conducting apoplast (large $\gamma^{apo} \cdot Lp^{apo}$), the osmotic hydraulic conductivity of the root can be low because of the low selectivity of the cell wall pathway ($\sigma_s^{apo} \approx 0$).

On the other hand, when water flow is driven by gradients in hydrostatic pressure, the water transport in the cell wall is fully efficient, i.e.:

$$Lp_{rh} = \gamma^{cc} \cdot Lp^{cc} + \gamma^{apo} \cdot Lp^{apo} \quad (5)$$

Equations (4) and (5) explain the large difference between $Lp_{rh}$ and $Lp_{ro}$ of up to two orders of magnitude when assuming a highly conducting apoplasmic pathway which exhibits a low selectivity for the nutrient salts used in this study.

Reflection coefficients ($\sigma_{sr}$) significantly lower than unity are easily explained within the model after transforming Equation (4) into:

$$\sigma_{sr} = \sigma_s^{cc} \cdot \frac{\gamma^{cc} \cdot Lp^{cc}}{Lp_{ro}} + \sigma_s^{apo} \cdot \frac{\gamma^{apo} \cdot Lp^{apo}}{Lp_{ro}} \quad (6)$$

from which it will be clear that $\sigma_{sr}$ is a weighted mean of the reflection coeffients of the cellular pathway ($\sigma_s^{cc} \approx 1$) and of the apoplasmic path ($\sigma_s^{apo} \approx 0$).

However, it should be kept in mind that the theory of parallel cellular water transport was applied to water relations of tissues without considering the effects of the serial arrangement of cells which leads to local interchange of flows (exchange between symplast and apoplast, for example at the endodermis). The existing models of root water transport are still not satisfying because they do not fully consider the complex structure of the root. Detailed numerical models may help further our understanding of basic components of water transport. The results presented in this publication, together with data on cellular water transport, will be used as the data basis for a cellular model of root water relations.

## Acknowledgments

The expert technical assistance of Mr B Stumpf is gratefully acknowledged. The authors also wish to thank Drs C A Peterson and S Hallgren for carefully reading the manuscript. This work was supported by a grant from the Deutsche Forschungsgemeinschaft, Sonderforschungsbereich 137.

## References

Birner T P and Steudle E 1993 Effects of anaerobic conditions on water and solute relations, and on active transport in roots of maize (*Zea mays* L.). Planta 190, 474–483.

Boyer J S 1985 Water transport. Annu. Rev. Plant Physiol. 36, 473–516.

Brundrett M C, Enstone D E and Peterson C A 1988 A berberine-aniline blue fluorescent staining procedure for suberin, lignin, and callose in plant tissue. Protoplasma 146, 133–142.

Frensch J and Steudle E 1989 Axial and radial hydraulic resistance to roots of maize (*Zea mays* L.). Plant Physiol. 91, 719–726.

Kedem O and Katchalsky A 1963 Permeability of composite membranes. Part 2: Parallel elements. Trans. Faraday Soc. 59, 1931–1940.

Melchior W and Steudle E 1993 Water transport in onion (*Allium cepa* L.) roots. Changes of axial and radial hydraulic conductivities during root development. Plant Physiol. 101, 1305–1315.

Miller D M 1985 Studies of root function in *Zea mays*. III. Xylem sap composition at maximum root pressure provides evidence of active transport into the xylem and a measurement of the reflection coefficient of the root. Plant Physiol. 77, 162–167.

Nonweiler T R F 1975 Flow of biological fluids through non-ideal capillaries. *In* Encyclopedia of Plant Physiology, New Series, Vol 1, Transport in plants I, Phloem transport, Appendix I. Eds. M H Zimmermann and J A Milburn. pp 474–477. Springer-Verlag, Berlin, Germany.

Perumalla C J and Peterson C A 1986 Deposition of Casparian bands and suberin lamellae in the exodermis and endodermis of young corn and onion roots. Can. J. Bot. 64, 1873–1878.

Peterson C A, Murrmann M and Steudle E 1993 Location of the major barriers to water and ion movement in young roots of *Zea mays* L. Planta 190, 127–136.

Peterson C A and Steudle E 1992 Lateral hydraulic conductivity of early metaxylem vessels in *Zea mays* L. roots. Planta 189, 288–297.

Steudle E 1992 The biophysics of plant water: compartmentation, coupling with metabolic processes, and flow of water in plant roots. *In* Water and Life: Comparative Analysis of Water Relationships at the Organismic, Cellular, and Molecular Levels. Eds. G N Somero, C B Osmond and C L Bolis. pp 173–204. Springer-Verlag, Heidelberg, Germany.

Steudle E and Frensch J 1989 Osmotic responses of maize roots. Water and solute relations. Planta 177, 281–295.

Steudle E, Oren R and Schulze E-D 1987 Water transport in maize roots. Measurement of hydraulic conductivity, solute permeability, and of reflection coefficients of excised roots using the root pressure probe. Plant Physiol. 84, 1220–1232.

*F. Baluška et al. (eds.), Structure and Function of Roots*, 215–222, 1995.


# Root hydraulic properties of spruce measured with the pressure probe

S.W. Hallgren[1], M. Rüdinger[2] and E. Steudle[2]
[1]*Department of Forestry, Oklahoma State University, Stillwater, Oklahoma, U.S.A. and* [2]*Lehrstuhl für Pflanzenökologie, Universität Bayreuth, Universitätstr. 30, D-95440 Bayreuth, Germany*

*Key words:* composite transport model, hydraulic conductivity, reflection coefficient

## Abstract

The root pressure probe was used for the first time to measure the hydraulic properties of entire root systems of young *Picea abies*. Hydraulic conductance was measured by osmotic and hydrostatic pressure relaxation techniques. Osmotic experiments were conducted by changing the nutrient solution and hydrostatic experiments by causing flow across the root with the pressure probe and with external pressure applied to the root system or to the cut stem of the excised root system. Usually, *Picea abies* root systems did not develop appreciable root pressure (< 0.02 MPA) and could be induced to reach a root pressure of 0.07 MPa by treating with $KNO_3$. In general, hydraulic conductance of the root system was large, but it was much smaller in the osmotic than in the hydrostatic experiments. Both hydrostatic techniques gave similar results. The results were explainable by a composite transport model of the root.

## Introduction

In the soil-plant-atmosphere continuum the capacity of the root system to conduct water and nutrients to the above ground parts is of critical importance to survival and productivity. Questions have been raised as to how much root system is sufficient (Carlson et al., 1988; Teskey et al., 1983), how roots adapt to drought (North and Nobel, 1991; Passioura, 1972) and the significance of root-shoot communication (Davies and Meinzer, 1990; Gowing et al., 1990; Tardieu et al., 1991). The answers will depend on a better understanding of root system hydraulics. Significant progress has been made in modelling root system hydraulics (Fiscus, 1975; Landsberg and Fowkes, 1978; Steudle 1992) and the pressure probe has permitted testing the models for a wide variety of plants (Steudle, 1989, 1993). Trees, especially conifers, have been studied only infrequently because of inherent difficulties such as the lack of root pressure. The research reported here concerns the development of techniques suitable for study of root hydraulics in conifers. Data were gathered by several different experimental approaches to test existing models.

## Material and methods

### *Plant material*

Two-year-old Norway spruce (*Picea abies* (L.) Karst.) seedlings from a local source were obtained from a private nursery in Bayreuth, FRG. They were transplanted in late March into 3 L pots filled with quartz sand (particle size: 0.6 to 1.2 mm) and watered every other day with a nutrient solution (composition: 0.5 m*M* $(NH_4)_2SO_4$, 0.5 m*M* $Ca(NO_3)_2$, 0.125, m*M* $CaCl_2$, 0.5 m*M* KCl, 0.1 m*M* $KH_2PO_4$, 0.3 m*M* $MgSO_4$, 5 $mgL^{-1}$ FeNaEDTA, 4.0 $\mu M$ $MnSO_4$, 0.4 $\mu M$ $CuSO_4$, 0.4 $\mu M$ $ZnSO_4$, 2.8 $\mu M$ $H_3BO_3$, and 0.1 $\mu M$ $Na_2MoO_4$ (Lutz and Breininger, 1986)). The seedlings were kept in a glasshouse where temperatures ranged from 18 to 28°C. Experiments began 6 to 8 weeks after transplanting when substantial new root growth had occurred. All root systems showed well developed mycorrhizae.

Surface area and dry weight were determined for seedlings following separation into needles, stems, fine (diameter: < 1 mm) and coarse (diameter: > 1 mm) roots. Roots were stained with toluidine blue before measurement. Surface area of roots was determined with an image analysis system based on a video camera and software from Skye Instruments LTD (Llandrindod Wells, UK) and of needles with a area meter (Delta-T Devices, Cambridge, England). Dry weight was determined for shoots and roots after 24 h at 80°C.

*Hydraulic measurements*

Before measurement seedlings were transferred to the laboratory and shoots placed in a black plastic bag for 12 h to stop transpiration and raise seedling water potential to near zero. The purpose of covering the seedlings was to prevent partial filling of the xylem with air upon cutting. Excised root systems in the original pots were watered with nutrient solution, placed in a pressure chamber and sealed around the stem at the ground line with a rubber stopper and molding clay. The seal enclosed 20 mm of stem and the stem was cut 25 mm above the seal for insertion in the pressure probe device (Fig. 1). The function of the root pressure probe has been explained in previous papers (Peterson and Steudle, 1993; Steudle et al., 1993; Zhu and Steudle, 1991). The root system was pressurized to 0.05 MPa for 12 h and allowed to exude sap to remove air from the xylem. The internal spaces of the Plexiglas part of the pressure probe device were flushed with distilled water as necessary during the experiment to remove air bubbles, because air would decrease the mechanical rigidity (the elastic coefficient) of the measuring system and dampen responses in pressure relaxation and other experiments.

*Pressure relaxation experiments*

The cut end of the stem of the root system was fixed to a root pressure probe (Steudle and Jeschke, 1983; Steudle et al., 1987) and 2 to 3 h allowed for root pressure to develop. Then the elastic coefficient ($\beta$) of the system (xylem plus equipment) was determined according to the following relation:

$$\beta = \frac{\Delta P_r}{\Delta V_S} \qquad (1)$$

$V_S$ is the volume of the measuring system including the xylem and the water in the pressure probe above the excised root. A known change in system volume $\Delta V_S$ was achieved by observing the movement of an oil-water meniscus in the capillary with a diameter of 410 $\mu$m, (Fig. 1) while rapidly moving a metal rod into or out of the pressure probe to a new position and rapidly returning it to the original position. The corresponding change in pressure ($\Delta P_r$) in the system was measured with a pressure transducer. Throughout the measurements with an individual plant which lasted up to 5 d, $\beta$ was measured often because it could change, mostly due to air entering the system at negative $P_r$.

Pressure relaxations were performed by instantaneously changing $V_S$ with the metal rod, maintaining the new volume by observing the oil-water meniscus and recording the change in pressure with a strip-chart recorder. The resulting root pressure relaxation (Fig. 2) was exponential with time. The half-time ($T_{1/2}{}^{W}$) and rate constant ($k_w$) for the pressure relaxation are related to each other and to the root surface area ($A_r$), $\beta$ and the hydraulic conductivity ($Lp_r$) of the root system as follows:

$$k_w = \frac{\ln(2)}{T_{1/2}{}^{w}} = Lp_r \cdot A_r \cdot \beta \qquad (2)$$

This relation was used to calculate $Lp_r$ from measured values of $k_w$, $\beta$ and $A_r$.

*Root exudation under pressure*

Root exudation under pressure has often been used with herbaceous plants (Fiscus, 1985; Weatherley, 1992; Mees and Weatherley, 1957) and occasionally with woody plants (Sands et al., 1982) to measure the hydraulic conductivity of roots. The root system was tightly sealed in a pressure chamber and air pressure was applied from a cylinder of pressurized air (Fig.1). The steady-state rate of exudation ($Q_V$ in $m^3s^{-1}$) was measured as a function of applied gas pressure ($\Delta P_{gas}$) by following the rise of the meniscus in a capillary with time. $Lp_r$ was obtained from the following relation:

$$Lp_r = \frac{Q_V}{\Delta P_{gas} \cdot A_r} \qquad (3)$$

The relation required the assumption that the contribution of the overall osmotic driving force ($\sigma_s \cdot \Delta\pi_S$) was small. To verify this assumption the osmotic pressure of the exudate was measured with an osmometer (Gonotec, Berlin, FRG). The hydraulic conductance of a root system ($Lp_r \cdot A_r$) should be independent of the direction of flow. Therefore, experiments were performed with the water in the capillary pressurized (Fig. 1) to force distilled water through the root system. In both types of experiments care was taken that the water flow had become constant (usually within ten minutes) before measured values were used in the analysis. The calculation of $Lp_r$ from the hydrostatic experiments

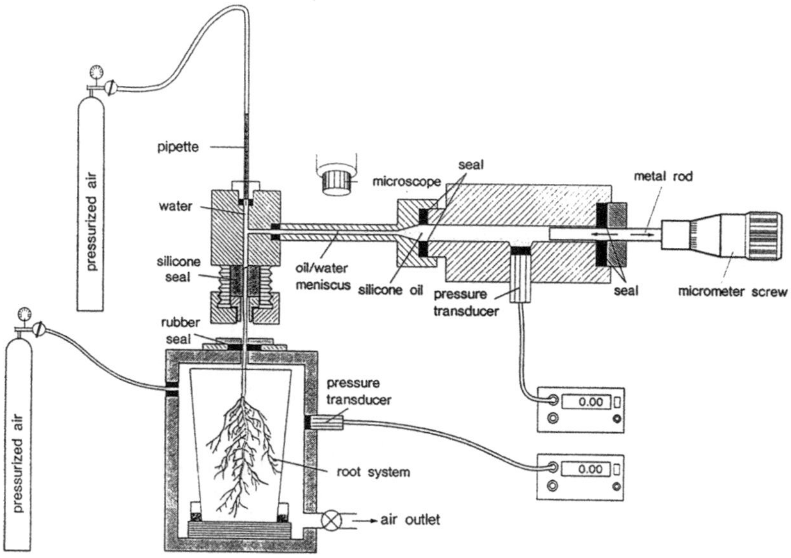


*Fig. 1.* Apparatus used in pressure relaxation and root exudation experiments. The excised root system was contained in a vessel which was pressurized as needed and the cut end inserted in a pressure probe device. Pressure could also be applied through the pipette to the cut end. The rise and fall of the meniscus in the pipette attached to the pressure probe device depended on whether pressure was applied to the root system or the cut end and indicated flow through the root system. Pressure relaxation experiments were conducted with the pipette removed and the pressure chamber open. The oil/water meniscus in the pressure probe device indicated flow across the root in pressure relaxation experiments where pressure was changed by rapid movement of a metal rod.

was based on the assumption that the hydraulic conductance of the xylem was much larger than the radial. In other words, the axial resistance of the root was negligible compared to the radial. Verification of this assumption was done in each experiment by cutting the main root and repeating the measurements to determine if the half-time for pressure relaxation decreased substantially after the cut.

*Osmotic measurements*

*Osmotic pressure relaxations*

First, water flow equilibration was established with the root system attached to the pressure probe and watered with nutrient solution. Then the soil solution was rapidly changed to a new solution containing an osmoticum (mannitol, KCl, $KNO_3$, etc.) in addition to the nutrients. The change in root pressure which was exponential with time was analysed in the same way as the pressure relaxations described above (Fig. 2C). Two different parameters were involved in the analyses: 1. the 'osmotic hydraulic conductivity' according to Eq. 1 and 2. the reflection coefficient of the root for a given solute. A good approximation of the latter is given by

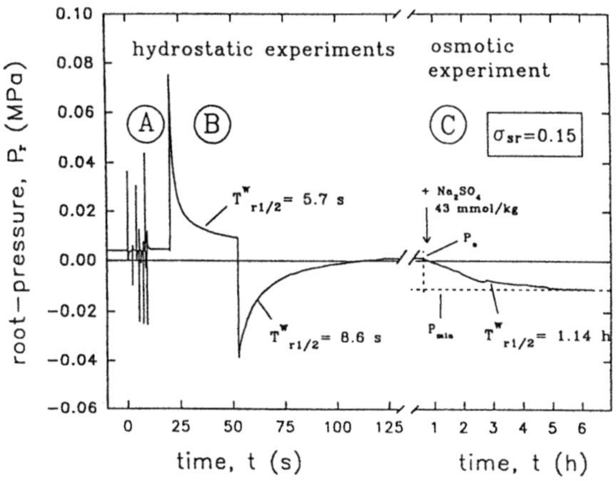


*Fig. 2.* Typical traces from the strip-chart recorder showing results of hydrostatic and osmotic experiments. Hydraulic conductivity ($Lp_r$) was calculated from measurements of the elastic coefficient ($\beta$) of the system (**A**), and the relaxation half-time ($T_{1/2}{}^{w}$) of hydrostatic (**B**) and osmotic (**C**) experiments. Root pressure decreased when a solute was added and increased when it was removed from the root-bathing solution. The reflection coefficient ($\sigma_{sr}$) was calculated by dividing the change in root pressure ($\Delta P_r$) by the change in the osmotic pressure of the soil solution ($\Delta\pi_S{}^{O}$).

(Steudle et al., 1987) :

$$\sigma_{sr} = \frac{\Delta P_{max}}{\Delta \pi_s^o} \tag{4}$$

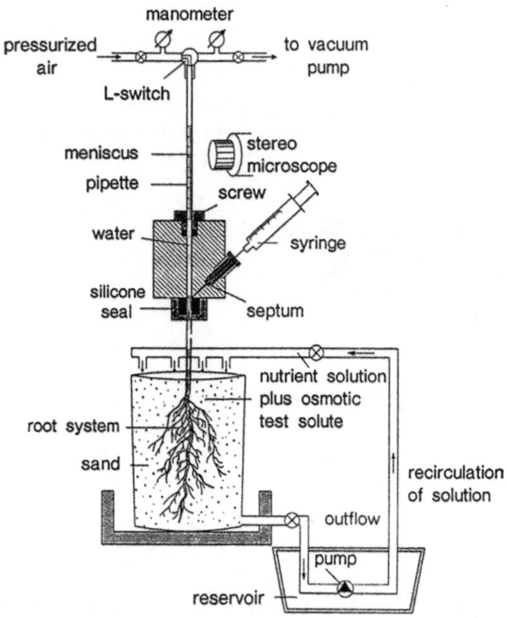


*Fig. 3.* Apparatus used in osmotic stop-flow experiments. Excised root systems were attached to a pipette connected to a vacuum pump and pressurized air. The root system was bathed with osmotic solutions and pressure or vacuum applied to the cut end to stop movement of the memiscus. Dyed oil was inserted with a syringe to develop a meniscus that was easy to observe.

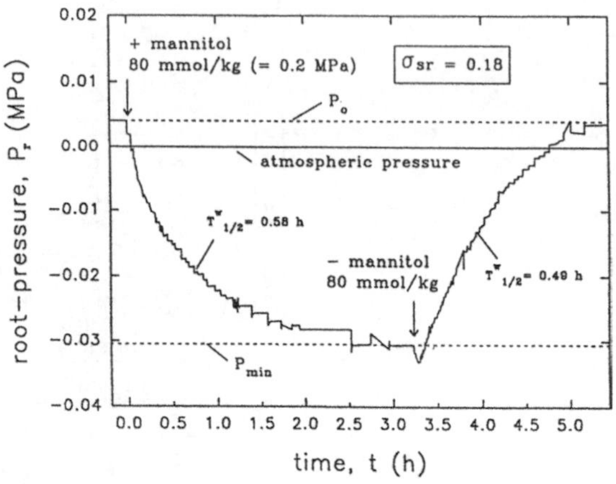


*Fig. 4.* Typical trace from strip-chart recorder showing results of stop-flow experiment. The excised root system was bathed in an osmotic solution and pressure or vacuum applied to the cut end to keep the meniscus in the pipette at a constant level. Addition and removal of a solute caused a fall and rise in the meniscus, respectively. A vacuum or pressure was applied as needed to stop movement of the meniscus. As in the osmotic relaxation experiments, $\sigma_{sr}$ was calculated as $\Delta P_r$ divided by $\Delta\pi_S{}^O$.

$\Delta P_{max}$ is the maximum change in root pressure and $\Delta\pi_s{}^O$ is the change in the osmotic pressure of the medium.

*Osmotic stop-flow experiments*

The reflection coefficient was also determined by a variation of the osmotic pressure relaxation experiment where water flow ($J_V$) was stopped by counterbalancing the effective osmotic pressure in the root created by the soil solution ($\sigma_{sr} \cdot \Delta\pi_S{}^O$) with an equivalent suction ($\Delta P_r$) applied to the cut end of the stem. The relation is described as follows:

$$\sigma_{sr} \cdot \Delta\pi_s^O = \Delta P_r \qquad (5)$$

Unstirred layer effects caused by a volume flow of water (sweep-away effects, Dainty, 1963) are avoided. The stop-flow experiment was performed by first establishing water flow equilibrium in the root with a pipette attached to the cut end of the stem (Fig. 3). A meniscus was held at a constant level with roots at atmospheric pressure and a slight overpressure was applied to the pipette. Then the soil solution was rapidly exchanged for new solution containing an osmotic solute (e.g. mannitol) which sucked water from the roots into the soil and caused the meniscus to fall. A vacuum was applied to the pipette to counterbalance the soil suction and hold the meniscus constant. The vacuum was measured by a manometer and compared to the water potential of the soil solution. The kinetics of the process was equivalent to that measured in the osmotic pressure relaxation experiments (Fig. 4); however, the system was open instead of closed as with the root pressure probe and tensions of about 0.1 MPa (1 bar) at maximum could be applied with the vacuum pump.

## Results

Root pressures were always close to zero. They ranged from $P_{ro}$ = -0.001 to 0.004 MPa (-0.01 to 0.04 bar or -0.1 to 0.4 m water column). Typical root pressure relaxation plots are shown for hydrostatic and osmotic experiments in Figures 2B and C, respectively. It is obvious in the figure that half-times of pressure relaxations were very different depending on the driving force. Osmotic gradients resulted in $T_{1/2}{}^W$ that were 1 to 2 order of magnitude greater than those caused by hydrostatic gradients. According to Eq. 2 this would result in a much lower osmotic $Lp_r$ than hydrostatic. The $T_{1/2}{}^W$ for pressure relaxations decreased sever-

*Table 1.* Results of hydrostatic experiments to measure hydraulic conductivity of excised root systems of spruce seedlings. (**A**) Hydrostatic experiments were conducted to determine $Lp_r$ by pressure probe, exudation and infiltration techniques. The plants were grown on either nutrient solution (NS) or deionized water (DW). Mean values are given ± SD and number of measurements are in brackets. (**B**) Osmotic experiments were conducted to determine $\sigma_{sr}$ and $Lp_r$ for several solutes

**A. Hydrostatic experiments**

| Tree number | Root surface area $A_r \cdot 10^3$ [$m^2$] | Hydraulic conductivity, $Lp_r \cdot 10^8$ [$m \cdot s^{-1} \cdot MPa^{-1}$] obtained by | | |
|---|---|---|---|---|
| | | Pressure probe relaxations | Exudation from pressurized root system | Infiltation of water into cut stem |
| 21 NS | 3.3 | 17 ± 13 (8) | - | - |
| 22 NS | 4.3 | 62 ± 77 (7) | 2.2 ± 0.14 (8) | 10 ± 0.42 (6) |
| 23 NS | 6.0 | 1.9 (1) | 0.28 ±0.032 (7) | 2.0 ± 0.24 (5) |
| 24 NS | 6.9 | 2.3 ± 1.4 (5) | 1.5 ± 0.15 (8) | 2.2 ± 0.09 (7) |
| 25 NS | 11.9 | 9.9 ± 3.3 (5) | 1.4 ± 0.46 (7) | 3.6 ± 0.23 (7) |
| 32 DW | 3.9 | 13 ± 8.4 (6) | - | - |

**B. Osmotic experiments**

| Tree number | Osmoticum | Concentration change in the medium (mmol·kg$^{-1}$) | Maximum change in root pressure $\Delta P_r$ (MPa) | Root reflexion coefficient $\sigma_{sr}$ | Osmotic experiment $Lp_r \cdot 10^8$ ($m \cdot s^{-1} \cdot MPa^{-1}$) | Ratio $\frac{Lp_r \text{ (hydr.)}}{Lp_r \text{ (osmot.)}}$ |
|---|---|---|---|---|---|---|
| 21 NS | $K_2SO_4$ (+) | + 45 | 0.020 | 0.18 | 0.054 | 305 |
| | (-) | -43 | -0.014 | 0.13 | 0.13 | 123 |
| | $Na_2SO_4$ (+) | + 43 | 0.019 | 0.18 | 0.012 | 1400 |
| 24 NS | $Na_2SO_4$ (+) | + 26 | 0.006 | 0.09 | 0.15 | 15 |
| | (-) | - 26 | -0.002 | (0.03) | 0.28 | 8 |
| | $Ca(NO_3)_2$ (+) | + 36 | 0.010 | 0.11 | 0.012 | 180 |
| | (-) | -36 | -0.010 | 0.11 | 0.012 | 190 |
| 25 NS | $Ca(NO_3)_2$ (+) | + 36 | 0.008 | 0.10 | 0.013 | 760 |
| | (-) | -30 | -0.007 | 0.10 | 0.059 | 170 |
| 32 DW | $Na_2SO_4$ (+) | + 45 | 0.018 | 0.17 | 0.021 | 610 |
| | $Ca(NO_3)_2$ (-) | -43 | -0.006 | (0.05) | 0.20 | 65 |

al fold when a major root was cut at the end of each experiment.

Results from the stop-flow experiments (Fig. 4) were consistent with those from the osmotic pressure relaxations done with the root pressure probe. Similar half-times of pressure relaxation were obtained from the two types of experiments ($T_{1/2}^W$ = 0.5 to 1.5 h) which resulted in similar osmotic $Lp_r$ when using typical values of $\beta$. When constant hydrostatic gradients were applied by pressurizing either the root systems or the cut surface of the stem steady-flow experiments gave values for $Lp_r$ similar to those obtained in the pressure relaxation experiments. Values for $Lp_r$ were in the range of 2 to $10 \cdot 10^{-8}$ $ms^{-1}$ $MPa^{-1}$ for steady flow experiments and 2 to $62 \cdot 10^{-8}$ $ms^{-1}$ $MPa^{-1}$ for pressure relaxation experiments (Table 1A).

In the osmotic experiments with both the pressure-probe and the stop-flow techniques reflection coefficients were calculated from the maximum changes in pressure for several solutes ($KNO_3$, mannitol, KCl, NaCl; Table 1B). Reflection coefficients were small ($\sigma_{sr}$ = 0.1 to 0.3) compared with values of the same solutes for cell membranes which are close to unity.

## Discussion

Root pressure phenomena are believed to be the result of root systems acting as osmometers through the accumulation of solutes in the xylem sap. It follows that any plant capable of accumulating solutes in the xylem should be also capable of developing root pressure and of exuding when the root is excised. Indeed, there are numerous examples of root pressure in herbaceous and agronomic plants. Kramer (1983) reported values for root pressure ranging from 0.05 to 0.6 MPa for these plants. The phenomenon has been also reported for deciduous trees but rarely for conifers (White et al., 1958). In deciduous trees, root pressures of as high as 0.2 to 0.3 MPa have been measured (Kramer, 1983; Steudle, 1994). Exudation from detopped conifer seedlings was found in several species (Lopushinsky, 1980; Sands et al., 1982). It appeared that exudation was enhanced by keeping the plants at low temperatures (Lopushinsky, 1980). Because of the low values of root pressure in conifers which were also found in this study, it may have been often overlooked because of the difficulties in measurement of the low values and of problems in carefully hydrating plants prior to the measurement. Low root pressures of conifers may result to some extent from a low reflection coefficient. The composite model for radial flow of water and solutes into the root may be used to explain how the low reflection coefficient can develop in roots where the reflection coefficient of individual cells is nearly unity (see discussion below).

In the research reported here, the root hydraulic conductivity was obtained by different types of hydrostatic experiments, root pressure relaxations and steady-state experiments, and the values were comparable for the different techniques (Table 1A). This means that the technique for determining $Lp_r$ from root pressure relaxations which was applied for the first time to the root of a coniferous tree yielded reasonable results. This is so because the axial hydraulic conductance of the roots was much larger than the radial despite the tracheid structure of the xylem and, thus, at least for spruce seedlings, the simple two compartment model of the root appeared to be valid.

The values of the 'hydrostatic $Lp_r$' obtained for spruce, ranged from 1.9 to 62 $\cdot 10^{-8}$ $ms^{-1}$ $MPa^{-1}$. These values agree reasonably well with previous findings. In a careful study on loblolly pine (*Pinus taeda*) Sands et al. (1982) found an overall 'hydrostatic $Lp_r$' of $1.4 \cdot 10^{-7}$ $ms^{-1}$ $MPa^{-1}$ which is close to our values. The hydraulic conductivity of older brown suberized roots was $7.6 \cdot 10^{-8}$ $ms^{-1}$ $MPa^{-1}$ and that of younger roots $2.0 \cdot 10^{-7}$ $ms^{-1}$ $MPa^{-1}$. They also reported a considerable axial hydraulic conductance measured with root segments which is also in line with the findings of this paper. Recent measurements of Douglas-fir gave values of $1 \cdot 10^{-8}$ $ms^{-1}$ $MPa^{-1}$ (recalculated from Coleman et al., 1990). The $Lp_r$ for oak was $1 \cdot 10^{-8}$ $ms^{-1}$ $MPa^{-1}$ which is also similar (Steudle, 1994).

Except for the recent study on oak roots mentioned above, studies of osmotic water flow and measurements of the 'osmotic $Lp_r$' of roots are rare in the literature for technical reasons. In the case of oak (*Quercus robur* and *Q. petraea*) it was found that the 'osmotic $Lp_r$' was by a factor of 40 to 400 smaller than the 'hydrostatic' which was interpreted by a different transport model. Similarly, for the spruce roots investigated in this paper 'osmotic $Lp_r$' was by 1 to 2 orders of magnitude smaller than the 'hydrostatic' which indicated the same differences in the transport model. Differences between osmotic and hydrostatic water flow have been also found for roots of some herbaceous species, although not for all. In herbaceous roots, the root $Lp_r$ has been compared with root cell Lp. It was found that during hydrostatic flow there was a predominant apoplastic by-pass around root protoplasts (perhaps also in the endodermis), whereas during osmotic flow the cell-to-cell path was preferred (Steudle, 1989; Steudle et al., 1993). The same should be true for spruce where differences between flows ($Lp_r$) were even larger. The physical explanation for the difference is quite simple: if there is an apoplastic path for water it should be quite ineffective in the presence of an osmotic gradient, since along the wall path the reflection coefficient should be virtually zero resulting in a very small overall driving force in the apoplast. By contrast, osmotic flow will fully develop along the cell-to-cell path. However, the cell-to-cell path has a high hydraulic resistance (many cell membranes to be crossed) so that the 'osmotic $Lp_r$' is low despite the high potential cell wall hydraulic conductivity. Thus, the apparent differences between 'osmotic' and 'hydrostatic $Lp_r$' become understandable.

The model which describes the root in terms of parallel pathways (cell-to-cell and apoplastic paths) has been termed the 'composite transport model of the root' (Steudle, 1989, 1992, 1993, 1994; Steudle et al., 1993). This model also explains the low reflection coefficients of roots which are even smaller for tree roots than for roots of herbaceous species (Peterson and Steudle, 1993; Steudle et al., 1993; Steudle, 1989,

1992, 1993, 1994). According to the model, the lower reflection coefficients of tree roots is due to the fact that the absolute value of $Lp_r$ is much smaller.

It has to be mentioned that in terms of the composite transport model the discrepancy between $Lp_r$'s from osmotic and hydrostatic water flow could also account for the well known effect that $Lp_r$ increases with increasing driving force. This effect has been known for a long time, but an appropriate explanation is still lacking. Mainly two models are discussed in the literature: (i) an effect of $J_V$ on the osmotic driving force across the root (dilution effect; Fiscus, 1975) and (ii) a valve-like mechanism which considers changes in the hydraulic conductance of plasmodesmata to be caused by hydrostatic pressure gradients (Passioura, 1988). The composite transport model offers another and straightforward explanation which is more reasonable. At low transpiration rates, the flow across the root is mostly osmotic in nature which results in a low $Lp_r$. The osmotic flow dominates because the osmotic driving force is large compared to the hydrostatic driving force at low transpiration and the cell-to-cell pathway has a much larger surface area for water uptake than the apoplastic pathway. With increasing transpiration, $Lp_r$ incorporates an increasing hydrostatic component which then also increases the root hydraulic conductance.

The much higher conductivity of the apoplastic pathway relative to the cell-to-cell pathway has already been demonstrated. When the cells were heat-killed in a small portion of the root (5 to 18% of the surface area) hydraulic conductivity increased 8-fold while the capacity to retain ions within the root dropped very low as indicated by root pressures near zero (Peterson and Steudle, 1993). An interesting test of the composite transport model would be to determine whether the curvilinear relation between $Lp_r$ and driving force is affected by eliminating the energy supply necessary to take up ions necessary for osmotic flow. For example, imposing anaerobic conditions on the roots would gradually deplete the roots of energy for active ion uptake simultaneously reducing the osmotic driving force. Eventually, only the hydrostatic driving force would produce flow and $Lp_r$ would not change as the driving force increased. This experiment should be done with plants lacking adaptations to flooding that can compensate for anaerobiosis through aerenchyma or alternative metabolic pathways.

The advantage for the plant of having a mechanism for changing hydraulic conductivity as the driving force for water uptake changes is obvious. As the demand for water increases in the shoot, root $Lp_r$ will also increase as tensions develop in the root xylem caused by transpiration. As the results demonstrate, the composite transport model functions despite a low solute permeability which is important for the plant to retain nutrients in the xylem.

## Acknowledgements

The research reported here was supported by a grant from EUROSILVA (project no. 39473c) to E S and a Fulbright Senior Scholar Award to S W H The expert technical assistance of Burkhard Stumpf and Libuš Badewitz is gratefully acknowledged.

## References

Carlson W C, Harrington C A, Farnum P and Hallgren S W 1988 Effects of root severing treatments on loblolly pine. Can. J. For. Res.18, 1376–1385.

Coleman M D, Bledsoe C S and Smit B A 1990 Root hydraulic conductivity and xylem sap levels of zeatin riboside and abscisic acid in ectomycorrhizal Douglas-fir seedlings. New Phytol. 115, 275–284.

Dainty J 1963 Water relations of plant cells. Adv. Bot. Res.1, 279–326.

Davies W J and Meinzer F C 1990 Stomatal responses of plants in drying soil. Biochem. Physiol. Pflanz. 186, 357–366.

Fiscus E L 1975 The interaction between osmotic- and pressure-induced water flow in plant roots. Plant Physiol. 55, 917–922.

Gowing D J G, Davies W J and Jones H G 1990 A positive root-sourced signal as an indicator or soil drying in apple, *Malus* × *Domestica*-Borkh. J. Exp. Bot. 41, 1535–1540.

Kramer P J 1983 Water Relations of Plants. Academic Press, Orlando. 489 p.

Landsberg J J and Fowkes N D 1978. Water movement through plant roots. Ann. Bot. (London) 42, 493–508.

Lutz H J and Breininger M Th 1986 Erste Ergebnisse von Auswaschungsversuchen mit Fichten (*Picea abies*) in Wasserkultur. Kali-Briefe 18, 267–273.

Lopushinsky W 1980 Occurrence of root pressure exudation in Pacific Northwest conifer seedlings. For. Sci. 26, 275–279.

Mees G C and Weatherley P E 1957 The mechanism of water absorption by roots. I. Preliminary studies on the effects of hydrostatic pressure gradients. Proc. Roy. Soc. London, Ser. B 147, 367–380.

North G B and Nobel P S 1991 Changes in hydraulic conductivity and anatomy caused by drying and rewetting roots of *Agave deserti* (Agavaceae). Am. J. Bot. 78, 906–915.

Passioura J B 1972 The effect of root geometry on the yield of wheat growing on stored water. Aust. J. Agric. Res. 23, 745–752.

Passioura J B 1988 Water transport in and to the root. Ann. Rev. Plant Physiol. Plant Mol. Biol. 39, 245–265.

Peterson C A and Steudle E 1993 Lateral hydraulic conductivity of early metaxylem vessels in *Zea mays* L. roots. Planta 189, 288–297.

Sands R, Fiscus E L and Reid C P P 1982 Hydraulic properties of pine and bean roots with varying degrees of suberzation, vascular

differentiation and mycorrhizal infection. Aust. J. Plant Physiol. 9, 959–969.
Steudle E 1989 Water flow in plants and its coupling to other processes: an overview. Methods Enzymol. 174, 183–225.
Steudle E 1992 The biophysics of plant water: compartmentation, coupling with metabolic processes, and water flow in plant roots. *In* Water and Life: Comparative Analysis of Water Relationships at the Organismic, Cellular, and Molecular Levels. Eds. G N Somero, C B Osmond and C L Bolis. pp 173–204. Springer-Verlag, Berlin.
Steudle E 1993a Pressure probe techniques: basic principles and application to studies of water and solute relations at the cell, tissue, and organ level. *In* Water Deficits: Plant Responses from Cell to Community. Eds. J A C Smith and H Griffith. pp 5–36. Bios Scientific Publishers, Oxford.
Steudle E 1994 Water transport across roots. Plant and Soil 167.
Steudle E and Jeschke W D 1983 Water transport in barley roots. Planta 158, 237–248.
Steudle E, Murrmann M and Peterson C A 1993 Transport of water and solutes across maize roots modified by puncturing the endodermis: further evidence for the composite transport model of the root. Plant Physiol. 103, 335–349.
Steudle E, Oren R and Schulze E-D 1987 Water transport in maize roots. Plant Physiol. 84, 1220–1232.
Tardieu F, Katerji N, Bethenod O, Zhang J and Davies W J 1991 Maize stomatal conductance in the field - its relationship with soil and plant water potentials, mechanical constraints and ABA concentration in the xylem sap. Plant Cell Environ. 14, 121–126.
Teskey R O, Hinckley T M and Grier C C 1983 Effect of interruption of flow path on stomatal conductance of *Abies amabilis*. J. Exp. Bot. 34, 1251–1259.
Weatherley P E 1982 Water uptake and flow in roots. *In* Encyclopedia of Plant Physiology, Vol. 12B. Eds. O L Lange, P S Nobel, C B Osmond, and H Ziegler. pp 79–109. Springer-Verlag, Berlin.
White P R, Schuker E, Kern J R and Fuller F H 1958 Root-pressure in gymnosperms. Science 128, 308–309.
Zhu G L and Steudle E 1991 Water transport across maize roots: simultaneous measurement of flows at the cell and root level by double pressure probe technique. Plant Physiol. 95, 305–315.

*F. Baluška et al. (eds.), Structure and Function of Roots*, 223–225, 1995.


# The development of melon roots under trickle irrigation: Effects of the location of the emitters

U. Shani[1], Y. Waisel[2] and A. Eshel[2]
[1]*Arava Experimental Station, Yotvata, Israel and* [2]*Department of Botany, Tel Aviv University, Tel Aviv Israel*

*Key words: Cucumis*, melon, root distribution, trickle irrigation

## Abstract

Desert agriculture is highly dependent upon trickle irrigation. Usually, plant roots that develop under high frequency trickle irrigation are shallow and densely crowded. As a result of that, plants become sensitive to temporal fluctuations in water availability, in temperature and in salinity. Thus, an attempt was made to establish a root system that will minimize the above mentioned hazards.

Winter melons (*Cucumis melo* cv. Haon), were grown under four irrigation treatments with tricklers placed at four different depths: surface, 15 cm, 30 cm and 45 cm. Root development was measured at harvest time, using the "wall profile" method. Distribution of the roots at various soil horizons, depended upon the depth of the emitters. Most roots have developed at the depth of 15–20 cm, in surface irrigation, and in the 15 cm deep treatments. Highest root densities were obtained at the depth of 25–30 cm, in the 30 cm deep emitters treatment, and at 35–40 cm where the emitters were 45 cm deep. Distribution of salt across the soil profile also depended upon the depth of the emitters; concentration of chloride at the soil surface was approximately 30–50 m*M*, in treatments with shallow emitters but reached a concentration of 1447 m*M* in the 45 cm treatment. Daily fluctuations in temperature were highest at the soil surface and negligible at the depth of 40 cm. As low root temperatures have an adverse effect on the water uptake capability of winter crops, plants with deeper roots should have an advantage. Indeed, melon plants of the 45 cm treatment remained green and vital throughout the winter, whereas plants with shallow roots developed scorched leaves and were severely damaged.

Thus, it seems possible to manipulate the architecture of the root systems of melon plants by regulating the depth of the emitters of the trickle irrigation system.

## Introduction

Modern desert agriculture is totally dependent upon irrigation. The most advanced irrigation system for such conditions, and the most efficient one, is the trickle irrigation. It is designed to deliver water in many points all over the field, at a low rate of a few liters per day (cf. Goldberg et al., 1971). The location and the distribution of the emitters, and the timing of irrigation, allows the grower to determine the precise distribution of water and nutrients in the soil profile (Bar-Yosef and Sagiv, 1980). Immediately around the emitter there is a soil volume which is constantly wet, and the salt concentration in it is similar to that of the irrigation water. At the periphery of this volume of soil, salts accumulate gradually, mostly at the soil surface (Shani, 1987). Such an irrigation system is widely used all over the world, but mostly in countries of the arid zones where water conservation is of prime importance. Nevertheless, even such a sophisticated irrigation system is not without shortcomings, especially when used with water of marginal quality or for growing off-season crops. Melon plants that are grown in the Arava valley of Israel in the autumn, often suffer from leaf scorches, and eventually wilt, especially when the water is rich in salts and when the plants are fully loaded with fruits (cf. Borochov-Neori and Borochov, 1991). This phenomenon was attributed to the insufficient development of the root systems under the decreasing temperatures of the soil, or from an undesirable distribution of the roots. In both cases the result was shortage of water.

Usually, the tricklers are placed on the soil surface and therefore, trickle irrigation induces the development of shallow and dense root systems (cf. Shani et al., 1993). Under such conditions the plants become sensitive to spatial and to temporal fluctuations in water availability, in soil temperature and in salinity (Shani, 1987; Waisel and Eshel, 1991). Therefore, in order to reduce the damage to the plants, high water content and low salt concentrations have to be maintained in the root zone. This requires precise timing of the irrigation pulses and their proper quantification, which are difficult to achieve under field conditions.

The question that was tackled in the reported investigation was if it would be possible to stimulate the development of a sufficiently extensive, a properly spread and an efficient root system, under such conditions of irrigation? Would it be possible to minimize the potential hazards of excessive accumulation of salts by the plants or of deterioration of their organs by a simple manipulation of the location of the tricklers?

## Experimental

A trickle irrigation system was installed in a sandy-loam field of melons (*Cucumis melo* cv. Haon) at Yotvata, Israel. The soil of this area is a sandy loam of the Arava type (cf. Shani et al., 1987), with a field capacity of 19% (V/V), and a wilting point of 6%. The tricklers were placed at four depths: surface, 15 cm, 30 cm and 45 cm. The plants were sown on these plots, during August 1991 and root development was measured three months later, at harvest time, using the "wall profile" method (Bohm and Kopke, 1977; Shani, 1987). The density of roots, at each horizon of the profile, was calculated from the number of roots that intersected a $m^2$ wall profile, with a $5 \times 5$ cm mesh (Newman, 1966). The plants were irrigated daily, during the experimental period, at a rate that compensated for the daily evaporation, as measured by a class A pan. The daily amount of irrigation water ranged between 3.6 L and 10 L $m^{-2}$. The average salt content of the irrigation water was approximately 20 m*M* $Cl^-$.

During that period, the average monthly temperatures fluctuated between 29.0°C and 13.1°C.

## Results

### *Root distribution*

Root distribution at the various soil horizons, was affected strongly by the emitters depth (Fig. 1). In the surface irrigation and in the 15 cm treatment, the highest density of roots was observed at the depth of 15–20 cm . The depth of highest root density increased with the drippers depth, reaching its highest values at the depth of 25–30 cm, in the 30 cm deep emitters treatment, and 35–40 cm where the emitters were 45 cm deep.

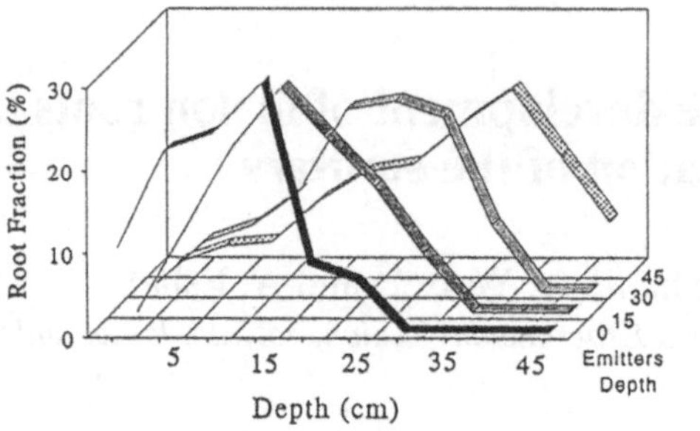


*Fig. 1.* Root distribution of melon plants (% of all roots in the profile) in four emitters-depth treatments.

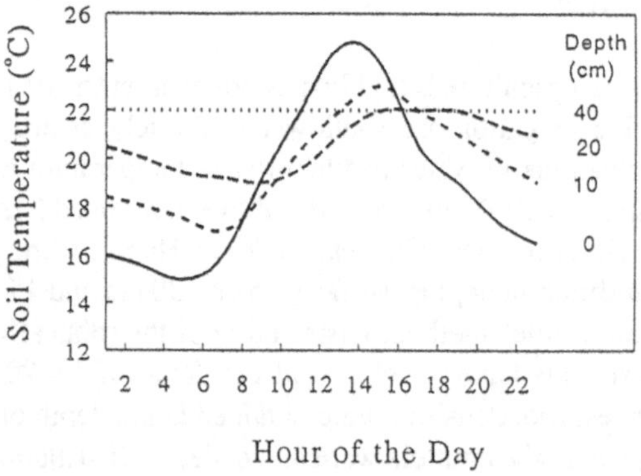


*Fig. 2.* An example of a daily time course of soil temperature at four depths in a field with a full plant canopy cover. Surface irrigation treatment. Measurements taken during mid-November.

### *Temperature*

Temperature profiles were very similar for the different irrigation systems, with the daily fluctuations being highest at the soil surface and much smaller at the depth of 40 cm (Fig. 2). As root zone temperature affects the water uptake, those plants with deeper roots should suffer less from water stresses. Indeed plants of the 45 cm emitters remained green and active till harvest time, whereas plants with shallow roots developed leaf scorches, were damaged and senescent early.

### *Yield*

Three factors have to be considered as affecting yield: water use, temperature and salinity. As linear relation-

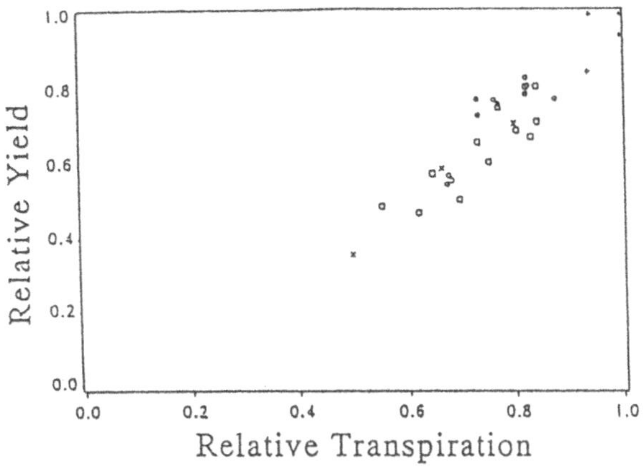


*Fig. 3.* Relative yield of melons (Y/Ymax) as a function of the rate of relative transpiration (Transpiration/Evaporation). The different symbols are the results of several experiments.

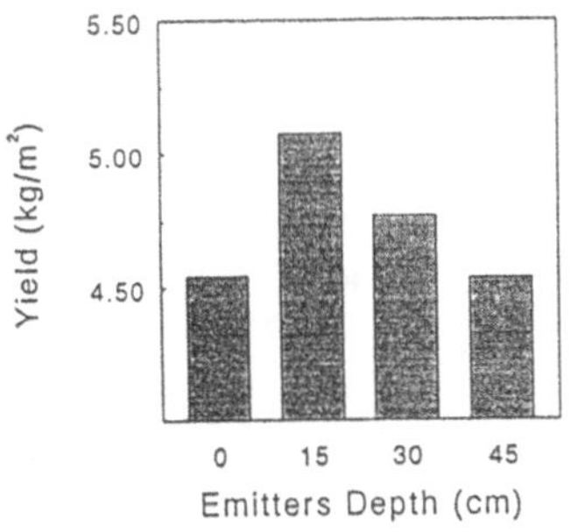


*Fig. 4.* Final yield (kg $m^{-2}$) in each of the four irrigation treatments.

ships were established between water use and the yield of the melons (Fig. 3), a high water use could have certain rewards under such conditions. Highest yields were obtained for the 15 cm emitter depth, where the soil temperature fluctuated between 18°C and 22°C only (Fig. 4). Thus, temperature alone was not the

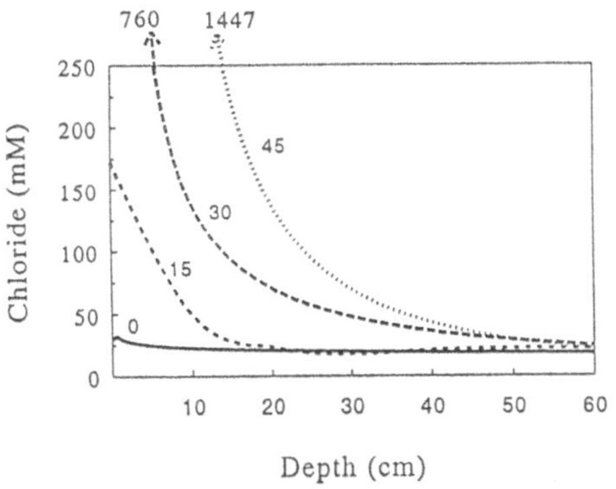


*Fig. 5.* Changes in chloride concentrations (m*M*) of saturated soil extracts with depth. Four irrigation treatments. Emitters depth indicated for each line.

controlling factor, and other effectors, such as salinity, must be considered.

A different salinity profile developed, in each of the four irrigation treatments, applied during the experiment (Fig. 5). Chloride concentration, in the saturated soil extract, at the soil surface, was 30–170 m*M* $Cl^-$, where the emitters were shallow. However, it reached values of 1447 m*M* Cl in the 45 cm treatment. No doubt that the development of such a salty layer on top of the soil has a dominant effect on the plants, especially during periods of excessive water use (cf. Nerson and Paris, 1984).

## Conclusions

It seems possible to manipulate the depth of the main part of the root system of melon plants by maneuvering the depth where water is discharged by the tricklers. The location of the emitters seems to affect mainly the surface soil salinity, and is probably the reason for the change in root distribution and for the development of a different root architecture. The assumption that temperature played a role in the collapse of melon plants, could not be corroborated.

## References

Bar-Yosef B and Sagiv B 1980 Growth of trickle irrigated tomato as related to rooting volume and uptake of N and water. Agron. J. 72, 815–822.

Bohm W and Kopke U 1977 Comparative root investigation with two profile wall methods. Z. Acker Pflanzenbau 144, 297–303.

Borochov-Neori H and Borochov A 1991 Response of melon plants to salt: Growth, morphology and root membrane properties. J. Plant Physiol. 139, 100–105.

Bowen G D 1991 Soil temperature, root growth and plant function. *In* Plant Roots: The Hidden Half. Eds. Y Waisel, A Eshel and U Kafkafi. pp 309-330. Marcel Dekker Inc., New York, USA.

Goldberg D, Gornat B and Bar Y 1971 The distribution of roots, water and minerals as a result of trickle irrigation. J. Am. Soc. Hortic. Sci. 96, 645–648.

Nerson H and Paris H S 1984 Effects of salinity on germination, seedlings growth and yield of melons. Irrig. Sci. 5, 265–273.

Newman E I 1966 A method of estimating the total length of roots in a sample. J. Appl. Ecol. 3, 139–145.

Shani U 1987 Soil, plant, water quality and drippers as criteria for trickle irrigation system design. Ph.D. Thesis, The Hebrew University, Jerusalem. 96 p.

Shani U, Hanks R J, Bresler E and Oliveira C A S 1987 Field method for estimating hydraulic conductivity and matric potential water content relations. Soil. Sci. Soc. Am. J. 51, 298–302.

Waisel Y and Eshel. A 1991 Multiform behavior of various constituents of one root system. *In* Plant Roots: The Hidden Half. Eds. Y Waisel, A Eshel and U Kafkafi. pp 39–52. Marcel Dekker Inc. New York, USA.

# ROOT-SHOOT INTERACTIONS

*F. Baluška et al. (eds.), Structure and Function of Roots*, 229–236, 1995.


# Effects of nitrogen source, nitrate concentration and salt stress on element and ion concentrations in transport fluids and on C and N flows in *Ricinus communis* L.

Andreas D. Peuke and W. Dieter Jeschke
*Julius-von-Sachs-Institut für Biowissenschaften, University of Würzburg, Mittlerer Dallenbergweg 64, D-97082 Würzburg, Germany*

*Key words:* N-nutrition, phloem, *Ricinus communis*, salt stress, transport, xylem

## Abstract

*Ricinus communis* L. was supplied with 0.2, 1.0, 4.0 m*M* nitrate or 1.0 m*M* ammonium and treated with a mild salt stress 40 m*M* NaCl (1.0 m*M* nitrate or ammonium). Between 41 and 51 days after sowing, element and ion concentrations in xylem and phloem sap were determined, and flows of C and N were modelled. Nutritional conditions particularly affected anion concentrations in the root-pressure xylem sap. Nitrate was the major N-compound in xylem sap of nitrate-fed, and amino acids in that of ammonium-fed plants. Lower nitrate was compensated mainly by chloride as an anion and by amino acids as a N-solute. Under salt treatment, $Na^+$ and $Cl^-$ levels increased, but a high selectivity of ion uptake into the xylem was observed. The phloem sap was less affected by nutritional conditions; only under stress conditions higher ion concentrations in the xylem, i.e. mainly of $Na^+$ and $Cl^-$, reflected in the phloem sap. Most of the N taken up was first transported to the shoot. In plants provided with adequate N, 70 – 77% of the N was incorporated into the shoot. This partitioning was shifted in favour of the shoot in salt-stressed, and in favour of the root, in N-limited plants, in which a net export of N from the shoot occurred. Salt stress and N-limitation decreased the photosynthetic and respiratory rates in *Ricinus* shoots, root respiration was stimulated by ammonium assimilation. Higher N assimilation in the root increased the proportion of C transported to the root, which was used there for respiration. Concomitantly more amino acids were translocated and led to higher recycling of carbon to the shoot via the xylem.

*Abbreviations:* dwt – dry weight, fwt – fresh weight

## Introduction

The type of the N source, $NH_4^+$ vs. $NO_3^-$, severely reflects on the flows of C, N and ions within the whole plant of *Ricinus communis* (Peuke and Jeschke, 1993) as does the rate of nitrate supply in *Pisum* (Duarte and Larsson, 1993) and in *Ricinus* (Peuke et al., 1994a). These studies are based on the ratio of C :N in xylem and phloem (Pate et al., 1979) or in the xylem only (Armstrong and Kirkby, 1979; Duarte and Larsson, 1993). The composition of xylem sap is strongly affected by the N source supplied in *Beta* and *Sorghum* (Arnozis and Findenegg, 1986), in *Phaseolus* (Allen et al, 1988) and in *Ricinus* (van Beusichem et al., 1988) and by low N supply in *Nicotiana* (Rufty et al., 1990) and in *Pisum* (Duarte and Larsson, 1993). The present paper aims to examine effects of the N source, of nitrate concentration and of external salt on ion and nutrient element concentrations in the xylem and phloem sap of *Ricinus* and to investigate how concentrations in the transport fluid reflect on the flows of C and N in xylem and phloem in the whole plant.

## Material and methods

### *Growth conditions*

Seeds of *Ricinus communis* were germinated in vermiculite moistened with 0.5 m*M* $CaSO_4$ and transferred after 12 or 13 days to quartz sand culture. The plants were daily supplied with nutrient solutions with different nitrate concentrations, nitrogen source and salt stress, based on Long Ashton solution (Hewitt, 1966):

| Nutrient condition | Nitrate | | | Ammonium |
|---|---|---|---|---|
| [m*M*] | 0.2 | 1.0 | 4.0 | 1.0 |
| compound | [$\mu M$] | | | |
| $K^+$ | 667 | 333 | 1332 | 333 |
| $Na^+$ | 341 | 170 | 382 | 170 |
| $Mg^{2+}$ | 300 | 150 | 600 | 150 |
| $Ca^{2+}$ | 457 | 333 | 1333 | 333 |
| $Cl^{-1}$ | 606 | 30 | 120 | 1100 |
| $SO_4^{2-}$ | 420 | 153 | 610 | 458 |
| $H_2PO_4^-$ | 300 | 150 | 300 | 150 |

Trace elements were added according to the Long Ashton solution, and the salt treatments were 40 m*M* with 1 m*M* $NO_3^-$ or $NH_4^+$. To prevent conversion of ammonium to nitrate the nitrification inhibitor dicyandiamide ($C_2H_4N_4$) was added (89.2 $\mu M$). Plants were grown in a greenhouse (15 – 25 °C; 45 – 70% relative humidity) with a supplementary illumination for 16h with Osram HQL 400 light that provided 300 – 500 $\mu$mol photons $m^{-2}$ $s^{-1}$.

*Harvesting and analyses*

At 41 and 51 days after sowing seven plants of each treatment (nine plants for 0.2 m*M* $NO_3^-$ and $NH_4^+$ +NaCl) were harvested and divided into roots and shoots. At each harvest in the late morning (10 to 12 a.m.) xylem was collected as root pressure exudate and phloem bleeding sap was obtained from incisions into the hypocotyl (Pate et al., 1979; Peuke and Jeschke, 1993). Additionally, during the experimental period xylem sap was obtained also by applying pneumatic pressure to the root system. Concentrations of elements and ions in the transport fluids were determined and flows of C and N were modelled (Peuke and Jeschke, 1993; Peuke et al., 1994a,b) according to the methods of Pate et al. (1979) and Jeschke et al. (1985) using increments of C and N in root and shoot and root respiration, each over the study period of 10 days, and the C:N ratios estimated xylem sap (averages of root pressure exudate and pressure-induced sap) and in phloem sap.

## Results

Growth of *Ricinus* was strongly influenced by the applied nutritional conditions. The mean fresh weight during the study period (41–51 days after sowing) increased strongly with nitrate supply (0.2 m*M*: 16.9 g to 4.0 m*M*: 99.0 g) and was not markedly influenced by salt treatment in nitrate-fed plants (27.7 g (40 m*M* NaCl) compared to 24.1 g (1.0 m*M* nitrate control)), but was substantially reduced in presence of ammonium and salt (7.0 g (40 m*M* NaCl) compared to 8.4 g (1.0 m*M* ammonium control)) (Peuke and Jeschke, 1993; Peuke et al., 1994a).

*Element and ion concentration in xylem sap*

Ion concentrations expressed in meq $L^{-1}$ in root pressure xylem sap were always higher than in the supplied nutrient solution with only few exceptions. Concentrations of sodium and potassium in the xylem were influenced more by their concentration in the medium than by the nitrate levels (Fig. 1). However, concentrations of magnesium and calcium increased with external nitrate levels. Nitrate, total N and amino acid N concentrations in the xylem reflected increased nitrate in the external medium, while concentrations of other anions were decreased. It is worth noting that the strongest, seventeen-fold nitrate concentration increase in xylem relative to the medium was to be observed at 0.2 m*M* external compared to a sevenfold elevation at 4.0 m*M* nitrate. At this low nitrate level chloride, phosphate and malate replaced missing quantities of nitrate as an anion in the xylem, replacement of nitrate in ammonium-fed plants was established by chloride and sulphate (Fig. 1). Generally higher ion and amino acid concentrations, as are required for generating a positive root pressure and a resulting flow, were found in xylem sap of salt-treated plants. This increase, however, was not found for nitrate, presumably because of competition by chloride at its high level. Concentrations of sodium and chloride increased 74 and 16 times respectively in xylem exudates of salt-treated plants compared to the controls. But, in this respect it is remarkable that ratios of xylem sap relative to external concentration were 36, 25 and 22 for $K^+$, $Mg^{2+}$ and $Ca^{2+}$, respectively, but only 0.4 for $Na^+$. Similarly these ratios were 40, 28 and 14 for $SO_4^{2-}$, $H_2PO_4^-$ and $NO_3^-$ but only 0.3 for $Cl^-$. This indicates highly specific overall transport processes to occur at particular sites of cortical root cell plasmalemma and xylem vessels, and an exclusion of sodium and chloride.

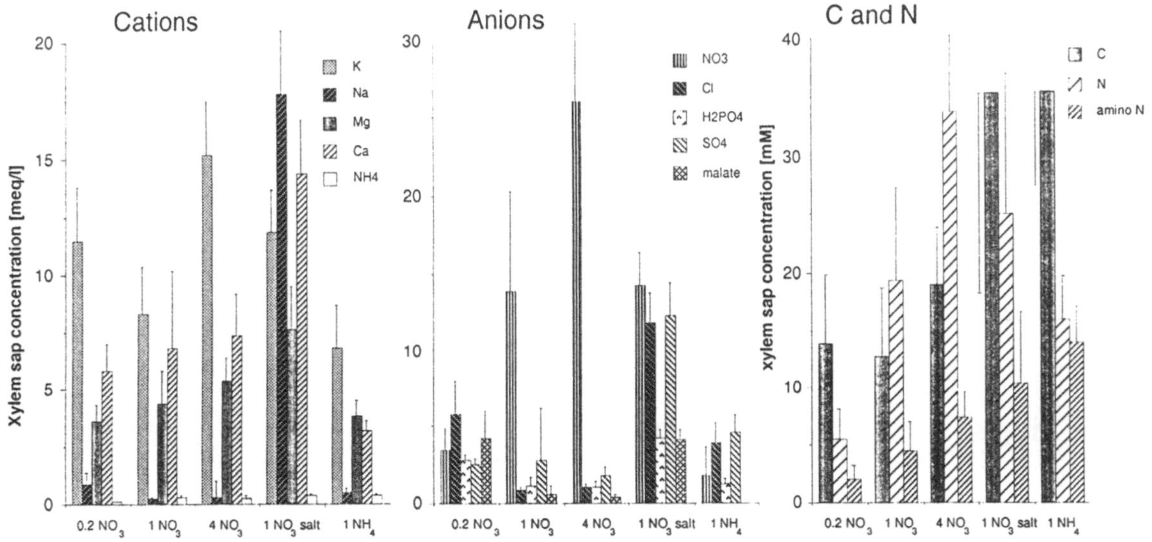


*Fig. 1.* Concentrations in meq $L^{-1}$ of ions and in mmol $L^{-1}$ of C, N and amino acid N (amino N) in root pressure xylem exudates of *Ricinus communis* grown under various nutritional conditions and harvested 41 to 51 days after sowing. The abscissa gives nutrient concentrations, e.g. '0.2 NO3' stands for 0.2 m*M* $NO_3^-$ and 'salt' denotes 40 m*M* NaCl. Error bars indicate standard deviation. Salt-treated, ammonium-fed plants did not produce root pressure exudate. Zero concentrations indicate that a solute was not detected.

In ammonium-fed plants, cation concentrations, with the exception of sodium, decreased slightly. Nitrate decreased to a very low level and, as mentioned before, sulphate and particularly chloride increased in the xylem exudates. Concentrations of amino acid nitrogen and carbon were increased. Ammonium itself could only be detected in very low levels in xylem saps and nutritional conditions had no effect on its concentration. From ammonium-fed plants no root pressure xylem sap was harvested.

*Element and ion concentration in phloem sap*
Besides carbon, mainly as sucrose, and amino acid nitrogen concentrations, also those of some ions, notably potassium, magnesium phosphate and malate, were higher in phloem than in xylem sap (Fig. 2), while those of nitrate and calcium, which are scarcely mobile in the phloem (Hall and Baker, 1972), were lower. Effects of varied external nitrate concentrations on most ions in the phloem exudate were relatively low, but amino acid nitrogen increased substantially in response to external nitrate levels, whereas chloride and malate in the phloem were increased due to the low nitrate supplied in N-deficient plants (Fig. 2). Ammonium nutrition led to somewhat higher carbon (sucrose) and to substantially elevated amino acid nitrogen and chloride concentrations in the phloem, but lowered those of potassium and malate. Sodium, chloride and to some extent also phosphate concentrations in phloem sap were increased due to salt stress (Fig. 2).

*Flows of N*
Most of the N taken up by the roots (60 – 160%) was transported via xylem to the shoot under all nutritional conditions and this flow was augmented due to recycling of phloem-borne N, but to different extents at the varied nutritional regimes (Fig. 3). In plants fed with 1.0 and 4.0 m*M* $NO_3^-$ and 1.0 m*M* $NH_4^+$ uptake rates were between 128 and 174 $\mu$mol $g^{-1}$ fwt/10 d, ammonium nutrition being slightly inhibiting, and the shoots incorporated 70 – 77% of the N taken up. Salt treatment reduced N-uptake by two-thirds in ammonium-fed plants compared to only one-third in nitrate-fed ones and the nitrogen partitioning was shifted in favour of the shoots. This was true also in ammonium-fed plants and if plants were additionally salt-stressed, there was a net transfer of nitrogen from the roots to the shoots. Regarding nitrogen partitioning, N deficiency (0.2 m*M* $NO_3^-$) had the opposite effect compared to salt stress or ammonium nutrition. Even though in these plants 60% of the nitrogen taken up by the roots was translocated first to the shoot via xylem, in the shoot there was a net loss in nitrogen, which was translocated to the root resulting in a higher phloem than xylem stream of nitrogen (Fig. 3).

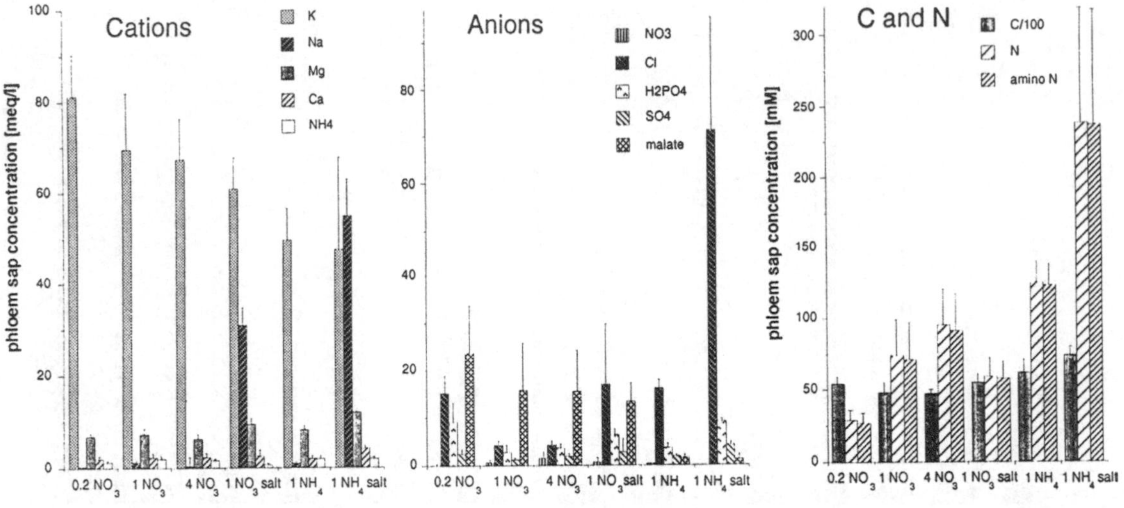


*Fig. 2.* Concentrations in meq $L^{-1}$ of ions and in mmol $L^{-1}$ of C, N and amino acid N (amino N) in phloem sap of *Ricinus communis* grown under various nutritional conditions and harvested 41 to 51 days after sowing. The abscissa gives nutrient concentrations, e.g. '0.2 $NO_3$' stands for 0.2 m*M* $NO_3^-$ and 'salt' denotes 40 m*M* NaCl. Error bars indicate standard deviation. Zero concentrations indicate that a solute was not detected.

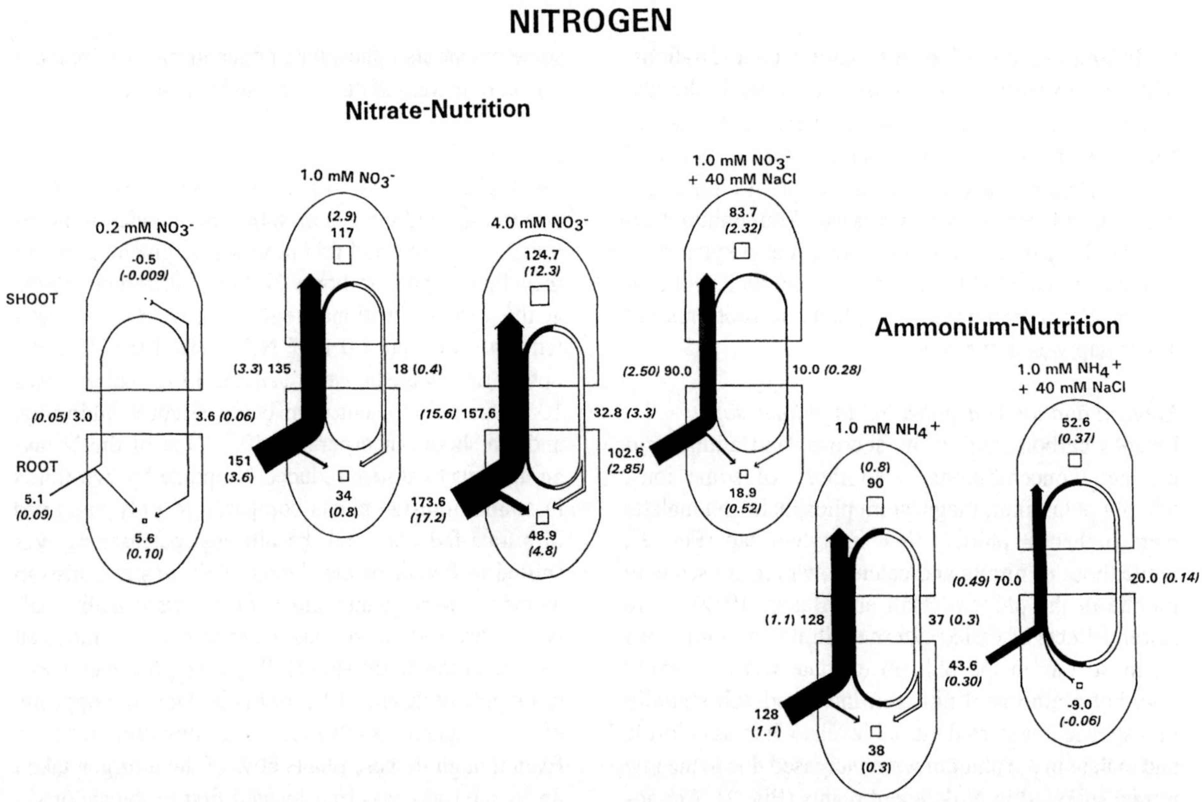


*Fig. 3.* Flow profile for uptake, transport and utilisation of nitrogen by whole plants of *Ricinus communis* L. over the study period 41–51 days after sowing. The plants were cultivated in quartz sand and were supplied with nutrient solutions which contained as nitrogen source either 0.2, 1.0, 4.0 m*M* $NO_3^-$, 1.0 m*M* $NO_3^-$ + 40 m*M* NaCl, 1 m*M*. $NH_4^+$ or 1 m*M* $NH_4^+$ + 40 m*M* NaCl. Widths of arrows (flow in xylem left with filled arrow, in phloem right with open arrow) and area of squares are drawn proportional to uptake, flow and deposition. The numbers indicate $\mu$mol $g^{-1}$ fwt $(10d)^{-1}$, in parenthesis mmol $plant^{-1}$ $(10d)^{-1}$. Data taken from Peuke and Jeschke (1993) and Peuke et al. (1994a).

*Table 1.* Selectivity ratios[1] of $K^+$: $Na^+$ and of $NO_3^-$ :$Cl^-$ in the xylem sap vs. the external solution in *Ricinus communis* under various nutritional conditions

| Condition | $K^+$ : $Na^+$ selectivity ratio[a] | $NO_3^-$ : $Cl^-$ selectivity ratio[a] |
|---|---|---|
| 0.2 m*M* nitrate | 6.8 | 1.8 |
| 1 m*M* nitrate | 17.7 | 0.52 |
| 4 m*M* nitrate | 13.2 | 0.78 |
| 1 m*M* nitrate + 40 m*M* NaCl | 80.7 | 48.1 |
| 1m*M* ammonium | 6.9 | — |

[a]The selectivity ratio was calculated as $([K^+])_{xylem}/([Na^+]/)_{external\ medium}$ or as $([NO_3^-]/Cl^-])/[Cl^-)_{xylem}/([NO_3^-]/[Cl^-])_{external\ medium}$ respectively.

*Flows of C*

Plants given adequate N (1.0 and 4.0 m*M* $NO_3^-$ or 1.0 m*M* $NH_4^+$) showed similar photosynthetic (about 4 mmol C $g^{-1}$ fwt/10d) and dark respiration rates (about 0.47 mmol C $g^{-1}$ fwt/10d) (Fig. 4). Salt stress, and particularly N deficiency, lowered these values. Under salt stress conditions, the inhibiting effects were more pronounced in ammonium-fed plants than in nitrate-fed ones. Remarkably, root respiration in ammonium-fed plants was doubled compared to the other treatments (0,82 mmol C $g^{-1}$ fwt/10d). In ammonium-fed and salt-treated plants relatively high recycling of C in the xylem was observed. Most of the carbon was incorporated into shoot tissues with exception of the N-limited plants in which the root was favoured. In those plants consequently the root/shoot ratio was increased (Peuke et al., 1994a).

## Discussion

Cation concentrations, except those of $Na^+$ in the xylem sap of *Ricinus communis* showed similar responses to the varied nutrient conditions, i.e. increases with elevated nitrate levels, including the change from ammonium to nitrate, and increases in response to salinity. In the case of $K^+$, however, also concomitant changes in external $K^+$ level reflected in the xylem sap concentrations. $Na^+$ concentrations were decreased by the above changes but were strongly increased by external salinity. Also, most anions except nitrate showed corresponding changes, i.e. concentrations were decreased with higher nitrate levels but increased in response to salinity.

Although the changes in nutritional conditions applied in the present paper were rather unique and seemingly unrelated, the responses in xylem sap composition revealed some related patterns, which may be attributed to four basic features, i.e. a) maintenance of charge balance, b) osmotic requirements, c) preferential xylem loading or selectivity during uptake of certain ions and d) metabolic responses. Maintenance of cation-anion balance (a) clearly led to the opposing changes in anion (decrease) and cation (increase) concentrations in response to higher nitrate levels and similarly to the change from ammonium to nitrate supply. Osmotic requirements (b), on the other hand, were the basis of the observed increases in anion and cation concentrations where the plant was salt-treated. In this case, additionally, the unchanged nitrate concentration at otherwise increased levels in the xylem required a charge balance by other anions. Preferential loading of ions into the xylem, or selectivity during uptake, (c), can be seen by relating concentrations in the xylem sap to those in the external solution by calculating the K/Na or $NO_3$/Cl selectivity ratios (Table 1). Selective xylem loading with K relative to Na, or with $NO_3$ relative to Cl, is strikingly evident from the respective ratios for the saline treatment, showing an eighty- or a forty-eight-fold preference for K or $NO_3$, respectively, which confirms the superior capacity of *Ricinus* in discriminating in favour of $K^+$ against $Na^+$ (Jeschke and Wolf, 1988a; Jeschke and Pate, 1991a). These ratios point to $K^+$ and $NO_3^-$ as the most favoured cation or anion in xylem transport, but they do not specify at which site selectivity is achieved, either during uptake or during release into the xylem vessels. A preference for $K^+$, i.e. a selectivity ratio greater than unity, is seen also under non-saline conditions and the ratio was also higher with ample than with limiting nitrate supply (Table 1). This observation points to an improved K/Na selectivity in response to sufficient nitrate supply and could be related to increased salinity tolerance in response to higher nitrate levels (Silberbush et al., 1988). A small preference for nitrate under non-saline conditions, on the other hand, was seen only under N-limiting conditions (Table 1). Similar changes in anion concentration due to ammonium nutrition, as found here, have been reported for other species (Allen and Raven, 1987; Allen and Smith, 1986; Arnozis and Findenegg, 1986).

Metabolic responses of the root (d) can be inferred from the xylem sap composition of nitrogenous solutes, since the ratio of reduced to total nitrogen (Table 2) in xylem sap may be taken as a measure of the proportional contribution of the root to nitrate reduction in the plant (Jeschke and Pate, 1991b; Pate, 1973; Smirnoff and Stewart, 1985). Nitrate reduction in the root was clearly decreased with increasing nitrate supply (Table 2) as reported for other species (Andrews, 1986), but remarkably it was increased in the salt-treated plants, while ammonium assimilation in the root approached 90% (see Allen and Raven, 1987; Allen and Smith, 1986; Arnozis and Findenegg, 1986; Van Beusichem et al., 1988). The shift of nitrate reduction towards the root in the presence of salt is remarkable, since this implies that reduction is shifted to the site of stress imposition. Nitrate reduction is accompanied by synthesis of organic acids and specifically malate (Naik and Nicholas, 1986; Raven and Smith, 1976) and it is noteworthy that malate concentrations (Fig. 1) increased in parallel with the extent of nitrate reduction in the root (Table 2). Malate levels were not, however, directly linked the relative rate of nitrate reduction but they increased more than proportionately in the presence of salt and under limiting $NO_3$ nutrition, as observed before (Föster and Jeschke, 1993; Jeschke and Wolf, 1988b). It appears as if substantial secretion of malate into the xylem vessels occurs in response to anion deficits as they occur under limiting $NO_3^-$ supply, or when an osmotically required increase in nitrate transport is inhibited by chloride competition in the presence of salt. Interestingly, no malate was found in xylem sap of ammonium-fed plants (Fig. 1), a finding which contrasts with reports by Allen and Rave (1987) and van Beusichem et al. (1988) but which corresponds with decreases in organic acids under ammonium nutrition (Epstein, 1972; Kirkby and Mengel, 1967) and with the proposed pH regulation in the case of ammonium assimilation in the root (Raven and Smith, 1976).

The ionic composition of phloem sap was less affected by nutritional conditions than was the xylem sap. However, in the presence of salt, when the xylem sap and hence the leaf apoplast contained high concentrations of $Na^+$ and $Cl^-$, these ions were loaded substantially also into the phloem (Fig. 2), as has been observed for $Na^+$ in *Ricinus* (Jeschke and Wolf, 1988a) and particularly in *Lupinus albus* (Jeschke et al., 1986). Chloride in the phloem sap was increased also when nitrate concentrations were low (0.2 m*M* $NO_3^-$ or ammonium nutrition). Interestingly, phosphate responded in parallel with chloride suggesting that conditions of low nitrate and ammonium feeding induced an anion deficit in the phloem, which was replenished by these two anions and, by malate in the presence of nitrate, but not with ammonium nutrition.

*Table 2.* Relative contribution of the root to total nitrate reduction or total ammonium assimilation in the plant as calculated from the ratio of inorganic to total N in the xylem sap

| Condition | Contribution of the root (%) |
|---|---|
| 0.2 m*M* nitrate | 37.1 |
| 1 m*M* nitrate | 24.7 |
| 4 m*M* nitrate | 22.2 |
| 1m*M* nitrate + 40 m*M* NaCl | 43.0 |
| 1 m*M* ammonium | 89.7 |

While C concentrations, mainly of sucrose, in the phloem were very stable, total N responded strongly (Fig. 2). Total and amino acid N in the phloem was very low in N-limited (0.2 m*M* $NO_3$) but high in ammonium-fed plants, indicating clearly that the observed growth reduction under ammonium nutrition (see Results) was not due to a limiting N availability.

Differences in the metabolic activity of the root alluded to above reflected on the flow patterns of C, since N assimilation inflicts a high demand of the root for assimilates as C-skeletons and as an energy source, if it is the site of N-assimilation. Thus, 33% of total photosynthate were translocated to the root at 1.0 m*M* nitrate nutrition and this percentage rose to 67% at 0.2 m*M* nitrate, to 42% in salt-treated plants and to 52% in presence of ammonium, conditions which all instigate assimilation in the root. Concomitantly, root respiration increased from 9.5% (1.0 m*M* nitrate) to 34% (0.2 m*M*), 17% (salt) and to 22% of total net photosynthesis (ammonium). Under more severe salt stress, root respiration of *Ricinus* was even higher (32%) (Jeschke and Pate, 1991a).

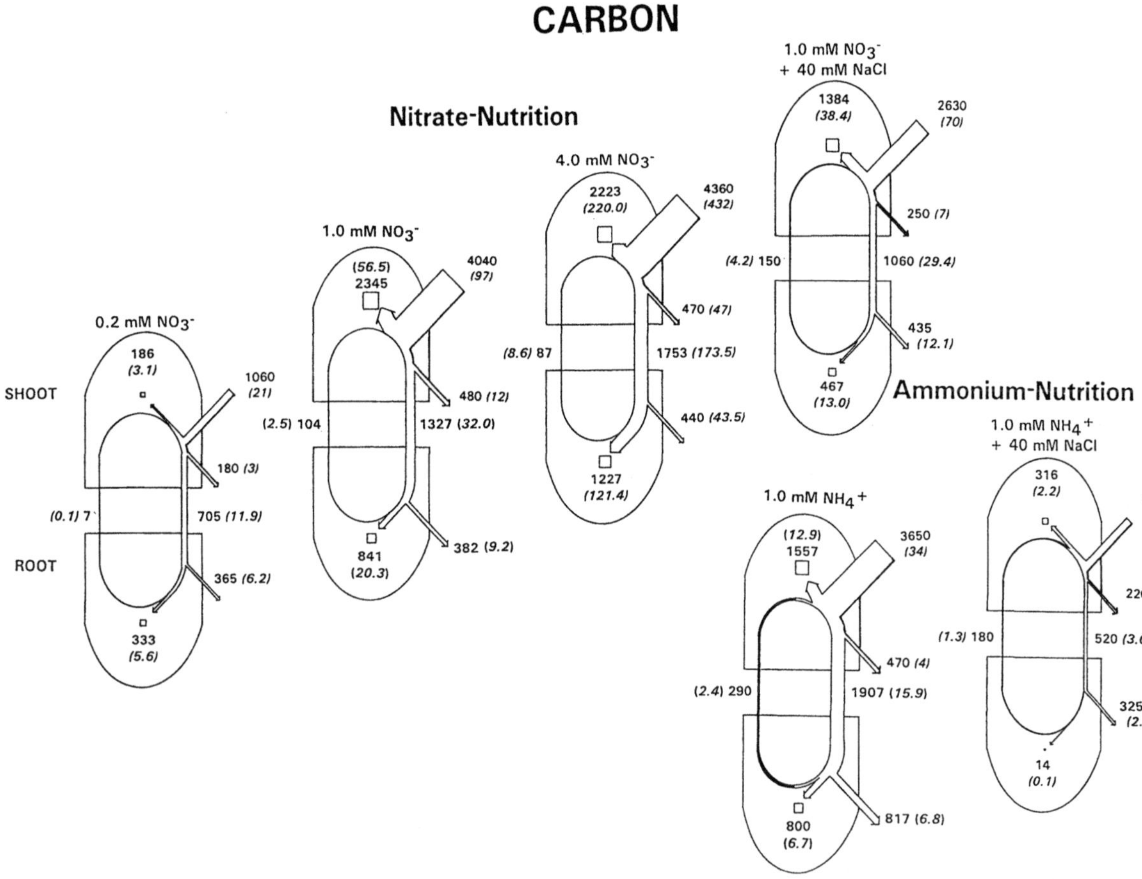


*Fig. 4.* Flow profile for uptake, transport and utilisation of carbon by whole plants of *Ricinus communis* L. over the study period 41–51 days after sowing. For other details, see Figure 1 legend. Data taken from Peuke and Jeschke (1993) and Peuke et al. (1994a).

Turning to the flow patterns of N, most of the N taken up by the root was first translocated to the shoot of *Ricinus* under all nutritional conditions. It would be expected, therefore, that flow rates in the xylem would directly reflect the concentrations of total N in the xylem. This was not always the case, however. For example, the xylem flow was one forty-fifth lower at 0.2 compared to 1.0 m*M* nitrate but N in the xylem decreased to one sixth only. A reason for lower expected translocation can be found in differences in transpiration, which was lower in the N-limited plants than at 1.0 m*M* nitrate. The leaf conductance was strongly decreased by salt treatment (1.0 m*M* nitrate: 353, 1.0 m*M* ammonium: 229 mol $m^{-2}$ $s^{-1}$) and slightly affected in N-limited plants (0.2 m*M*: 423 compared to 666 mol $m^{-2}$ $s^{-1}$ in 1.0 m*M* nitrate) (Peuke et al., 1994b). Another reason appears to be that root pressure xylem exudate increases, relative to the in situ xylem sap, specifically in ions, such as nitrate, readily secreted into the xylem vessels.

Figures 3 and 4 allow a comparative view of C and N flows under greatly varied conditions and one aspect should be noted, i.e. the relative magnitude of C and N streams. While in *Ricinus* supplied with 1.0 or 4.0 m*M* nitrate, C flows in the xylem were lower than those of N, they were higher under N-limiting conditions, in presence of salt or with ammonium nutrition. Higher C than N flows in the xylem accompany nitrogen assimilation in the root and were found in $N_2$-fixing legumes (Pate et al., 1979), in *Ricinus* under saline conditions (Jeschke and Pate, 1991a) and in *Pisum*, a species which predominantly reduces nitrate in its root (Pate, 1973) at low and in contrast to *Ricinus* also at high nitrate supply (Duarte and Larsson, 1993).

**Acknowledgements**

This paper was supported by the Sonderforschungsbereich (SFB) 251 of the Deutsche Forschungsgemeinschaft (DFG). The help of A Hilpert in technical assistance, and Dr Kaiser in anion analysis is acknowledged.

## References

Allen S and Raven J A 1987 Intracellular pH regulation in *Ricinus communis* grown with ammonium or nitrate as N source : The role of long distance transport. J. Exp. Bot. 38, 580–596.

Allen S, Raven J A and Sprent J I 1988 The role of long-distance transport in intracellular pH regulation in *Phaseolus vulgaris* grown with ammonium or nitrate as nitrogen source, or nodulated. J. Exp. Bot. 39, 513–528.

Allen S and Smith J A C 1986 Ammonium nutrition in *Ricinus*: Its effect on plant growth and the chemical composition of the whole plant, xylem and phloem saps. J. Exp. Bot. 37, 1599–1610.

Andrews M 1986 The partitioning of nitrate assimilation between root and shoot of higher plants. Plant Cell Environ. 9, 511–519.

Armstrong M J and Kirkby E A 1979 Estimation of potassium recirculation in tomato plants by comparison of potassium and calcium accumulation in the tops with their fluxes in the xylem stream. Plant Physiol. 63, 1143–1148.

Arnozis P A and Findenegg G R 1986 Electrical charge, balance in the xylem sap of beet and sorghum plants with either $NO_3$ or $NH_4$ nitrogen. J. Plant Physiol. 125, 441–449.

Duarte P J P and Larsson C-M 1993 Translocation of nutrients in N-limited, non-nodulated pea plants. J. Plant Physiol. 141, 182–187.

Epstein E 1972 Mineral nutrition of Plants: Principles and Perspectives. John Wiley, New York.

Förster J C and Jeschke W D 1993 effects of potassium withdrawal on nitrate transport and on the contribution of the root to nitrate reduction in the whole plant. J. Plant Physiol. 141, 322–328.

Hall S M and Baker D A 1972 The chemical composition of *Ricinus* phloem exudate. Planta 106, 131–140.

Hewitt E J 1966 Sand and water culture methods used in the study of plant nutrition. Commonw. Bur. Hortic. Plant Crops G.B. Tech. Commun. 22.

Jeschke W D, Pate J S and Atkins C A 1986 Effects of NaCl salinity on growth, development, ion transport and ion storage in white lupin (*Lupin albus* L. cv. Ultra). J. Plant Physiol. 124, 257–274.

Jeschke W D and Pate J S 1991a Modelling of the uptake, flow and utilization of C, N and $H_2O$ within whole plants of *Ricinus communis* L. based on empirical data. J. Plant Physiol. 137, 488–498.

Jeschke W D and Pate J S 1991b Modelling of the partitioning, assimilation and storage of nitrate within root and shoot organs of castor bean (*Ricinus communis* L.). J. Exp. Bot. 42, 1091-1103.

Jeschke W D and Wolf O 1988a Effect of NaCl salinity on growth, development, ion distribution, and ion translocation in castor bean (*Ricinus communis* L.). J. Plant Physiol. 132, 45–53.

Jeschke W D and Wolf O 1988b External potassium supply is not required for root growth in saline conditions: Experiments with *Ricinus communis* L. grown in a reciprocal split root system. J. Exp. Bot. 39, 1149–1167.

Jeschke W D, Atkins C A and Pate J S 1985 Ion circulation via phloem and xylem between root and shoot of nodulated white lupin. J. Plant Physiol. 117, 319–330.

Kirkby E A and Mengel M 1967 Ionic balance in different tissues of the tomato plant in relation to nitrate, urea, or ammonium nutrition. Plant Physiol. 42, 6–14.

Naik M S and Nicholas D J D 1986 Malate metabolism and its relation to nitrate assimilation in plants. Phytochemistry 25, 571–576.

Pate J S 1973 Uptake, assimilation and transport of nitrogen compounds by plants. Soil Biol. Biochem. 5, 109–119.

Pate J S, Layzell D B and McNeil D L 1979 Modeling the transport and utilization of carbon and nitrogen in a nodulated legume. Plant Physiol. 63, 730–737.

Peuke A D and Jeschke W D 1993 The uptake and flow of C, N and ions between roots and shoots in *Ricinus communis* L. I. Grown with ammonium or, nitrate as nitrogen, source. J. Exp. Bot. 44, 1167–1176.

Peuke A D, Hartung W and Jeschke W D 1994a The uptake and flow of C, N and, ions between roots and, shoots in *Ricinus communis* L. II. Grown with low and high nitrate supply. J. Exp. Bot. 45 (*In press*).

Peuke A D, Jeschke W D and Hartung W 1994b The uptake and flow of C, N and ions between roots and, shoots in *Ricinus communis* L. III. Long distance transport of abscisic acid depending on nitrogen nutrition and salt stress. J. Exp. Bot. 45 (*In press*).

Raven J A and Smith F A 1976 Nitrogen assimilation and transport in vascular land plants in relation to intracellular pH regulation. New Phytol. 76, 415–431.

Rufty T W Jr, MacKown C T and Volk R J 1990 Alterations in nitrogen assimilation and partioning in nitrogen stressed plants. Physiol. Plant. 79, 85–95.21

Silberbush M, Golan-Goldhirsh A, Heimer Y, Ben-Asher J and Lips S H 1988 Response of peanuts (*Arachis hypogaea* L.) grown in saline nutrient solution to potassium nitrate. J. Plant Physiol. 132, 229–233.

Smirnoff N and Stewart G R 1985 Nitrate assimilation and translocation by higher plants: Comparative physiology and ecological consequences. Physiol. Plant. 64, 133–140.

Van Beusichem M L, Kirkby E A and Baas R 1988 Influence of nitrate and ammonium on the uptake, assimilation, and distribution of nutrients in *Ricinus communis*. Plant Physiol. 86, 914–921.

*F. Baluška et al. (eds.), Structure and Function of Roots*, 237–241, 1995.


# Abscisic acid in the ear of winter wheat: Transport from the roots, the leaves, or biosynthesis in situ?

V. Borkovec and S. Procházka
*Department of Nuclear Methods and Department of Botany and Plant Physiology, University of Agriculture, Zemědělská 1, 613 00 Brno, Czech Republic*

*Key words:* cytokinins, endogenous level, genotype, kernels, leaf, transport

## Abstract

The role of abscisic acid (ABA) in the formation of yields of winter wheat was studied from the following points of view: transport of $^{14}$C-ABA from the flag leaf to the other parts of plant in the period prior to anthesis, effect of ABA and cytokinins on the transport of $^{14}$C-sucrose to the kernels, and endogenous level of (+)-S-ABA in the flag leaf and in the ear before and after anthesis. Experiments revealed that the transport of $^{14}$C-ABA from the flag leaf is directed to the internode and the ear. The application of ABA inhibited the transport of $^{14}$C-sucrose from the nutrient solution to the developing kernels simultaneously with the number of kernels. The interactions between ABA and cytokinins in the period prior to anthesis could be related to the effects on the respiration system which could consequently influence the transport of assimilates towards sink tissues and play an important role in yield formation. The endogenous level of ABA in the ear corresponded with its level in the flag leaf up to the period of kernel maturation: from this point the level sharply rose in the leaf whilst in kernels it started to decrease. The results indicate that the flag leaf could be a more important source of ABA than the roots and that ABA synthesised there could be easily transported to the developing ear.

## Introduction

Abscisic acid (ABA) is the plant hormone described in textbooks as a substance of an inhibitory nature which is involved in the regulation of many physiological processes — the acceleration of fruit and leaf abscission, dormancy onset, control of long distance transport of minerals, changes in hydraulic conductivity, stomata apperture, being examples of its wide spectrum of effects (Addicott and Van Stevenick, 1983). As is the case with other plant hormones, various sites of its biosynthesis were identified in plants: chloroplasts of mature leaves, roots and according to some findings sinks are also considered to be the site of its biosynthesis. Although transport occurs predominantly in the phloem, xylem sap analyses yielded substantial amounts of ABA too (Hartung and Davies, 1992).

From the extent of literature which is available it is evident that it is necessary to understand how the capacity of ABA synthesis in the shoots relates to the capacity of roots to produce ABA and thereby sense soil drying. It seems possible that the capacity for turgor maintenance in different plant parts may have a crucial influence on the sensitivity of ABA biosynthesis in roots and leaves and thus on whole plant drought response (Hartung and Davies, 1992). Wolf et al. (1990) proposed a model for flows and utilisation of ABA in *Lupinus* where the main source is leaves from which ABA is transported basipetally via phloem to the roots and, after some addition from root biosynthesis, ABA is transported in the xylem acropetally. Jackson (1993) reconciles the opposite statements when explaining the effects of flooding with the hypothesis that most of, if not all, the increases in ABA in the roots, xylem sap or shoot originated in the shoot. It is possible that at least some of the extra ABA in leaves and sap may represent an accumulation of ABA no longer exported in the phloem to active sinks the growth which are suppressed by flooding.

Free ABA content is highest in the developing seeds and it is generally relatively low or even absent at maturity, though in several species appreciable levels of ABA are consigned to the mature seed (Black, 1983). Modifications of ABA accumulation are brought about by environmental and genetic factors from which water deficit and ABA-deficient mutants are used for further experimentation most frequently. Experiments of Haeder and Beringer (1981) and Ober and Setter (1990) confirm that elevated levels of ABA of wheat and maize seeds are brought about by drought. The question of the origin of ABA in seeds could help to resolve the experiments with ABA deficient mutants. Arabidopsis ABA-deficient mutants (aba) reciprocally crossed with the wild type produced homozygous mutant embryos on a mother plant which were able to produce ABA (Karssen et al., 1983). In tomato, both endosperm and embryo are sites of ABA synthesis, as revealed by crossing with the *sitiens* mutant. On the other hand, the genotype of the mother determines the ABA content of the testa, which is of maternal origin. The level of ABA here peaks sooner than that in the other parts of the seed (Groot et al., 1991). There is other evidence, as in maize, for the dual origin of ABA in developing seeds, since in the stress situation referred to above, the higher level of ABA deposited in the seeds during their development arises partially from increased translocation from the leaves (Black, 1992; Ober and Setter, 1990).

The aim of this work was to present data pointing out the possibility that the flag leaf of winter wheat plays the most important role for the export of ABA in the period shortly prior to and after anthesis, and that the role of ABA in this period is in close relationship with cytokinins.

## Material and methods

### *Transport and distribution of $^{14}$C-ABA*

$^{14}$C-ABA (Amersham, U.K., specific radioactivity 947 MBq.mmol $L^{-1}$, radioactive concentration 1,48 KBq $\mu L^{-1}$) was applied on the flag leaf blade of winter wheat plant cv. Košútka pretreated with a 0.5% Tween solution. The plants were sprayed with $N^6$-(*m*-hydroxybenzyl) adenosine (HBA) $10^{-5}$ mol $L^{-1}$. Control plants were sprayed with water only. After a transport period of 48 hours, the plants were divided in to the individual parts (see Fig. 1) and, after drying to constant weight, combusted in a flow of oxygen. Radioactivity was estimated using a PACKARD 2000 CA scintillation counter. Data were corrected automatically for quenching.

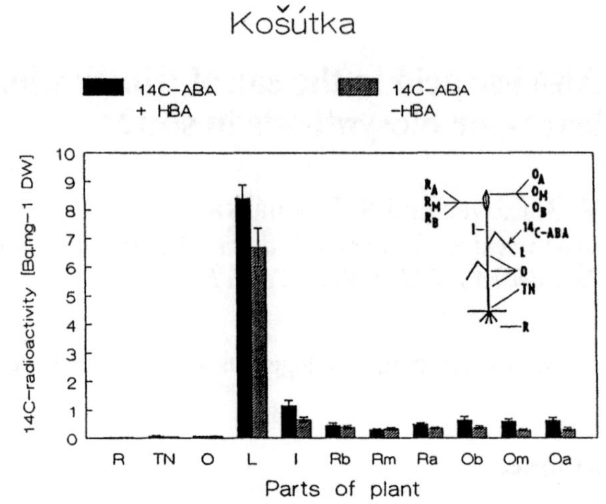


*Fig. 1.* The effect of HBA on the $^{14}$C-ABA transport from flag leaf of cv. Košútka in the period 7 days to anthesis, as measured by its mean concentration in various plant parts. (Explanation of symbols: R-roots, TN-tillering node, O-other vegetative parts bellow flag leaf, L-flag leaf, I-internode, $R_b$,$R_m$,$R_a$, rachis basal, medial and apical part, $O_b$, $O_m$, $O_a$-other parts of the ear — basal, medial and apical section). Bars indicate standard errors.

### *The effect of cytokinins and ABA on the transport of $^{14}$C-sucrose*

The effect of exogenous application of cytokinins (BA $10^{-6}$mol $L^{-1}$, $N^6$-(*m*-hydroxybenzyl) adenosine (HBA) $10^{-6}$ mol $L^{-1}$, or spray of HBA $10^{-5}$mol $L^{-1}$ and their interaction with ABA ($10^{-4}$mol $L^{-1}$) in the winter wheat plants cv. Slavia was studied. For the experiments, field-grown plants in the period 10 days prior to anthesis were used. Ears with the 1st. internode were cultured on nutrient media according to the method of Waters et al. (1984), with $^{14}$C-sucrose as the indicator of assimilate transport. The phytohormones were applied into the nutrient media (except the spray of HBA) for the first 5 days. $^{14}$C-sucrose was used throughout the whole experiment. The radioactivity of the individual parts was determined after another 14 days as described earlier.

### *The level of ABA in flag leaf and ear*

The endogenous level of ABA was determined using a monoclonal antibody raised against (+)-S-ABA (Monoclonal antibody Centre, Cambridge) according to the method of Quarrie et al. (1988). For the experiments, field-grown plants of two winter wheat cultivars differing in the number of kernels (Viginta-average number 18.2 kernels per ear, and Zdar — average number 33.3

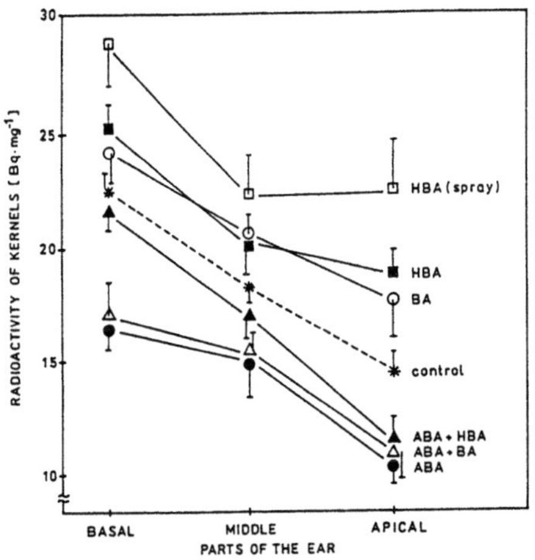


*Fig. 2.* The effect of treatment with different phytohormones 10 days before anthesis on the transport of $^{14}$C-sucrose from nutrient solution to the growing kernels of winter wheat cv. Slavia (means Bq.mg$^{-1}$ of dry matter, ±SE, 14 day transport period).

kernels per ear) were used. The calibration curve and appropriate values of the ABA level were calculated using PACKARD 2000 CA SECURIA programme. Data from the flag leaf and parts of the ear in the period prior to, during, and after anthesis up to the complete maturity of kernels were related to fresh weight.

## Results

### *Transport of $^{14}$C-ABA in winter wheat plants*

Experiments on the transport and distribution of $^{14}$C-ABA after its application to the flag leaf of winter wheat cv. Košútka in the period 7 days prior to anthesis revealed (Fig.1) that within 24 hours after application the radioactivity was distributed solely to the growing internode, rachis and spikelets. To the part of plant below the 1st internode, i.e. the rest of the 2nd and the 3rd internode, leaves, tillering node and roots, only traces of label were transported. The distribution to these parts was not affected by the treatment with HBA, whilst if distributed to the parts situated above the treated leaf a significantly higher accumulation of radioactivity was observed under the effect of HBA.

### *Effect of ABA on assimilate transport*

When ABA was applied at the concentration of $10^{-4}$ mol L$^{-1}$ to the internode of winter wheat cv. Slavia prior to anthesis in the experimental setting where only detached internodes with ear were cultivated on nutrient media, a significant decrease of the transport of $^{14}$C-sucrose to all parts of the ear was observed (Fig. 2). Simultaneously, a reduction in the number of kernels and their mass was observed (data not shown). The effect of HBA ($10^{-6}$mol. L$^{-1}$) gave the opposite result and was connected with the increase of the transport of $^{14}$C-sucrose to the ear, the number of kernels in ear and their mass. Simultaneous application of ABA and HBA resulted in a partial decrease of the inhibitory effect of ABA from all aspects of the study.

### *The endogenous level of abscisic acid*

The analysis of the endogenous level of (+)-S-ABA in the flag leaf in the period prior to anthesis (days 7 to 1) (Figs. 3,4) revealed differences between the two cultivars under study. The highest level was found in the cultivar with the low number of kernels (Viginta) and vice versa. In the ear, similar differences were evident at anthesis. In the weeks following anthesis, the fluctuation of the level of ABA was similar in both cultivars and no significant differences relating to the type of cultivar were detected. The oscillation of the ABA level corresponded in most cases to air humidity with the lower level of ABA being associated with a higher air humidity (data not shown). Thirty-five days after anthesis, when maturation and leaf senescence started, the ABA level in the flag leaf in all cultivars rose several times independently of air humidity and rainfall, and this increase was detected in the glummas and rachis too. In kernels, the level did not follow this tendency and decreased to a very low level. It would mean that, as the transport of assimilate from flag leaf ceases, and assuming that the main bulk of ABA is transported from this source, it ceases in the kernels too and does not increase its level.

## Discussion

The similarity of the distribution of $^{14}$C-ABA with that of $^{14}$C-sucrose suggests that ABA could easily be transported from the flag leaf to the developing sinks in a manner similar to sucrose, i.e. alongside the main stream of assimilates via the phloem. The low amount of radioactivity transported to the roots suggests that the main transport direction will be from the leaves to the 1st internode and ear, i.e. to the parts which are still growing. As the flag leaf represents a powerful source of ABA, it is probable that the main portion of ABA in the ear comes from this source. The HBA

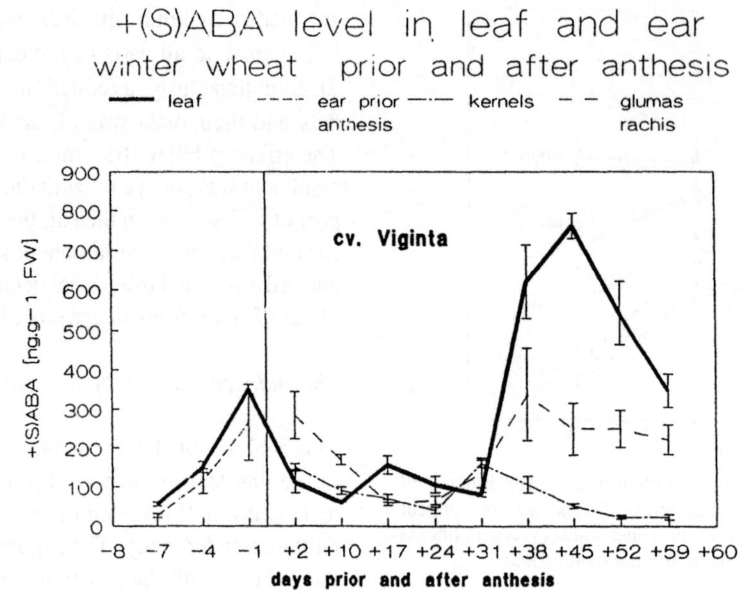


*Fig. 3.* The endogenous level of ABA (means, ±SE) in various parts of the winter wheat plant cv. Viginta in the period prior to and after anthesis.

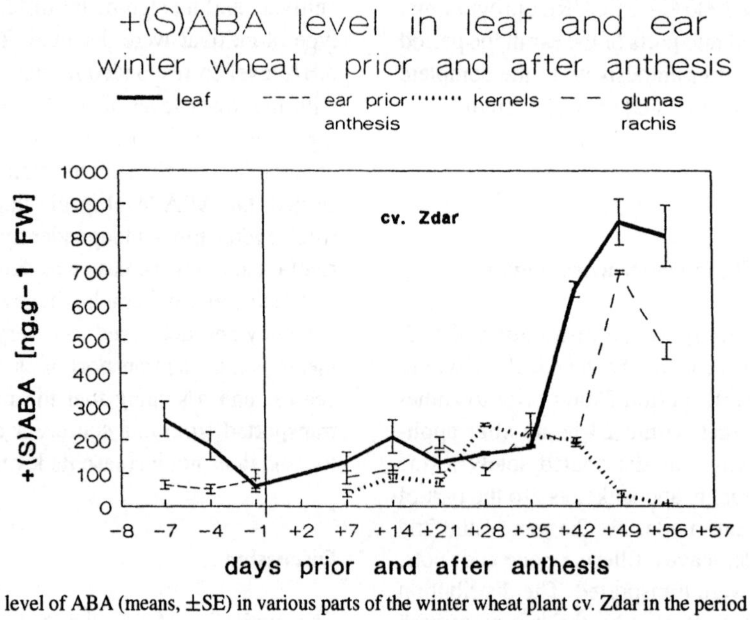


*Fig. 4.* The endogenous level of ABA (means, ±SE) in various parts of the winter wheat plant cv. Zdar in the period prior to and after anthesis.

treatment increased the flow of $^{14}$C-ABA to the ear and internode in this cultivar. Assuming that HBA would affect the endogenous flow of ABA, this experiment could serve as an explanation for the regulatory role of HBA on the flow of endogenous ABA from the leaf to the ear in this cultivar. In spite of the relatively long transport period (2 days) it is reasonable to assume that the labelled substance was $^{14}$C-ABA, as our previous experiments with pea proved that, even after a 24 hour experimental period, more than 50 % of the applied substance was still $^{14}$C-ABA (Borkovec et al., 1984). As the interaction between ABA and cytokinins in the process of opening stomata is well known (Blackman and Davies, 1984), it is probable that, in this experimental arrangement also, the stimulation of the entire transpiration stream could occur.

The transport of $^{14}C$-sucrose to the developing kernels was significantly reduced by the application of ABA in the period prior to anthesis, simultaneously with the reduction of the number of kernels. This inhibitory effect of ABA was explained in the reduction of the flow of assimilate to the developing ear. This reduction probably limited the availability of assimilate for maturing pollen grains too (Waters et al., 1984). The use of $^{14}C$-labelled sucrose supported the idea of a limited transport of sucrose to the kernels. The cytokinins could positively influence the number of endosperm cells in the process of kernel formation. Stimulative effect of HBA on the transport of $^{14}C$-sucrose to the developing ear and reversion of the inhibitory effect of ABA could occur on this basis, as indicated by the higher mass of kernels observed as a result of HBA treatment (data not shown). Another explanation of this process could be that it occurs at the level of transpiration, as ABA significantly reduced the uptake of the labelled nutrient media. A similar interaction was described in tomato (Blackman and Davies, 1984), where cytokinins are considered to be an antagonist of ABA in the closing of stomata.

The results of the transport and endogenous level of ABA support the idea that ABA in the winter wheat plant is synthesized predominantly in the flag leaf, i.e. in the organ which exerts a great influence on the processes occurring in the ear in the period immediately prior to and after anthesis, and probably only a minor part is of root origin. However, as the site of ABA biosynthesis could be any organ of the plant (or part of it) that contains plastids, the interpretation meets difficulties in situations where the main source is for some reason limited or when stress conditions are exerted. The finding of a decisive answer could be helped by studies at a molecular level, e.g. detection of RNAs responsible for ABA biosynthesis directly in the tissue under study.

## References


Addicott F T and Van Stevenick R F M 1993 The significance of abscisic acid in the life of plants. *In* Abscisic Acid. Ed. F T Addicott. pp 581–586. Praeger New York, USA.

Black M 1983 Abscisic acid in seed germination and dormancy. *In* Abscisic Acid. Ed. F T Addicott. pp 331–364. Praeger, New York, USA.

Black M 1992 Involvement of ABA int he physiology of developing and mature seeds. *In* Abscisic Acid: Physiology and Chemistry. Eds. W J Davies and H G Jones. pp 99–124. Bios Scientific Publishers, Oxford, UK.

Blackman P G and Davies W J 1984 Age related changes in stomatal response to cytokinins and abscisic acid. Ann. Bot. 54, 121–125.

Borkovec V, Lehmann H and Procházka S 1984 Influence of auxin-like substances upon the transport of $^{14}C$-ABA in long pea epicotyl segments. part II. Metabolism of $^{14}C$-ABA during its IAA-promoted transport in etiolated segments of pea (*Pisum sativum* L.). Biol. Plant. 26, 358–363.

Groot S P C, van Yperen I I and Karssen C M 1991 Strongly reduced levels of endogenous levels of abscisic acid in developing seeds of tomato mutant *sitiens* do not influence in vitro accumulation of dry matter and storage proteins. Physiol Plant. 81, 73–78.

Haeder H E and Beringer H 1981 Influence of potassium nutrition and water stress on the content of abscisic acid in developing grains and flag leaves of wheat during grain development. J.Sci. Food Agric. 32, 552–556.

Hartung W and Davies W J 1992 Drought induced changes in physiology and ABA. *In* Abscisic Acid: Physiology and Chemistry. Eds. W J Davies and H G Jones. pp 63–80. Bios Scientific Publishers, Oxford, UK.

Jackson M B 1993 Are plant hormones involved in root to shoot communication. Adv. Bot. Res. 19, 104–187.

Karssen C M, Brinkhorst-van der Swan D L C, Breekland A E and Koorneef M 1983 Induction of dormancy during seed development by endogenous abscisic acid: studies on abscisic acid deficient genotypes of *Arabidopsis thaliana* (L.). Heynh. Planta 157, 158–165.

Ober E S and Setter T L 1990 Timing of kernel development in water-stressed maize: water potentials and abscisic acid concentrations. Ann.Bot. 66, 665–672.

Quarrie S A, Whitford P N, Appleford N E J, Wang T L, Cook S K, Henson L E and Loveys B R 1988 A monoclonal antibody to (S)-abscisic acid: its characterisation and use in a radioimmunoassay for measuring abscisic acid in crude extracts of cereal and lupin leaves. Planta 183, 330–339.

Waters S P, Martin P and Lee B T 1984 The influence of sucrose and abscisic acid on the determination of grain number in wheat. J. Exp. Bot. 85, 829–840.

Wolf O, Jeschke W D and Hartung W 1990 Long distance transport of abscisic acid in salt stressed *Lupinus albus* plants. J. Exp. Bot. 41, 593–600.

*F. Baluška et al. (eds.), Structure and Function of Roots,* 243–251, 1995.


# Hormones and root-shoot relationships in flooded plants - an analysis of methods and results

Michael B. Jackson, Paul A. Attwood, Robert W. Brailsford, David Coupland, Mark A. Else, Philippa J. English and Jacky E. Summers
*Department of Agricultural Sciences, University of Bristol, AFRC Institute of Arable Crops Research, Long Ashton Research Station, Bristol BS18 9AF, UK*

*Key words:* environmental stress, ethylene, flooding, 1-aminocyclopropane-1-carboxylic acid (ACC), hormones, root to shoot communication, xylem sap

## Abstract

Two aspects of root to shoot communication in flooded plants are discussed (i) the formation of porous aerenchyma that enhances the passage of oxygen, and other gases, from shoots to roots and (ii) the movement of the ethylene precursor 1-aminocyclopropane-1-carboxylic acid (ACC) from roots to shoots in the transpiration stream, and the effect of this on ethylene production and epinastic curvature in the shoots. For aerenchyma studies a highly sensitive photoacoustic laser detector for ethylene was used to avoid interference associated with other methods of ethylene measurement that require tissue excision. ACC concentrations in xylem sap were measured by physico-chemical means to ensure correct identification and account for processing losses. Solute concentrations, e.g., abscisic acid (ABA), in xylem sap are shown to be distorted by temporary contamination caused by the method used to collect sap. Concentrations of solutes in xylem sap (e.g., ACC) are also altered by changes in sap flow brought about by conventional methods of sap collection or by experimental treatments such as flooding the soil. Ways of for overcoming these problems are described together with a summary of preliminary results.

## Introduction

Plant physiologists are well-aware of the interconnectedness of growth and development in roots and shoots, a relationship founded principally on the division of labour arising from the very different environmental resources and absorbing properties of above- and below-ground parts. Only when water and minerals taken in by roots are exchanged for assimilates obtained photosynthetically by the shoots can the plant function successfully as a whole organism. This competitive nutritional interdependent has been expressed in mathematical terms by several authors (e.g., Davidson, 1969; Hunt, 1975; Thornley, 1972) and underpins the predictability of root:shoot dry weight ratios (Wilson, 1988) and their return to near normal levels if some roots are removed to upset the balance (Fig. 1).

However, it is a mistake to believe that roots and shoots influence each other solely by supplying assimilates, minerals and water. There are other forms of root-shoot communication that also profoundly affect patterns of growth and development. These are especially evident when soil conditions deteriorate sufficiently for shoot and root morphology to adjust, sometimes dras-

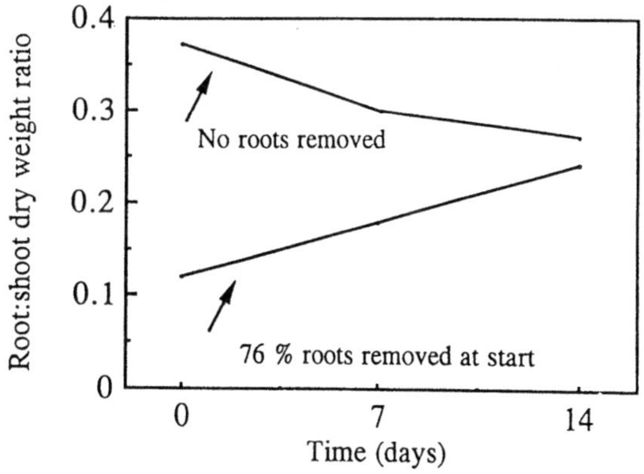


*Fig. 1.* Recovery of the root:shoot dry weight ratio over 14 d following removal of approximately three quarters of the root system from 2-week-old seedlings of barley (*Hordeum vulgare*) grown in solution culture. (M.B. Jackson and J.V. Lake, unpublished).

tically, to the change in circumstances. These developmental events may commence within a few hours and be sustained for days. The location of these responses is sometimes remote from the point of stress perception, implying the action of communicating messages, possibly hormonal. Such messages may be 'positive' (increased movement from stressed to responding parts), 'negative' (decreased movement from stressed to responding parts), 'accumulative' (decreased movement out of the responding parts), or (more improbably) 'debit' messages resulting from increased loss of some physiologically active material from the responding tissues to the stressed parts (Jackson, 1993). The morphological changes these messages induce are of interest not only because they provide an experimental handle on how roots and shoots communicate but because, in themselves, these modifications are of adaptive significance for the plants.

Two aspects of communication between roots and shoots in plants subjected to soil flooding discussed here are (i) formation of highly porous tissue (aerenchyma) that facilitates gas exchange between shoots and roots, and (ii) changes in hormone 'output' from roots to shoots that affects epinastic curvature, stomatal closure and other aspects of leaf morphology. Such changes in morphology help decrease the injury that flooding might otherwise cause.

**Aerenchyma formation**

The stems and roots of most wetland plants are preadapted to the flooded soils in which they usually grow, by forming, constitutively, an extensive system of interconnected gas-filled space. This aerenchyma is created either by cell separations resulting from complex patterns of differential cell expansion and division (so-called schizogenous aerenchyma), or by the collapse of files of existing cortical cells (lysigenous aerenchyma). An example of schizogenous aerenchyma can be seen in the shoots of the submersed aquatic weed *Potamogeton pectinatus* (Fig. 2a), while roots of willow (*Salix viminalis*) contain schizogenous aerenchyma in 1 d-old tissue (Fig. 2b) and lysigenous aerenchyma in older parts of the root (Fig. 2c). In wetland plants, aerenchyma in roots and shoots is often connected, allowing ready passage of oxygen from the aerial parts to the roots (Gaynard and Armstrong, 1987). Not surprisingly, therefore, close correlations have been found between the maximum depth of rooting in flooded soil and the extent of aerenchyma in the roots (Justin and Armstrong, 1987). Only in *Trapa natans* (Menegus et al., 1992) can roots survive or grow for long without oxygen. Roots of all other species so far documented, including those with anoxia tolerant shoots such as rice (Vartapetian et al., 1977), require oxygen from the soil, or from aerenchyma, for their survival.

Marsh or riparian species are not the only ones to form aerenchyma. Maize (*Zea mays*) can also develop highly aerenchymatous roots and leaf bases, and does so as a response to poorly aerated environments. This aerenchyma is lysigenous, and, in the roots, is generated by a collapse of certain cortical cells identifiable by being shorter and wider than more peripheral cells that do not collapse, and by having thinner walls and less ordered cytoplasmic microtubules (Baluška et al., 1993). A loss of vacuolar integrity may precipitate cell collapse (Campbell and Drew, 1983). Little is known of the biochemistry and enzymology of cell wall breakdown that takes place during lysis. Some increases in the levels of cellulase [$\beta$-(1→4) glucanase] have been measured (authors' unpublished results) and this enzyme could enhance wall dissolution by degrading points where cereal-wall glucans contain several contiguous $\beta$-(1→4)-linked glucose molecules.

The promotion of aerenchyma in maize roots by waterlogging the soil, or by stopping the aeration of the roots of solution-grown plants, is mediated by ethylene (Jackson, 1989). This is reminiscent of ethylene-promoted cell collapse in abscission zones of leaves and fruit (Sexton et al., 1985) and of the action of various extra-cellular inducers of apoptosis or programmed cell death in animal tissues (Williams and Smith, 1993). There is some evidence that ethylene also promotes aerenchyma in roots of species other than maize. For examples it promotes aerenchyma formation in poorly aerated roots of willow (*Salix viminalis*) (Attwood, Newman and Jackson, unpublished), *Pinus serotina* (Topa and Mcloed, 1988) and in at least one cultivar of rice (Justin and Armstrong, 1991), but the response is not found in all species where lysigenous aerenchyma develops e.g., lodgepole pine or sitka spruce (Coutts and Philipson, 1987) and the rice cultivar RB3 (Jackson et al., 1985). Unlike some other ethylene-mediated processes, such as inhibition of pea stem extension (Burg and Burg, 1967) or the promotion of fruit ripening (Burg and Burg, 1965), aerenchyma development is not retarded by carbon dioxide (Brailsford and Jackson, unpublished). This indifference to carbon dioxide is an important observation because

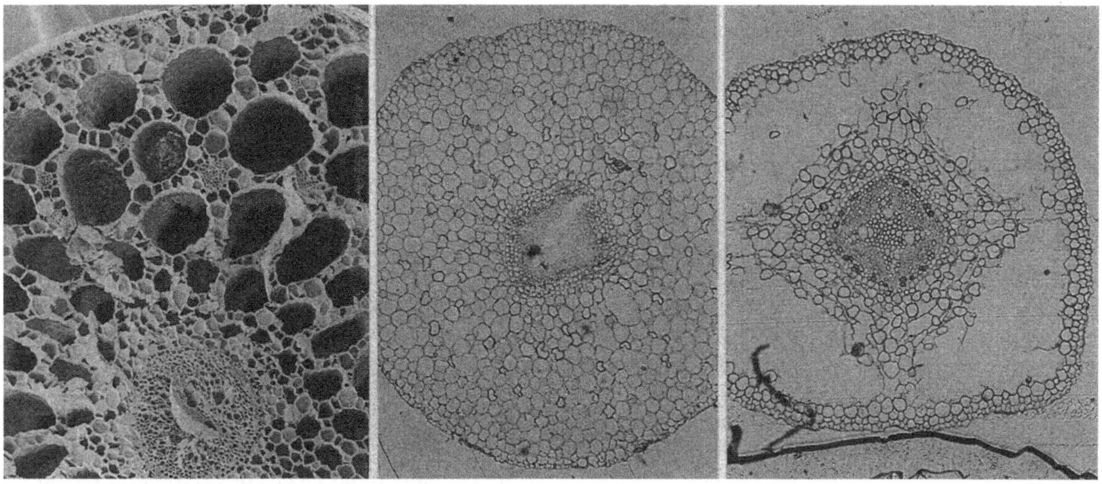

*Fig. 2.* Schizogenous and lysigenous aerenchyma. **Left** - schizogenous aerenchyma in a mature stem of the submersed aquatic dicot *Potamogeton pectinatus*. **Middle** - schizogenous aerenchyma in 1 d-old root tissue of the basket willow, *Salix viminalis*. **Right** - lysigenous aerenchyma in approximately 7 d-old root tissue of willow.

flooded soil may contain equilibrium partial pressures of over 10 kPa carbon dioxide (Jackson, 1979).

One of several lines of evidence implicating ethylene in aerenchyma development in poorly aerated roots of maize has been that partial oxygen shortage stimulates ethylene production (Atwell et al., 1988; Jackson et al., 1985a). Conventional flame ionization gas chromatography used in this work is approximately as sensitive to ethylene as most plants (minimum partial pressure circa 0.001 Pa). Nevertheless, even at this level of sensitivity, excised segments of roots must first be incubated in small, closed vials for at least 1 h to raise ethylene concentrations to easily measurable levels. Such a method raises the prospect of artifactual results created by effects of wounding, cutting and handling or re-orientating the roots, and by gaseous changes in the incubation vials such as oxygen depletion and carbon dioxide enrichment. Many of these problems have been overcome by using a laser-driven photoacoustic detector for ethylene that is over 1000 times more sensitive than flame ionization detectors (Harren et al., 1990). This has enabled ethylene evolution by individual roots of intact seedlings of maize to be monitored in situ, thus keeping perturbation of the roots to a minimum. Ethylene release into a slowly flowing humidified air passing over such roots, decreased over 6 h in response to the small perturbation caused by transferring the seedling to the through-flow chamber (Fig. 3A). Thereafter, ethylene production was very slow and decreased only slightly over the subsequent 20 h. If, 6 h after transfer to the chamber, oxygen partial pressure was decreased to 0 kPa from that in air (20.8 kPa), ethylene production was arrested, and did not recover when roots were returned to air 16 h later (Fig. 3B). This confirmed that oxygen is needed for ethylene biosynthesis in these roots. Decreasing oxygen supply to 3 kPa or 5 kPa had the opposite effect. Instead of depressing ethylene production, rates increased, compared to those in air (Fig. 3C, D). If these hypoxic roots were returned to air after 16 h, ethylene production then fell-back to levels typical of roots kept continuously in air (Fig. 3C, D; Brailsford et al., 1993). Carbon dioxide (10 kPa) did not interfere with this positive response to small oxygen partial pressures. Thus, unlike the behaviour of most other tissues and species reported in the literature, maize roots increase their production of ethylene when partially deficient in oxygen with a time lag of less than 2 h, the response being reversible by air and unaffected by carbon dioxide.

The biochemical basis of the stimulation of ethylene biosynthesis by a restricted oxygen supply is not clear and several possibilities have been put forward. The effect is associated with increased concentrations of ACC (Atwell et al., 1988) and the extra ethylene is derived from this ACC (Jackson et al., 1985). Clearly the amounts of ACC available of oxidation to ethylene are increased in hypoxic maize roots. One possible explanation is that less ACC is sequestered as vacuolar malonyl ACC (MACC) thus releasing more ACC for

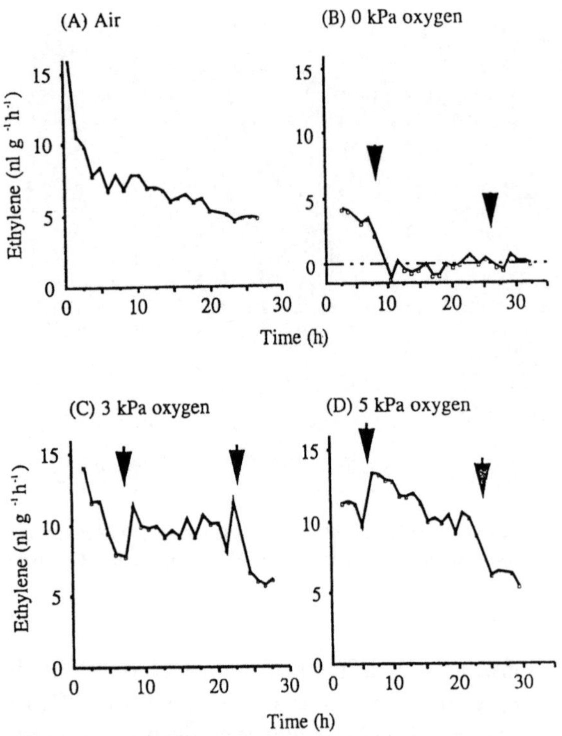


*Fig. 3.* Effect of different oxygen partial pressures on ethylene production by individual primary roots of intact 3 d-old seedlings of *Zea mays*. Arrows indicate the starting and finishing times for the various oxygen treatments.

oxidation to ethylene. However, such an effect does not take place in hypoxic seedlings of rice or *Echinochloa oryzoides* (Pearce et al., 1992). A second suggestion for explaining the increase in ACC production under partial oxygen shortage is that competition for S - adenosylmethionine (the precursor of ACC), by spermidine and spermine biosynthesis is weakened by hypoxia. This hypothesis too has now largely been discounted following analyses showing no marked diminution of these polyamines in hypoxic maize roots (Jackson and Hall, 1993). However, levels of the diamine putrescine are enhanced by oxygen deficiency, and exogenous putrescine increases ethylene production when administered to maize roots. If endogenous putrescine can act in this way, the effect could help explain at least some of the additional ethylene production. A more complex explanation centres on the possibility that small partial pressures of oxygen outside the roots give rise to zones of complete oxygen depletion *within* root axes. Indirect evidence (Bradford and Yang, 1980; Wang and Arteca, 1992) suggests these anoxic zones may then synthesise more ACC than better-aerated tissues as a consequence of increased expression of ACC synthase genes (Zaraembinsky and Theologis, 1993). It is envisaged that the extra ACC produced in anoxic zones diffuses into nearby cells that contain sufficient oxygen to permit conversion to ethylene by the ethylene forming enzyme complex (ACC oxidases). As little as 1 kPa oxygen is needed for ethylene biosynthesis in maize roots (Brailsford et al., 1993; Jackson et al., 1985a) A schematic outline of the spatial requirements of this hypothesis has been published elsewhere (Jackson, 1994). Anoxia is likely to develop in the apex, where rates of respiration and tissue density are particularly high, and the central stele which is surrounded by suberized endodermal cells that are thought to be less permeable to oxygen than other, less thickened cell types. Direct measurements of oxygen within maize roots, using microelectrodes (Armstrong et al., 1993) encourage the idea that cells in the apex and stele are the most likely to become anoxic.

## Hormones and root to shoot communication in flooded plants

The idea that hormones from roots influence shoot development dates back at least to the work of Went (1938) and Chibnall (1939). Many surgical-type experiments, have since been reported which supported this view (e.g., Miginiac and Scotta, 1985; Smith and Wareing, 1964; Went, 1943) but there remains a lack of convincing evidence that marked changes in amounts of hormone moved from roots to shoots in the transpiration stream commonly take place, or that any changes which do take place are of physiological significance. Both problems are interrelated since reliable estimates of the former are required as a basis for testing physiological significance to hormone supply to the shoots (Jackson, 1993). The intention, stated or implied, lying behind much of published work is to estimate how much hormone is present in the transpiration stream as it enters the shoot from the root. Unfortunately, sampling transpiration fluid *directly* is difficult since severing xylem elements to obtain sap destroys the negative pressures that drive transpiration in light-grown plants. Consequently, xylem sap obtained from severed organs (e.g., a de-topped root system) is almost always used. However, there are two principal reasons why measurements in xylem sap obtained in this way may give misleading estimates of what is present in true transpiration fluid. The first is contamination caused by the wounding that inevitably accompanies cutting, and handling plant tissue prior to collecting sap. This

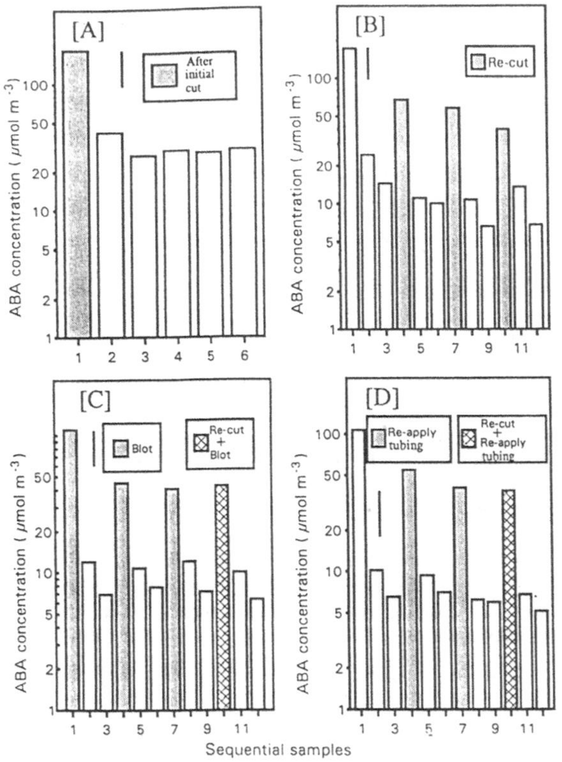


*Fig. 4.* Abscisic acid contamination in xylem sap from de-topped root systems of tomato plants. Effects of **[A]** initial severing of shoot from the root system (first 100 $\mu$L sample taken within 30 s); **[B]** re-cutting the hypocotyl stump three times, each re-cutting being separated by two intervening samplings of 200 $mm^3$ sap; **[C]** three sequential blottings of the cut surface of the hypocotyl without re-cutting, followed by re-cutting and blotting together; **[D]** removing and re-applying sap collection tubing on three occasions, the first two without re-cutting of the hypocotyl, the third in association with re-cutting the hypocotyl. Means of six or seven replicates. Sap samples were 100 $mm^3$ in [A] and 200 $mm^3$ in [C, D and E]). Vertical bars are least significant differences ($p \leq 0.05$). Taken from Else et al., 1994.

can lead to an overestimate of true hormone levels. The second reason is the effect of changes in sap flow, through dilution, on the concentrations measured. The major cause of changed sap flow is the removal of the shoot. Once the shoot is excised, fast transpiration is replaced by relatively slow osmotically-driven sap flow out of the roots. Flow rates also differ between stressed and non-stressed plants both as whole plants and after de-capitation. We have examined these problems with a view to minimising them in our studies of flooded plants.

*Contamination*

Contamination by wounding can be a particular problem if the first few droplets of sap issuing from a de-topped root system are used in the hope they will faithfully represent the contents of the transpiration stream moments before the shoot was excised. We used de-topped root systems of tomato plants placed in a pressure chamber at 0.3 MPa within 30 secs of removing the shoot to study the effects on ABA levels of wounding and other physical perturbations associated with obtaining sap. The first 100 $mm^3$ of sap, obtained within less than 0.5 min, contained much more ABA than in five subsequent samples of 100 $mm^3$ taken over approximately 10 min (Fig. 4A). This suggests that the initial sample may not be unchanged transpiration fluid but, instead, is enriched with ABA as a result of the methods used to obtain sap. This was confirmed by finding that after removing the de-topped root system from the pressure chamber and re-cutting the surface of the hypocotyl stump prior, sap ABA was again temporarily increased. Re-cutting the same hypocotyl three times, at intervals, gave essentially similar results (Fig. 4B). Increases and decreases in air pressure within the pressure chamber associated with the disassembly and re-cutting were found not to be the cause; damaged cells at the cut surface of the hypocotyl being suspected instead. However, blotting the cut surface with filter paper failed to remove the contamination. On the contrary, blotting also appeared to contaminate the sap with ABA, even in the absence of recutting (Fig. 4C). This result eliminated the cut surface as the ABA source. Since re-cutting *and* blotting did not give more ABA than blotting alone (Fig. 4D), the act of cutting the tissue was thereby eliminated as the source of the problem. The application of rubber tubing, needed to collect the sap issuing from the cut stump, remained to explain the ABA contamination. When this possibility was tested by removing and re-applying the tubing several times, but without any re-cutting of plant tissue, the familiar short-lived increases in xylem sap ABA were repeated (Fig. 4D). Gentle radial pressures from the collecting tubing are thus held responsible for contaminating the initial few microlitres of xylem sap with ABA. The effect could be reproduced readily simply by squeezing the hypocotyl slightly between finger and thumb. Our conclusion is that far from giving the most reliable estimates of the concentration of hormone in the transpiration stream, the first 100–200 $mm^3$ of xylem sap are contaminated and will lead to an over estimate of transpiration stream ABA. ACC, sucrose and osmolarity are also affected in a similar way to ABA (Else et al., 1991, 1993).

*Effect of sap flow rate*

Despite the contamination present in the initial 100–200 $mm^3$ of sap there remains a possibility that concentrations in subsequent samples may comprise unchanged transpiration fluid. If this is true, hormone concentrations should not alter if speed of sap flow is changed. In other words, varying the pressure around de-topped root systems to push sap more quickly or more slowly out of the stelar pipework of the hypocotyl and root system should not alter the concentration of a solute such as ABA. However, contrary to this expectation, concentrations were found to be much altered by flow rate. Abscisic acid concentrations in sap flowing at approximately 0.6 $mm^3$ $s^{-1}$ (0.05 MPa pressure) were much higher than in sap flowing at the faster rate of 2.75 $mm^3$ $s^{-1}$ (0.3 MPa). Samples obtained after the initial contaminated ones cannot, therefore, be captured transpiration fluid because concentration was susceptible to change by different rates of sap flow.

The observation that hormone concentration alters with sap flow rate has a significance beyond simply demonstrating the unreliability of measurements in sap from de-topped plants as estimates of concentrations present in the transpiration stream. It has a profound influence on how we should interpret the large body of published data on hormone concentrations measured in sap samples from excised plant organs such as roots. It also suggests alternative and more reliable methods for estimating concentrations in the transpiration stream. These implications have been discussed elsewhere (Jackson, 1991, 1993, 1994) and can be summarised as follows.

(1) - Most published measurements of hormone concentrations in xylem sap are probably overestimates of those of the transpiration stream of intact plants because the sap analyzed usually flowed at rates much slower than transpiration.

(2) - Even when estimates concentrations in the transpiration stream are correct, differences in concentration between plants do NOT demonstrate that roots are exporting more or less hormone, as is often claimed. The results may sometimes only reflect differences in the extent of dilution arising from faster or slower whole plant transpiration.

(3) - Export from roots is more realistically estimated as the amount of solute delivered per unit time; concentration data alone will not provide this information dependably. Delivery rates can be obtained by multiplying concentration in sap by its flow rate. This is the preferred criterion for determining the existence of hormone-based messages passing from roots to shoots.

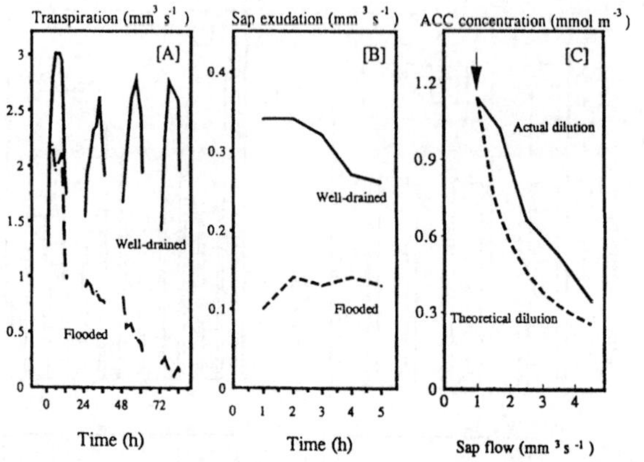


*Fig. 5.* Effects of soil flooding on tomato plants. **[A]** Day-time transpiration over 4 d. **[B]** Rates of osmotically-driven sap flow from de-topped plants after 24 h flooding. **[C]** Concentration of 1-aminocyclopropane-1-carboxylic acid (ACC) in sap induced to flow at an increasing rate by pressurizing the roots of plants flooded for 48 h. Arrow shows the rate of transpiration of intact plants. The dashed line shows the expected dilution if strictly in proportion to flow rate. Means of six or seven plants. (Else, Davies and Jackson, unpublished.)

*Estimating xylem sap concentrations and delivery rates in whole plants*

Credible estimates of concentrations and delivery rates in the transpiration stream can be obtained only after the relationships between sap flow, concentration and delivery have been established experimentally for the species under study and environmental conditions of interest (e.g., flooding, drought, salinity etc.). We have sought to determine these relationships for flooded and non-flooded tomato plants (*Lycopersicon esculentum*) as a necessary step to determining (i) the likely concentrations of the ethylene precursor ACC in the transpiration stream of intact plants, (ii) the amounts actually delivered from the roots. (iii) the timing of any increases in positive ACC message in relation to physiological changes such as epinastic curvature in the shoots.

Decreases in transpiration during soil flooding (Fig. 5A) suggest that hormone concentrations in the transpiration stream will increase as a result of less dilution, even if hormone output from roots remains unchanged. Two other problems are (i) osmotically-driven sap flow issuing from de-topped plants is very much slower than whole plant transpiration, (ii) osmotically driven sap flows more slowly from roots of flooded plants than from roots of well-drained plants (Fig. 5B). To overcome these three inter-related dilution problems, we commenced by examining the changes in concentration of ACC when sap flow rates were increased by pressurizing de-topped root systems. Any problems of chemical specificity of the assay, interference from contaminants, losses during extraction and processing arising from the conventional Lizada and Yang technique of ACC assay (Lizada and Yang, 1979) were overcome by using internal standards, HPLC purification, gas chromatography and a nitrogen/phosphorus detector (Hall et al., 1993). The range of sap flow through the roots was adjusted to include the daytime transpiration rate. Curvilinear relationships were obtained (e.g., Fig. 5C for ACC). These curves serve to emphasise that hormones and related solutes in xylem sap are increasingly diluted as sap flow rate rises. They also help in estimating concentrations present in the transpiration stream of whole plants. This estimation is done simply by reading across from the transpiration rate to intersect the fitted curve (arrow in Fig. 5C), and then down to concentration. ACC output from roots to shoots of whole plants is then obtained by multiplying this concentration by the rate of transpiration.

If the relationship between concentration and sap flow rate had been strictly proportional, the calculation would be valid using concentrations in sap flowing at any convenient rate. However, this convenience was not open to us because the concentration was *not* proportional to sap flow (Fig. 5C). Thus, calculated deliveries will only estimate whole plant deliveries closely if the sap analyzed flows at or close to the rate of transpiration.

We have taken the above approach to study changes in the delivery of ACC from roots of tomato plants during 36 h of soil flooding. ACC delivery into the shoots was found to begin within 6 h and to coincide with the start of petiole epinasty, an ethylene-promoted growth process (Fig. 6). The petioles of tomato plants readily oxidise ACC to ethylene using the ethylene forming enzyme (ACC oxidase). Plants genetically engineered to produce less than normal amounts of ethylene by the insertion of an ACC oxidase gene (pTOM 13 cDNA clone) in reverse orientation ('anti-sense' plants) exhibit less epinastic curvature when flooded even though the delivery of ACC from the flooded roots is similar to that from wild type roots (English et al., 1993). Anti-sense plants are as sensitive to ethylene as the wild type but produce less ethylene from ACC. These results support the view of earlier workers (Bradford and Yang, 1980) that the extent of epinastic curvature in flooded tomato plants is limited by the amount of additional ACC passing into the tissues

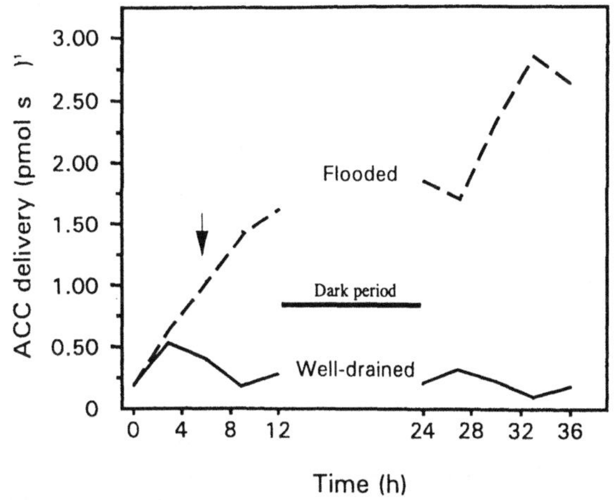


*Fig. 6.* Delivery rates of ACC from roots into the shoots of flooded and well-drained tomato plants over 36 h. The arrow indicates the time at which epinastic curvature commenced. Delivery rates were calculated by multiplying transpiration rate by concentration of ACC in sap induced to flow from de-topped root systems at rates similar to those of whole plant transpiration (Else, Davies and Jackson, unpublished).

from the roots, and by the capacity of the shoots to oxidise this ACC to ethylene. The levels of ACC oxidase activity in petioles have been found to increase in flooded plants. This may be important to secure maximum conversion rates of root-sourced ACC. As expected, enhancement of ACC oxidase activity by flooding was smaller in plants anti-sensed for ACC oxidase (English, Roberts, Lycett and Jackson, unpublished).

## Conclusions

Plant hormones and their precursors are produced in extremely small amounts, making their identification and measurement difficult in small samples. These problems are exacerbated by the susceptibility of biosynthetic pathways to the perturbations inherent in most experimental procedures. Furthermore, once a hormone or precursor is discharged into a transport system, such as the transpiration stream, it is subject to dilution as flow of the carrying solvent adjusts to environmental and experimental circumstances. These problems have been addressed by developing a combination of techniques that include (i) physico-chemical measurement of ACC, (ii) highly sensitive photoacoustic laser detection of ethylene, (iii) harnessing the effect of sap flow on concentration to estimate correctly hormone fluxes from roots to shoots, and (iv) establishing when samples are free of the effects of the wounding which inevitably accompanies even the most careful handling of plants. By these means we have linked enhanced ethylene production in partially oxygen deficient maize roots with aerenchyma development. We also related the timing of increased delivery of ACC from roots of flooded tomato plants to increased ethylene production and epinastic growth by the petioles.

## References


Armstrong W, Cringle S, Brown M and Greenway H 1993 A microelectrode study of oxygen distribution in the roots of intact maize seedlings. *In* Interacting stresses on plants in a changing climate. Eds. M B Jackson and C R Black. pp 287–304. Springer-Verlag, Berlin.

Atwell B J, Drew M C and Jackson M B 1988 The influence of oxygen deficiency on ethylene biosynthesis, 1-aminocyclopropane-1-carboxylic acid levels and aerenchyma formation in roots of *Zea mays*. Physiol. Plant. 72, 15–22.

Baluška F, Brailsford R W, Hauskrecht M, Jackson M B and Barlow P W 1993 Differential morphogenesis in the maize root cortex: involvement of microtubules and phytohormones during post-mitotic cell growth in relation to aerenchyma and other responses to environmental stress. Bot. Acta 106, 394–403.

Bradford K J and Yang S F 1980 Xylem transport of 1-aminocyclopropane-1-carboxylic acid, an ethylene precursor, in waterlogged plants. Plant Physiol. 65, 322–326.

Brailsford R W, Voesenek L A C J, Blom C W P M, Smith A R, Hall M A and Jackson M B 1993 Ethylene production by primary roots of *Zea mays* L. in response to sub-ambient partial pressures of oxygen. Plant, Cell Environ. 16, 1071–1080.

Burg S P and Burg E A 1965 Ethylene action and the ripening of fruits. Science 148, 1190–1196.

Burg S P and Burg E A 1967 Molecular requirements for the biological activity of ethylene. Plant Physiol. 42, 144–152.

Campbell R and Drew M C 1983 Electron microscopy of gas space (aerenchyma) formation in adventitious roots of *Zea mays* L. subjected to oxygen shortage. Planta 157, 350–357.

Chibnall A C 1939 Protein Metabolism in the Plant. Yale University Press, New Haven (p 266 in reprinted edition, 1964).

Coutts M P and Philipson J J 1987 Structure and physiology of sitka spruce roots. Proc. Roy. Soc. Edinburgh 93b, 131–144.

Creelman R A, Bell E and Mullet E 1992 Involvement of a lipoxigenase-like enzyme in abscisic acid biosynthesis. Plant Physiol. 99, 1258–1260.

Davidson R L 1969 Effect of root/leaf temperature differentials on root/shoot ratios of some pasture grasses and clover. Ann. Bot. 33, 561–569.

Else M A, Davies W J and Jackson M B 1991 The delivery of hormones and other solutes from roots to shoots of tomato plants: contamination of xylem sap in early samples from de-topped plants. J. Exp. Bot. 42 (suppl.) p.13.

Else M A, Davies W J, Whitford P N, Hall K C and Jackson M B 1994 Concentrations of abscisic acid and other solutes in xylem sap from root systems of tomato and castor oil are distorted by wounding and variable sap flow rates. J. Exp. Bot. 45, 317–323.

English P J, Lycett G W, Roberts J A and Jackson M B 1993 The use of antisense transgenic tomato plants to study the role of ethylene in responses to waterlogging. *In* Cellular and Molecular Aspects of the Plant Hormone Ethylene. Eds. J C Pech, A Latche and C. Balague. pp 262–262. Kluwer Academic Publishers, Dordrecht, The Netherlands.

Gaynard T J and Armstrong W 1987 Some aspects of internal aeration in amphibious plants. *In* Plant Life in Aquatic Environments. Ed. R M M Crawford. pp 303–312. Blackwell Scientific, Oxford, UK.

Hall K C, Else M A and Jackson M B 1993 Determination of 1-aminocyclopropane-1-carboxylic acid (ACC) in leaf tissue and xylem sap using capillary column gas chromatography and a nitrogen/phosphorus detector. Plant Growth Regul. 13, 225–230.

Harren F J M, Bijnen F C G, Reuss J, Voesenek L A C J and Blom C W P M 1990 Intercavity photoacoustic measurements with $CO_2$ waveguide laser: detection of $C_2H_4$ as trace trace gas at ppt level. Applied Physics B50, 137–144.

Hunt R C 1975 Further observations on root-shoot equilibria in perennial rye grass (*Lolium perenne*). Ann. Bot. 39, 745–755.

Jackson M B 1979 Rapid injury to peas by soil waterlogging. J. Sci. Food Agric. 30, 143–152.

Jackson M B 1989 Regulation of aerenchyma formation in roots and shoots by oxygen and ethylene. *In* Cell Separation in Plants. Physiology, Biochemistry and molecular Biology. Eds. D J Osborne and M B Jackson. pp 263–274. Springer-Verlag, Berlin, Germany.

Jackson M B 1991 Regulation of water relationships in flooded plants by ABA from leaves roots and xylem sap. *In* Abscisic Acid, Physiology and Biochemistry. Eds. W J Davies and H G Jones. pp 217–226. Bios Scientific, Oxford, UK.

Jackson M B 1993 Are plant hormones involved in root to shoot communication? Adv. Bot. Res. 19, 103–187.

Jackson M B 1994 Root to shoot communication in flooded plants. Involvement of abscisic acid, ethylene and 1-aminocyclopropane-1-carboxylic acid in adaptive responses of shoots and roots. Agron. J. (*In press*).

Jackson M B and Hall K C 1993 Polyamine production and action in roots of *Zea mays* L. in relation to aerenchyma development. Ann. Bot. 72, 569–575.

Jackson M B, Fenning T M and Jenkins W 1985 Aerenchyma (gas space) formation in roots of rice (*Oryza sativa* L.) is not controlled by ethylene and or small partial pressures of oxygen. J. Exp. Bot. 36, 1566–1572.

Jackson M B, Young S F and Hall K C 1988 Are roots a source of abscisic acid for the shoots of flooded pea plants? J. Exp. Bot. 36, 1631–1637.

Jackson M B, Fenning T M, Drew M C and Saker L R 1985 Stimulation of ethylene production and gas-space (aerenchyma) formation in adventitious roots of *Zea mays* by small partial pressures of oxygen. Planta 165, 486–492.

Justin S H F W and Armstrong W 1987 The anatomical characteristics of roots and plant responses to soil flooding. New Phytol. 106, 465–495.

Justin S H F W and Armstrong W 1991 Evidence for the involvement of ethene in aerenchyma formation in adventitious roots of rice (*Oryza sativa* L.). New Phytol. 118, 49–62.

Lizada C C and Yang S F 1979 A simple and sensitive assay for 1-aminocyclopropane-1-carboxylic acid. Anal. Biochem. 100, 140–145.

Menegus F, Cattaruzza C, Scaglioni L and Ragg E 1992 Effects of oxygen level on metabolism and development of seedlings of *Trapa natans* and two ecologically related species. Physiol. Plant. 86, 168–172.

Miginiac E and Sotta B 1985 Organ correlations affecting flowering in relation to phytohormones. Biologia Plant. 27, 373–381.

Parry A D and Horgan R 1992 Abscisic acid biosynthesis in roots. 1. The identification of potential abscisic acid precursors and other carotenoids. Planta 187, 185–191.

Pearce D M E, Hall K C and Jackson M B 1992 The separate effects of oxygen shortage and increases in carbon dioxide or ethylene on shoot extension and ethylene biosynthesis in dark-grown seedlings of rice (*Oryza sativa*) and barnyard grass (*Echinochloa oryzoides*). Ann. Bot. 69, 441–447.

Sexton R, Lewis, L N, Trewavas, A J and Kelly P 1985 Ethylene and abscission. *In* Ethylene and Plant Development. Eds. J A Roberts and G A Tucker. pp 173–196. Butterworths, London.

Shih-Ying H and Van Toai T T 1991 Abscisic acid induces anaerobic tolerance in corn plant. Plant Physiol. 97, 593–597.

Smith H and Wareing P F 1964 Gravimorphism in trees. 3. The possible implication of a root factor in the growth and dominance relationships of the shoots. Ann. Bot. 28, 297–309.

Thornley J H M 1972 A balanced quantitative model for root: shoot ratios in vegetative plants. Ann. Bot. 36, 431–441.

Topa M A and Mcloed K W 1988 Promotion of aerenchyma formation in *Pinus serotina* seedlings by ethylene. Can. J. Forest Sci. 18, 276–280.

Vartapetian B B, Andreeva I N, Kozlova I and Agapova I P 1977 Mitochondrial ultrastructure in roots of mesophyte and hydrophyte at anoxia and after glucose feeding. Protoplasma 91, 243–256.

Wang T-W and Arteca J M 1992 Effects of low $O_2$ root stress on ethylene biosynthesis in tomato plants (*Lycopersicon esculentum* Mill. cv Heinz 1350). Plant Physiol. 135, 631–634.

Went F A 1943 Effect of the root system on tomato stem growth. Plant Physiol. 18, 44–55.

Williams G T and Smith C A 1993 Molecular regulation of apoptosis:genetic controls on cell death. Cell 74, 777–780.

Wilson J B 1988 A review of evidence on the controls of root:shoot ratios, in relation to models. Ann. Bot. 61, 433–449.

Zarembinsky T I and Theologis A 1993 Anaerobiosis and plant growth hormone induce two genes encoding 1-aminocyclopropane-1-carboxylate synthase in rice (*Oryza sativa* L.). Mol. Biol. Cell 4, 363–373.

Zeevaart J A D, Rock C D, Fantauzzo F, Heath T G and Gage D A 1991 Metabolism of ABA and its physiological implications. *In* Abscisic Acid, Physiology and Biochemistry. Eds. W J Davies and H G Jones. pp 39–52. Bios Scientific, Oxford, UK.

*F. Baluška et al. (eds.), Structure and Function of Roots*, 253–256, 1995.


# Translocation of nutritional and growth regulating substances in vitro to the developing roots and shoots of cucumber explants

E. Abbas[1], V. Borkovec[2], S. Procházka[1] and L. Havel[1]
[1]*Department of Botany and Plant Physiology and* [2] *Department of Nuclear Methods, University of Agriculture, Zemědělská 1, 613 00 Brno, Czech Republic*

*Key words: Cucumis sativus* L., tissue, plant hormones, target sites, explants uptake, translocation

## Abstract

Hypocotyl and cotyledon explants of cucumber *Cucumis sativus* L. cv. Regina were subcultured on MS-medium supplemented with labelled and non-labelled nutrients and plant hormones /$^{14}$C-sucrose, $^{14}$C-napthaleneacetic acid (NAA), $^{14}$C-benzyladenine (BA), $^{3}$H-abscisic acid (ABA)/ for 21 or 28 days followed by two additional days on basal medium /MS-hormone-free/ with the aim of following their distribution in the developing organs.The results indicate that, although the regenerated plant is an isolated system in which the source-sink is developing, the roots are target for both $^{14}$C-NAA and $^{3}$H-ABA. The developing shoots and calluses, on the other hand, are targets for $^{14}$C-sucrose and $^{14}$C-BA. A higher accumulation of $^{14}$C-sucrose, $^{14}$C-NAA, $^{14}$C-BA and $^{3}$H-ABA was observed in the regenerating explants compared to non-regenerating ones. The uptake of $^{14}$C-NAA and sucrose occurred at, a higher rate compared to the uptake of $^{14}$C-BA and $^{3}$H-ABA indicating, that similarly to intact plants, the movement of cytokinine is much less than the transport of sucrose and NAA.

## Introduction

Hypocotyl and cotyledon explants of cucumber (*Cucumis sativus* L.) are a convenient experimental system to study organ formation because they rapidly product organs in the presence of exogenous hormones. Adventitious regeneration direct from explant tissue has been reported a number of times. Custers and Bergervoet (1980) observed that adventitious shoot formation from hypocotyls of cucumber seedlings was possible on medium containing kinetin at 10 mg $L^{-1}$ and IAA at 0.01 mg $L^{-1}$.

Few studies have been done on the uptake and distribution of nutritive and growth regulating substances in plant tissue cultures. Madore (1990) studied the uptake of exogenously supplied $^{14}$C-sucrose into photosynthetic and non-photosynthetic tissues of variegated leaves of *Coleus blumei* Benth. It was observed that, in green tissues, label from exogenous sugars was recovered as raffinoee-series sugars; in white tissues, exogenous sugars were hydrolysed and metabolised to amino acids and organic acids. Smulders et al. (1986) studied auxin transport and metabolism during in vitro flower bud development in tobacco. They reported that polar development as an result of polar transport of auxin was found e.g. in the adventitious growth of roots and buds.

Van der Krieken et al. (1988) studied the effect of BA on metabolism of BA in the early stages of in vitro flower bud formation in tobacco. There was only a small difference in the ratio between BA and its conjugates at inductive and non-inductive BA concentrations.

The aim of the present study was to estimate the uptake and distribution of $^{14}$C-sucrose, $^{14}$C-naphthaleneacetic acid (NAA), $^{14}$C-benzyladenine (BA) and $^{3}$H-abscisic acid (ABA) using hypocotyl and cotyledon explants of cucumber.

## Material and methods

Seeds of cucumber (*Cucumis sativus* L.) cv. Regina were surface sterilized by dipping for 20 second in 70% ethanol, then washed vigorously with 0.2% diconit (dichloroisocyanuric acid, sodium salt) for 20 min.. After six rinses with distilled water, seeds were transferred into culture tubes containing 5mL of solid

MS-medium (Murashige and Skoog, 1962), and germinated at 25°C in the light. After 7 days, hypocotyl or cotyledon explants were subcultured for regeneration on MS- medium containing labelled substances ($^{14}$C-sucrose, $^{14}$C-NAA, $^{14}$C-BA or $^{3}$H-ABA) for 21 or 28 days followed by 2 additional days on basal medium (MS- hormone-free). In the basic arrangement there were four explants per culture flask. The incubation was in the light (OSRAM tubes L 36W/20, coolwhite Hellweiss, 3800 lx) at 23°C. After the incubation period the explants were divided into the developing parts, callus, root, shoot and the rest of the explants, dried to a constant weight at 50°C. Samples of the individual explants were combusted in a flow of oxygen according to the method Kala and Peśka (1980) and the radioactivity was determined by the liquid scintillation method using a PACKARD CA 2000 spectrometer with automatic correction for quenching.

The following labelled substances were used: $^{14}$C-sucrose (Institute of Radioisotopes, Prague, Czech Rep., specific activity 12 G Bq.mmol$^{-1}$, radioactive concentration 800 KBq. L$^{-1}$). $^{14}$C-NAA (Amersham International PLC, U.K., specific activity 2.02 GBq. mmol$^{-1}$, radioactive concentration 42.85 KBq. L$^{-1}$). $^{14}$C-BA (Amersham International PLC, U.K., specific activity 2.11 GBq.mmol$^{-1}$, radioactive concentration 27.7 KBq. L$^{-1}$). $^{3}$H-ABA (Amersham International PLC, U.K., specific activity 2.55 TBq. mmol$^{-1}$, radioactive concentration 92.5 KBq. L$^{-1}$).

## Results

The results presented in Table 1 show that cucumber explants exposed to inductive medium for 21 days induced the formation of callus, root and shoot on. 70%, 80% and 40% of hypocotyl explants and on 65%, 80% and 45% of the cotyledon explant H, respectively. After a 28 day exposure, 70%, 85% and 40% of the hypocotyl explants responded by producing callus, root and shoot, respectively. The cotyledon explants induced the production of callus, root and shoot on 75%, 80% and 55%, respectively.

Figures 1A and B reveal that $^{14}$C-sucrose was transported to all the developing parts of the regenerating explant. The radioactivity was markedly higher in the developing shoot of the explants than in the other parts, (i.e. developing callus and roots). On the other hand, in both non-regenerating hypocotyl and cotyledon explants the radioactivity was found to be the low-

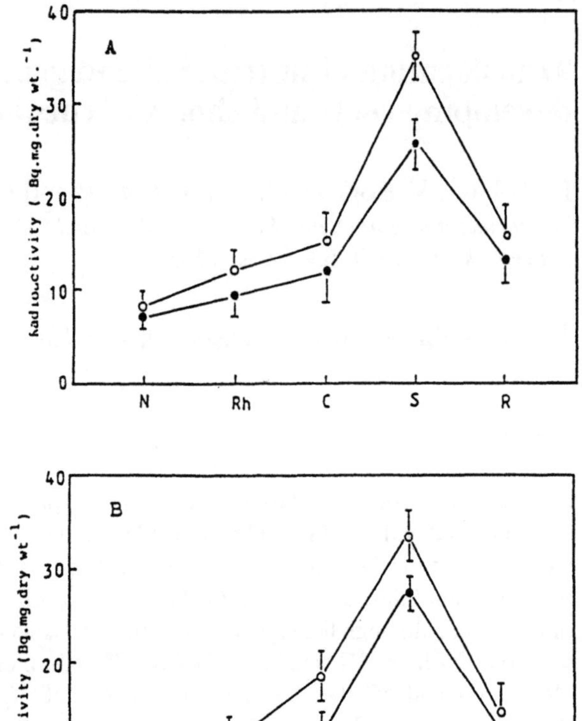


*Fig. 1.* Uptake and distribution of $^{14}$C-sucrose in cucumber hypocotyl (**A**) and cotyledon (**B**) explants, (**C**) uptake and distribution of $^{14}$C-NAA in cucumber hypocotyl explants. (Explanation of the symbols: N - parts of non-regenerating explants, Rh - parts of regenerating hypocotyl explants, Rc-parts of developing cotyledon explants, C - parts of developing callus, S - parts of developing shoots, R - parts of developing roots. ● transport period 21 days, o transport period 28 days).

*Table 1.* Induction of callusing and organogenesis from hypocotyl and cotyledon of cucumber explants on MS-medium containing IAA (2 mg $L^{-1}$); kinetin (2 mg $L^{-1}$) and BA (1 mg $L^{-1}$) plus $^{14}$C-sucrose

| Treatment (Days) | Percent cultures response | | |
|---|---|---|---|
| | Callus Production % | Root Production | Shoot Production |
| Hypocotyl | 70 | 80 | 40 |
| 21 | | | |
| Cotyledon | 65 | 80 | 45 |
| Hypocotyl | 70 | 85 | 40 |
| 28 | | | |
| Cotyledon | 75 | 80 | 55 |

Each value represents the results of twenty explants.

est. The same distribution pattern was maintained in both hypocotyl and cotyledon explants.

In the case of $^{14}$C-NAA, it was found (Fig. 1C) that the uptake of $^{14}$C-NAA was much lower in non-regenerating explants in contrast to regenerating cotyledon and hypacotyl (data not shown but compare Figure 1A and 1B for sucrose) explants, respectively. In general, it can be said that the transport of labelled NAA was intensive into the developing roots and calluses. In contrast, low values of radioactivity were found in the developing shoots. As regards the uptake and distribution of $^{14}$C-BA, it is clearly evident (Fig. 2A) that radioactivity from $^{14}$C-BA was markedly higher in the regenerating explants than non-regenerating explants. The highest accumulation of BA was found in callus cells and values of radioactivity were lower in shoot and particularly in the root. The results shown in Figure 2B reveal that the developing calluses and roots were the site of the highest accumulation of $^{3}$H-ABA but low values of radioactivity were found in the developing shoots.

In general, the uptake and distribution, of $^{14}$C-sucrose, $^{14}$C-NAA, $^{14}$C-BA and $^{3}$H-ABA were, from the point of view of minimal and maximal values, similar in both hypocotyl and cotyledon explants (data not shown but compare Figure A and B for sucrose).

## Discussion

The results of the translocation of $^{14}$C-sucrose, $^{14}$C-NAA, $^{14}$C-BA, $^{3}$H-ABA in both regenerating hypocotyl and cotyledon explants showed clearly that

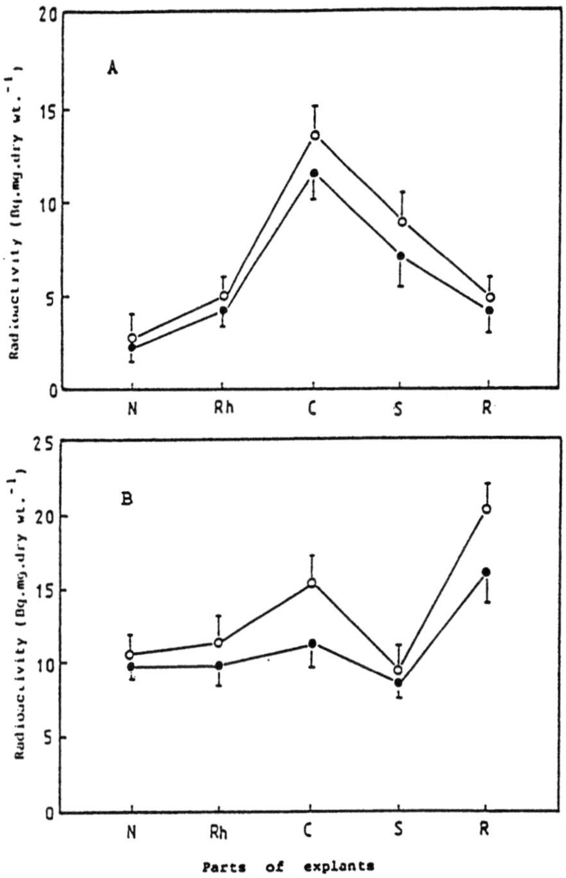


*Fig. 2.* Uptake and distribution of $^{14}$C-BA in cucumber hypocotyl explants (For explanation of symbols see Figure 1).

in the regenerating tissues the uptake of all these substances was significantly higher compared to non-regenerating explants. With this experimental arrangement it is difficult to say whether this observation could be explained as the cause or consequences, but as this increased uptake was observed with all substances it seems to depend predominantly on the level of differentiation of the cultures. In this respect, Rose and Thorpe (1973) and Thorpe and Murashige (1970) observed that starch accumulation occured prior to the formation of meristemoide and shoot primordia and that there was a need for a continuous supply of free sugars from the medium for organagenesis.

When studying the distribution of $^{14}$C-sucrose, it was observed that labelled sucrose accumulated predominantly in the developing shoot of the explants. Higher values of radioactivity of $^{14}$C-NAA and $^{3}$H-ABA were found in the developing roots and the major part of labelled $^{14}$C-BA accumulated in the developing calluses and shoots. It indicates that the transport of these substances to the developing organs of the culture is regulated by different mechanisms probably related to the location of their use (sucrose) or their action in the plant in vivo.

The $^{14}$C-NAA and $^{14}$C-BA used in the study were used in relatively equal specific radioactivities (i.e. GBq) and radioactive concentrations (Bq. $L^{-1}$), $^{14}$C-sucrose was used in a rather higher specific radioactivity and concentration. The specific radioactivity of $^{3}$H-ABA was of a more than three orders higher magnitude (TBq.mmol$^{-1}$) and higher radioactive concentration compared to previous labelled substances. The comparison of uptake of all labelled substances from this point of view suggests that the highest uptake was that of $^{14}$C-NAA, followed by $^{14}$C-BA, $^{14}$C-sucrose and $^{3}$H-ABA.

The uptake of $^{14}$C-NAA seemed thus to occur at a much higher rate compared to other labelled substances (BA, sucrose and ABA). The explanation of this event is probably not only in the above-mentioned differences but also in a high lipid solubility of NAA and pH dependent uptake and thus easier transport in the system. This is similar to observations by Barendse et al. (1986) on thin-layer tissue cultures from flower stalks of tobacco. They observed that the uptake of NAA occurred at a much higher intensity compared to BA (more than 10-fold in the first 24 hours, when both hormones were supplied at equal concentrations).

The results of this study suggest that organ differentiation is followed by selective and active uptake of substances of nutritional and regulating nature and that early in organogenesis the source /medium/ - sink /explant/ system is formed.

## References


Barendse G W M, Croes A F, Krieken U D and Smulders M S M 1986 Metabolism of hormones in thin-layer tissue cultures from flower stalks of tobacco. *In* Proceedings of the International Symposium Conjugated Plant Hormones, Structure, Metabolism and Function. pp 123–127. Gera, GDR.

Custers J N and Bergervoet J H W 1980 In vitro adventitious bud formation on seedling and embryo explante of *Cucumis sativus* L. Cucurbit Genetic Coop. 3, 2–4.

Kala A and Peáka J 1980 *In* Utilizatian of Nuclear Methods and Ionising Radiation in Genetics, Plant breeding and Physiology. pp 167–169. University of Agriculture Brno.

Madore M A 1990 Carbohydrate metabolism in photosynthetic and nonphotosynthetic tissues of variegated leaves of *Coleus blumei* Benth. Plant Physiol. 39, 617–622.

Murashige T and Skoog F 1962 A revised medium for rapid growth and bioassays with tobacco tissue cultures. Physiol. Plant. 19, 473–497.

Ross M K and Thorpe T A 1973 Physiological gradients and shoot initiation in tobacco callus. Plant Cell Physiol. 14, 473–480.

Smulders M J M, Croes A F and Robben A J 1986 Auxin transport and metabolism during in vitro flower bud development in tobacco. *In* Proc Int. Symp. Conjugated Plant Hormones, Structure, Metabolism, Function. pp 87–91. Gera, GDR.

Thorpe T A and Murashige T 1970 Some histochemical changes underlying shoot initiation in tobacco callus cultures. Can. J. Bot. 48, 277–285.

Van der Krieken W M, Croes A F, Barendse G and Wullems G J 1988 Uptake and metabolism of benzyladenine in the early stage of flower bud develapment in vitro in tobacco. Physiol. Plant. 74, 113–118.

*F. Baluška et al. (eds.), Structure and Function of Roots*, 257–267, 1995.


# Root respiration and its importance for the carbon balance of beech saplings (*Fagus sylvatica* L.) in a montane beech forest

Dirk Gansert
*Systematisch-Geobotanisches Institut, Universität Göttingen, Untere Karspüle 2, D–37073 Göttingen, Federal Republic of Germany*

*Key words:* carbon balance, *Fagus sylvatica* L., field experiments, gap, root respiration, sunflecks

## Abstract

Root respiration of 10-year-old beech saplings (*Fagus sylvatica* L.) grown in the understorey (UND) and in a natural gap (GAP) of a mature beech forest in the Solling mountains, FRG, was investigated from April until December, 1990. Respiration rates of fine, medium and coarse roots were measured in situ by a PC-controlled cuvette system. Fine root respiration rates were in the range of 0.5 – 9.8 nmol $CO_2$ $gDW^{-1}$ $s^{-1}$ at both sites, but respiration rates of UND saplings were higher, compared to those of GAP saplings. The dependence of respiratory activity on soil temperature proved to be highly significant ($p < 0.001$) for both plots, following a quasi-Arrhenius type curve. Fine root respiration rates of UND saplings were highly significantly, negatively correlated with the water content of the attached organic material, whereas respiration rates of GAP saplings did not show such a correlation. Further, a significant correlation ($p < 0.01$) between mycorrhizal biomass and respiration rate was detected at the UND site, but not at the GAP site. Medium and coarse root respiration rates were very similar and no significant differences between the two sites were detected. Maximum respiration rates of 3.1 nmol $CO_2$ $gDW^{-1}$ $s^{-1}$ were reached in the middle of July. Due to low light intensities in the under storey, daily net $CO_2$ assimilation rates of UND saplings were much smaller than those of GAP saplings. At both sites, net $CO_2$ assimilation rates varied more than respiration rates and thus the carbon balance of beech saplings was more affected by the rate of carbon fixation than by the rate of respiratory carbon loss.

## Introduction

Plant growth beneath the closed canopy of a mature deciduous forest is severely limited by light intensities which are only a fraction of full sunlight (Chazdon, 1986, 1988; Endler, 1993; Poulson and Platt, 1989). In a mature beech forest, less than 5% of the radiation above the canopy is available for assimilation by the herb layer (Eber, 1972; Ehrhardt, 1988; Schulte, 1993). Consequently, successful competition for light requires a highly efficient conversion of light into biomass, i. e. a high ratio of carbon gain to carbon loss for the plant as a whole. Studies on the aboveground growth patterns of shade-tolerant tree species showed large variations in height and lateral growth, leaf exposure, leaf area index (LAI) and other phenological parameters, when light intensity was altered by an opening in the canopy (Canham, 1988, 1989; Pearcy, 1983). Saplings of European beech (*Fagus sylvatica* L.) growing in the understorey (UND) show a clear suppression of lateral shoots and the main shoot grows plagiotrop, whereas saplings of the same age in a natural gap (GAP) distribute larger amounts of biomass into lateral shoots and growth is predominantly orthotrop (Roloff 1986; Schmaltz 1964; Stickan et al., 1992; Suner and Röhrig, 1980).

Photosynthetic gas exchange measurements of beech saplings revealed quantum yields of 0.04 (Stickan and Zhang, 1992) to 0.07 mol $CO_2$ mol $Photon^{-1}$ (Rees, personal communication), which is about twice as high as quantum yields of sun leaves of mature trees (Schulte, 1993; Schulze, 1970).

Further, light saturation of photosynthesis ($I_s$) of shaded beech saplings was measured in a range of 170 to 230 $\mu$mol Photons $m^{-2}$ $s^{-1}$ (Stickan and Zhang,

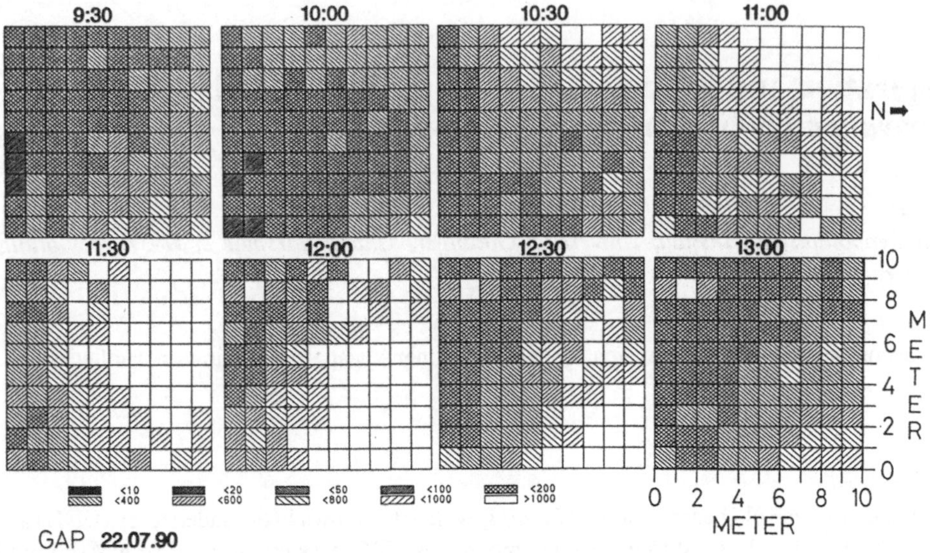


*Fig. 1.* Distribution of PPFD at the GAP site derived from the light measurement field on 22nd of July, 1990. Each square represents a half hour mean value of $\mu$mol Photons $m^{-2}$ $s^{-1}$ calculated from 1800 single measurements. At 11 a.m. a sunfleck occurs in the north-western part of the field, which extends and moves southward. At 1 p.m. the sunfleck has vanished.

1992) which was about one third of $I_s$-values derived from the light-crown of mature trees (Larcher, 1969; Schulte, 1993; Schulze, 1970). Both low $I_s$-values and high quantum yields should be considered as adaptive reactions of the photosynthetic apparatus to insufficient light intensities (Boardman, 1977; Bolhar-Nordenkampf, 1985).

Hence morphological and physiological features of the aboveground organs of beech saplings ensure an efficient use of low light intensities for carbon fixation.

On the other hand, only limited information is available on the respiratory activity of understorey saplings and studies made so far concentrate exclusively on the investigation of leaf respiration (Masarovicova, 1980, 1988). Data on the respiratory carbon loss of the roots of tree saplings in a low-light climate is lacking, though this factor may contribute considerably to a plant's overall carbon loss.

Based on field experiments done in 1990, data on the respiratory activity of beech sapling roots and its dependence on climatic variables is presented here. Priority has been given to the investigation of fine root respiration rates, due to their high physiological activity in water and nutrient absorption.

Simultaneous measurements of rates of net photosynthesis and root respiration of saplings grown in a natural gap and in the understorey will give information about differences in the C-input : C-output ratio which indicates contrasting carbon economies of shaded and unshaded plant individuals.

## Materials and methods

### *Study area*

The study area is situated in the Solling mountains, Lower Saxony, Germany (51°46'N, 9°35'E) at 500 m above sea level. The annual mean temperature is 6.4°C and the annual sum of precipitation is about 1048 mm. The soil type is Typic Dystrochrept with a humus moder type and mean pH($H_2O$)-values ranging from 3.8 to 4.2 (Ellenberg et al., 1986).

Two areas, each 600 $m^2$ were chosen in a 140 to 150-year-old natural beech forest. The density of trees is approximately 240 $ha^{-1}$ with an average height of 30 m.

One plot (UND) was chosen beneath a closed canopy, the other plot was established within a natural gap (GAP) created by a storm in 1972. The average number of seedlings plus saplings at the UND site was only 1 $m^{-2}$, whereas the density at the GAP site was about 7 $m^{-2}$. As the subject of investigation, ten-year-old saplings were chosen which had reached an average height of 70 cm (GAP) and 35 cm (UND), respectively.

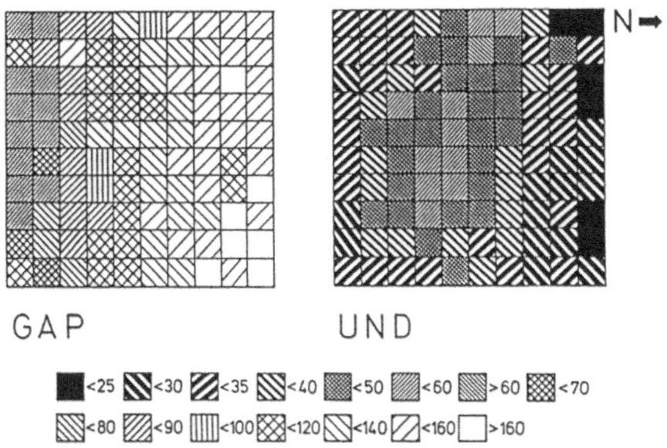


*Fig. 2.* Daily mean value of PPFD ($\mu$mol Photons $m^{-2}$ $s^{-1}$) at the GAP and the UND site on 22nd of July, 1990. At the UND site PPFD did not exceed 70 $\mu$mol Photons $m^{-2}$ $s^{-1}$ but at the GAP site more than 160 $\mu$mol Photons $m^{-2}$ $s^{-1}$ were reached.

## *Methods*

### *a) Light measurement system*

To monitor the considerable spatial heterogeneity of the photosynthetic photon flux density (PPFD) a 'light measurement field' was installed at the centre of each plot. Each field consisted of 100 self constructed quantum sensors which were distributed regularly over an area of 10 by 10 m. The signal from each sensor was stored every second by a Pocket Computer (SHARP). Half hour mean values were calculated for each sensor and also stored. This procedure allowed the continuous registration of varying light conditions with special regard to sunflecks over a 3.5 day period (Stickan and Hentzelt, 1991).

Using the mean values as a data base for the creation of light plots by applying Statistical Analysis Software (SAS), the diurnal course of PPFD at each site could be mapped. Figure 1 gives an example of a drifting sunfleck at the GAP site on 22nd of July. Figure 2 demonstrates the daily mean PPFD at the GAP and the UND site on the same day. As an average of several days with diffuse radiation, about 8% of the above canopy light intensity was transmitted at the UND site and 20% was transmitted at the GAP site during the period of closed canopy.

### *b) Respiration cuvette system*

The respiratory activity of different parts of the root system, i.e. fine ($\phi \leqslant 2$ mm), medium (2 mm $< \phi \leqslant 5$ mm) and coarse roots ($\phi > 5$ mm) was measured continuously by a PC-controlled cuvette system developed by Prof Kakubari at the University of Shizuoka, Japan. Twelve PVC cuvettes were connected in parallel with a PC-controlled value timing unit by PVC tubes ($\phi$ = 5 mm inside) up to 30 m long. The dimensions of a single PVC cuvette were 16.9×10×4.3 cm and it had a volume of 238 $cm^3$. At one narrow side two PVC tubes were inserted for the aeration of the interior. The opposite side was provided with a slit ($\phi$ = 6 mm) through which live fine roots, still attached to the sapling, could be inserted. The slit was sealed hermetically with an inert sealing compound. Additionally, each cuvette was equipped with an NTC-temperature sensor. After the insertion of a sample (fine roots, organic material or mineral soil) the cuvette was hermetically sealed with a thick PVC lid and buried at the depth at which the sample originated from.

The $CO_2$ evolution rate of an enclosed sample was measured by calculating the difference between the $CO_2$ content of fresh air - collected from just above the forest floor at a remote undisturbed place - pumped into the cuvette and the $CO_2$ content of the air leaving the cuvette. The duration of a single $CO_2$ accumulating period was 50 minutes. Previous experiments had shown that the rate of $CO_2$ evolution during this period was neither restricted by excessive $CO_2$ concentrations in the cuvette nor accelerated due to preceding aeration. Experiments were run for 3 days and thus, respiration data of 86 $CO_2$ accumulating periods were recorded. All data necessary for converting $CO_2$ concentration data into mass flow data (i.e. mass flow of air, time of accumulation, time of cuvette measurement) was logged automatically by an NEC Personal Computer.

### *c) Calculation of fine root respiration*

Respiration rates of soil material were related to the dry mass of the samples, but respiration rates of fine roots were exclusively related to the dry weight of biomass of the samples i.e. without the fraction of necromass (see below). Fine root respiration was calculated as the difference between the respiration rate of a soil sample ($O_f$, $A_h$) taken from the immediate vicinity of the selected root and the respiration rate of the root plus attached soil material. For this purpose, the dry weight of the attached soil material had to be determined by treatment in the laboratory (see below).

Most of the investigated fine roots grew in the organic layer and were excavated carefully with a small fork. Care was taken not to tear off distal ramifications so as to avoid wound respiration.

*d) Fine root sample treatment*

Respiration experiments were conducted between April and December, 1990. As a rule, one sample per tree was investigated, but in some cases the respiration rates of three fine root samples of a single tree were measured simultaneously. At the end of an experiment, the enclosed fine root samples were detached with a pair of scissors and treated in the laboratory as follows:

1. The roots were washed under a binocular with a gentle jet of water and rinsed out soil material was collected quantitatively. Subsequently, soil matter was dried at 105°C for 24 hours.
2. Mycorrhiza species were determined following Agerer (1986, 1988) and Agerer et al. (1986) and collected quantitatively. Mycorrhizal necromass was also collected, but without regard to different species. Each fraction was dried in a freeze-drier (Alpha II–12, Christ) for 24 hours and its dry weight determined.
3. Non mycorrhized fine roots were separated into dead and living root fragments, considering the white colour of the central cylinder and its elasticity as live criteria. The texture of the root tissue, its colour and the condition of lateral root tips were used as supplementary criteria (Murach, 1984).

*e) Medium and coarse root respiration*

To measure medium and coarse root respiration, a small section was layed bare and washed carefully with a weak antiseptic solution (Orthocid 83: 5 g $L^{-1}$; Tetracycline: 100 mg $L^{-1}$; Tween 20: 2 mL $L^{-1}$). Subsequently the fraction was hermetically sealed in a cylindrical PVC cuvette (28.3 $cm^3$) and covered entirely with moist soil material. After the measurement period the sections were cut off and dry weight was determined. The percentage of dry weight of live tissue of such a section was calculated from the histochemical investigation of cross sections, applying staining methods for starch (Essiamah and Eschrich, 1985; Neemann and Stickan, 1991).

*f) Photosynthetic measurements*

During the vegetation period, assimilation and respiration rates of leaves of the GAP and UND saplings were investigated by a minicuvette gas exchange system using the IRGA technique (Walz, Effeltrich, FRG). Experiments were carried out with special regard to the influence of sunflecks on the net assimilation rate. For this purpose, sunflecks were simulated using halogen light of high radiant energy (200W) (Rees and Stickan, 1991).

*Table 1.* Biometric data of 10-year-old beech saplings grown in the understorey (UND) and in a natural gap (GAP). Mean values of six saplings are given which were harvested quantitatively at each site. Root systems were excavated with the aid of rinsing water minimizing loss of biomass

| | | Local-site | |
|---|---|---|---|
| | | UND | GAP |
| Leaf area | $dm^2$ | 7.0 | 35.0 |
| Leaf biomass | gDW | 2.6 | 14.4 |
| Mass of main shoot | " | 3.0 | 22.0 |
| Mass of lateral shoots | " | 3.3 | 35.1 |
| Mass of main root | " | 3.8 | 19.9 |
| Mass of lateral roots | " | 0.5 | 11.4 |
| Biomass of fine roots | " | 1.0 | 6.5 |
| Total mass | " | 14.2 | 109.3 |
| Root:Shoot ratio | | 1.7 | 1.9 |

DW = dry weight.

*g) Statistics*

Statistical Analysis Software (SAS) was applied for statistical calculations. The correlation of respiration rates with different independent variables was investigated by nonlinear regression analysis (Sachs, 1992, 489 ff). A student 't' test was used to determine significant ($p < 0.01$) differences between the respiration rates of UND saplings and those of GAP saplings.

## Results

*a) Biometric data*

Table 1 gives the average biometric data of 10-year-old beech saplings grown at the GAP site and in the understorey (UND), respectively. At both sites, live fine roots make up about 20% of the entire root mass. Although total dry weight of GAP saplings is about 8 times higher than that of UND saplings, root : shoot ratios are very similar. The specific leaf area which will be defined here as leaf area per gram plant dry weight ($cm^2$ $gDW^{-1}$) is 50 for UND saplings and 32 for GAP saplings, respectively.

*b) Carbon assimilation*

Light saturation curves of net photosynthesis of leaves from UND saplings corresponded with those of shade leaves from mature trees and reached maximum PPFD values at the end of May (177 $\mu$mol Photons $m^{-2}$ $s^{-1}$). Net photosynthesis of leaves from GAP saplings was light saturated by a PPFD of 258 $\mu$mol Photons $m^{-2}$ $s^{-1}$ (Rees and Stickan, 1991).

Net photosynthetic rates under light saturation ($A_{max}$) of GAP saplings exceeded those of UND saplings during the entire vegetation period and reached their maximum in July (6.7 $\mu$mol $CO_2$ $m^{-2}$ $s^{-1}$). The maximum value for UND saplings was 4.4 $\mu$mol $CO_2$ $m^{-2}$ $s^{-1}$, which was also reached in July (Rees, personal communication).

Experiments on the significance of sunflecks for carbon fixation were carried out at 10 minute intervals under light saturated conditions. It was detected that leaves of UND saplings reached half the maximum net photosynthetic rates after 3 minutes of illumination, whereas leaves of GAP saplings reached half the maximum values in half the time. About 29% of the net assimilation rate of GAP saplings under limited PPFD (20–50 $\mu$mol Photons $m^{-2}$ $s^{-1}$) was due to the occurence of an artificial sunfleck i. e. when light saturation was generated for 10 minutes. For UND saplings, this fraction was only 10%. Thus, leaves of UND saplings showed no clear disposition to the efficient use of sunflecks. It seems more likely that a profound adaptation of the photosynthetic apparatus to continuously low light intensities is more important for UND saplings than short-term reactions to sunflecks (Rees and Stickan, 1991).

*c) Fine root respiration*

Figure 3 gives a typical graph of the respiration rate of a fine root sample at the GAP and UND site over three days at the end of May, 1990. The mean rate was 3.3 ± 1.7 nmol $CO_2$ $gDW^{-1}$ $s^{-1}$ for the former and 6. 2 ± 0. 9 nmol $CO_2$ $gDW^{-1}$ $s^{-1}$ for the latter, whereas mean values of temperature were very similar (GAP: 10.8 ± 1.3 °C, UND : 10.1 ± 1.3 °C). The respiration rates corresponded with the seasonal temperature curves which indicates a strong dependence of respiration on temperature.

This relationship could best be described by a quasi-Arrhenius type curve (Janecek et al., 1989): $r(T) = \exp(a \cdot (T-b))$ with r = respiration rate (mgC $gDW^{-1}$ $d^{-1}$) and T = temperature (°C). The parameters ‘a’ and ‘b’ were calculated by applying nonlinear regression analysis of the Statistical Analysis Software (SAS). Based on the respiration data throughout the year, correlation coefficients of 0.91 (n=24) for the UND site and 0.82 (n=20) for the GAP site were found ($p < 0.001$) (Fig. 4a,b). For the vegetation period from May until October correlation coefficients were 0.78 (n=19) for the former and 0.72 (n=16) for the latter ($p < 0.001$).

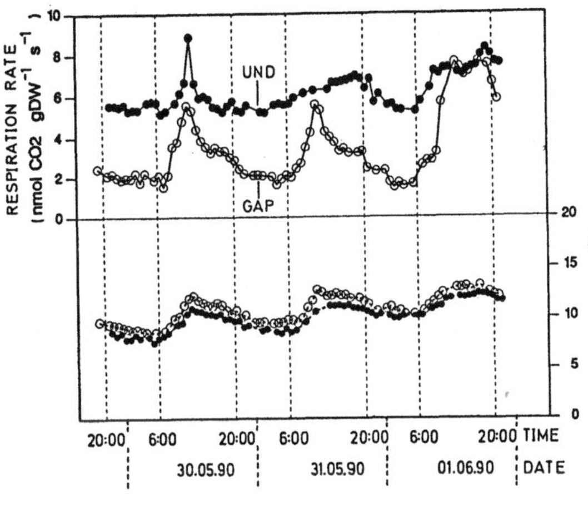


*Fig. 3.* Respiration rates of two fine roots at the UND and at the GAP site over a three-day period at the end of May. Both graphs coincided with the temperature graphs (UND: n=70, r=0.82; GAP: n=73, r=0.85).

In Figure 5, a comparison is made between the respiration rates of UND and GAP fine roots from April until December. Despite nearly identical temperature curves at both sites from April until July, the respiration rates at the UND site were higher than those at the GAP site.

During the vegetation period, a highly significant negative correlation of root respiration and water content of the attached organic material was detected at the UND site (Fig. 6a), whereas at the GAP site no such correlation was found (Fig. 6b).

Beside temperature and water content as climatic variables, the influence of ectomycorrhizal infection on the respiration rates was considered. As a rule, a single root sample was infected by the hyphae of several basidiomycetes e.g. *Xerocomus chrysenteron* ((Bull. ex St. Amans) Quel.), *Lactarius subdulcis* (Bull. ex Fr.) and *Russula ochroleuca* ((Pers.) Fr.). Hence, the mycorrhizal biomass, as a portion of the total respiring biomass of a sample was used as the reference quantity. At the UND site (Fig. 7a) a significant correlation ($p < 0.01$) of mycorrhizal biomass and respiration rate was detected (r=0.63, n=19) but this was not the case at the GAP site (r=0.5, n=14) (Fig. 7b).

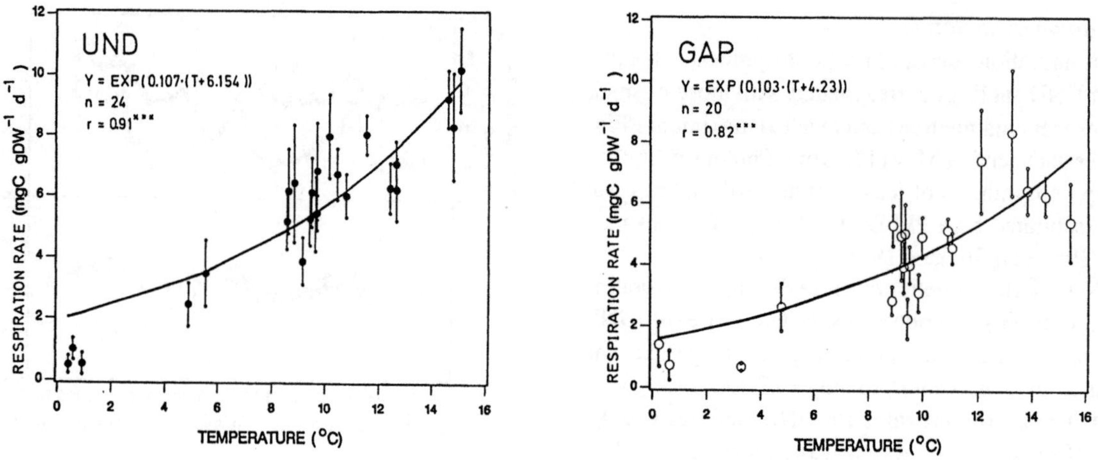


*Fig. 4.* Dependence of fine root respiration on temperature. Each dot (circle) represents a mean value of a three-day period which consists of 70 to 80 single measurements. Respiration rates were investigated from April until December, 1990.

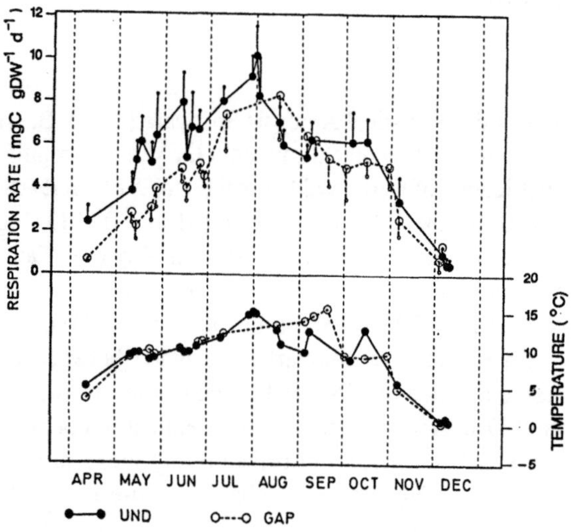


*Fig. 5.* Annual graph of fine root respiration of beech saplings grown in the understorey (UND) and in the natural gap (GAP). Root respiration at the UND site was higher than at the GAP site during the first half of the vegetation period, although temperatures were very similar.

*d) Respiration of medium and coarse roots*

Medium and coarse root respiration rates were found to be very similar. Furthermore, no significant differences between samples from the gap and the understorey could be determined. During spring, the mean respiration rates of 4 medium roots, which had been measured at each plot over 2 three-day periods, did not exceed 0.9 nmol $CO_2$ $gDW^{-1}$ $s^{-1}$ (T=3.9 ± 1.7 °C). Often at night no respiratory activity could be detected. During daytime, the respiration rates of five samples increased with increasing temperature, whereas the others did not follow such a temperature course. Here, respiration rates were more or less alternating with no clear pattern. In mid-summer, a mean rate of 1. 5 ± 0.2 nmol $CO_2$ $gDW^{-1}$ (T=14.1 ± 1.5°C) was typical and daytime respiration did not differ significantly from night time respiration.

In Figure 8, an example of coarse root respiration at the end of July is presented. Although temperature shows a slight decrease at both sites during the experiment, the respiration rates (GAP and UND) were increasing. The mean values of 1.6 ± 0.4 (GAP) and 1.2 ± 0.2 nmol $CO_2$ $gDW^{-1}$ $s^{-1}$ (UND) were not significantly different from each other. The maximum respiration rate of this root class was measured in the middle of July and had a mean value of 3.1 ± 0.3 nmol $CO_2$ $gDW^{-1}$ $s^{-1}$ (T=13 ± 1°C). In comparison with fine root respiration rates, medium and coarse roots showed much lower respiratory carbon losses, but these were more constant during the diurnal course and even throughout the year.

*e) Input: output ratios for carbon*

In Table 2, mean values of daily net assimilation and fine root respiration per plant are given for each month during the vegetation period 1990. The calculation of net assimilation considers varying light saturation curves (Schulte, 1993; Schulze, 1970) of GAP and UND saplings as well as changes in light intensities at both plots during this period. Due to low light intensities at the UND site a net assimilation rate of only 0.3 mg C $plant^{-1}$ $d^{-1}$ was calculated for September. During October, leaf respiration rates of UND saplings often exceeded assimilation rates and thus, a negative

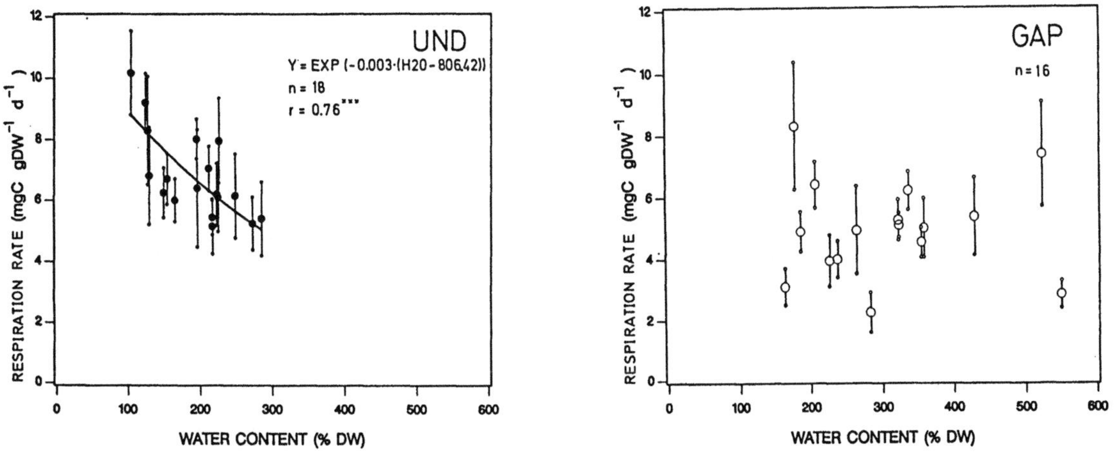


*Fig. 6.* Dependence of fine root respiration on the water content of the attached organic soil material ($O_f$ + $O_h$) during the vegetation period in 1990. Water content is given in per cent of dry weight of the organic material.

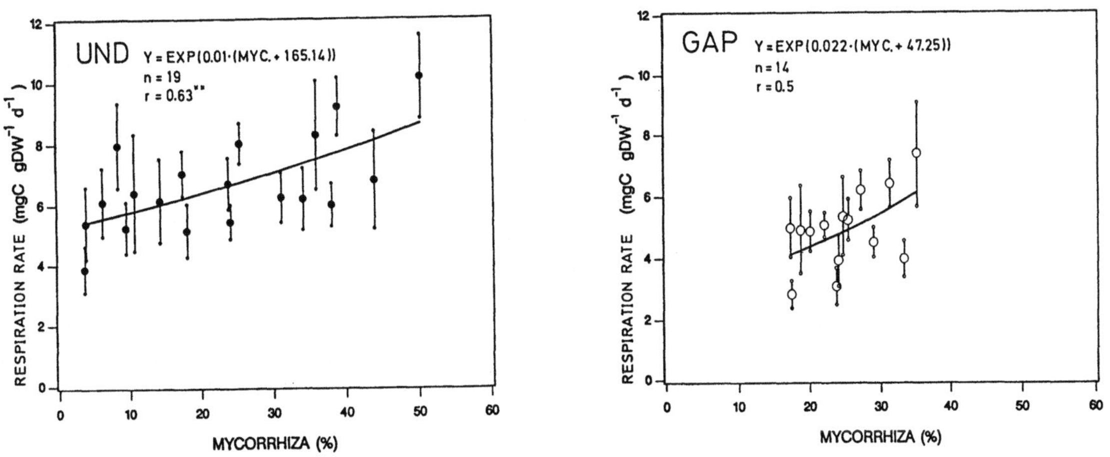


*Fig. 7.* Dependence of fine root respiration on the mycorrhizal infection rate during the vegetation period in 1990. The infection rate is given in per cent of dry weight of respiring biomass. Root samples were infected by the hyphae of several ectomycorrhiza forming basidiomycetes.

*Table 2.* Net assimilation rates ($C_{in}$) and fine root respiration rates ($C_{out}$) of a representative UND and GAP sapling calculated as a mean value for each month during the vegetation period in 1990

| | Local site UND | | | GAP | | |
|---|---|---|---|---|---|---|
| | Net assimilation rate | Fine root respiration rate (mgC plant$^{-1}$ d$^{-1}$) | $C_{in}$ : $C_{out}$ | Net assimilation rate | Fine root respiration rate (mgC plant$^{-1}$ d$^{-1}$) | $C_{in}$ : $C_{out}$ |
| May | 56 ± 15 | 5.4 ± 1 | 10 | 548 ± 95 | 19.7 ± 4.6 | 28 |
| Jun | 111 ± 19 | 6.7 ± 1 | 17 | 560 ± 113 | 30.1 ± 3.2 | 19 |
| Jul | 55 ± 21 | 8.9 ± 1 | 6 | 740 ± 141 | 48 ± 2.4 | 15 |
| Aug | 51 ± 22 | 6.5 ± 0.7 | 8 | 494 ± 113 | 54 ± 3.7 | 9 |
| Sep | 0.3 ± 17 | 5.8 ± 0.6 | 0 | 262 ± 100 | 39 ± 3.6 | 7 |
| Oct | -2 ± 18 | 6.2 ± 0.1 | 0 | 241 ± 140 | 33 ± 1.5 | 7 |

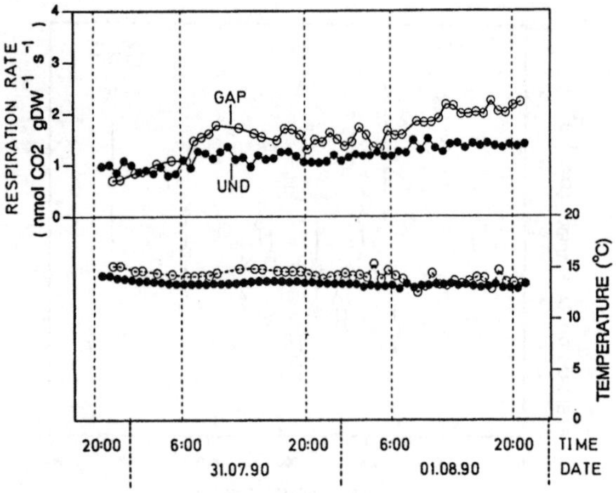


*Fig. 8.* Respiration rates of two coarse root segments at the GAP and at the UND site over a two-day period at the end of July, 1990. Although temperature graphs showed a slight decrease, respiration rates of both increased during this period.

mean value was calculated, whereas the net assimilation of GAP saplings reached a rate of 241 mgC plant$^{-1}$ d$^{-1}$. Mean respiration rates of fine roots were derived from several three-day measurement periods each month. The comparison of C-input : C-output ratios of both sites revealed higher values for the GAP saplings each month, especially in May and July. These large differences are due to the high ratios of daily net assimilation (GAP : UND) compared to those of daily fine root respiration, which underlines the importance of light for UND saplings.

## Discussion

The comparative study of photosynthesis and respiration as the two basic processes in carbon metabolism in saplings grown in the understorey and in a natural gap proved to be a useful tool for the interpretation of contrasting intraspecific growth patterns.

The optimized adaptation of leaves to low light intensities has been emphasized as a requirement for efficient growth (Canham, 1988, 1989; Chazdon, 1986, 1988; Gross, 1982), but it remains open, whether insufficient light alone is responsible for growth limitation. Therefore, emphasis was paid on the investigation of respiratory patterns of the root systems of GAP and UND saplings. Different respiration rates with varying dependencies on environmental variables may additionally strengthen divergent growth patterns.

Basically, in situ measurements of root respiration can only yield approximate data, because any disturbance of the root and its environment e.g. alterations in soil moisture, nutrient supply, $O_2 : CO_2$ ratio, microbial density etc. may cause deviations from the natural respiratory state. Chapman (1979) suggested, that washed root material had up to five times higher respiration rates than estimated from his regression models. Our measurements on washed fine roots which had been embedded in sterilized and moistened silica sand however, revealed a distinct decrease in respiration and almost no reaction to temperature variations.

On the other hand, indirect measurement of fine root respiration rates as described here, must consider the respiratory activity of microbes, living on the root surface and in the adhering soil material, i.e. the rhizosphere (Campbell and Greaves, 1990; Newman, 1985). The investigation of harvested beech fine roots showed that the ratio of microbial density of the rhizosphere to root-free organic soil material ranged from 9 to 22 (Carlberg, 1992), but this was not in accordance with similar ratios of the respective respiration rates measured in our laboratory (Gansert, unpubl. data). Thus, it is likely that the respiration rate as calculated here, was not overestimated due to an underestimation of rhizosphere respiration.

At both sites, root respiration showed a close dependence on soil temperature throughout the year. This was also valid during the vegetation period, when the temperature varied only 7°C. Although the relationship between temperature and respiration rate has been well documented (Larcher, 1980; Long and Woodward, 1988 ; Lyr et al., 1992; Precht et al., 1973), it should be kept in mind, that root respiration in the field is affected by several other interacting variables such as water and nutrient supply (Lambers, 1985; Lambers et al., 1991; Palta and Nobel, 1989), root growth activity (Veen, 1981; Waisel and Eshel, 1991) and internal carbohydrate levels (Amthor, 1989; Farrar and Williams, 1991). Hence, a clear correlation of temperature with root respiration will only be possible, when such variables have no important influence.

The correlation of fine root respiration and water content at the UND site and the absence of such a relationship at the GAP site points to an influence of water content on root respiration. During the vegetation period, the water content of the organic layers ($O_f$ plus $O_h$) at the UND site was 30% lower than at the GAP site and water shortage occured for several days during July and August. This phenomenon was probably caused by the water uptake of mature trees, which accumulate a large part of their fine roots in these layers (Ellenberg et al., 1986; Rapp, 1991).

Ladefoged (1939) observed that beech fine roots originating from thick taproots showed an increase in growth rate during periods of drought, whereas those originating from thin horizontal roots of the same tree were often supressed. He assumed, that the former could be sufficiently supplied with water from deeper soil layers and, thus, an increased growth rate could be due to higher temperatures. This assumption may be of importance for the interpretation of increased fine root respiration rates of UND saplings during dry periods, because these roots often originated from the main root. Other authors, however, observed a decrease of fine root biomass during the dry summer months and related this finding to the prevailing low soil moisture (Deans, 1979; Göttsche, 1972; Hoffmann, 1972). On the other hand, Comeau and Kimmins (1989), Nisbet and Mullins (1986) and Santantonio and Hermann (1985) determined an increased production of fine roots on dry sites which is in accordance with Gales (1977, 1979) who reported that the root : shoot ratio increased in response to drought. From these reports can be concluded that further information on the interaction of root respiration and water availability is of great importance for the understanding of carbon economy under various climatic conditions.

A significant relationship between rates of fine root respiration and rates of mycorrhizal infection could only be detected for UND saplings but not for GAP saplings. This may be due to the small range of mycorrhizal infection of fine roots from this site. Additionally, results from laboratory experiments (Gansert, unpubl. data) showed that respiration rates of non mycorrhized fine roots of 3-year-old beech saplings were of the same magnitude as rates of mycorrhized roots in the field. Harley (1969, pp 98) emphasized that 'on a dry weight basis, uninfected and mycorrhizal roots of *Fagus sylvatica* respired at much the same sort of rate and no explanation of function or difference of behaviour could really be linked with their rates of respiration'. Similarly, Kramer and Hodgson (1954) pointed out that the oxygen consumption of mycorrhizal roots of *Pinus taeda* was about 1.4 times as great as that of non mycorrhizal roots on a fresh-weight basis, but was considerably smaller on a dry-weight or surface-area basis. Thus, it is of importance to look much closer at the relationship between fine root respiration and mycorrhizal infection under field conditions.

With respect to the ratios of net assimilation and those of fine root respiration of the GAP and UND saplings during the vegetation period it should be concluded that limited net assimilation rates due to insufficient light intensities are largely responsible for limited growth of UND saplings. Additionally, the relationships of respiration rates and water content as well as mycorrhizal infection rate for the UND saplings indicate an increased sensitivity of fine root respiration to other abiotic and biotic variables. Thus, it is likely that under detrimental environmental conditions like water shortage in dry summer periods the growth rate of understorey beech saplings will be retarded, whereas that of GAP saplings remains unaffected.

## Acknowledgements

Financial support was received from the Bundesministerium für Forschung und Technologie (Research program 'Stabilitätsbedingungen von Waldökosystemen', Forschungszentrum Waldökosysteme, Universität Göttingen: Project No P.6.3.5.1.2). I wish to thank Walter Stickan for providing the data of the light measurement fields, his critical support and encouragement. I would also like to thank Christoph Leuschner who critically read the manuscript.

## References

Agerer R 1986 Studies on Ectomykorrhiza II: Introducing remarks on characterization and identification. Mycotaxon Vol. XXVI, 473–492.

Agerer R 1988 Colour Atlas of Ectomycorrhizae. Einhorn Verlag Eduard Dietenberger GMBH, Schwäbisch Gmünd, FRG.

Agerer R, Brand F and Gronbach E 1986 Die exakte Kenntnis der Ektomykorrhizen als Voraussetzung für die Feinwurzeluntersuchungen im Zusammenhang mit dem Waldsterben. AFZ 20, 497–503.

Amthor J S 1989 Respiration and Crop Productivity. Springer, Berlin, Heidelberg, New York, FRG. 204p.

Boardman N K 1977 Comparative photosynthesis of sun and shade plants. Ann. Rev. Plant Physiol. 28, 355–377.

Bolhar-Nordenkampf H R 1985 Shoot morphology and leaf anatomy in relation to photosynthesis. *In* Techniques in Bioproductivity and Photosynthesis. Eds. J Coombs, D O Hall, S P Long and J M O Scurlock. pp 107–117. Pergamon Press, Oxford, UK.

Campbell R and Greaves M P 1990 Anatomy and community structure of the rhizosphere. *In* The Rhizosphere. Ed. J M Lynch. Environmental and Applied Microbiological Series. John Wiley and Sons, Chichester, UK.

Canham Ch D 1988 Growth and canopy architecture of shade-tolerant trees: response to canopy gaps. Ecology 69, 786–795.

Canham Ch D 1989 Different responses to gaps among shade-tolerant tree species. Ecology 70, 548–550.

Carlberg S 1992 Vergleichende Untersuchungen der Mikroflora von wurzelfreiem Boden und der Rhizosphäre von Jungbuchen. Diplomarbeit, Inst. Pflanzenpathologie und Pflanzenschutz, Universität Göttingen, FRG. 90p.

Chapman S B 1979 Some interrelationships between soil and root respiration in lowland calluna heathland in southern England. J. Ecol. 67, 1–20.

Chazdon R L 1986 Light variation and carbon gain in rain forest understorey palms. J. Ecol. 74, 995–1012.

Chazdon R L 1988 Sunflecks and their importance to forest understorey plants. *In* Advances in Ecological Research Vol 18. Eds. M Begon, A M Fitter, E D Ford and A Macfadyen. pp 1–63. Academic Press London, New York, UK.

Comeau P G and Kimmins J P 1989 Above - and below-ground biomass production of lodgepole pine on sites with differing soil moisture regimes. Can. J. For. Res. 19, 447–454.

Deans J D 1979 Fluctuations of the soil environment and fine root growth in a young Sitka spruce plantation. Plant and Soil 52, 195–208.

Eber W 1972 Über das Lichtklima von Wäldern bei Göttingen und seinen Einfluß auf die Bodenvegetation. Scripta Geobot. 3. Goltze Verlag, Göttingen, FRG. 150p.

Ehrhardt O 1988 Der Strahlungshaushalt eines Buchenwaldes und dessen Abwandlung während der verschiedenen phänologischen Entwicklungsphasen. Ber. Forschungszentrums Waldökosysteme, Univ. Gött., Reihe A, 45, 1–170.

Ellenberg H, Mayer R and Schauermann J 1986 Ökosystemforschung, Ergebnisse des Sollingprojekts 1966–1986. Eugen Ulmer, Stuttgart, FRG, 507 p.

Endler J A 1993 The color of light in forests and its implications. Ecol. Monogr. 63, 1–27.

Essiamah S and Eschrich W 1985 Changes of starch content in the storage tissues of deciduous trees during winter and spring. IAWA Bull. n.s. Vol. 6, 97–106.

Farrar J F and Williams M L 1991 The effects of increased atmospheric carbon dioxide and temperature on carbon partitioning, source - sink relations and respiration. Plant, Cell Environ. 14, 819–830.

Gales K 1977 A study of the effect of drought on root growth and root-shoot relationships in *Lolium perenne*. L. Ph. D. thesis, University of London.

Gales K 1979 Effects of water supply on partitioning of dry matter between roots and shoots in *Lolium perenne*. J. Appl. Ecol. 16, 863–877.

Göttsche D 1972 Die Verteilung von Feinwurzeln und Mykorrhizen im Bodenprofil eines Buchen- und Fichtenbestandes im Solling. Mitt. Bundesforschungsanst. Forst Holzwirtsch. 88, 1–120.

Gross L J 1982 Photosynthetic dynamics in varying light environments : a model and its application to whole leaf carbon gain. Ecology 63, 84–93.

Harley J L 1969 The Biology of Mycorrhiza. Leonard Hill, London, UK. 334p.

Hoffmann G 1972 Wachstumsrhythmik der Wurzeln und Sproszachsen von Forstgehölzen. Flora 161, 303–319.

Janecek A, Benderoth G, Lüdeke M K B, Kindermann J and Kohlmaier G H 1989 Model of the seasonal and perennial carbon dynamics in deciduous-type forests controlled by climatic variables. Ecol. Modell. 49, 101–124.

Köstler J N, Brückner E and Biebelriether H 1968 Die Wurzeln der Waldbäume. Parey Verlag, Hamburg, Berlin, FRG. 284p.

Kramer P J and Hodgson R H 1954 Differences between mycorrhizal and non-mycorrhizal roots of Loblolly pine. VIIIth Int. Bot. Cong. Paris, 13, 133–134.

Ladefoged K 1939 Untersuchungen über die Periodizität im Ausbruch und Längenwachstum der Wurzeln bei einigen unserer gewöhnlichsten Waldbäume. Forstl. Forsogsvaes. Dan. 16, 1–256.

Lambers H 1985 Respiration in intact plants and tissues : its regulation and dependence on environmental factors, metabolism and invaded organisms. *In* Encyclopedia of Plant Physiology 18. Eds. R Douce and D A Day. pp 418–473. Springer, Berlin, Heidelberg, New York, FRG.

Lambers H, Van der Werf A and Konings H 1991 Respiratory patterns in roots in relation to their functioning. *In* Plant Roots: The Hidden Half. Eds. Y Waisel and A Eshel. pp 229–263. Marcel Dekker, Inc. New York, USA.

Larcher W 1969 The effect of environmental and physiological variables on the carbon dioxide gas exchange of trees. Photosynthetica 3, 167–198.

Larcher W 1980 Physiological Plant Ecology. Springer, Berlin, Heidelberg, New York, FRG. 252p.

Long S P and Woodward F I 1988 Plants and Temperature. Symp. Soc. Experimental Biology, 42, Cambridge, UK.

Lyr H, Fiedler H J and Tranquillini W 1992 Physiologie und Ökologie der Gehölze. Gustav Fischer Verlag, Jena, Stuttgart, FRG. 620p.

Masarovicova E 1980 Photosynthesis, photorespiration and mitochondrial respiration on *Fagus sylvatica* L. seedlings : effects of temperature and oxygen concentration. Photosynthetica 14, 321–325.

Masarovicova E 1988 Comparative study of growth and carbon uptake in *Fagus sylvatica* L. trees growing under different light conditions. Biol. Plant. (Praha) 30, 285–293.

Murach D 1984 Die Reaktion der Fichtenfeinwurzeln (*Picea abies* Karst.) auf zunehmende Bodenversauerung. Gött. Bodenkdl. Ber. 77, 1–126.

Neemann G and Stickan W 1991 Untersuchungen zum Kohlenstoff-Haushalt von Jungbuchen - Assimilatspeicherung in Abhängigkeit von der Einstrahlung. Ber. Forschungszentrums Waldökosysteme, Univ. Gött. Reihe B, 22, 152–155.

Newman E I 1985 The rhizosphere: carbon sources and microbial populations. *In* Ecological Interactions in Soil. Ed. A H Fitter. pp 107–121. Blackwell Scientific Publications, Oxford, UK.

Nisbet T R and Mullins C E 1986 A comparison of live and dead fine root weights in stands of Sitka spruce in contrasting soil water regimes. Can. J. For. Res. 16, 394–397.

Palta J A and Nobel P S 1989 Root respiration for *Agave deserti:* influence of temperature, water status and root age on daily patterns. J. Exp. Bot. 40, 181–186.

Pearcy R W 1983 The light environment and growth of $C_3$ and $C_4$ species in the understory of a Hawaiian forest. Oecologia (Berlin) 58, 26–32.

Poulson Th L and Platt W J 1989 Gap light regimes influence canopy tree diversity. Ecology 70, 553–555.

Precht H, Christophersen J, Hensel H and Larcher W 1973 Temperature and Life. Springer, Berlin, Heidelberg, New York, FRG. 779p.

Rapp C 1991 Untersuchungen zum Einflusz von Kalkung und Ammonium-sulfat-Düngung auf Feinwurzeln und Ektomykorrhizen eines Buchenaltbestandes im Solling. Ber. Forschungszentrums Waldökosysteme, Univ. Gött. Reihe A, 72, 1–293.

Rees U and Stickan W 1991 Untersuchungen zum C-Haushalt von Jungbuchen - $CO_2$ -Assimilation in Abhängigkeit von der Einstrahlung. Ber. Forschungszentrums Waldökosysteme, Univ. Gött., Reihe B Bd. 22, 138–141.

Roloff A 1986 Morphologische Untersuchungen zum Wachstum und Verzweigungssystem der Rotbuche (*Fagus sylvatica* L.). Mitt. Dtsch. Dendrol. Ges. 76, 5–47.

Sachs L 1992 Angewandte Statistik. Springer, Berlin, Heidelberg, New York, FRG. 846p.

Santantonio D and Hermann R K 1985 Standing crop, production and turnover of fine roots on dry, moderate and wet sites of mature Douglas-fir in western Oregon. Ann. Sci. For. 42, 113–142.

Schmaltz J 1964 Untersuchungen über den Einflusz von Beschattung und Konkurrenz auf junge Buchen. Diss. Univ. Göttingen, FRG.

Schulte M 1993 Saisonale und interannuelle Variabilität des $CO_2$-Gaswechsels von Buchen (*Fagus sylvatica* L.) - Bestimmung von C-Bilanzen mit Hilfe eines empirischen Modells. Shaker Verlag Aachen, FRG. 164p.

Schulze E D 1970 Der $CO_2$ -Gaswechsel der Buche (*Fagus sylvatica* L.) in Abhängigkeit von den Klimafaktoren im Freiland. Flora 159, 177–232.

Stickan W and Hentzelt H 1991 Untersuchungen zum Kohlenstoffhaushalt von Jungbuchen - Der Einsatz preisgünstiger Si-Fotodioden zur Bestimmung der photosynthetisch wirksamen Strahlung. Ber. Forschungszentrums Waldökosysteme, Univ. Gött., Reihe B 22, 133–137.

Stickan W and Zhang X 1992 Seasonal changes in $CO_2$ and $H_2O$ gas exchange of young European beech (*Fagus sylvatica* L.). Trees 6, 96–102.

Stickan W, Gansert D, Neemann G and Rees U 1992 Kohlenstoff- und Wasserhaushalt von Jungbuchen und Krautschicht im Verlauf einer Bestandesentwicklung. Ber. Forschungszentrums Waldökosysteme, Univ. Gött., Reihe B 31, 591–604.

Suner A and Röhrig E 1980 Die Entwicklung der Buchennaturverjüngung in Abhängigkeit von der Auflichtung des Altbestandes. Forstarchiv 51, 145–149.

Veen B N 1981 Relation between root respiration and root activity. Plant and Soil 63, 73–76.

Waisel Y and Eshel A 1991 Plant Roots : The Hidden Half (Soils, Plants and the Environment). Marcel Dekker, Inc. New York, USA. 944 p.

# ROOTS UNDER STRESS CONDITIONS

*F. Baluška et al. (eds.), Structure and Function of Roots*, 271–276, 1995.


# Impact of the exodermis on infection of roots by *Fusarium culmorum*

Susan A. Kamula, Carol A. Peterson and Colin I. Mayfield
*Department of Biology, University of Waterloo, Waterloo, ON, Canada N2L 3G1*

*Key words:* Casparian bands, exodermis, *Fusarium culmorum*, hypodermis, passage cells, short cells, suberin lamellae

**Abstract**

Patterns of infection with *Fusarium culmorum* (W G Smith) Saccardo were observed in seedling roots of barley (*Hordeum vulgare* L.), wheat (*Triticum aestivum* L.), maize (*Zea mays* L.) and asparagus (*Asparagus officinalis* L). Apical regions of the main roots were not infected. Since penetration into the root occurred several days after inoculation and the roots were growing during the experiment, these regions had apparently not been in existence long enough to be infected. In older regions of barley, wheat and asparagus, hyphae entered through the tips of lateral roots. In barley and wheat, which had not developed any suberin lamellae in their subepidermal layer, infection occurred randomly over the remainder of the root. In maize, the fungus penetrated the epidermis at many sites but did not breach the exodermis in which all cells possessed both Casparian bands and suberin lamellae. Maize roots, therefore, sustained only minimal infections. In asparagus, the fungus grew through the short (passage) cells but never the long cells of the exodermis. In doing so, it penetrated cells possessing Casparian bands but lacking suberin lamellae. The results support the hypothesis that suberin lamellae provide effective barriers to the growth of *F. culmorum* hyphae.

## Introduction

Deposition of suberin in plant cell walls is known to deter fungal infection (Kolattukudy, 1984). Thus, the suberin in the Casparian bands and suberin lamellae of the exodermis, a specialized layer in the roots of many angiosperm species (Perumalla et al., 1990; Peterson and Perumalla, 1990), may play a role in preventing or delaying fungal invasion. Since the exodermis is the outermost layer of the cortex, it is well positioned to protect a large area of the organ. To determine the potential importance of the exodermis as an antifungal layer, we compared the pattern of infection of four species by pathogen *Fusarium culmorum* (W G Smith) Saccardo. According to Perumalla et al. (1990), barley (*Hordeum vulgare* L.) and wheat (*Triticum aestivum* L.) have nonexodermal roots. Maize (*Zea mays* L.) and asparagus (*Asparagus officinalis* L.) have exodermal roots (Peterson et al., 1982; Peterson and Perumalla, 1990). The exodermis of maize is similar, comprised of cells of uniform length and structure (Kroemer, 1903). At maturity, all cells of the exodermis have both Casparian bands and suberin lamellae (Ferguson and Clarkson, 1975; Peterson, et al., 1982). The exodermis of young roots of asparagus, on the other hand, is dimorphic and consists of long and short cells (Kroemer, 1903). At maturity, the long cells and possibly some of the short cells have both Casparian bands and suberin lamellae but the remainder of the short cells have only Casparian bands (Hussey, 1982; Kamula, 1991). Inclusion of a species with a dimorphic exodermis made it possible to test the effects of Casparian bands and suberin lamellae on fungal growth separately.

## Materials and methods

Caryopses of barley (*Hordeum vulgare* L.) and wheat (*Triticum aestivum* L.) were purchased from Carolina Biological Supply, Burlington, NC, USA ; caryopses of maize (*Zea mays* L. cv. Seneca Chief) and seeds of asparagus (*Asparagus officinalis* L. var. Mary Washington) were obtained from Ontario Seed Company, Ltd., Waterloo, ON, Canada. All propagules were surface sterilized in 5.2% sodium hypochlorite (commercial Javex bleach) for five min and soaked overnight in cold, running tap water. Plants were grown in vermiculite, initially watered to saturation and subsequently every 2–4 d with tap water, in a Conviron PGW-36 growth chamber (Controlled Environment, Winnipeg, MN, Canada) at 25 °C during 16 h light (intensity 215 $\mu$E $s^{-1}$ $m^{-2}$) and 23 °C during 8 h darkness, relative humidity 75–85%. Maize, barley and wheat grown for 5 d and asparagus grown for 21 d were used in the experiments.

A culture of *Fusarium culmorum* (W.G. Smith) Saccardo was provided by Dr Robert Hall, Department of Environmental Biology, University of Guelph, ON, Canada, who had isolated it from the crown of a winter wheat plant in Elora, ON, Canada. The fungus was grown on potato-dextrose agar (Difco Laboratories, Detroit, MI, USA) at room temperature in a dimly lit area for two months. A spore suspension was prepared by flooding the mycelium with sterile, distilled water and adjusting the volume to give $2 \times 10^5$ spores $mL^{-1}$.

To begin an experiment, seedlings were gently uprooted and their root systems washed and laid horizontally in large, moist, sterile chambers. The primary roots of each species were inoculated with *F. culmorum* by pipetting 2 mL of the spore suspension over them. This resulted in an inoculum ranging between 14 and 38 spores $mm^{-1}$ root length. Roots which served as controls were placed in similar surroundings and treated with water. The chambers were covered with aluminum foil so that the roots were in darkness and the shoots emerged from the side. The containers were incubated at room temperature and the contents were wetted with sterile distilled water as necessary to maintain a moist environment around the roots. To determine the time course of infection, three seedlings of each species were inoculated as described above and individual roots were harvested 0.5, 3, 7, 9, 14 and 21 d later. This experiment was repeated three times. A further comparison of the infection patterns in the four plant species was made by inoculating three seedlings of each and harvesting them after 21 d. Entire roots were analyzed from both experiments. They were cut into segments 10–30 mm long which were cleared and stained with chlorazol black E (CBE) according to the method of Brundrett et al. (1984) with the following modifications. An autoclave was used (20 min at 121° C) for both clearing and staining steps, the concentration of CBE was 0.03%, and the roots were mounted in 70% glycerin. With this procedure, fungal walls, lignified plant walls and Casparian bands stained black. Root segments were mounted on microscope slides with props at the ends of the coverslips to prevent squashing the specimens. To determine the location of suberin lamella development, cross sections were taken near the apices of exodermal species and stained with Sudan red 7B which partitions into lipids, imparting a red colour to them (Brundrett et al., 1991). The experiment was repeated three times.

All experimental specimens were observed with a Zeiss Axiophot microscope. Colour photographs were taken on Kodak Ektachrome 200 ISO, 35 mm slide film, black and white on Ilford Pan F 50 ISO, 35 mm negative film. Black and white prints were made from black and white negative film or from internegatives made from colour slides using Ilford Pan F 50 ISO, 35 mm negative film.

## Results

No signs of contamination or infection were apparent on control roots. The distinctive macroconidia of *Fusarium culmorum* were clinging to the surfaces of all treated roots and the majority of the conidia had germinated within 3 d. By 7 days, copious hyphae were present on the surfaces of the roots and were beginning to invade them. Penetration sites were at midlength and more proximal regions of the roots, never in the apical regions. Several distinct patterns of infection were discerned.

In barley and wheat, infection occurred randomly in the midlength to basal regions of the roots (average total root length about 170 mm). Hyphae grew through epidermal cells ( Fig. 1) and subsequently many runner, branched and coiled hyphae proliferated throughout the cortex (Fig. 2). Infection also occurred in some young lateral roots, often through their root caps (Fig. 3). Occasionally, black staining of the host tissue near an infection site was observed (Fig. 1) indicating that suberin and/or lignin had been deposited as a wound response.

In maize roots, the infection process was arrested at the exodermis. This layer became mature 40 mm from the main root tip and very close to the tips of the laterals. Hyphae frequently penetrated the epidermis in mid and basal regions of the main roots, as well as in the young laterals. However, growth did not proceed into the cortex (Fig. 4). From 9 to 21 days after inoculation, masses of hyphae had formed in some epidermal cells but the fungus had still not penetrated further into the root (Fig. 5). Furthermore, hyphae did not grow through the tips of the lateral roots. The pathogen, therefore, produced only a minimal infection of the root, i.e. in the epidermis.

The exodermis of asparagus roots also influenced the infection process. In these roots, Casparian bands and suberin lamellae had formed 10 mm from the main root apex. The dimorphic nature of the exodermis was apparent in clearings of control roots (Fig. 6). Fungal invasion of the root proceeded through the epidermis overlying the short cells and then through the short cells themselves, forming hyphal coils (Fig. 7). As an example of the extent of short cell penetration, in samples taken 14 days post inoculation, hyphae were present in 41% of the 115 short cells tallied. In contrast, penetration of a long cell was never observed anywhere in the roots. The fungus also gained entry by growing through the lateral roots, either through the short cells of the exodermis or directly through the root tips.

## Discussion

A consistent feature of the invasion of all four plant species with *Fusarium culmorum* was that it did not occur in young regions of the roots. This result is not typical of *Fusarium* infections, which often proceed through the root cap (Pennypacker, 1981; Smith and Peterson, 1985). In the present investigation, this result was probably a consequence of the relatively long time required by the fungus to invade the root. Since the roots were elongating by tip growth throughout the experiment, the youngest regions would not have been in existence long enough to be infected. In species with an exodermis (maize and asparagus), the layer was mature before the fungus succeeded in gaining entry into the roots. The maturation of the exodermis in rapidly growing maize roots occurs about 120 mm from the root tip but is within 5 mm of the tip in roots which have ceased growing (Perumalla and Peterson, 1986). The intermediate maturation distance of 40 mm obtained in the present study was a consequence of a reduced growth rate caused by forcing the roots to grow horizontally. In the case of maize and asparagus, the apical, uninfected zone correlated with the region of immature exodermis, in which a large area of plasmalemma is exposed to the soil solution (Kamula, 1991). However, if the fungus were capable of a more rapid infection than the strain used in the present study, the length of the uninfected zone would be reduced and invasion in zones with an immature exodermis would be anticipated.

A comparison of the infection patterns of various species indicates that the exodermis plays a role in the invasion process and that the type of exodermis is critical for the outcome. Specifically, the failure of the fungus to penetrate the uniform exodermal cells of maize and the long cells of asparagus indicates that suberin lamellae were the structures responsible for prevention of its ingress. The Casparian band had no such effect, since it was present in the short cells of asparagus through which the hyphae grew. These results are reasonable because the Casparian band is located in the anticlinal (radial and transverse) walls of the cells whereas the lamellae also cover the tangential walls through which the fungus penetrates. Our observations agree with those of Gallaud (1905) who noted that *Fusarium* grew through the short cells but not the long cells in the exodermis of *Cyripedium barbatum.* Noelle (1910) also described penetration of short cells (as well as unsuberized, immature hypodermal cells) by fungal hyphae as a general phenomenon in conifers. Shishkoff (1986) found that hyphal penetration by *Pythium coloratum* on *Allium cepa* roots occurred with much greater frequency through unsuberized short and long cells, although she did find a few cases where the fungus had grown through the suberized wall of a long cell. The question of whether or not the exodermis constitutes a resistant layer was also approached by Perry and Evert (1983) and Brammall and Higgins (1988) with *Solanum tuberosum* and *Lycopersicon esculentum,* respectively. It is not clear from the literature whether the exodermis of these plants is dimorphic or uniform (Shishkoff, 1987). In the infected zones, most of the exodermal walls in *S. tuberosum* and some of the exodermal walls in *L. esculentum* lacked suberin lamellae and none of the electron micrographs illustrate hyphal growth into an exodermal cell with a lamella. Therefore, it is possible that the invasion of the exodermis which led to the formation of lignitubers in the case of *S. tuberosum* (Perry and Evert, 1983), or further invasion of the cortex in *L. esculentum* (Brammall and Higgins, 1988), occurred through cells lacking suberin lamellae. Although the evidence is somewhat scattered at the present time, much of it indicates that suberin lamellae are effective in preventing growth of pathogenic fungi through exodermal cells.

Whether or not a root system is successfully attacked by a pathogenic fungus depends not only on the exodermis but also on the existence of other sites of invasion. For example, *Fusarium* spp. frequently

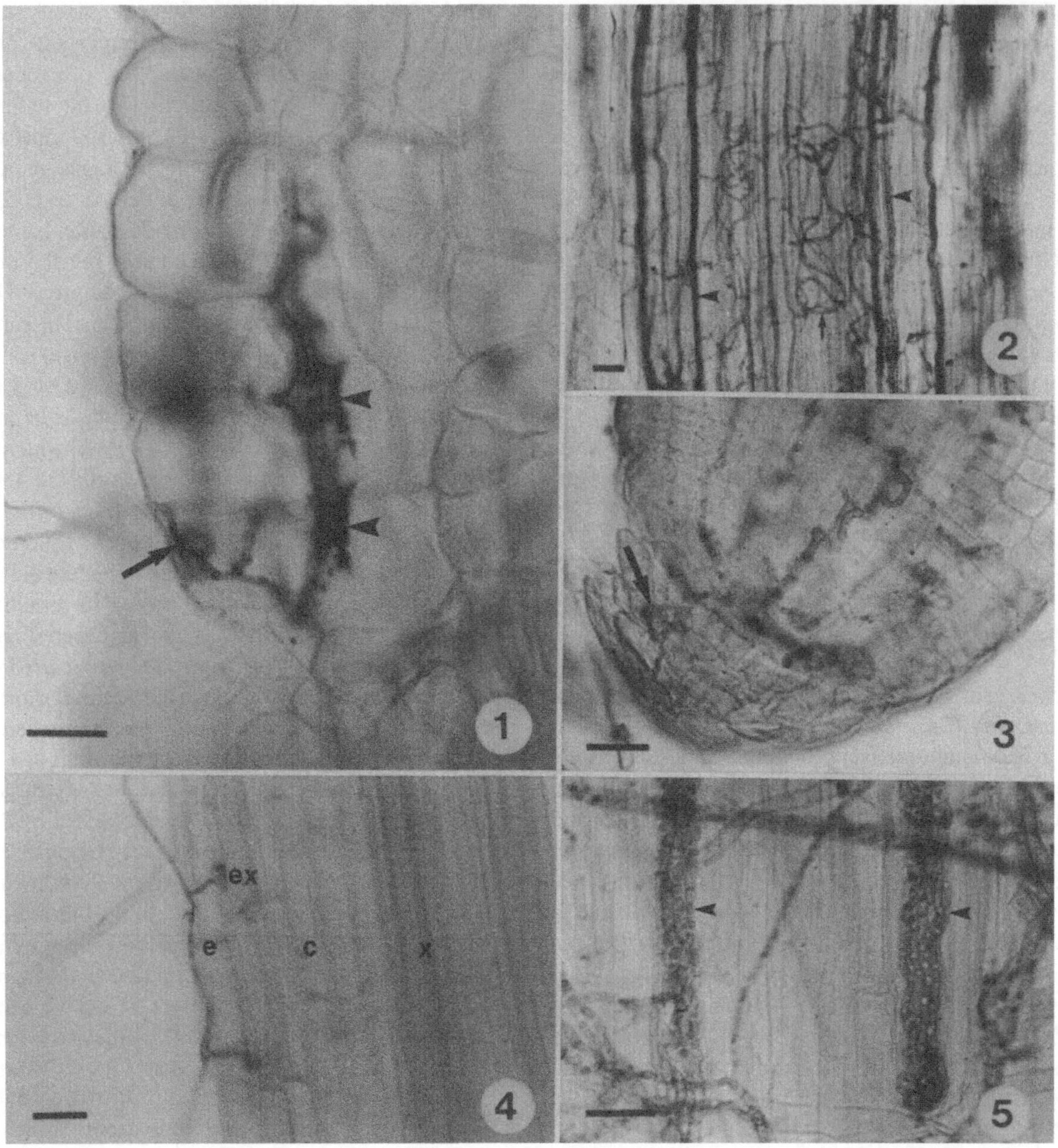


*Fig. 1–5.* Whole roots which had been infected with *Fusarium culmorum*, fixed in FAA, cleared with KOH and stained with chlorazol black E. Scale *bars* = 25 μm.

*Fig. 1.* Median longitudinal, optical section of barley root with *F. culmorum* hyphae in epidermal cell (arrow). The cell layer under the epidermis underwent a wound reaction (arrowheads).

*Fig. 2.* Paradermal, optical section through the cortex of a heavily infected wheat root. Runner hyphae (arrowheads) occupy the intercellular spaces and branched hyphae (arrow) have grown into cortical cells.

*Fig. 3.* Median longitudinal, optical section of a young, lateral root of barley. Hyphae of *F. culmorum* had grown through the root cap (arrow) and continued into the root.

*Fig. 4.* Median longitudinal, optical section of a maize lateral root with *F. culmorum* hyphae in its epidermis (e). The pathogen did not grow into the exodermis (ex), central cortex (c) or xylem (x).

*Fig. 5.* Paradermal, optical section of a maize main root with extensive hyphal coiling in some epidermal cells (arrowheads).

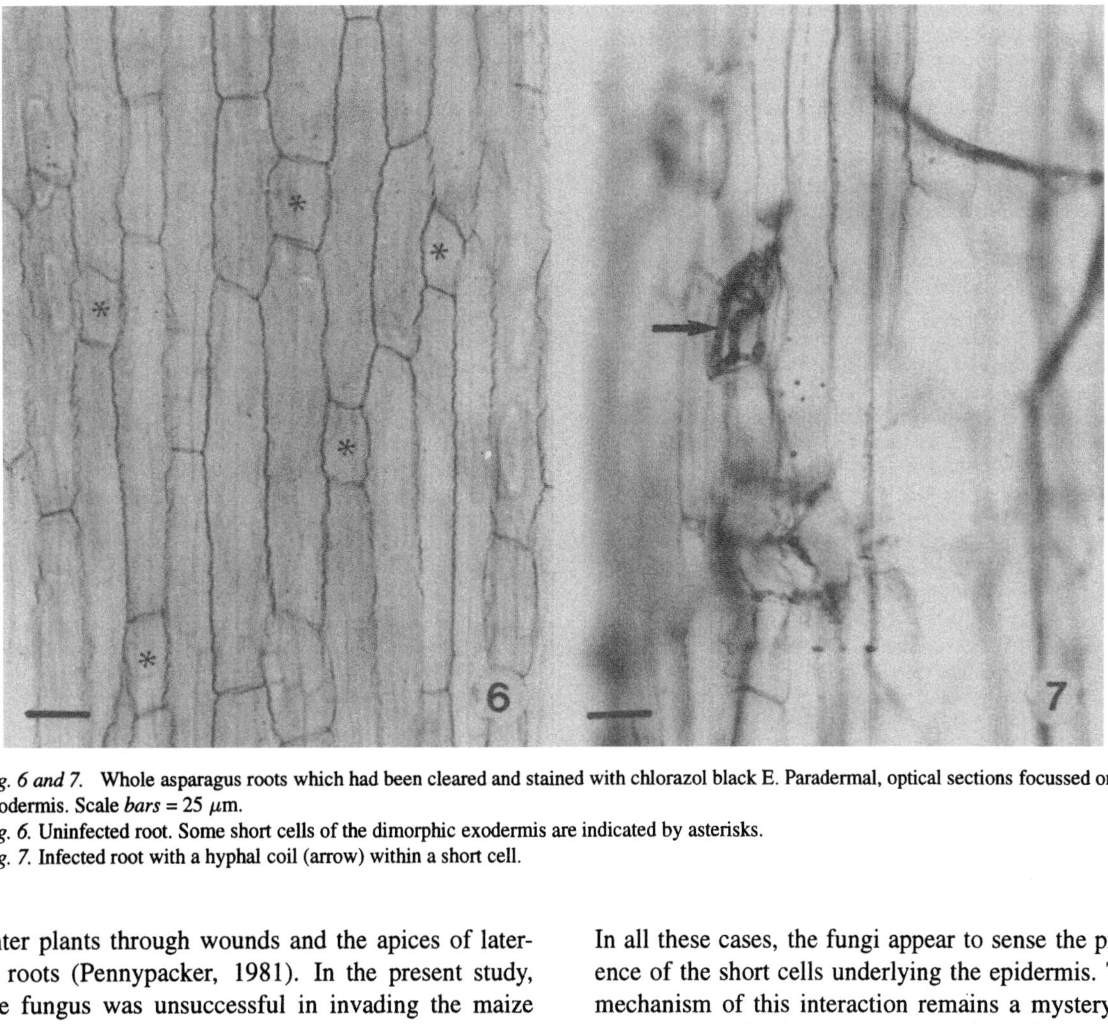


*Fig. 6 and 7.* Whole asparagus roots which had been cleared and stained with chlorazol black E. Paradermal, optical sections focussed on the exodermis. Scale *bars* = 25 μm.
*Fig. 6.* Uninfected root. Some short cells of the dimorphic exodermis are indicated by asterisks.
*Fig. 7.* Infected root with a hyphal coil (arrow) within a short cell.

enter plants through wounds and the apices of lateral roots (Pennypacker, 1981). In the present study, the fungus was unsuccessful in invading the maize root through these other potential sites (for unknown reasons) and the interior of the root remained free of infection. However, in the case of asparagus, invasion occurred through both young lateral roots and short cells so that the root was thoroughly colonized. Thus, the presence of suberin lamellae in the exodermis can only provide protection to a limited area of the root and successful pathogens can use other sites of entry.

It is interesting to note that when *F. culmorum* attacked the roots of asparagus, it entered the epidermal cells overlying the short cells of the exodermis. Shishkoff (1986) found that zoospores of *Pythium coloratum* tended to encyst on the epidermis over the short cells of *Allium cepa* and that after germination, a large fraction of their hyphae grew through the short cells. There are several reports that endomycorrhizal fungi also enter the roots through the short cells (see Shishkoff, 1986 lit. cit.; Peterson, 1992 and lit. cit.). In all these cases, the fungi appear to sense the presence of the short cells underlying the epidermis. The mechanism of this interaction remains a mystery. It may have evolved to attract mycorrhizal fungi to an appropriate entry site, and was later used to advantage by pathogenic fungi.

## Acknowledgements

This research was supported by grants to C A P and C I M from the Natural Sciences and Engineering Research Council of Canada. We thank Ms Daryl Enstone for reviewing the manuscript and preparing the plates.

## References

Brammall R A and Higgins V J 1988 A histological comparison of fungal colonization in tomato seedlings susceptible or resistant to *Fusarium* crown and root rot disease. Can. J. Bot. 66, 915–925.

Brundrett M C, Kendrick B and Peterson C 1991 Efficient lipid staining in plant material with Sudan red 7B or Fluorol yellow 088 in polyethylene glycol-glycerol. Biotechnic Histochem. 66, 111–116.

Brundrett M C, Piché Y and Peterson R L 1984 A new method for observing the morphology of vesicular-arbuscular mycorrhizae. Can. J. Bot. 62, 2128–2134.

Ferguson I B and Clarkson D T 1975 Ion transport and endodermal suberization in the roots of corn. New Phytol. 75, 69–79.

Gallaud I 1905 Etudes sur les mycorrhizes endophytes. Rev. Gén Bot. 17, 313–325.

Hussey R B 1982 Interactions between V-A mycorrhizal fungi and *Asparagus officinalis* L. roots. M Sc Thesis, University of Guelph, Guelph, ON, Canada.

Kamula S A 1991 The significance of suberin lamella development in the dimorphic exodermis. MSc Thesis, University of Waterloo, Waterloo, ON, Canada.

Kolattukudy P E 1984 Fungal penetration of defensive barriers in plants. *In* Structure, Function and Biosynthesis of Plant Cell Walls. Eds. W M Dugger and S Bartnicki-Garcia. pp 302–343. American Society of Plant Physiologists, Rockland, MD.

Kroemer K 1903 Wurzelhaut, Hypodermis und Endodermis der Angiospermenwurzel. Biblioth. Bot. 12, 1–159.

Noelle F W 1910 Studien zur vergleichenden Anatomie und Morphologie der Koniferen wurzeln mit Rücksicht auf die Systematik. Bot. Zeitung 68, 169–266.

Pennypacker B W 1981 Anatomical changes involved in the pathogenesis of plants by *Fusarium*. *In Fusarium*: Diseases, Biology, and Taxonomy. Eds. P E Nelson, T A Toussoun and R J Cook. pp 400–408. Penn. State Univ. Press, University Park and London.

Perry J W and Evert R F 1983 Histopathology of *Verticillium dahliae* within mature roots of Russet Burbank potatoes. Can. J. Bot. 61, 3405–3421.

Perumalla C J and Peterson C A 1986 Deposition of Casparian bands and suberin lamellae in the exodermis and endodermis of young corn and onion roots. Can. J. Bot. 64, 1873–1878.

Perumalla C J, Peterson C A and Enstone D E 1990 A survey of angiosperm species to detect hypodermal Casparian bands. I. Roots with a uniseriate hypodermis and epidermis. Bot. J. Linn. Soc. 103, 93–112.

Peterson C A, Emanuel M E and Wilson C 1982 Identification of a Casparian band in the hypodermis of onion and corn roots. Can. J. Bot. 60, 1529–1535.

Peterson C A and Perumalla C J 1990 A survey of angiosperm species to detect hypodermal Casparian bands. II. Roots with a multiseriate hypodermis or epidermis. Bot. J. Linn. Soc. 103, 113–125.

Peterson R L 1992 Adaptations of root structure in relation to biotic and abiotic factors. Can. J. Bot. 70, 661–675.

Shishkoff N 1986 The dimorphic hypodermis of plant roots: its distribution in the angiosperms, staining properties, and interaction with root-invading fungi. PhD Thesis, Cornell University, Ithaca, NY.

Shishkoff N 1987 Distribution of the dimorphic hypodermis of roots in angiosperm families. Ann. Bot. 60, 1–15.

Smith A K and Peterson R L 1985 Histochemical features of wall appositions in *Asparagus* root meristems infected by *Fusarium*. Can. J. Plant Pathol. 7, 28–36.

*F. Baluška et al. (eds.), Structure and Function of Roots*, 277–284, 1995.


# Ultrastructural responses of root caps to the herbicides chlorsulfuron and metsulfuron methyl

K.A. Fayez, I. Gerken and U. Kristen[1]
*Institut für Allgemeine Botanik, Universität Hamburg, Ohnhorststrasse 18, D-22609 Hamburg, Germany*
[1] *Corresponding author*

*Key words:* herbicides, chlorsulfuron, metsulfuron methyl, root cap ultrastructure, root growth, *Pisum sativum, Zea mays*

## Abstract

Herbicide residues may affect seedlings during early stages of their development. We studied this possibility by the use of light and electron microscopy after incubation of germinating seeds of *Pisum sativum* L. and *Zea mays* L. with different concentrations of chlorsulfuron and metsulfuron-methyl. By in vitro experiments, we have shown that both herbicides caused growth reduction of the very young roots, and severe ultrastructural alterations and injuries of the root caps of both species. Chlorsulfuron caused increase of electron-dense material in the vacuoles, cytoplasmic degeneration even in the inner secretory cell layers of the cap, and disruption of the amyloplast envelopes with release of the statolithic starch grains. In the initial cell complex of the root cap, the herbicides caused the formation of large concentric aggregates of the rough ER and wall disformations in the cells adjacent to this complex. Scanning electron microscopic observations revealed a decrease of the slime layer ensheathing the root cap and the subapical root surface.

We conclude that even in early stages of seed germination, both herbicides seriously affect the gravity perception centre (consisting of the statocytes), and the secretory tissue of the root caps, thus probably disturbing the processes of gravitropism and the protective slime secretion of the roots.

## Introduction

The development of modern herbicides has recently been directed to a short half life time in the environment to avoid a carry-over of herbicide residues to a second crop year following spring application. However, the persistence data reported for the family of sulfonylurea herbicides, especially for chlorsulfuron, metsulfuron-methyl, sulfometuron-methyl and amidosulfuron (Anderson and Dulka, 1985; Smith, 1986; Smith and Aubin, 1992; Walker and Welch, 1989) indicates a carry-over of herbicide residues to a second season. This may be especially true for tropical and subtropical regions where crop rotation is usually 2 or even 3 times per year.

For chlorsulfuron it has been reported that growth reduction of beans, maize and many other crops occurred up to 36 months from the application of the herbicide to the soil (Dyer and Fay, 1984; Nicholls et al., 1987). Thus, it is suggested that herbicide residues would affect seedlings, following herbicide application, possibly during very early stages of their development. Therefore, we studied this possibility by light and electron microscopic observations after incubation of in vitro germinating seeds of *Pisum sativum* L. and *Zea mays* L. with different concentrations of chlorsulfuron and metsulfuron-methyl.

The objectives of the present laboratory study were to determine the herbicide-induced subcellular alterations of the root tips of very young seedlings possibly leading to the root growth reduction observed. Our investigations were previously restricted to the root cap in order to focus on the secretory tissue and on the gravity perception centre of this organ.

## Materials and methods

Seeds of *Zea mays* L. cv. Aztek and of *Pisum sativum* L. cv. Kleine Rheinländerin were pretreated with Thiram or 10% NaOCl for 10 min for seed surface sterilisation, were then set up under clean bench condition in Petri dishes, filled with boiled tap water (1 mm high), for soaking and germination until break-through of the radicule. Seeds with a root length of 2 mm were selected and transfered into sterilized Petri dishes filled with distilled water containing 10% B5 medium without phytohormones (1 mm high). Two control media were used; one with 10% B5 medium only, the other with 10% B5 medium plus 0.5% acetone. This acetone concentration, which had no influence on the root growth, was necessary to dissolve the herbicides.

For the experiments, two sulfonylurea herbicides were selected: 1-(2-chlorophenyl sulfonyl)-3-(4-methoxy-6-methyl-1.3.5-triazine-2-yl) urea (chlorsulfuron; Sigma, Deisenhofen, Germany) and methyl-2-[3-(4-methoxy-6-methyl-1,3,5-triazine-2-yl) sulfonylurea] benzoate (metsulfuron-methyl; Sigma, Deisenhofen, Germany). The herbicides were added to the culture medium (except the controls) to achieve final concentrations of 1, 10 and 50 mL $L^{-1}$ chlorsulfuron or metsulfuron-methyl, respectively. After an incubation for 6 days at 22 °C in the dark, the root lengths were measured to calculate the mean percentage of growth reduction in relation to the control.

For the fluorescence and transmission electron microscopic investigations, root tips (2 mm long) were fixed with 2.5% glutaraldehyde in phosphate buffer (at 20 °C for 3 h) after an incubation period of only 24 h to study very early stages of root development under herbicide influence. The root tips were then post-fixed in phosphate-buffered 1% osmium tetroxide for 2 h at 20 °C, rinsed with 0.05 *M* phosphate buffer, dehydrated in an acetone series and embedded in Spurr's medium (Spurr, 1969). Ultrathin sections (60 nm thick) were treated with lead citrate and examined with the Philips electron microscopes EM 201c and EM 420.

For fluorescence microscopy, semithin sections (0.7 $\mu$m thick) of the embedded root tips were heat fixed to glass slides, and the embedding medium was removed with 2% ethanolic NaOH as described by Lane and Europa (1965). Periodic acid Schiff's (PAS) reaction for fluorescent staining of cell walls was carried out according to Schröder et al. (1980). PAS fluorescence was observed using a Zeiss fluorescence microscope (Zeiss 15 colour filter, excitation 546 nm, beam splitter 580 nm, barrier 590 nm).

For scanning electron microscopy, the root tips were fixed as described above. Following fixation, the specimens were dehydrated in an acetone series on ice, critical-point dried using liquid $CO_2$, coated with gold, and observed in a Philips SEM 505 electron microscope.

## Results

Both herbicides applied caused drastic reduction of root growth of *Zea mays* and *Pisum sativum* seedlings. After 6 days of germination, the root length of *Zea mays* was reduced to 97% by 10 mL $L^{-1}$ and to 30% by 50 mL $L^{-1}$ chlorsulfuron, and to 94% by 10 mL $L^{-1}$ and to 28% by 50 mL $L^{-1}$ metsulfuron-methyl in comparison with 100% growth of the control seedlings (Fig. 1).

These values were calculated as the mean of 45 seedlings each. The standard deviations were 12.3 and 5.5% or 14 and 10.9%, respectively. A concentration of 1 mL $L^{-1}$ of the herbicides did not influence the root growth of *Zea mays*. For *Pisum sativum*, the root length was reduced to 58, 42 and 34% of the control by 1, 10 or 50 mL $L^{-1}$ chlorsulfuron, respectively, and to 64, 40 and 29% by 1, 10 or 50 mL $L^{-1}$ metsulfuron-methyl after 6 days of germination (Fig. 2). These values were calculated as the mean of 42 seedlings each with a range of standard deviations from 6.3 to 9.8%.

Mainly, two types of structural alterations were observed in the root caps of the herbicide-treated *Zea mays* and *Pisum sativum* seedlings. One type concerned the central tissue of the root cap predominantly consisting of statocytes, and the other was found in the peripheral region where the secretory tissue is located.

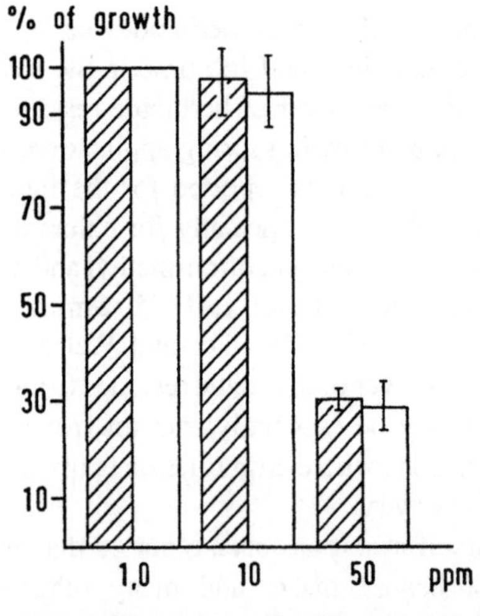


*Fig. 1.* Effect of different concentrations (ppm) of chlorsulfuron (■) and metsulfuron methyl (□) on the root length (growth) of *Zea mays.*

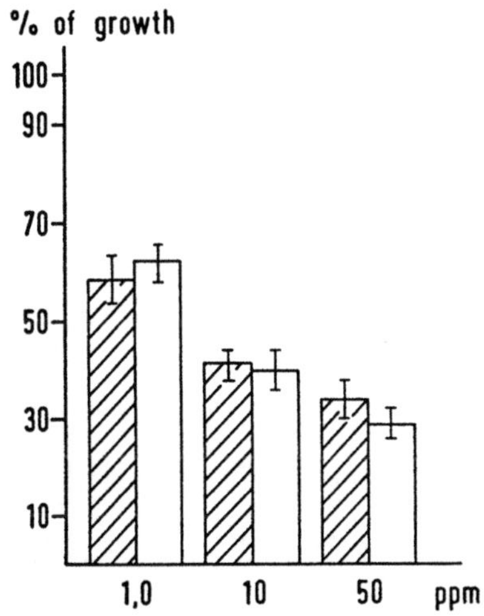


*Fig. 2.* Effect of different concentrations (ppm) of chlorsulfuron (■) and metsulfuron methyl (□) on the root length (growth) of *Pisum sativum.*

After 24 h of treatment with 1% chlorsulfuron or metsulfuron-methyl, the cytoplasm of the statocytes (Fig. 4) was more electron-dense than in the corresponding control cells (Fig. 3), the number and size of vacuolar compartments were increased, the statolithes seemed to be shrunken, and the vacuoles contained dots of electron-dense material (Fig. 4). Mitosis was sporadically observed in the statocytes (Fig. 4). However, treatment with 10 or 50 mL $L^{-1}$ chlorsulfuron or metsulfuron-methyl for 24 h caused damage of the cytoplasm and degeneration of the organelles with release of starch grains from the statolithic amyloplasts which lost their envelopes (Fig. 5).

The observation of the secretory tissue revealed the most obvious injuries among all regions of the root cap, probably due to the most exposed situation of the peripheral cell layers which had direct and intense contact with the herbicides. The tip region of the root cap was especially affected. The two or three outermost cell layers were totally damaged as demonstrated by fluorescence microscopy of the PAS-stained cell walls (compare Figs. 6 and 7). Most of the secretory cells of the root cap treated with 10 mL $L^{-1}$ chlorsulfuron or metsulfuron-methyl for 24h were degenerated (Fig. 8) whereas the corresponding control cells showed active dictyosomes, normal mitochondria and a well developed ER (Fig. 9).

The damage of the secretory tissue was reflected by the changed surface feature of the root tips as shown by the use of scanning electron microscopy. The SEM micrographs (Figs. 10 and 11) clearly demonstrate the differences between the control (Fig. 10) and the herbicide-treated root tips (Fig. 11). The surface of the subapical region of the untreated root tip was covered with slime produced by the intact root cap (Fig. 10), whereas the root tip treated with 10 mL $L^{-1}$ chlorsulfuron or metsulfuron-methyl seemed to be naked in this region (Fig. 11). The features of the herbicide-treated root caps also differed from the control in that the sloughed cells were visible due to the lack of secretory material after herbicide treatment (compare Figs. 12, 13 and 14).

In addition to the herbicide-induced changes mentioned above, two further features were observed in the root caps of *Zea mays*: An obvious zig-zag shape of some cell walls in the lateral, peripheral root cap tissue (Fig. 15) which was never present in the control, and a frequent occurrence of large concentric aggregates of rough ER in the initial cell complex of the root cap after 24 h treatment with 50 mL $L^{-1}$ chlorsulfuron (Fig. 16).

## Discussion

The influence of sulfonylurea herbicides on the growth and development of roots is relatively well known. It has been reported that the number of cell divisions in maize roots was drastically reduced after treatment with chlorsulfuron (Ray, 1982a, b; Rost, 1984). A similar effect of chlorsulfuron was demonstrated with pea roots (Rost, 1984). It is not surprising that chlorsulfuron and also metsulfuron-methyl interfere with the processes of cell growth and cell proliferation because both herbicides inhibit the synthesis of the amino acids valine, leucine and isoleucine by blocking the acetolactat synthase pathway (Larossa and Schloss, 1984; Scheel and Casida, 1985). To our knowledge, root caps have not yet been examined in relation to the action of sulfonylurea herbicides although this tissue has very important functions, e.g. perception of gravity (Iversen and Rommelhoff, 1978; Volkmann, 1974; Volkmann and Sievers, 1975), mechanical protection of the meristematic root tip region and protective slime secretion (Bacic et al., 1986). Disturbance of even one of these functions would produce severe consequences for root growth and development.

Our observations exhibited that chlorsulfuron and metsulfuron-methyl just affected those parts of the root cap which are responsible for all three functions mentioned. The cytoplasms of the statocytes was obviously damaged by relatively low concentrations

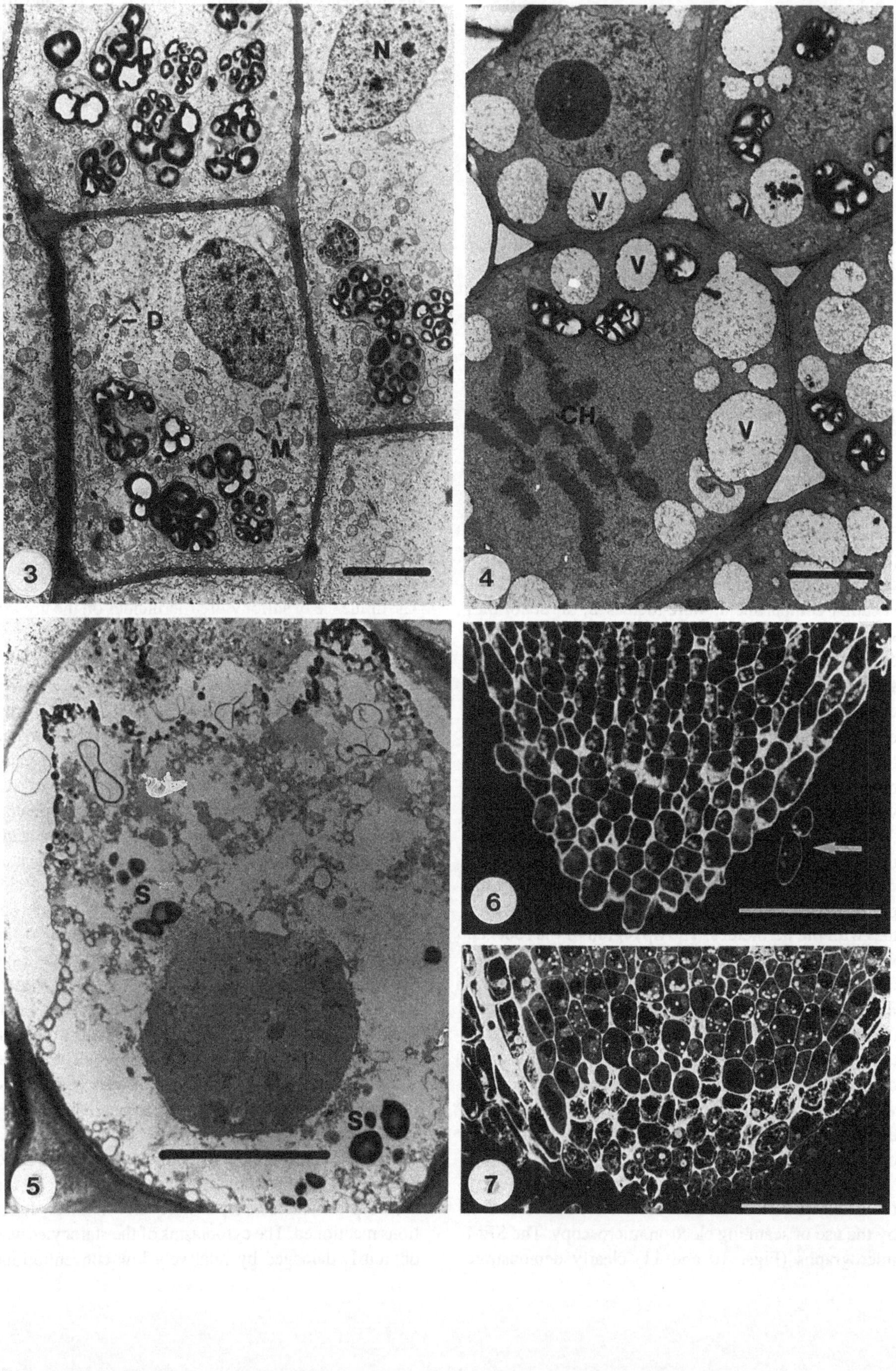
N
D
N
M
3
V
V
CH
V
4
S
S
5
6
7

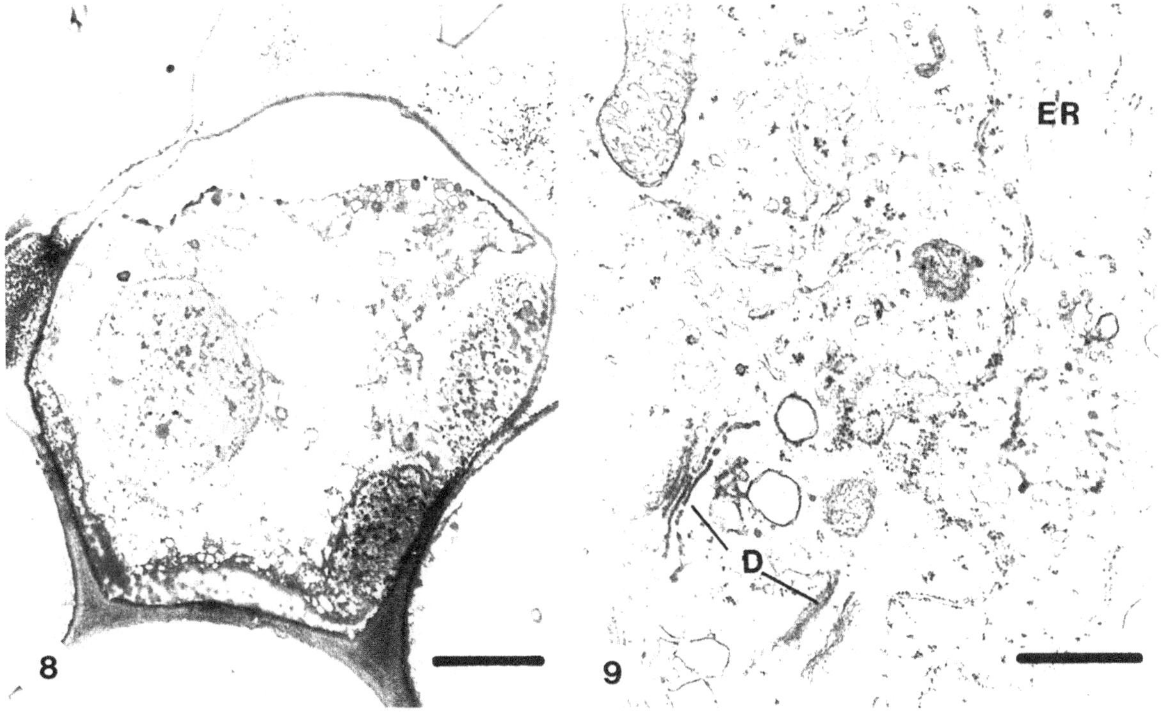


*Fig. 8 and 9.* Electron micrographs of the secretory tissue of the root cap of *Pisum sativum*.
*Fig. 8.* Treatment with 10 ppm chlorsulfuron for 24 h. The cytoplasm and the organelles of the secretory cells are totally damaged. *Bar* = 5 $\mu$m.
*Fig. 9.* Control. The cytoplasm shows a normal feature; The ER, the dictyosomes (D) and the other organelles are well developed. *Bar* = 1 $\mu$m.

of the herbicides. The endoplasmic reticulum usually forming large paralelly arranged cisternae (Volkmann and Sievers, 1975) was destroyed, and the statolithic starch grains, normally surrounded by the amyloplast envelope, was released into the cytosol. Under such conditions, the root would no longer be able to correctly respond to gravity. Barlow (1974) has clearly shown that removal of the root cap abolished the geotropism but also induced cell division in the remaining portion of the root tip. This led to a generation of a new cap in *Zea mays* and *Pisum sativum* (Barlow and Hines, 1982). However, regeneration of the statenchyme with recovery of the geotropism seems unlikely in an herbicide-polluted environment.

Normally, the outer cell layers of the root cap secrete polysaccharide slime by the activity of the Golgi apparatus (Northcote and Picket-Heaps, 1966; Paull and Jones, 1976; Rougier, 1981; Piche et al., 1985). Thereafter, the cells of these layers are separated from each other by partial cell wall digestion and are sloughed from the root cap (Clowes and Woolston, 1978). In the herbicide-treated seedlings, the secretory cells were degenerated before sloughing. They showed different stages of damage from destruction of their endomembrane system, including the Golgi appa-

*Fig. 3–7.* Electron micrographs of the statenchym in the root cap of *Pisum sativum*.
*Fig. 3.* Control. Longitudinal section. The statocytes contain the characteristic large, enveloped starch grains. Numerous mitochondria (M) and dictyosomes (D) are also present. N = nucleus. *Bar* = 5 $\mu$m.
*Fig. 4.* Treatment with 1 mL $L^{-1}$ chlorsulfuron for 24 h. Cross section. The cytoplasm is more electron-dense than in the control (Fig. 3). The number of vacuolar compartments (V) is increased and most of the vacuoles contain electron-dense material. The envelope of the amyloplasts (statolithes, which seemed to be shrunken) is still present. One cell is in the stage of mitosis (C = chromosomes). *Bar* = 5 $\mu$m.
*Fig. 5.* Treatment with 10 mL $L^{-1}$ chlorsulfuron for 24 h. Cross section. The cytoplasm and the organelles are damaged. The starch grains, still visible, are no longer ensheathed by membranes. *Bar* = 5 $\mu$m.
*Fig. 6 and 7.* Fluorescence micrographs of the root cap tip of *Pisum sativum*. Longitudinal semithin sections stained with PAS. *Bars* = 100 $\mu$m.
*Fig. 6.* Control. The apex shows a regular arrangement of the peripheral secretory cells and two sloughed cells (arrow).
*Fig. 7.* Treatment with 10 mL $L^{-1}$ chlorsulfuron for 24 h. The cells of the apical root cap tissue are damaged.

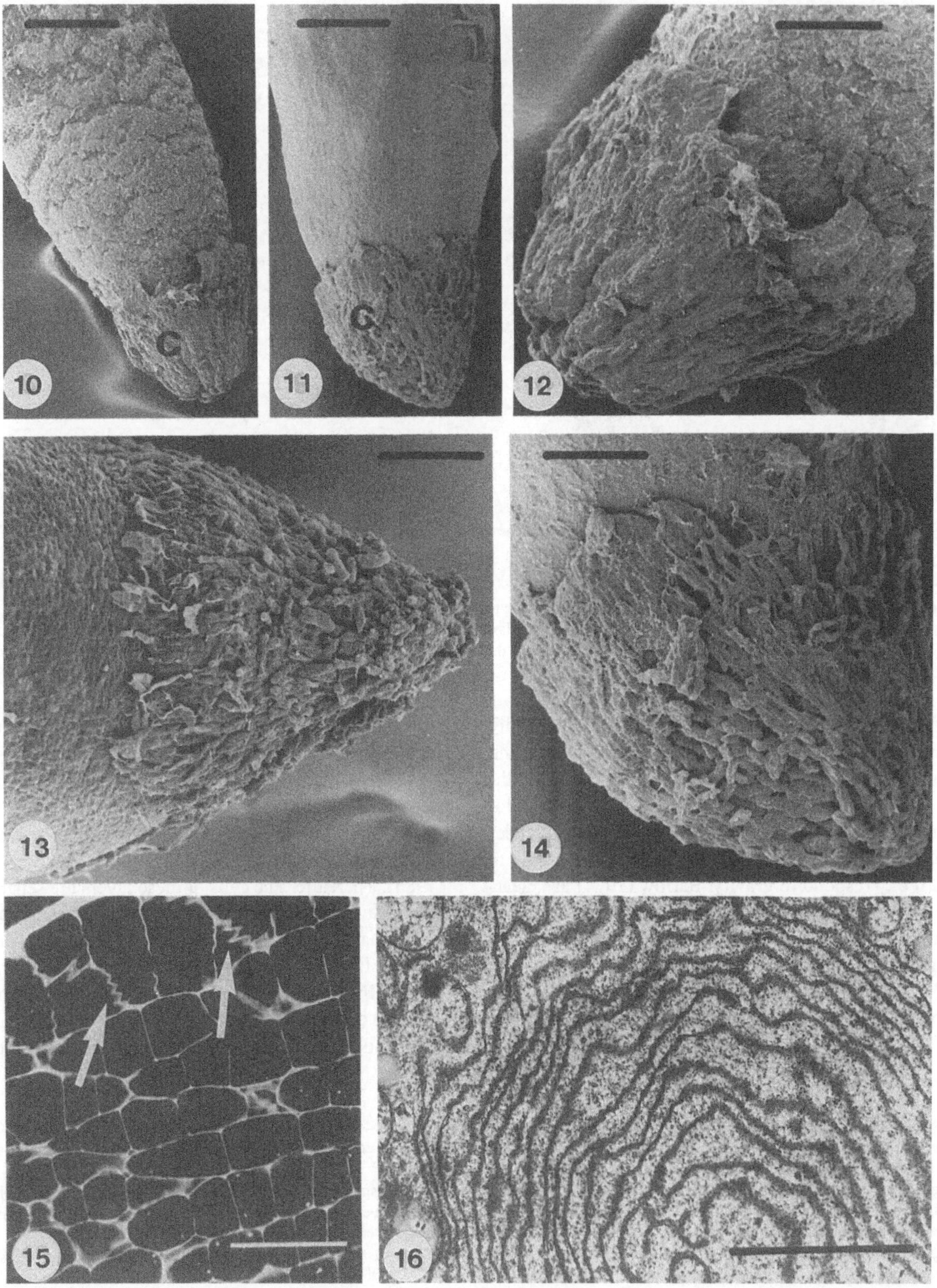
C
C
10
11
12
13
14
15
16

ratus, up to total degeneration of their cytoplasm. As a consequence of such severe injuries of the outer root cap cells, the production and the secretion of the protective root cap slime must be inhibited. Comparison of the scanning electron micrographs from untreated and herbicide-treated maize root tips clearly supported this suggestion.

Moreover, the herbicide action may have reached the initial cell complex of the root cap as indicated by the formation of concentric complexes of rough ER in the maize roots after treatment with 50 mL $L^{-1}$ chlorsulfuron. Similar ER complexes have been reported to occur in the cortex of maize roots with drought or cold stress and have been related to the synthesis of stress proteins (Ciamporova and Mistrik, 1993). However, it is still unclear whether these complexes reflect enhanced or reduced protein synthesis since they have been observed in very active cells (Kristen, 1977) as well as in cells of resting tissue (Bergfeld and Schopfer, 1984).

Both herbicides used in this study have been developed to inhibit predominantly the growth of dicotyledonous weeds. The injuries of the maize root cap show that they also seriously affect monocotyledonous plants. However, the main finding of our study is that low concentrations of herbicides affect seedlings in very early stages of their development. It is likely that early root cap injuries will negatively influence later stages of plant development, thus causing growth retardation until maturity of crops.

## Acknowledgements

We gratefully acknowledge the technical assistance of Irmhild Wachholz, Christine Schumacher (electron microscopy) and Christa Adami (photo lab), and the fruitful discussion with Ute Joos and Jan van Aken.

## References

Anderson J J and Dulka J J 1985 Environmental fate of sulfometuron methyl in aerobic soils. J. Agric. Food Chem. 33, 596–602.

Bacic A, Moody S F and Clarce A E 1986 Structural analysis of secreted root slime from maize (*Zea mays* L.). Plant Physiol. 80, 771–777.

Barlow P W 1974 Recovery of geotropism after removal of the root cap of *Zea mays*. J. Exp. Bot. 25, 1137–1146.

Barlow P W and Hines E R 1982 Regeneration of the root cap of *Zea mays* L. and *Pisum sativum* L.: a study with the scanning electron microscope. Ann. Bot. 49, 521–529.

Bergfeld R and Schopfer P 1984 Transitory development of rough endoplasmic reticulum aggregates during embryo maturation in seeds of mustard (*Sinapis alba* L.). Eur. J. Cell Biol. 34, 27–33.

Ciamporova M and Mistrik I 1993 The ultrastructural response of root cells to stressful conditions. Environ. Exp. Bot. 33, 11–26.

Clowes F A L and Woolston R E 1978 Sloughing of root cap cells. Ann. Bot. 42, 83–89.

Dyer W E and Fay P K 1984 The effect of chlorsulfuron soil residues on 11 crops, 36 month after herbicide application. West. Soc. Weed Sci. Res. Rep. Weed Science Society of America, Champaign, IL. 226p.

Iversen T H and Rommelhoff A 1978 The starch statolith hypothesis and the interaction of amyloplasts and endoplasmic reticulum in root geotropism. J. Exp. Bot. 29, 1319–1328.

Kristen U 1977 Auffällige ER-Konfigurationen in Protein-Polysaccharidschleim-sezernierenden pflanzlichen Drüsen. Planta 133, 161–167.

Lane B P and Europa D L 1965 Differential stayning of ultrathin sections of epon-embedded tissue for light microscopy. J. Histochem. Cytochem. 13, 579–582.

Larossa R A and Schloss J V 1984 The sulfonylurea herbicide sulfometuron methyl is an extremely potent and selective inhibitor of acetolactat synthase in *Salmonella typhimurium*. J. Biol. Chem. 259, 8753–8757.

Nicholls P H, Evans A A and Walker A 1987 The behaviour of chlorsulfuron and metsulfuron in soils in relation to incidents of injury to sugar beet. Proc. British Crop. Prot. Conf. Weeds 2, 549–556.

Northcote D H and Pickett-Heaps J D 1966 A function of the Golgi apparatus in polysaccharide synthesis and transport in root cap cells of wheat. Biochem. J. 98, 159–167.

Paull R E and Jones R L 1976 Studies on the secretion of maize root cap slime. IV. Evidence for the involvement of dictyosomes. Plant Physiol. 57, 249–256.

Piche Y Ackerley C A and Peterson R L 1985 Localization of L-fucose in the root cap and slime of *Zea mays* L. utilizing fluores-

*Fig. 10–14.* Scanning electron micrographs of root tips of *Zea mays*.

*Fig. 10.* Control. The root tip surface is totally covered with root cap slime. C = root cap. *Bar* = 200 μm.

*Fig. 11.* Treatment with 10 mL $L^{-1}$ chlorsulfuron for 24 h. Root cap slime is less present on the surface of the root cap (C) and absent from the subapical root surface. *Bar* = 200 μm.

*Fig. 12.* Control. The peripheral and the sloughed cells of the root cap are not clearly visible because of their enclosure by the root cap slime. *Bar* = 100 μm.

*Fig. 13 and 14* Treatment with 10 mL $L^{-1}$ metsulfuron methyl (Fig. 13) or 10 mL $L^{-1}$ chlorsulfuron (Fig. 14). Root caps of *Zea mays*. Most of the peripheral and sloughed cells are visible because of the absence or shortage of root cap slime. *Bar* = 100 μm.

*Fig. 15.* Fluorescence micrograph of a longitudinal section through the subapical, peripheral region of a *Zea mays* root cap. Treatment with 50 mL $L^{-1}$ chlorsulfuron for 24 h. Some cell walls are zig-zag (arrows). *Bar* = 50 μm.

*Fig. 16.* Electron micrograph of the initial cell complex of the root cap. Treatment with 50 mL $L^{-1}$ chlorsulfuron for 24 h. The cell shows a large concentric aggregate of rough ER. *Bar* = 4 μm.

cent lectin conjugates and colloidal gold-lectin techniques. Plant Sci. 40, 179–184.

Ray T B 1982a The mode of action of chlorsulfuron: a new herbicide for cereals. Pestic. Biochem. Physiol. 17, 10–17.

Ray T B 1982b The mode of action of chlorsulfuron: the lack of direct inhibition of plant DNA synthesis. Pestic. Biochem. Physiol. 18, 262–266.

Rost T L 1984 The comparative cell cycle and metabolic effects of chemical treatments on root tip meristem. III. Chlorsulfuron. J. Plant Growth Regul. 3, 51–63.

Rougier M 1981 Secretory activity of the root cap. *In* Encyclopedia of Plant Physiology. New Series 13B. Ed. F Loewus and W Tanner. pp 542–574. Springer, Berlin, Heidelberg, New York.

Scheel D and Casida J E 1985 Sulfonylurea herbicides: growth inhibition in soybean cell suspension cultures and in bacteria correlated with block in biosynthesis of valine, leucine or isoleucine. Pestic. Biochem. Physiol. 23, 398–412.

Schröder H E, Rossinsky U and Müller W 1980 An established routine method for differential staining of epoxy-embedded tissue sections. Microsc. Acta 83, 111–116.

Smith A E 1986 Persistence of the herbicides [$^{14}$C] chlorsulfuron and [$^{14}$C] metsulfuron methyl in prairie soils under laboratory conditions. Bull. Environ. Contam. Toxicol. 37, 698–704.

Smith A E and Aubin A J 1992 Degradation of the sulfonylurea herbicide [$^{14}$C] amidosulfuron (HOE 075032) in Saskatchewan soils under laboratory conditions. J. Agric. Food Chem. 40, 2500–2504.

Spurr A R 1969 A low-viscosity epoxy resin embedding medium for electron microscopy. J. Ultrastruct. Res. 26, 31–43.

Volkmann D 1974 Amyloplasten und Endomembranen: Das Geoperzeptionssystem der Primärwurzel. Protoplasma 79, 159–183.

Volkmann D and Sievers A 1975 Wirkung der Inversion auf die Anordnung des Endoplasmatischen Retikulums und die Polarität der Statocyten in Wurzeln von *Lepidium sativum*. Planta 127, 11–19.

Walker A and Welch S J 1989 The relative movement and persistence in soil of chlorsulfuron, metsulfuron-methyl and triasulfuron. Weed Res. 29, 375–383.

*F. Baluška et al. (eds.), Structure and Function of Roots*, 285–291, 1995.


# Biophysics of the growth responses of pea roots to changes in penetration resistance

A.G. Bengough, C.J. Mackenzie and H.E. Elangwe
*Soil-Plant Dynamics Group, Cellular and Environmental Physiology Department, Scottish Crop Research Institute, Dundee DD2 5DA, Scotland, UK*

*Key words:* cell expansion, *Pisum sativum* L., root growth pressure, root penetration, soil strength, temperature


## Abstract

Two experiments were performed in simplified soil-less systems to study how roots respond to changes in mechanical impedance. In the first the increases in root force and diameter that occur when a pea root was impeded mechanically inside a hole with rigid conical walls were determined. The experiment was performed at 8°C and at 25°C, and the root growth pressures generated were calculated during periods of 12, 24, 48 and 72 hours. The maximum growth pressures generated were approximately the same at both temperatures, although the maximum pressure was achieved approximately twice as quickly at 25°C than at 8°C, being reached within 15–20 hours. In the second set of experiments a new technique was developed to measure simultaneously the elongation rate and the force exerted by the roots of seedlings grown in moist air. A constant force was exerted by a force transducer on a pea radicle using a system of pulleys, and the elongation rate of the pea root was monitored using a linear variable differential transformer (LVDT). The changes in root elongation rate were recorded that occurred in response to increases and decreases in the applied force. Root elongation rate decreased by more than 50% within 30 min of increasing the applied force by 100 mN. A similarly fast, but smaller increase in growth rate occurred when the force was removed. The interpretation of results from both studies will be discussed in terms of a modified form of the Lockhart model of growth.


## Introduction

Root elongation is often slowed in compacted and in dry soil due to the physical resistance to root extension. When a root encounters a particle of soil it has to either exert a force sufficient to dislodge the particle, or change its direction of growth to grow around the particle, if an alternative lower resistance pathway exists. The axial penetration resistance, $Q_r$, of soil is given by

$$Q_r = F_r / A_r \qquad (1)$$

where $F_r$ is the axial force exerted by the root tip, and $A_r$ is the cross-sectional area of the root.

Greacen (1986) and Greacen and Oh (1972) suggested that if water is not limiting cell expansion, the elongation rate, R, of a root growing against a mechanical impedance can be described by a modified form of the Lockhart Equation:

$$R = m_r \cdot (P - Y - Q_r) \qquad (2)$$

where P is the turgor pressure inside the root cells, $m_r$ is the extensibility of the length of growing tissue, and Y is the yield threshold of the cell walls. However, it is not known to what extent P, Y, and $m_r$ are themselves dependent on the value of $Q_r$, and it is likely that $m_r$ and Y are related (Passioura and Fry, 1992). Root tips contain a progression of cells at different stages of expansion, and in fully expanded cells the cell walls will have hardened such that the yield threshold balances the turgor pressure.

The maximum growth pressure that a root can exert is given by setting R=0 in equation (2), so that

$$Q_{rmax} = P - Y \qquad (3)$$

$Q_{rmax}$ has been determined by measuring the force exerted by a root pressing into a hole in a rigid block of plaster of Paris or ceramic material. The cross-sectional area of the root is normally measured at the end of the experiment, so that $Q_{rmax}$ can be calculated using Equation (1). However, the diameter of roots increases generally with mechanical impedance (Bengough and Mullins, 1990), and so as both root force and diameter are increasing with time, it is not normally possible to have a continuous record of $Q_r$.

Values for $Q_{rmax}$ vary markedly even within a cultivar, and Eavis, Taylor and Ratliff (1969) recorded a range of values for pea seedlings from 0.6 to 2.6 MPa. For pea seedlings, mean values of $Q_{rmax}$ have been measured as 1.2 and 1.3 MPa (Eavis et al., 1969; Taylor and Ratliff, 1969) at about 30°C, and 0.3 to 0.5 MPa at 20°C (Misra et al., 1986). In an attempt to explain these differing pressures, Greacen (1986) hypothesized that the larger maximum growth pressures were due to the warmer temperature used by Eavis et al. (1969) and Taylor and Ratliff (1969). Under normal conditions, warmth increases the rate of translocation of sugars to the root tip and the rate of cell production/division (Barlow, 1993), although it also increases the rate at which these sugars are broken down by respiration, and the rate of dilution of the cell contents by growth. Temperature can also affect the mechanical properties of plant cell walls (e.g. Pritchard et al., 1990a). It is possible that temperature could influence the maximum growth pressure, either through its effect on the cell walls, or through an effect on the osmotic balance when root growth is halted by a mechanical impedance.

In the field, roots encounter zones of soil of different strength, and can grow into cracks and pores, and so it is important to understand how roots respond to changes in mechanical impedance. The effects of changes in soil strength have only been investigated in a few studies (Bengough and Young, 1993; Goss and Russell 1980), and it has not been possible to obtain simultaneous measurements of root force and elongation rate. Detailed data on changes in growth rate and on interactions with factors such as temperature will provide better insight into the mechanisms involved and clarify the validity of Equation (2) as a model for the effects of mechanical impedance on root elongation.

The first aim of this paper was therefore to investigate the effect of temperature on the maximum growth pressure exerted by pea radicles, and how this pressure changes with time.

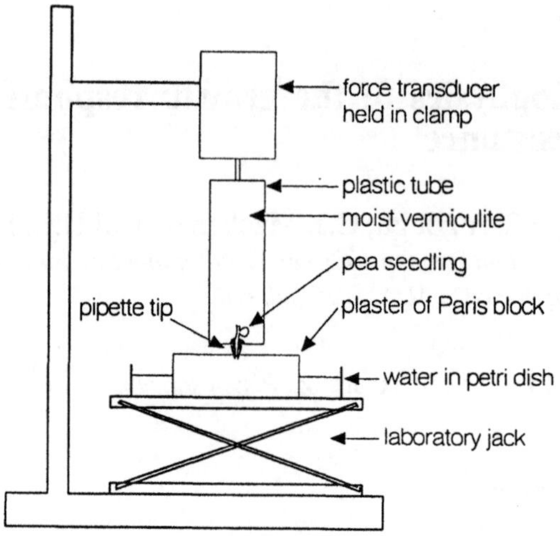


*Fig. 1.* Apparatus for measuring the root growth pressure when a root is grown against a fixed obstruction.

The second aim was to investigate what happens to the root elongation rate when the force applied to the root tip is changed.

## Materials and methods

### *Seedling preparation*

Pea seeds (*Pisum sativum* L. c.v. Helka) were surface sterilized with a saturated solution of calcium hypochlorite, washed in distilled water, and germinated in the dark on moist filter paper. Different seed batches were used for Experiments 1 and 2, and the seeds were soaked for 5 hours in Experiment 1 prior to germination. Temperature was maintained constant (± 1°C) using a growth cabinet.

Seedlings were selected with radicles between 10 and 20 mm long and transferred to a polythene sample tube that had been adapted specially (Fig. 1). The radicle was inserted through a hole in the base of the sample tube so that about 2–3 mm of root protruded. The root was anchored in place with a small quantity of plaster of Paris. Moist vermiculite was placed around the seedling.

### *Experiment 1*

The sample tube was attached to a force transducer while the root tip was placed inside the top of a 6 mm deep plastic cone made from a disposable pipette tip (Fig. 1). The base of the pipette tip was embedded in a plaster of Paris block which sat in a petri dish containing deionised water, to keep the air surrounding the

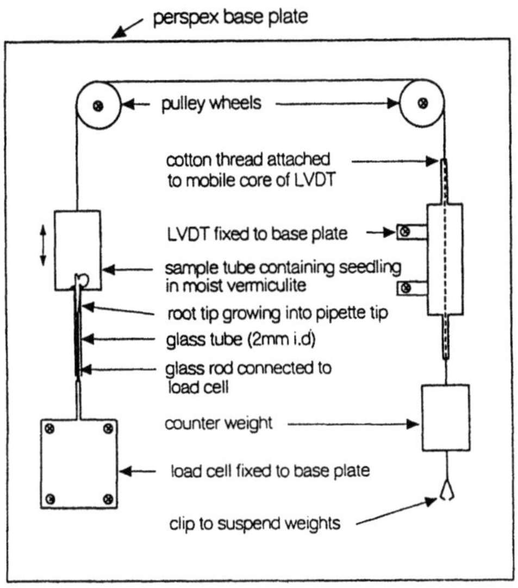


*Fig. 2.* Pulley arrangement for applying a force to the root tip and recording the force and elongation.

root saturated with water-vapour. The force exerted by the root on the cone was registered by the force transducer and the readings were recorded every 10 minutes using a data logger. The experiment was performed in a growth cabinet in the dark at 8°C and at 25°C. An individual run for a single seedling lasted for a period of either 12, 24, 48 or 72 hours with the force exerted by the root being recorded at 10 min intervals. At the end of each run the root diameter was measured. A total of 24 replicates were used at 8°C, and 32 at 25°C.

*Experiment 2*

A detailed description of the experimental arrangement is given in Bengough and MacKenzie (1994). The sample tube was suspended over two pulleys and connected by cotton thread to a Linear Variable Displacement Transformer (LVDT) and a counter-weight (Fig. 2). A glass tube (2 mm internal diameter) was fixed onto the sample tube for the root to grow down. The root tip was allowed to grow into a plastic pipette tip, the end of which was plugged with a little plaster of Paris, and attached via a rod to a force transducer. The elongation of the root was registered on the LVDT, while the force exerted by the root tip was registered by the force transducer. The difference in mass between the root holder and the counter-weight determined the force applied to the root tip, and readings of displacement and force were recorded at 5 min intervals by the data logger. To investigate the effect of increasing or decreasing the force experienced by a growing root

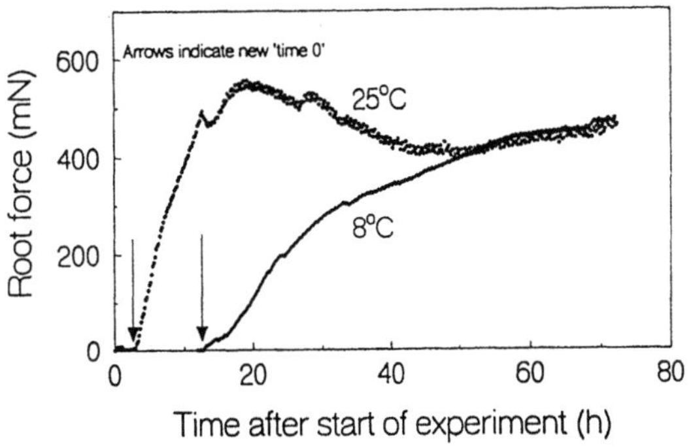


*Fig. 3.* Root force as a function of time for two individual pea roots, one at 8° C and one at 25° C.

tip, a 10 g weight was added or removed from the counter weight after the experiment had been running for about 24 h, and the experiment was then left to run for a further 24 hours. The treatment in which the force on the root tip was increased will be referred to as the '0-10g' treatment, and '10-0g' refers to the treatment in which the force on the root tip was decreased. The experiment was performed in a growth cabinet in the dark at 15°C.

*Measurement of root diameter and calculation of the final root growth pressure*

Diameters were measured at the end of each experiment using a binocular microscope with graticule scale, either along the intact root (Experiment 1), or on thin cross-sections of root cut using a razor blade (Experiment 2). The average diameter was determined for the section of root 2–5 mm behind the root tip in both experiments and used to calculate the cross-sectional area by assuming that the root was cylindrical. The 'final' root growth pressure was calculated as the last value of force exerted by the root at the end of the experiment divided by the root cross-sectional area.

## Results

*Experiment 1*

Figure 3 shows the increase of force with time for two pea roots, one at 8°C and one at 25 °C. In general, the force increased more rapidly for the roots at 25°C than at 8° C. The initial period before the force started to increase corresponded to the time taken for the root tip to grow down to the base of the cone. The time when the force just started to increase was taken as time zero (indicated by arrows in Fig. 3). The final root growth pressure was calculated for each run and

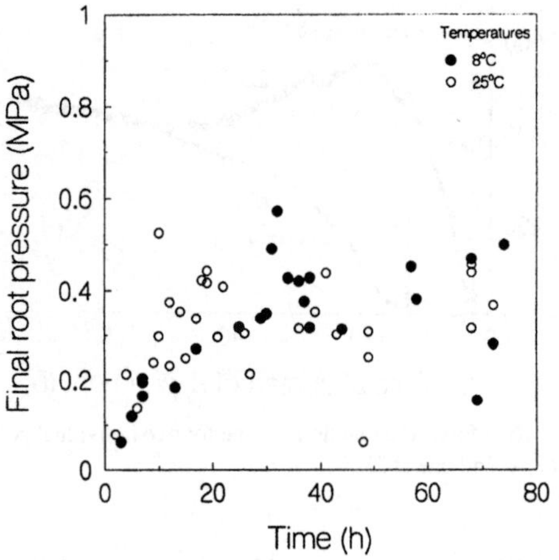


*Fig. 4.* Root growth pressure for pea radicles as a function of time at 8° C and at 25° C.

plotted against the time taken to reach that pressure (Fig. 4). The maximum growth pressure was reached in 15–20 h at 25°C compared with more than 30 h at 8°C. However, the maximum value reached did not differ between the temperatures. There was a slight tendency for the root growth pressure to increase with increasing root diameter at 8° C (Fig. 5a), and they were positively correlated (r=0.478, $p$<0.05). However at 25° C the growth pressure was independent of diameter (Fig. 5b) and there was no significant correlation (r=-0.028, $p$>0.05).

*Experiment 2*

The mean force and elongation rates are shown in Figure 6. The time when the weights were changed over was taken as time '0'. In the 0-10 g treatment the elongation rate decreased from about 10 mm/day to less than 5 mm/day within 0.5 h. There followed a gradual decline in elongation rate during the remaining 24 hours of the experiment.

In the 10–0 g treatment, the elongation rate increased within 0.5 h from 3 to 4 mm/day to 6 mm/day, and continued to increase gradually to 8 mm/day by the end of the experiment.

## Discussion

*Experiment 1*

It is interesting to speculate on what occurs inside the root during the initial increase in force at the start of each experiment. There will be some initial compression of tissue in the root as the advance of the root tip is stopped by the base of the cone. The root tissues are likely to compress faster at warmer temperatures simply because of the faster speed of axial growth. It is also possible that the value of P-Y in Equation 3 is increasing due to an increase in turgor resulting from either the passive accumulation of imported solutes, or to some form of osmotic adjustment (e.g. Atwell, 1990; as suggested by Greacen and Oh, 1972). The yield threshold in the axial direction could also be decreasing, for example, if cell wall loosening occurred initially while cell elongation was prevented by the obstruction.

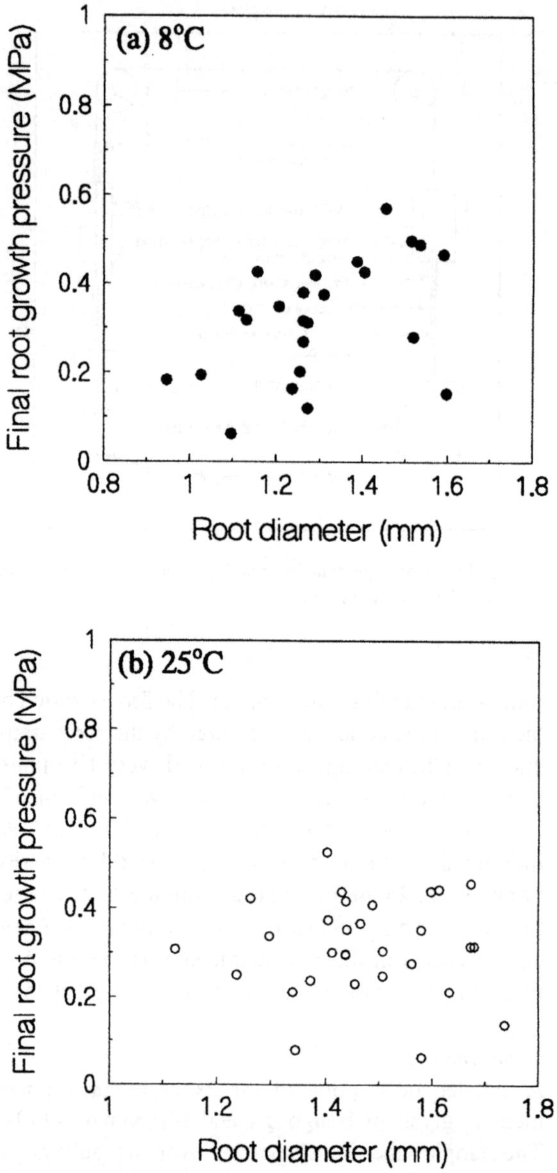


*Fig. 5.* Root growth pressure as a function of root diameter at (**a**) 8° C and at (**b**) 25° C.

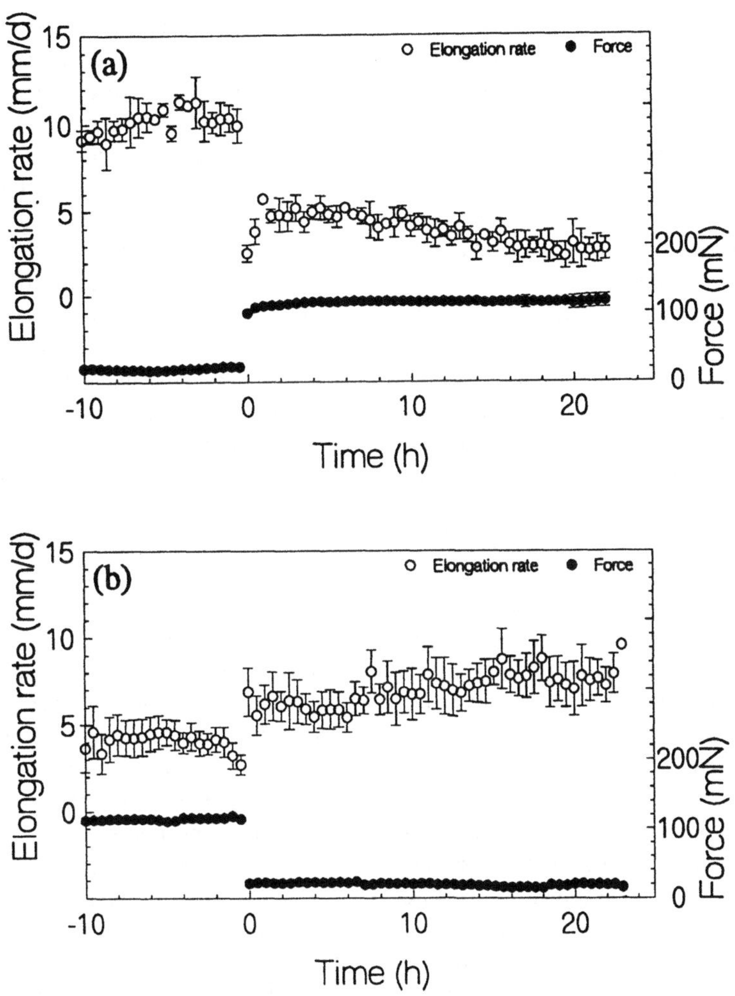


*Fig. 6.* Root force and elongation rate for the (**a**) 0–10 g and (**b**) 10–0g treatments. Points indicate mean values of 3–4 replicates ±1 standard error.

After the initial period of 10–15 h at 25°C and 30–35 h at 8°C, the growth pressures recorded were generally between 0.3 and 0.5 MPa. In the majority of runs at 25°C, the force exerted by the root passed through a maximum after about 20–25 h and decreased to approximately 80% of the maximum value by the end of the experiment (e.g. see Fig. 3). In contrast, there was no such decrease at 8°C. This may be due to continued solute import causing a transient increase in turgor or, alternatively, to an age (or thermal time) effect. Misra, et al. (1986) found that the maximum root growth pressure was 0.08 MPa smaller for pea seedlings that were 4 days old than for seedlings that were 2 days old. Nearly all studies of root growth pressure are performed using seedlings of only a few days old and so such age effects deserve further investigation. It could also be that after a prolonged period during which growth has stopped, the cell walls become stiffer. Veen (1982) found that reorientation of microfibrils occurred in mechanically impeded roots, suggesting that the axial yield threshold would increase.

The maximum growth pressures measured were similar to those determined by Misra et al. (1986) for peas, but less than half of those determined by Eavis et al. (1969) and Taylor and Ratliff (1969). As temperature did not have a major effect on the maximum growth pressure of peas in this experiment, it seems unlikely that this could explain the large differences between the growth pressures measured in these different studies. Root diameter and hence the cross-sectional area was determined 1, 4, 4.25, and 2–5 mm behind the root tip in the studies of Eavis et al. (1969),

Taylor and Ratliff (1969), Misra et al. (1986), and in the current studies respectively. As the root diameter often increases with distance from the root tip, Eavis et al. (1969) would calculate relatively larger values of root growth pressure due to the smaller root cross-sectional area at 1 mm from the tip.

It is debatable which is the most appropriate position to measure the root diameter. In soil, the length of root apical to the zone of elongation is pushed through the soil by the elongating cells. The resistance of the soil to deformation by the root tip will create mechanical stresses within the root from the root cap including the elongating zone, and these stresses will act both radially and axially. In this paper we have considered the axial force per unit area acting on the cross-sectional area of the root determined from the average diameter between 2 and 5 mm behind the root tip. This region was selected as being the region where the rate of cell elongation is normally fastest (Eavis, 1967) and so corresponding to where the pressures are generated that drive the apical region of the root through the soil. In soil the radial forces exerted by the root may also help the root to penetrate axially, by creating a zone of stress relief in front of the root tip (e.g. Abdalla et al., 1969).

Another approach to examine possible effects of temperature on the maximum root growth pressure is to consider the value of P-Y determined in pressure probe studies in the absence of mechanical impedance. Pritchard et al. (1990b) found that in wheat roots P was the same (=0.77 MPa) at 27°C and at 15°C, while Y was 0.07 MPa at 25°C and $\leqslant$ 0.2 MPa at 15°C. For maize P increased from 0.59 to 0.82 MPa when the temperature was decreased from 20°C to 5°C, and for a temperature decrease from 30°C to 15°C, Y was unchanged at 0.23 MPa (Pritchard et al., 1990a). These results suggest that, for *unimpeded* wheat and maize roots, P-Y was between about 0.4 and 0.8 MPa and was not decreased by cooler temperature. This work supports the finding of this paper, that the pressure available for growth is not affected within this temperature range. In both studies (Pritchard et al., 1990a,b) the extensibility, $m_r$, decreased at cooler temperatures.

*Experiment 2*

In this experiment an axial force was applied directly to the root tip independent of its elongation rate, whereas in soil a root tip has to push against the reaction force of deforming soil both axially and radially. The roots grew straight for at least 10 to 20 mm down the tube against forces of up to 120 mN, and so the displacement recorded by the LVDT gave a valid measure of the root elongation rate. Bending of the root may occur if larger forces are applied or if the length of root is increased.

Two separate stages can be identified in the response of roots to a change in the force applied. The first stage is a fast response which occurs within 30 min of a change in the applied force. This change in growth rate is due to the change in the pressure available to drive cell elongation (=P-Y-$Q_r$), caused by $Q_r$ increasing or decreasing. The subsequent slower alteration in elongation rate during a period of hours after the force changeover could be due to changes in turgor pressure, the cell wall properties ($m_r$, Y), or in the rate of cell production in the meristem. Perhaps the simplest explanation is that either $m_r$ decreases axially in response to mechanical impedance or, alternatively, that Y increases axially during impedance. This hypothesis could also account for why the 'fast' increase in elongation rate on reducing the force applied to a root that had been impeded mechanically was smaller than the 'fast' decrease in elongation rate that occurred when the force on a previously unimpeded root was increased. The question remains, however, as to why Greacen and Oh (1972) found no change in $m_r$ or in Y with increasing mechanical impedance.

The forces applied to the root tip in this experiment were equivalent to axial root penetration resistances of less than 0.1 MPa. This is less than one quarter of the maximum growth pressures that were measured for seedlings in Experiment 1, and may suggest that the relation between elongation rate and axial root penetration resistance is non-linear, especially at large root penetration resistances. However, caution must be exercised in extrapolating these results to soil, where radial pressures are very important to the process of root penetration.

## Acknowledgements

We are grateful to the Scottish Society for Crop Research for providing travel funds to enable this paper to be presented, and to the referees for thoughtful and constructive comments. This work was funded by the Scottish Office Agriculture and Fisheries Department.

## References


Abdalla A M, Hettiaratchi D R P and Reece A R 1969 The mechanics of root growth in granular media. J. Agric. Eng. Res. 14, 263–248.

Atwell B J 1990 The effect of soil compaction on wheat during early tillering. II. Concentrations of cell constituents. New Phytol. 115, 37–41.

Barlow P W 1993 The response of roots and root systems to their environment - an interpretation derived from an analysis of the hierarchical organization of plant life. Env. Exp. Bot. 33, 1–10.

Bengough A G and Mullins C E 1990 Mechanical resistance to root growth: A review of experimental techniques and root growth responses. J. Soil Sci. 41, 341–358.

Bengough A G and Young I M 1993 Root elongation of seedling peas through layered soil of different penetration resistances. Plant and Soil 149, 129–139.

Bengough A G and MacKenzie C J 1994 Simultaneous measurement of root force and elongation for seedling pea roots. J. Exp. Bot. 45, 95–102.

Eavis B W, Ratliff L F and Taylor H M 1969 Use of a deadload technique to determine axial root growth pressure. Agron. J. 61, 640–643.

Goss M J and Russell R S 1980 Effects of mechanical impedance on root growth in barley (*Hordeum vulgare* L.). III. Observations on the mechanism of response. J. Exp. Bot. 31, 577–588.

Greacen E L 1986 Root response to soil mechanical properties. Trans. 13th Int. Congr. Soil Sci. 5, 20–47.

Greacen E L and Oh J S 1972 The physics of root growth. Nature, New Biol. 235, 24–25.

Misra R K, Dexter A R and Alston A M 1986 Maximum axial and radial growth pressures of plant roots. Plant and Soil 95, 315–326.

Passioura J B and Fry S C 1992 Turgor and cell expansion: Beyond the Lockhart Equation. Aust. J. Plant Phys. 19, 565–576.

Pritchard J, Barlow P W, Adam J S and Tomos A D 1990a Biophysics of the inhibition of the growth of maize roots by lowered temperature. Plant Physiol. 93, 222–230

Pritchard J, Wyn Jones R G and Tomos A D 1990b Measurement of yield threshold and cell wall extensibility of intact wheat roots under different ionic, osmotic and temperature treatments. J. Exp. Bot. 41, 669–675.

Taylor H M and Ratliff L F 1969 Root growth pressures of cotton, peas, and peanuts. Agron. J. 61, 398–402.

Veen B W 1982 The influence of mechanical impedance on the growth of maize roots. Plant and Soil 66, 101–109.

*F. Baluška et al. (eds.), Structure and Function of Roots*, 293–298, 1995.


# Tolerance of meristematic root cells to desiccation in seeds; protection of the nucleus

Donato Chiatante
*Dipartimento di Scienze Animali, Vegetali e dell'Ambiente, Università degli Studi del Molise, via Cavour 50, 86100 Campobasso, Italia*

*Key words:* desiccation, germination, maturation, nucleus, root

## Abstract

Variation in the content of a particular nuclear protein (QP47) during seed maturation and the first phases of germination was investigated. In particular, the content QP47 in the nuclei increases during maturation and decreases during germination. The sub-nuclear distribution of QP47 during maturation and germination was examined by immunofluorescence with a polyclonal antibody and revealed that QP47 may be arranged along the chromosome. Despite the striking similarity between QP47 and LEA (Late Embryogenesis Abundant) proteins with regard to variations of content, many differences emerged; therefore, the function of QP47 remains to be determined. It is hypothesized that the presence of this nuclear protein might serve to protect the DNA molecule of embryonic cells during desiccation of maturing seeds.

## Introduction

In the majority of higher plants, tolerance to protoplasmic dehydration is restricted to the seed (Bartels et al., 1990). The only exception is represented by a small group of poikilohydric angiosperms (resurrection plants) whose leaves and other organs can exist in a completely dry but viable state for long periods (Bartels et al., 1990; Piatkowski et al., 1990). Within hours of rehydration, all the plant tissues resume full physiological activity.

In the case of maize, castor beans, barley and kidney-bean seeds, tolerance to desiccation is acquired at a later stage during embryogenesis (Bartels et al., 1988; Bewley et al., 1989; Bochicchio et al., 1988; Crouch and Sussex, 1981) but begins before desiccation (Bartels et al., 1988). This fact suggests that acquisition of tolerance to desiccation is a developmentally programmed event. In this regard, many genes have been shown to be expressed only in late embryogenesis before desiccation, and in the case of cotton and *Brassica napus* these genes encode LEA (Late Embryogenesis Abundant) proteins (Harada et al., 1989). The mRNAs encoding LEA proteins are most abundant in dry seeds (Galau et al., 1986) and decrease as the seeds germinate (Dure III et al., 1989). The decrease in the amount of LEA proteins during germination would explain the rapid loss of tolerance to desiccation shown by the seedlings (Bartels et al., 1990). It is likely that LEA proteins are common to all seeds that undergo desiccation during maturation (Dure III et al., 1989).

LEA genes can be precociously induced when immature rape seed embryos excised from seeds are cultured on media containing ABA (abscisic acid), and maximal accumulation of LEA mRNA coincides with the stage of embryogenesis during which ABA reaches peak levels (Finkelstein et al., 1985). The correlation of Lea mRNAs accumulation with ABA level and dessication has been observed also in a class of LEA genes in cotton; in fact, it has been suggested that in cotton the ABA level regulates the expression of a class of LEA genes (Galau et al., 1987). This raises the question of whether RAB (Responsive to Abscisic acid) and LEA genes are homologous. In this regard, in rape seed, a gene family induced by ABA and desiccation specifies an mRNA which gives a predicted sequence sharing similarities with a cotton LEA protein and containing structural elements in common with RAB and desiccation-induced proteins (Harada et al., 1989).

With regard to the mechanism of action, the survival of cells during desiccation seems to be due to the amount of specific gene products or of their specific subcellular locations (Piatkowski et al., 1990 ). LEA proteins may play their protective role by means of their ability to sequester a considerable amount of ions, thus avoiding the formation of an extremely high ionic environment in the dehydrating cytoplasm (Dure III, 1993).

It is known that the nucleus is not affected by desiccation; in fact, it starts its activities within minutes of the start of imbibition. Because of the peculiar molecular organization of the nucleus, it would be interesting to ascertain the protective mechanisms devised to survive desiccation and to maintain the integrity of the DNA molecule. In my laboratory, a novel nuclear protein named QP47 has been purified which might play such a role (Chiatante and Onelli, 1993). This paper analyses the behaviour of this protein in *Pisum sativum* during seed maturation and the first phases of germination. Furthermore, the subcellular localization of QP47 has been examined in an attempt to throw light on the relationship between QP47 and the DNA molecule.

## Materials and methods

### *Extraction and purification of nuclei*

Seeds of *Pisum sativum* L. cv. Lincoln were surface sterilized with 10% (v/v) sodium hypochlorite for 30 min, rinsed in running tap water for 1 h and left at room temperature for different periods of time. After 24 h, the seeds were put on moist agriperlite until use. Meristematic root tips (0 - 2 mm) were excised by scalpel and collected in extraction buffer (0.4 m*M* sucrose, 2.5% (w/v) Ficoll 40,000, 25 m*M* Tris/HC1 pH 7.6, 10 m*M* $MgCl_2$, 10 m*M* $\beta$-mercaptoethanol and 30% (v/v) glycerol).

Differentiated tissue was obtained by cutting a 0.5 cm segment 0.5 cm from the root tip. The tissue was homogenized and nuclei were purified according to Chiatante et al. (1991).

Nuclei were also prepared from root meristems of embryos during seed maturation. The procedure for extracting fresh nuclei was the same as described above.

### *Gel electrophoresis and protein phosphorylation of QP47*

SDS gel electrophoresis was carried out according to Chiatante et al. (1991)

### *QP47 purification*

Purified nuclei were freed from glycerol by centrifugation, washed with buffer containing 0.4 m*M* sucrose, 25 m*M* Tris/HC1 pH 7.6, 10 m*M* $MgCl_2$, 10 m*M* $\beta$-mercaptoethanol and pelleted by centrifugation (20,000 *g* for 30 min). The pellet of nuclei was resuspended in lysis buffer containing 25 m*M* Tris/HC1 pH 7.6, 10 m*M* $MgCl_2$, 10 m*M* $\beta$-mercaptoethanol 0.5 *M* NaC1. Lysis was continued for 2 h at 4°C with gentle stirring. The lysate was centrifuged at 20,000 *g* for 30 min, the pellet was discarded and the supernatant was saturated with ammonium sulphate. The precipitated proteins were collected by centrifugation (20,000 *g* for 45 min) and dialysed overnight against lysis buffer without NaC1. The volume of the extract was reduced by freeze-drying and this preparation was used as 'crude nuclear preparation'. Aliquots of the crude nuclear preparations were used for gel electrophoresis separation of the proteins.

### *Immunodetection of QP47 by fluorescence*

Nuclei to be used for immunodetection were obtained from root tips fixed in 4% paraformaldehyde at 4°C for 20 min in buffer containing 10 m*M* Tris/HC1 pH 7.6.

The fixed tissue was homogenized and nuclei were purified as described in Chiatante et al. (1991), except that the glycerol concentration used for preparation of the buffer and gradient was 10% instead of 30%. The only exception was the 2 *M* sucrose cushion of the gradient which still contained 30% glycerol. The fixed nuclei formed a pellet on the 36 - 54% Percoll gradient instead of on the 2 *M* cushion as in the case of the fresh nuclei. Fixed nuclei were collected, diluted with an equal volume of Tris/HC1 pH 7.6 and centrifuged at 20,000 *g* for 10 min. The supernatant was discarded and the pellet of nuclei was resuspended in 300 $\mu$L of serum prepared by Vectastain ABC elite kit (Vector Laboratories, Inc.). After 20 min of incubation at 4°C, the nuclei were centrifuged at 20,000 *g* for 10 min, resuspended in 300 $\mu$L of QP47 antibody (diluted 1:100) and incubated overnight in a horizontal shaker. Nuclei were centrifuged, washed with 10 m*M* Tris/HCl pH 7.6, and resuspended in the same washing buffer. Twenty $\mu$L of nuclear preparation was spread on a slide and dried. The dried slides were incubated with a secondary anti-rabbit biotinylated antibody (Ab) then stained with Texas Red-Streptavidin diluted 1:200 in 10 m*M* Tris/HCl pH 7.6, 10 m*M* EDTA, 100 m*M* NaCl. This same buffer was used to wash the slides. The slides were incubated with DAPI (4',6-diamidino-2-phenylindole, 2$\mu$g mL$^{-1}$) for 7 min, washed and mounted in 95% glycerol. Photographs were made using a Leitz Aristoplan microscope.

## Results

We had shown previously that a considerable amount of QP47 was present in nuclei of meristematic root cells after 6 h of imbibition (Chiatante and Onelli, 1993). In the same paper, the comparison of the QP47 content at different times of germination demonstrated that the amount of this nuclear protein must decrease to an undetectable level before cell proliferation can be resumed. These results suggested the existence of a negative correlation between the content of this protein and the metabolic activity of the nucleus. When cell proliferation was arrested in seedlings grown in conditions of starvation (no sugar present in the medium) or osmotic stress (presence of 10% of Polyethyleneglycol 8,000) we did not observe any increase in the level of QP47 (data not shown). This result suggested that the synthesis and degradation of QP47 might be two physiological events confined respectively to seed maturation and to early seed germination. A similar behaviour is typical of LEA proteins which are present in the dry seed and accumulate in the embryo cells during seed maturation. For this reason, it seemed of interest to discover whether QP47 was also synthesized during maturation. In this regard Figure 1 shows the quantitative level of QP47 in nuclei of meristematic root cells during different phases of seed maturation and germination. It seems clear that QP47 was synthesized at a later stage of embryogenesis, although the data do not establish any correlation between the metabolic activity of the embryo cells and the time of appearance of QP47.

An experiment of western blotting indicated that the protein with the molecular weight of 47 kDa found in root meristems during seed maturation was homologous to QP47 purified from germinating seedlings. In fact, the 47 kDa protein was recognized (data not shown) by the QP47 polyclonal antibody prepared previously (Chiatante and Onelli, 1993).

An indication concerning the possible physiological role of QP47 might be provided by the localization of this protein within the nucleus and from information concerning its relationship with the DNA molecule. For this purpose, QP47 polyclonal Ab had been used previously to study the immunolocalization of this protein in nuclei during seed germination (Chiatante and Onelli, 1993). In the present paper, the immunolocalization of QF47 in nuclei during seed maturation has been investigated (Fig. 2). When the pattern of fluorescent labelling were compared in the nuclei of germinating or maturing seeds, it was possible to observe some differences. In fact, during maturation the fluorescence seemed to be particularly concentrated on the nucleolus (Fig. 2) whereas, during germination, it seemed to be diffused uniformly over all the nucleus (see figure in Chiatante and Onelli, 1993). In both cases, the QP47 fluorescence seemed to be organized in short threads closely resembling the filamentous chromosomes. However, when the DAPI fluorescence was compared with QP47 fluorescent thread structures in the same nucleus, no correspondence was found.

When the conserved sequences of the repeat unit of some known LEA proteins were compared with the only stretch of sequence of QP47 identified so far (Chiatante and Onelli, 1993) no homology was found.

## Discussion

The differential accumulation of LEA mRNAs in seeds and seedlings of *Brassica napus* suggests that the expression of these genes is regulated during the transition from embryogenesis to germination (Harada et al., 1989). In fact, these mRNAs are accumulated during late embryogenesis, are stored in dry seed, and become undetectable in seedlings within 24 h of imbibition (Harada et al., 1989). Genes, mRNAs, and proteins with properties similar to rape seed LEA have been recently described in wheat, rice, cotton, barley, and carrots (Dure III, 1993; Dure III et al., 1989; Hong et al., 1988; Litts et al., 1987; Mundy and Chua, 1988). Such a wide diffusion suggests that LEA proteins are likely to be common to all seeds that undergo desiccation during maturation (Dure III et al., 1989).

The findings presented in this paper concerning the variations of QP47 content suggest the existence of a similarity between the behaviour of this protein in pea nuclei of root meristems and LEA proteins in the cyto-

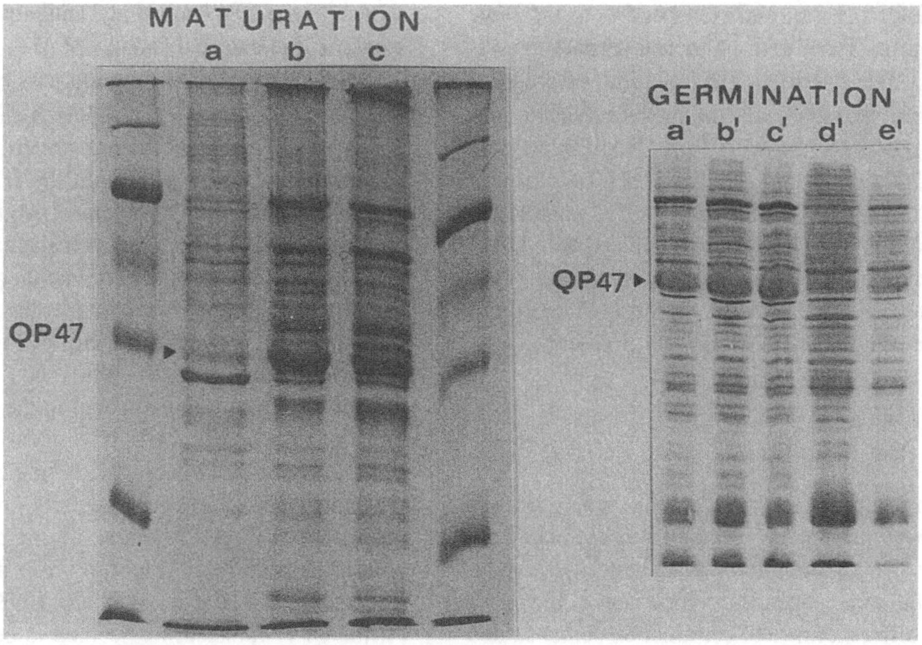


*Fig. 1.* Quantitative variations of QP47 during seed germination and maturation. The three different stages of maturation correspond to the beginning (**a**), middle (**b**) and end (**c**) of desiccation period following the completion of embryogenesis (respectively 25, 35 and 45 days after anthesis). The different lanes during germination represent, respectively, the nuclear extract 6 (a'), 16 (b'), 24 (c'), 48 (d'), and 72 (e') hours from the beginning of imbibition. The arrows point to the position of QP47 nuclear protein. The lanes on the left and right side of the left-hand figure represent molecular weight markers (29, 45, 66, 195, 207 kDa).

plasm of the embryo in other higher plants. Contrary to the hypothesis of QP47 being a LEA protein is the lack of homology in the sequence, as well as a number of differences of properties, such as molecular weight and isoelectric point (Chiatante and Onelli, 1993). A better comparison will be possible only when the QP47 sequence is completely known; it should also be borne in mind that LEA proteins with different sequences have been reported.

There are a number of genes (RAB) responsive to ABA levels which seem to be involved in the acquisition of tolerance to desiccation. These genes are active in different plant organs, including the seeds where their products might be responsible for the survival of embryo cells after a period of severe desiccation. Although a certain degree of similarity between some RAB and LEA genes has been found, it is accepted that differences of specificities exist between these two classes of genes. We have not yet studied the similarities between QP47 and RAB gene products; however, the administration of ABA to pea seedlings during the first stages of germination did not increase the amount of QP47 in the nuclei, even though the treatments delayed its degradation. This last result has been interpreted as the consequence of a more general effect of the plant hormone on the transition of embryo cells from the state of quiescence to the state of active proliferation (Levi et al., 1993).

The protective effect of LEA proteins on the cytoplasm seems to be based upon the sequestration of ions that prevent the formation of too high an ionic environment during desiccation (Dure III, 1993). This effect would be the direct consequence of the particular conformational structure of these proteins. Nothing is known about the function of the RAB gene products in protecting the cells from desiccation.

The abundance of LEA proteins in the cytoplasm of the embryo cells of many higher plants during maturation seems to be a condition required by these proteins in order to perform their protective function. However, LEA proteins have so far never been found in nuclei, and nothing is known about the mechanism by which the nucleus survives desiccation. The fact that QP47 is abundant in pea nuclei prior to desiccation suggests that it may serve a protective function. The manner in which the nucleus is protected against desiccation could differ from that of the cytoplasm. For example, protection could be provided by increas-

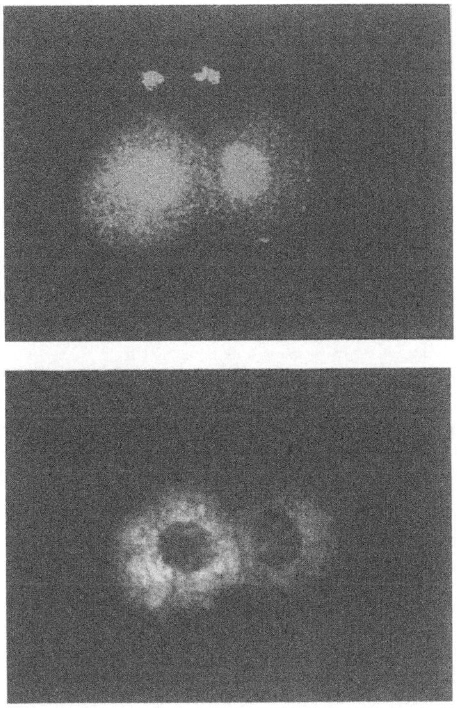

*Fig. 2.* Immunofluorescent staining of QP47 inside the nuclei purified from meristematic root tips during seed maturation. The nuclei have been stained by DAPI to point out the chromosome positions (**B**) and are counter-stained by Texas Red to localize the position of QP47 (**A**). It is evident there are difference between the patterns of the two types of fluorescence. The QP47 protein seems to be more concentrated in a nuclear compartment which includes the nucleoulus. This distribution changes during seed maturation.

ing the state of compactness of the chromatin. If so, then QP47 would have a physiological function different from that revealed for LEA and RAB proteins. Therefore, the sequence of QP47 could be different. Future investigations should consider the possibility that QP47 might have some properties in common with proteins involved in the regulation of metabolism and the conformational organization of the DNA.

In this regard, a better understanding of the relationship between this protein and the DNA molecule could be provided by studying its sub-nuclear distribution. The data presented here suggest that QP47 could be arranged along the chromosomes, although it is not clear why the protein is more concentrated in the nucleolus during maturation. An arrangement of QP47 along the chromosome during desiccation could serve a protective purpose with regard to the DNA molecule. The nuclear distribution of the protein at different times of maturation and germination should also be investigated.

In conclusion, although the function of QP47 in the nucleus remains unclear, there is evidence that this protein may serve to protect the chromatin during the period of desiccation in the dry seed.

## Acknowledgements

This research has been supported by National Research Council of Italy, Special Project Raisa, Sub-project, Paper XXX. I wish to thank Antonio Grippo for photographic assistance.

## References

Bartels D, Schneider K, Terstappen G, Piatkowski D and Salamini F 1990 Molecular cloning of abscisic acid - modulated genes which are induced during desiccation of the resurrection plant *Craterostigma plantagineum*. Planta 181, 27–34.

Bartels D, Singh M and Salamini F 1988 Onset of desiccation tolerance during development of the barley embryo. Planta 175, 485–492.

Bewley J D, Kermode A R and Misra S 1989 Desiccation and minimal drying treatments of seeds of castor bean and *Phaseolus vulgaris* which terminate development and promote germination cause changes in protein and messenger RNA synthesis. Ann. Bot. 63, 3–7.

Bochicchio A, Vazzana C, Raschi A, Bartels D and Salamini F 1988 Effect of desiccation on isolated embryos of maize. Onset of desiccation tolerance during development. Agronomy 8, 29–36.

Chiatante D, Brusa P, Levi M and Sparvoli E 1991 Nuclear proteins during the onset of cell proliferation in pea root meristems. J. Exp. Bot. 42, 45–50.

Chiatante D and Onelli E 1993 Nuclear proteins and the onset of cell proliferation in root meristems of *Pisum sativum*: QP47 a novel acidic protein. Seed Sci. Res. 3, 35–42.

Crouch M L and Sussex I M 1981 Development and storage - protein synthesis in *Brassica napus* L. embryos in vivo and in vitro. Planta 153, 64–74.

Dure III L, Crouch M, Harada J, Ho T-H, Mundy J, Quatrano R, Thomas T and Sung Z R 1989 Common amino acid sequence domains among the LEA proteins of higher plants. Plant Mol. Biol. 12, 475–486.

Dure III L 1993 A repeating 11-mer amino acid motif and plant desiccation. Plant J. 3, 363–369.

Finkelstein R R, Tenbarge K M, Shumway J E and Crouch M L 1985 Role of ABA in maturation of rape seed embryos. Plant Physiol. 78, 630–636.

Galau G A, Hughes D W and Dure III L 1986 Abscisic acid induction of cloned cotton late embryogenesis abundant (LEA) mRNAs. Plant. Mol. Biol. 7, 155–170.

Galau G A, Bijaisoradat N and Hughes D W 1987 Accumulation kinetics of cotton late embryogenesis abundant mRNAs and storage protein mRNAs; coordinate regulation during embryogenesis and the role of abscisic acid. Dev. Biol. 123, 198–212.

Harada J J, De Lisle A J, Baden C S and Crouch M L 1989 Unusual sequence of an abscisic acid inducible mRNA which accumulate late in *Brassica napus* seed development. Plant Mol. Biol. 12, 395–401.

Levi M, Brusa P, Chiatante D and Sparvoli E 1993 Cell cycle reactivation in cultured pea embryo axes. Effect of abscisic acid. In Vitro Cell Dev. Biol. (*In press*).

Litts J C, Colwell G W, Chakerian R L and Quatrano R S 1987 The nucleotide sequence of a cDNA clone encoding the wheat Em protein. Nucl. Acids Res. 15, 3607–3618.

Mundy J and Chua N-H 1988 Abscisic acid and water stress induce the expression of a novel rice gene. EMBO J. 7, 2279–2286.

Piatkowski D, Schneidter K, Salamini F and Bartels D 1990 Characterization of five abscisic acid - responsive cDNA clones isolated from the desiccation - tolerant plant *Craterostigma plantegineum* and their relationship to other water stress gene. Plant Physiol. 94, 1682–1688.

*F. Baluška et al. (eds.), Structure and Function of Roots*, 299–304, 1995.


# Inhibition of root growth by salinity stress: Toxicity or an adaptive biophysical response?

Peter M. Neumann
*Plant Physiology Laboratory, Faculty of Agricultural Engineering, Technion — Israel Institute of Technology, Haifa 32000, Israel*

*Key words:* adaptation, cell-wall hardening, $CaCl_2$, NaCl, root inhibition, *Zea mays* L.

## Abstract

This review considers mechanisms underlying the inhibition of root elongation growth by salinity stress. The first section considers effects of salinity on quasi steady state elongation growth, morphology and mature cell size, in maize (*Zea mays* L) primary roots. The following sections review evidence indicating that the inhibition of root elongation growth by salinity need not be a toxic consequence, e.g. of competitive displacement of essential $Ca^{2+}$ from plasmamembrane binding sites in the expanding tip tissues. Thus growth inhibitory levels of salinity did not compete with $Ca^{2+}$ for initial attachment to plasmamembrane binding sites, did not reduce capacity for trans-membrane proton transport, and did not reduce capacity for osmotic adjustment or turgor maintenance in growing root tips supplied with adequate external calcium. As an alternative to the ion toxicity hypothesis, it is suggested that salinity induces regulated biophysical restraints to cell wall expansion, which in turn inhibit root expansion growth. Finally, the possibility that regulated reductions in cell, root and overall plant size, have adaptive advantages for prolonging plant survival in drying salinized soils, is considered.

## Introduction

Salinity in the rhizosphere can inhibit growth and yields of glycophytic crop plants in arid and semi arid regions of the world. Thus excessive accumulation of salt in the apoplasmic and/or cytoplasmic spaces of mature tissues of hydroponically grown plants may accelerate leaf dessication and plant death (see reviews by Munns, 1993; Rengel, 1992). However, several recent reports indicate that under drying soil conditions, salinization can actually prolong the survival of several species of agriculturally important plants (Devitt et al., 1993; Richards, 1992). There is little doubt that an earlier plant response to salinization is the inhibition of ongoing expansion growth in roots and leaves. This early onset of growth inhibition could be an expression of the toxic consequences of salt accumulation in the growing tissues, or, alternatively, an actively regulated plant stress response. This article reviews evidence indicating that salinity induced changes in root structure and growth rate may, at least initially, be viewed as metabolically regulated biophysical responses with possible adaptive value for plant survival.

## Growth responses to salinity

The rapid elongation of plant roots is primarily a consequence of the massive expansion of young cells produced by ongoing divisions in the meristem. Excessive salinity has been reported to inhibit both root cell production and cell expansion (Kurth et al., 1986, Zidan et al., 1990). Some indicative details of the typical inhibitory effects of salinity on growth, cell length and morphology of maize (*Zea mays* L) primary roots are given in Figure 1A and Table 1. Figure 1A shows the rapidly established inhibitory effects of salinization (with 100 mol $m^{-3}$ NaCl and with extra calcium) on root elongation growth. For comparison, Figure 1 B shows the inhibitory effects of salinity on elongation of the first true leaf. Table 1 shows that the inhibition of root elongation by salinity is also associated with reductions in length of mature epidermal cells. Moreover these salinity induced reductions in both root

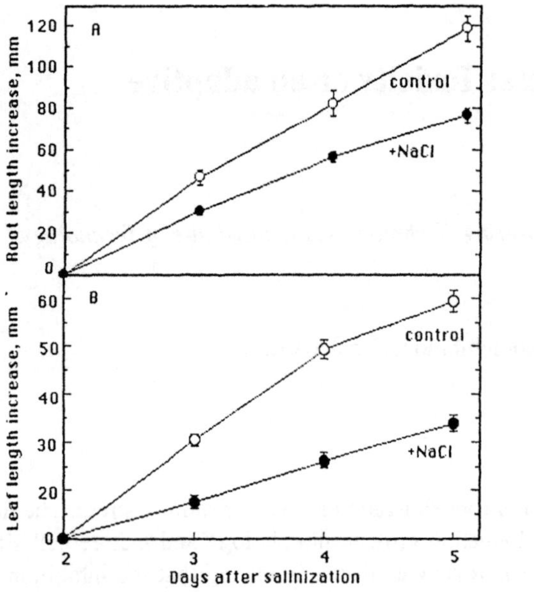


*Fig. 1.* **A**. Elongation of intact maize primary roots from 2d after salinization of 0.1 strength nutrient solution containing extra calcium 10.5 mol $m^{-3}$ $CaCl_2$) with 100 mol $m^{-3}$ NaCl. Plants hydroponically grown in a chamber as previously described by Zidan et al. (1990). Means of 14 control plants and 20 salinized plants. Vertical bars indicate ±SE. **B**. Shows elongation of first true leaves grown similarly. Means of 15 control plants and 14 salinized plants. Vertical bars indicate ± SE. The lengths of mature epidermal cells produced by the first true leaves after 2d in saline growth conditions with additional calcium were reduced from 139±14 to 83±7 μm, means ± SD, n = 81±2 cells (Neumann and Etzion, unpublished).

*Table 1.* Effects of salinity and additional calcium on growth parameters of primary roots of maize seedlings. Root length and diameter are means, ±SD, n = 10 after 4d growth in aerated nutrient solutions (NS) ± 100 mol $m^{-3}$ NaCl ± 10 mol $m^{-3}$ $CaCl_2$ (Adapted from Evlagon et al., 1990). Cell lengths based on assay of 67 ± 3 cells from 5 roots in each treatment (means ±SD, adapted from Zidan et al., 1990)

| | Root length (cm) | Root diameter (mm) | Mature cell length (μm) |
|---|---|---|---|
| NS | 18.1±12.3 | 0.82±0.06 | 195±13 |
| NS+NaCl | 8.3± 0.0 | 1.42±0.05 | 104± 7 |
| NS+Ca | 17.1± 1.1 | 0.93±0.04 | 190±12 |
| NS+NaCl+Ca | 13.7± 1.3 | 0.78±0.04 | 160± 8 |

length and cell length were greater when calcium concentrations were at 0.5 mol $m^{-3}$ (basic concentration in 0.1 strength nutrient solution) rather than 10.5 mol $m^{-3}$ (0.1 strength nutrient solution with extra Ca, as in Zidan et al., 1990). Salinity and calcium had other interactive effects on the morphology of the roots: Thus roots of plants salinized with NaCl and additional calcium were thinner than control roots, whereas the roots of salinized plants grown without additional calcium were thicker. These changes were consistent with salinity induced changes in lengths a n d diameters of cortical cells in maize and cotton roots observed by Kurth et al. (1986) and Azaizeh et al. (1992). The fact that raising the calcium concentration (and activity) in dilute nutrient solutions used for hydroponic culture of salinized plants can increase root (and leaf) growth, is well established (e.g. Lehaye and Epstein, 1969; Maas and Grieve, 1987; Zidan et al., 1991).

Salinity induced changes in root length, mature cell length and root diameter could be caused by a) toxic effects of salt on the metabolism of expanding cells b) reductions in water availability for cell expansion caused (osmotically) by a high external salt concentrations, or, c) induction of metabolically regulated, and possibly adaptive, plant responses. These possibilities are considered below.

## Toxic effects of salinity

Cramer et al. (1985) put forward the attractive hypothesis that a primary response of root cells to salt stress involves competitive displacement by sodium of essential calcium ions from attachment sites at the outer plasmamembrane surface. This would then lead to reductions in membrane integrity and function e.g. loss of sodium exclusion capacity, increased leakage of essential solutes, turgor losses and consequent growth inhibition. However subsequent reports indicated that salinity induced calcium release from internal membranes of root cells rather than plasmamembrane surfaces (Lynch et al., 1987; Lynch and Lauchli, 1988). Moreover even high sodium concentrations did not appear to compete effectively with calcium ions in initially saturating plasma membrane binding sites in maize roots (Zidan et al., 1991).

Additional evidence against the hypothesis that salinity necessarily leads to loss of plasmamembrane integrity and function came from a study of the effects of salinity on the capacity of root epidermal cells to acidify their surrounding cell walls (Zidan et al., 1990). Wall acidification (e.g. via plasmamembrane proton pumping ATPases) is, according to the acid growth hypothesis (Rayle and Cleland, 1992) one of the essential prerequisites for cell expansion growth. Disruption of transmembrane proton transport and wall acidification, as a result of membrane disruption by salinity, would therefore lead automatically to a restriction of expansion growth. However, capacity for progressive wall acidification by growing epidermal cells in maize root tips was not inhibited by growth in NaCl solutions with or without extra calcium. Similar results were obtained when root tip tissues were assayed for capacity to change the colour of bromocresol purple pH indicator in agar gels. Presumably therefore, this membrane function was relatively unaffected by levels of salinity which clearly inhibited growth.

Thus membrane disruption. as a result of competitive displacement by sodium of calcium bound to the plasmamembrane surface, did not appear to be a primary response of growing maize root cells to salinity. This conclusion is consistent with the fact that the membranes of animal cells routinely function well in high sodium environments, despite the equivalent importance, and low concentration, of calcium ions. A probable explanation for the beneficial effect of additional calcium on growth of salinized roots is that high NaCl concentrations reduce the ionic activity of calcium in dilute nutrient solutions (via ion to ion interactions) to sub-optimal levels (Maas and Grieve, 1987; Zidan et al., 1991): Raising the calcium concentration in dilute nutrient solutions containing NaCl therefore restores calcium activity to physiologically normal levels. However, additional growth inhibitory factors are involved since root and leaf growth were still inhibited by salinity, even when calcium levels in the external solution were greatly increased (Evlagon et al., 1990; Kurth et al., 1986; Maas and Grieve, 1987; Zidan et al., 1990 and Fig 1). Similarly, calcium ion activities in salinized calcareous soils of semi-arid and arid regions are expected to be higher than those in dilute nutrient solutions, so that external calcium deficiencies are unlikely to be the primary cause of salt stress responses in the field (Maas and Grieve, 1987; Zidan et al., 1991). Salinity might however induce root changes by interfering with possible cell messenger roles of calcium e.g. calcium flux across the plasmamembrane or cytoplasmic calcium activity (reviewed by Rengel, 1992).

*Table 2.* Effects of salinity and additional calcium on osmotic potential gradient ($\Delta\pi$) between tip tissues from root elongation zone and external nutrient solution. Root tip segments (5mm length) were frozen and thawed prior to psychrometric determination of bulk tissue osmotic potentials. Osmotic potential of external nutrient solutions (NS) were subtracted to give $\Delta\pi$. Means, ±SD, n = 9. Adapted from Neumann et al. (1994) and Azaizeh and Neumann (unpubl. data)

| | 2d | 4d | 8d |
|---|---|---|---|
| | $\Delta\pi$(-*MPa*) | | |
| NS | 0.71±0.05 | 0.78±0.07 | 0.79±0.04 |
| NS+NaCl | 0.58±0.12 | 0.58±0.07 | 0.60±0.04 |
| NS+Ca | 0.70±0.07 | 0.79±0.08 | 0.70±0.08 |
| NS+NaCl+Ca | 0.65±0.11 | 0.65±0.09 | 0.70±0.04 |

## Osmotic effects of salinity

Salinization causes the osmotic potential of external water sources to be more negative and could therefore decrease water availability for growth. Root growth inhibition following salinization might therefore be related to an effect of salinity on capacity to maintain osmotic potential gradients between growing cells and external water sources (cf. Bernstein, 1971). This could result from induced leakage of essential solutes, as discussed above, or through insufficient availability of the compatible solutes required for osmotic adjustment. The fact that osmotic potential gradients and cell turgor pressures in fully expanded maize root cells and tissues were not greatly affected after several days exposure to growth inhibitory levels of salinity with additional calcium, argues against this possibility (Evlagon, 1990; Azaizeh et al., 1992). There is less information about the capacity of the immature and rapidly expanding root tip tissues to maintain osmotic potential gradients and turgor pressure under saline conditions. The results in Table 2 show that in the absence of additional calcium, the expanding tip tissues of salinized primary roots showed a consistent loss of capacity to maintain osmotic potential gradients. However when calcium levels were increased to adequate levels, the immature tip tissues responded to salinization in the same way as the mature tissues i.e. bulk osmotic potential gradients were relatively unaffected. Moreover recent micro-pressure probe and psychrometric measurements, did not reveal reductions in cell, or bulk, turgor pressures in root tips of salinized

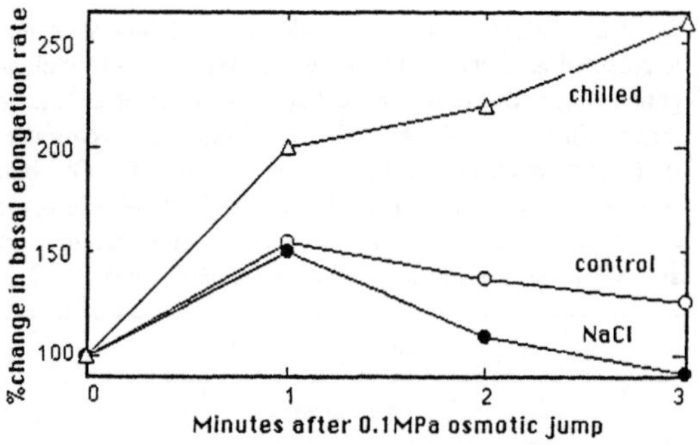


*Fig. 2.* Percent changes in basal elongation rate of root tip tissues of intact maize seedlings induced by an 0.1 MPa osmotic jump. Basal elongation rate of chilled roots 6+2 $\mu$m min$^{-1}$ (means, ±SD for 6 roots). Basal elongation rate of non salinized control roots 27±7 and of 3d salinized roots 22±6 $\mu$m min$^{-1}$ (means ±SD). Data adapted from Neumann and Leon (1992) and Neumann et al. (1994).

plants provided with extra calcium in the nutrient solution (Neumann et al., 1994). Thus, provided calcium activity is maintained at adequate levels, the growing tip tissues of maize primary roots appear able to maintain osmotic potential gradients and turgor pressures under saline conditions.

## Effects of salinity on water transport capacity and cell wall hardening

The much used model equations of Lockhart (1965) suggest that expansion growth in plants could be inhibited by hydraulic restraints to water uptake and/or by wall mechanical restraints to cell expansion (i.e. cell wall hardening). Both these parameters are affected by cell metabolism. Several reports indicate that salinity can reduce the hydraulic conductivity in mature roots of maize (e.g. Evlagon et al., 1990; Azaizeh et al., 1992). However the immature tissues of the growing root tips in maize are hydraulically isolated from the rest of the root (Steudle and Frensch, 1989) and salinity may not affect their hydraulic conductivity in the same way. The fact that cell turgor pressures along growing root tips were not significantly reduced by saline conditions, suggests that the hydraulic conductivity of water uptake pathways in the root tips was not reduced. This conclusion is further supported by the observation that comparative rates of osmotically induced water transport out of root tips of salinized maize seedlings, are no lower than in control plants (Neumann et al., 1994).

The fact that water uptake capacity, osmotic potential gradients and turgor pressures in the root tip tissues can be maintained while growth is reduced, leads, by default,to the conclusion that salinity inhibits expansion growth by reducing cell wall extension capacity. Support for this conclusion came from direct estimates of extension capacity in the root tips of salinized or control plants. Figure 2 shows the comparative kinetics of percent changes in instantaneous elongation rates of root tips of whole plants following dilution of the external nutrient solution, so as to induce an 0.1 MPa osmotic jump. The extensibility characteristics of the control roots are compared with those of control roots which were chilled for 3 min at 11°C prior to osmotic jump treatment at the same temperature, and with those of salinized roots. This osmotic jump method (Neumann and Leon, 1992; Neumann et al., 1994) involves microscopically following changes in the position of the root tip, at 1 min intervals. Microscopic measurement of instantaneous growth rates on the basis of minute by minute changes in root position are clearly less definitive than measurements based on longer time intervals of hours or days. However the approach allows accurate comparisons of the relatively large short term changes in basal growth rate induced by the osmotic jump treatment. These are in turn related to the extension capacity of the expanding cell walls. The control roots in Figure 2 showed an initial increase in comparative elongation which was followed by a progressive decline. The roots of salinized plants showed the smallest initial increase and subsequently the most rapid decline in elongation. Moreover the extension rates of the salinized roots were significantly lower than those of the controls throughout the osmotic jump treatment. The chilled roots responded to the osmotic jump treatment with an initial increases in comparative elongation but there was no sign of the subsequent decline seen with the other treatments. Chilling had no effect on the mechanical extensibility of killed cell wall preparations (Cleland, 1971) and one possible explanation for the different behaviour of the chilled roots is that metabolic processes involved in the initiation of wall hardening responses to the osmotic jump, were inhibited at low temperatures (Neumann and Leon, 1992). Conversely the comparative extensibility of the salinized roots was lowest because the saline environment promoted wall hardening processes (Neumann et al., 1994). Green et al. (1971) concluded that metabolically regulated wall hardening and softening processes were involved in regulating growth responses of algal cells to sudden imposition and removal of osmotic stresses. Similar conclusions were drawn by Kuzmanoff and Evans (1981) in order to explain rapid root growth responses to osmotic stress. Rapid wall hardening responses to salinity have also been directly measured by extensiometric assay in growing maize leaf tissues (Neumann, 1993). Models and potential mechanisms for metabolically controlled changes in cell wall mechanical properties in higher plants were recently presented by Passioura and Fry (1992).

## Possible adaptive advantages of growth inhibitory responses to salinity

Some of the long term root responses to salinity stress reviewed here are similar to previously reported root responses to water deficits (Cutler et al., 1977; Fraser et al., 1990; Sharp et al., 1988) and osmotic stress (Pritchard et al., 1991). Thus these stresses were also shown to induce apparent cell wall hardening, reductions cell production rate or reductions in mature cell size. It is interesting to consider whether the inhibition of cell expansion in response to salinity, osmotic and water stress could have an adaptive role in increasing plant survival: For example the small root and leaf cells produced will have higher surface to volume ratios than the relatively larger cells in unstressed plants. A relative increase in plasmamembrane surface area per unit cell volume could result in a relative increase in solute uptake capacity per cell. Moreover smaller cell size could increase the comparative effectiveness of the accumulated solutes i.e. same number of solute molecules per reduced volume (Cutler et al., 1977). A consequence of these differences would be an increase in capacity for turgor maintenance. Similarly, the production of thinner roots and a shortening of the root elongation zone under saline conditions would reduce sugar requirements for new cell wall production and could result in increased sugar availability for osmotic adjustment and maintenance respiration activities (Bernstein, 1971; Fraser et al., 1990; Pritchard et al., 1991; Sharp et al., 1988; Yeo, 1983; Zidan et al., 1990). Finally it is worth noting that salinity induced reductions in cell expansion will also lead to reductions in rates of root and shoot area increase. Such reductions can optimise long term survival of plants in drying saline soils under field condition, by prolonging the period of availability of limited soil water reserves (Devitt et al., 1993; Passioura et al., 1993; Richards, 1992).

In conclusion salinity rapidly inhibits growth and alters the structure of maize roots, even when external calcium activity is maintained at adequate levels. These changes do not appear to be the result of toxic effects of NaCl e.g. on membrane associated activities such as capacity for wall acidification, osmotic adjustment, turgor maintenance, or water transport, in the growing root tip tissues. Rather root growth inhibition appears to be related to a regulated hardening of the growing cell walls which, in turn, helps limit root tip expansion rates. The plant growth responses initiated in response to salinity stress i.e. reductions in root cell length, root diameter and overall plant size, may have adaptive advantages for prolonging survival in drying saline soils.

## Acknowledgement

This research was supported by the fund for the promotion of research at the Technion.

## References

Azaizeh H, Gunse B and Steudle E 1992 Effects of NaCl and $CaCl_2$ on water transport across cells of maize (*Zea mays* L.) seedlings. Plant Physiol. 99, 886–894.

Bernstein L 1971 Osmotic adjustment of plants to saline media.1. Steady state. Am. J. Bot. 48, 909–918.

Cleland R 1971 the mechanical behavior of isolated *Avena* coleoptile walls subjected to constant stress. Properties and relation to cell elongation. Plant Physiol. 47, 805–811.

Cramer G R, Lauchli A and Polito V S 1985 Displacement of $Ca^{2+}$ by $Na^{+}$ from the plasmalemma of root cells. A primary response to salt stress? Plant Physiol. 79, 209–211.

Cutler J M, Rains D W and Loomis R S 1977 The importance of cell size in water relations of plants. Physiol. Plant. 40, 255–260.

Devitt D A, Bowman D C and Schulte P J 1993 Response of *Cynodon dactylon* to prolonged water deficits under saline conditions. Plant and Soil 148, 239–251.

Evlagon D, Ravina I and Neumann P M 1990 Interactive effects of salinity and calcium on hydraulic conductivity, osmotic adjustment and growth in primary roots of maize seedlings. Isr. J. Bot. 39, 239–247,

Fraser T E, Kuhn Silk W and Rost T L 1990 Effects of low water potential on cortical cell length in growing regions of maize roots. Plant Physiol. 93, 648–651.

Green P B, Erickson R O and Buggy J 1971 Metabolic and physical control of cell elongation rate. Plant Physiol. 47, 423–430.

Kurth E, Cramer G R, Lauchli A and Epstein E 1986 Effects of NaCl and $CaCl_2$ on cell enlargement and cell production in cotton roots. Plant Physiol. 82, 1102–1186.

Kuzmanoff K M and Evans M L 1981 Kinetics of adaptation to osmotic stress in Lentil (*Lens culinaris* Med), roots. Plant Physiol. 68, 244–247.

LaHaye P A and Epstein E 1969 Salt tolerance by plants: enhancement with calcium. Science 166, 395–396.

Lockhart J A 1965 An analysis of irreversible plant cell elongation. J. Theor. Biol. 8, 264–275.

Lynch J and Läuchli A 1988 Salinity affects intracellular calcium in corn root protoplasts. Plant Physiol. 87, 351–356.

Lynch J, Cramer G R and Läuchli A 1987 Salinity reduces membrane associated calcium in corn root protoplasts. Plant Physiol. 83, 390–394.

Maas E V and Grieve C M 1987 Sodium induced calcium deficiency in salt stressed corn. Plant Cell Environ. 10, 559–564.

Munns R 1993 Physiological processes limiting plant growth in saline soils:some dogmas and hypotheses. Plant Cell Environ. 16, 15–24.

Neumann P M and Leon D 1992 Physical restraints underlying short term inhibition by auxin of root elongation in intact maize seedlings. Plant Growth Regul. 11, 119–123.

Neumann P M 1993 Rapid and reversible modifications of extension capacity of cell walls in elongating maize leaf tissues responding to root salinization. Plant Cell Environ. 16, 1107–1114.

Neumann P M, Azaizeh H and Leon D 1994 Hardening of root cell walls: a growth inhibitory response to salinity stress. Plant Cell Environ. 17, 303–309.

Passioura J B and Fry S C 1992 Turgor and cell expansion: beyond the Lockhart equation. Austr. J. Plant Physiol. 19, 565–576.

Passioura J B, Condon A G and Richards R A 1993 Water deficits, the development of leaf area and crop productivity. *In* Water Deficits, Plant Responses from Cell to Community. Eds. J A C Smith and H Griffiths. pp 253–264. Bios Scientific Publishers Ltd., Oxford, UK.

Pritchard J, Wyn Jones R G and Tomos A D 1991 Turgor, growth and rheological gradients of wheat roots following osmotic stress. J. Exp. Bot. 42, 1043–1049.

Rayle D L and Cleland R E 1992 The acid induced growth theory of auxin- induced cell elongation is alive and well. Plant Physiol. 99, 1271–1274.

Rengel Z 1992 The role of calcium in salt toxicity. Plant Cell Environ. 15, 625–632.

Richards R A 1992 Increasing salinity tolerance of grain crops: is it worthwhile? Plant and Soil 146, 89–98.

Sharp R E, Hsaio T C and Silk W K 1988 Growth of the maize primary root at low water potentials 1. Spatial distribution of expansive growth. Plant Physiol. 93, 1337–1346.

Steudle E and Frensch J 1989 Osmotic responses of maize roots. Water and solute relations. Planta 177, 281–295.

Yeo A R 1983 Salinity resistance: Physiologies and prices. Physiol. Plant. 58, 214–222.

Zidan I, Azaizeh H and Neumann P M 1990 Does salinity reduce growth in maize root epidermal cells by inhibiting their capacity for cell wall acidification? Plant Physiol. 93, 7–11.

Zidan I, Jacoby B, Ravina I and Neumann P M 1991 Sodium does not compete with calcium in saturating plasmamembrane sites regulating $^{22}Na$ influx in salinized maize roots. Plant Physiol. 96, 331–334.

*F. Baluška et al. (eds.), Structure and Function of Roots,* 305–310, 1995.


# Distribution of calcium in tomato plants in response to heat stress and plant growth regulators

Z. Starck, A. Siwiec and D. Chotuj
*Department of Plant Physiology, Warsaw Agricultural University, Rakowiecka 26/30, 02538 Warsaw, Poland*

*Key words:* $\beta$-naphthoxy acetic acid (NOA), calcium, gibberellin, heat, tomato

## Abstract

Two cultivars of tomato with contrasting response to elevated temperature were compared: sensitive - Roma and tolerant - Robin. Experiments were done on fruit explants and on rooted cuttings with small fruits. In both cases $^{45}Ca$ was poorly transported to the fruits. Nevertheless in fruit explants elevated temperature (40°C) increased $^{45}Ca$ import into the fruits in both cultivars. In the compared cuttings, treated or not treated with growth regulators and at various temperatures, the greatest differences were observed in the amount of $^{45}Ca$ transported to the fruits. Sensitive Roma cuttings scantily supplied their fruits with $^{45}Ca$ both under optimal temperature and heat stress. In plants previously treated with NOA + $GA_3$ high temperature increased $^{45}Ca$ transport to the fruits. Robin cuttings inversely responded to heat stress by transporting a much higher portion of $^{45}Ca$ to the fruits, both in control and NOA + $GA_3$ cuttings.

The diversity of $^{45}Ca$ distribution during elevated temperature in cuttings, but not in fruit explants of both cultivars seems to be connected with an ability to control calcium supply to fruit or at least to prevent its decrease; this mechanism is perhaps located outside the cluster.

## Introduction

Physiological functions of $Ca^{2+}$ in plants are diverse. In addition to its role as a cellular messenger, effects of $Ca^{2+}$ on integrity of membranes, rigidity of cell wall and maintenance of cell-to-cell contact are reported. Membrane-associated $Ca^{2+}$ prevents ion leakage caused by stress (Poovaiah, 1993; Simon, 1978). Physical and chemical signals alter the status of $Ca^{2+}$ distribution in the cell (Gilroy et al., 1990; Palta, 1990). The change in internal and extracellular calcium concentration triggers a cascade of biochemical and physiological events leading to the plant response. Calcium seems to be the primary transducer of hormonal and environmental signals e.g. salinity to plant response (Rengel, 1992). Such signals as touch, cold shock and fungal elicitors rapidly increase $Ca^{2+}$ in the cytoplasm of tobacco (Knight et al., 1991).

Exposure of cells to suboptimal temperature results in membrane injury (Santarius and Weis, 1988; Steponkus, 1981). Calcium magnifies the heat tolerance of the membrane; it counteracts heat-induced efflux of betacyanin in roots of red beet (Toproven and Glinka, 1976) and increases protoplasmic viscosity, (Alexandrov, 1967, cited in Steponkus, 1981). Ben-Arie (1982) reported that $Ca^{2+}$ and, even more, spermidine stabilize microsomal membrane fluidity, preventing its natural enhancement due to increasing temperature. Optimal $Ca^{2+}$ content and proper distribution in individual organs prevents the incidence and severity of physiological disorders that are caused in many cases by unfavorable external conditions (Poovaiah, 1993). Salt toxicity almost instantly reduces the amount of $Ca^{2+}$ transported to the leaves accompanied by a drop in the apoplast of leaf cells and a rise in $Na^{+}$ (Rengel, 1992).

Special consideration should be paid to physiological disorders in soft fruits, which may be causally emitted with local deficiencies of calcium (Bangerth, 1979). In spite of their very low $Ca^{2+}$ content, fruits are sensitive to its decrease. Low calcium concentrations in apple fruits accompany bitter pit (Fucumoto et al., 1987) or pit breakdown (Tomala and Dilley, 1990) and are observed with blossom-end rot in tomato fruits (Ehret and Ho, 1986). The above physiological diseases are often described under adequate $Ca^{2+}$ supply of the whole plant, indicating some perturbation in its distribution, especially its supply to the fruits (Adams and Ho, 1992). In the case of salinity, transport of $Ca^{2+}$ to the tomato fruits is reduced (Adams and Ho, 1992; Ho et al., 1993). Klein and Ferguson (1987) reported lower sensitivity of pear fruit to heat stress after treatment with $Ca^{2+}$.

As Poovaiah (1993) suggested, further investigations concerning adjustment of $Ca^{2+}$ distribution within the whole plant are needed. It would enable better understanding of the role of calcium in fruit physiology under environmental stresses.

In our previous experiments, treatment of tomato flowers with growth regulators (NOA + $GA_3$) diminished negative effects of high temperature on photosynthate translocation and import into developing fruits (Starck and Cieśla, 1989). As response to supraoptimal temperature fruit supply with $^{14}C$-photosynthates (unpublished) or $^{14}C$-sucrose (Starck and Witek-Czupryńska, 1993) in many cases coincided with the changes in their $Ca^{2+}$ content. If elevated temperature depressed C - import to the fruits, calcium content also decreased. In contrast in the case of plant heat tolerance, calcium level increased. The diversity of two compared cultivars (Roma and Robin) with different heat tolerance seems to be connected, among others with their ability to direct more $Ca^{2+}$ to the fruits during or immediately after heat treatment (Starck and Witek-Czupryńska, 1993). Therefore, the aim of the present experiments is to check if elevated temperature affects current uptake of $^{45}Ca$ and its distribution in the fruit explants of rooted tomato cuttings with particular attention to its delivery to fruit.

## Material and methods

Two experiments were done with cultivars of tomato plants with contrasting response to high temperature, Roma (very sensitive) and Robin (tolerant). Plants were grown in the greenhouse with complete fertilizer. At the beginning of flowering, half of the plants was treated with growth regulators. Flowers were sprayed twice a week with $\beta$-naphthoxyacetic acid - NOA (50 mg $\times$ $dm^3$), Gibrescol - $GA_3$ (30 mg $\times$ $dm^3$) and 0.1% Tween 20 as in our previous experiments (Starck and Cieśla, 1989). Experimental series are denoted as not treated control (C) and treated with growth regulators (NG). Both experiments were done in three replications (three plants per treatment).

*Experiment I*

Part of the clusters with three small fruits (each about 20 g F.W.) detached from plants that were treated and not treated with growth regulators (C and NG) were placed in 5 $cm^3$ of 1 m$M$ $^{45}CaCl_2$ (150 kBq per explant) in darkness for 24h under optimal or elevated temperature. The temperature regimes for Roma were 42°C during 10h and 35°C during 14h and for Robin they were respectively 42°C and 38°C. In both temperature regimes air was saturated with water. After that, part of the peduncles that were immersed in radioactive solutions ($^{45}CaCl_2$) were thoroughly rinsed and fruit explants were separated into fruits, pedicels, peduncles and calyxes. Samples were ashed and dissolved in HCl and their radioactivities were estimated by scintillation couning (Beckman LS 9000).

*Experiment II*

At the onset of flowering, Roma and Robin cuttings were divided into C and NG series, as in Experiment I. When small fruits developed (about 10 g F.W. of Roma and Robin cvs.), some of them were removed, leaving only two in each cluster. The uptake of $^{45}Ca$ was estimated by the whole cuttings (one cluster, one leaf below and a short piece of the stem with roots).

All cuttings of both cultivars with C and NG series were labelled with $^{45}Ca$ introduced to the nutrient solution (200 kBq per cutting) during 5h under natural light and about 20°C. After plant labelling, one fruit with calyx and peducle as well as one leaflet and piece of roots were harvested to estimate their radioactivity. After that, plants of both series and cultivars were placed for 20h in two temperature regimes, 17–22°C (optimal) and 40°C (heat stress), in complete darkness with a water-saturated atmosphere.The radioactivities of each organs were estimated as in Experiment I. Statistical analyses were performed with the use of t-Student test.

## Results

*Experiment I*

Total uptake of $^{45}Ca$ by Roma fruit-explants in NG-series, under optimal temperature was about twice higher than in nontreated controls (Table 1). In both cases heat stress drastically increased $^{45}Ca$ uptake, but differences were much smaller if radioactivity was calculated per $cm^3$ of water uptake during 24h exposure. Explants of cv. Robin took up more $^{45}Ca$ under elevat-

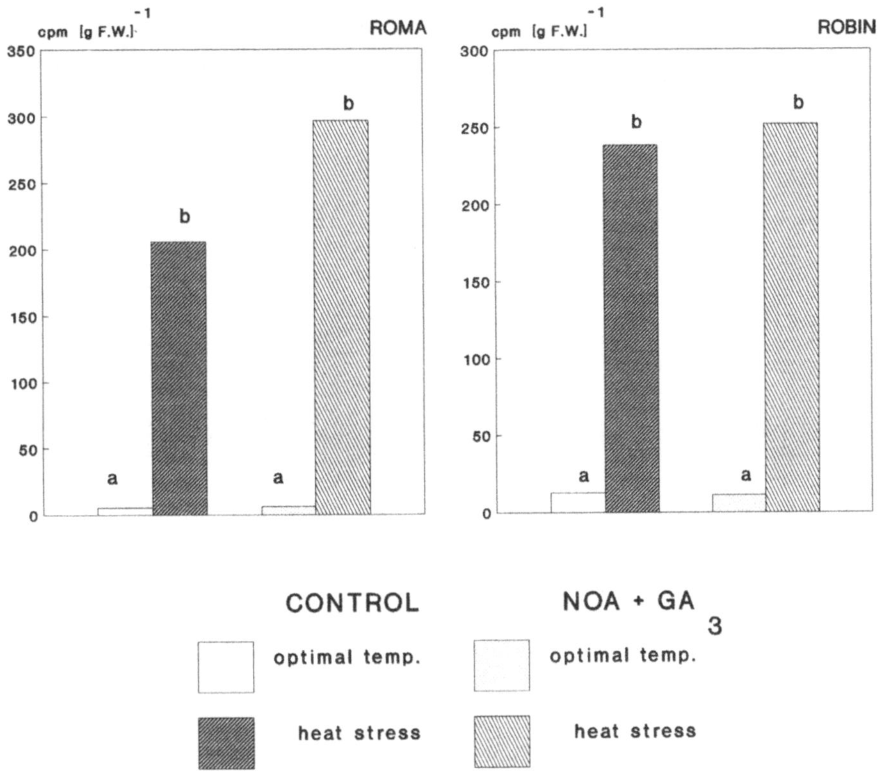


*Fig. 1.* Translocation of $^{45}Ca$ to the fruits in tomato explants.

*Table 1.* Uptake of $^{45}Ca$ by tomato fruit-explants (Exp. I)

| Radioact. | Temperature regime | cv. Roma Control | cv. Roma NOA+GA$_3$ | cv. Robin Control | cv. Robin NOA+GA$_3$ |
|---|---|---|---|---|---|
| $10^3$ cpm per explant | Optimal | 1034±132 | 2232±360 | 2430±270 | 2400±260 |
| | Heat | 3133±92 | 2965±619 | 3195±641 | 3374±240 |
| $10^3$ cpm cm$^{-3}$ | Optimal | 595±123 | 1119±102 | 1095±154 | 1190±130 |
| | Heat | 1016±116 | 857±222 | 881±65 | 936±197 |

*Table 2.* Export of $^{45}Ca$ from the roots (in percentage) and $^{45}Ca$-distribution in tomato cuttings (total export assumed as 100 percent) (Exp. II)

| | cv. Roma | | | | cv. Robin | | | |
|---|---|---|---|---|---|---|---|---|
| | Control | | NOA+GA$_3$ | | Control | | NOA+GA$_3$ | |
| | Opt. | Heat | Opt. | Heat | Opt. | Heat | Opt. | Heat |
| Export from roots | 68 | 61 | 42 | 48 | 11 | 20 | 74 | 47 |
| **Distribution** | | | | | | | | |
| Cluster | 13.4 | 12.4 | 15.1 | 13.1 | 3.7 | 22.2 | 3.5 | 9.7 |
| Blades | 1.6 | 7.6 | 0.9 | 8.1 | 0 | 0.1 | 1.4 | 1.0 |
| Stem + petioles | 85.0 | 80.0 | 84.0 | 78.8 | 96.3 | 77.7 | 95.1 | 89.3 |

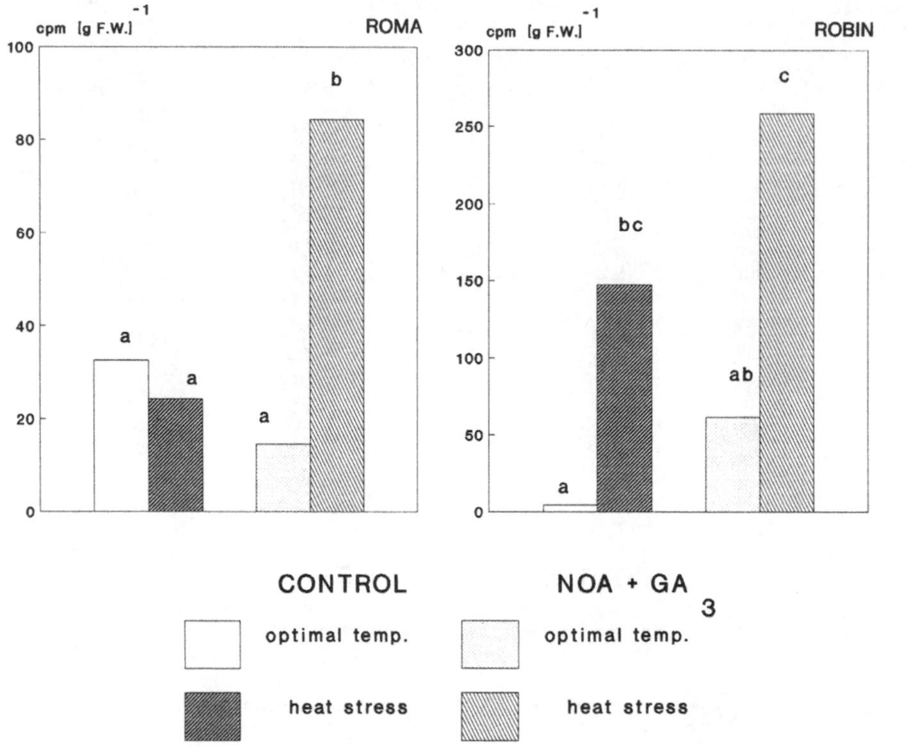


*Fig. 2.* Translocation of $^{45}Ca$ to the fruits in tomato cuttings.

ed temperature, but only due to higher water absorption (Table 1). Translocation of $^{45}Ca$ to the fruits was very low. Nevertheless, the greatest explant response to heat stress in both cultivars concerned the quantity of $^{45}Ca$ transported to the fruits. Heat stress strongly stimulated $^{45}Ca$ movement to Roma fruits in the NG and in control plants (Fig. 1). Robin explants were affected by elevated temperature in a very similar way as Roma (Fig. 1). Heat shock did not strongly affect the supply of $^{45}Ca$ to the calyx (data not presented) in spite of its higher transpiration as compare with fruits (Starck and Witek-Czupryńka, 1993).

*Experiment II*

Uptake of $^{45}Ca$ calculated per g root F.W. by Roma cuttings under optimal temperature conditions was similar in the C and NG series ($17 \times 10^3$ and $14 \times 10^3$ cpm), in contrast to enhanced calcium uptake by Robin NG-series (control and NG series, respectively: $16 \times 10^3$ cpm and $23 \times 10^3$ cpm).

Plant samples harvested immediately after $^{45}Ca$ exposure had very low radioactivity in each series of Roma and Robin fruits as well as in blades (data not presented).

Export of $^{45}Ca$ from the roots during 20h in the two cultivars was differently changed in both series (Table 2). In Roma, upward $^{45}Ca$ movement was higher under control series than in NG cuttings both under optimal and high temperature conditions. In Robin, $^{45}Ca$ export from the roots in control series, was very low under optimal temperature and was almost two times higher under heat. In NG series $^{45}Ca$ export from the roots was enhanced by growth regulator treatment but was depressed by elevated temperature.

As concerns $^{45}Ca$ distribution in Roma the stem with petioles consisted most to whole plant radioactivity. Less $^{45}Ca$ was transported to the cluster and very little to the blades, especially under optimal temperature. As with the other physiological parameters, the effect of elevated temperature on $^{45}Ca$ partitioning differed somewhat in Robin. Stem with petiole were also the most radioactive organs but the contribution of clusters was relatively low under optimal temperature (Table 2).

In spite of dramatically poor $^{45}Ca$ transport to the fruits in both cultivars, distinctive differences in response to the elevated temperature were observed (Fig. 2). The contribution of fruits to $^{45}Ca$ estimated in the whole cuttings fluctuated from 0.035% to 0.3% after optimal temperature treatment and from 0.2% to 2% after heat stress in both cultivars. Robin cuttings strongly responded to heat shock treatment by enhanced transport of currently uptake $^{45}Ca$ to the fruits. The reaction of Roma cuttings was smaller and occurred only in the NG series (Fig. 2, note the differences of the scales). Nevertheless the whole cluster of Roma contributed equally to total plant $^{45}Ca$ in all series (Table 2). Most $^{45}Ca$ and in a similar amount was contained in the peduncle with pedicel (data not presented). Calcium supply to the fruits was very low in NG-cuttings under optimal temperature, but was enhanced only in that series by heat stress. Robin peduncle and pedicel had variable radioactivity in each series, probably as the result of differential $^{45}Ca$ export from the roots (see Table 2).

## Discussion

The concept of calcium action in regulation of plant processes is based on its effect on membrane integrity. Some stimuli lead to changes in $Ca^{2+}$ fluxes between external or internal reservoirs and the cytoplasm (Carafoli, 1987). In a wide range of elevated temperature $Ca^{2+}$ is essential for maintaining thermal stability of the membranes, preventing their leakage (Toprover and Glinka, 1976). Some papers deal with tomato fruit physiological disorders, that accompany local calcium deficiency (Brown and Ho, 1993; Minamide and Ho, 1993).

In the opinion of Schulte-Baukloh and Fromm (1993) calcium is a necessary factor in the regulation of phloem loading and unloading. According to these authors as a general tendency, the sieve element cytoplasm has a higher concentration of $Ca^{2+}$ in the sink leaf than the source leaf, with very small differences in its content in the cell wall.

Susceptibility of tomato plants to elevated temperature does not seem to be connected either with $Ca^{2+}$ uptake or its transport to the shoot. On the other hand the above presented results suggest, either adaptative $Ca^{2+}$ distribution between individual organs or only its supply to the fruits during heat stress. In tomato plants $Ca^{2+}$ supply to the fruits is generally very low (Bangerth, 1979; Ho, 1989; Starck and Witek-Czupryńska, 1987), as it is presented in this paper. Nevertheless, permanent flux of $Ca^{2+}$ to the fruits may have a pivotal role to maintain an optimal level of calcium in the cytoplasm of fruit cells, as factor increasing tolerance to high temperature.

In rooted cuttings, during heat stress import of $^{45}Ca$ to the fruits was drastically enhanced in heat tolerant Robin plants. A similar response in sensitive Roma, was recorded only in the NG series, with reduced susceptibility to elevated temperature and higher sink strength of their fruits (Starck and Cieśla, 1989). This is perhaps connected with an enlarged auxin and/or gibberellin level. Under greenhouse conditions fruit set and growth of Roma is very poor, presumably due to its hormonal inbalance. This cultivar responds positively to flower treatment with growth regulators, especially under unfavourable temperature conditions.

In contrast to rooted cuttings, in fruit explants high temperature affected $Ca^{2+}$ uptake and fruit supply with this ion in a very similar way in both cultivars: increase of $^{45}Ca$ uptake and its flux to the fruits was reported. It suggests, that in a detached cluster after suppression of the other organs the specific reaction of the cultivars compared declines. Therefore, precise regulation of current $Ca^{2+}$ supply to the fruits seems to be located outside the cluster or at least outside the fruit. This hypothesis could not be fully evaluated in the present investigations. Further studies are required to understand the role of $Ca^{2+}$ as a possible factor contributing to the mechanism of fruit cell protection against heat injury in tomato.

## References


Adams P and Ho L C 1992 The susceptibility of modern tomato cultivars to blossom-end rot in relation to salinity. J. Hortic. Sci. 67, 827–839.

Bangerth F 1979 Calcium-related physiological disorders of plants. Annu. Rev. Phytopathol. 17, 97–122.

Ben-Arie R, Lurie S and Mattoo A K 1982 Temperature - dependent inhibitory effects of calcium and spermine on ethylene biosynthesis in apple discs correlate with changes in microsomal membrane microviscosity. Plant Sci. Lett. 24, 239–247.

Brown M M and Ho L C 1993 Factors affecting calcium transport and basipetal IAA movement in tomato fruit in relation to blossom - end rot. J. Exp. Bot. 44, 1111–1117.

Carafoli E 1987 Intracellular calcium homeostasis. Annu. Rev. Biochem. 56, 395–433.

Ehret D L and Ho L C 1986 Translocation of calcium in relation to tomato fruit growth. Ann. Bot. 58, 679–688.

Fukumoto M, Nagai K, Yoshioka H and Aoba K 1987 Mechanism of the development of a calcium-related disorder (bitter pit) in apple. JARQ 20, 248–252.

Gilroy S, Read N D and Trewavas A J 1990 Elevation of cytoplasmic calcium by caged calcium or caged inositol triphosphate initiates stomatal closure. Nature 346, 769–771.

Ho L C 1989 Environmental effects on the diurnal accumulation of $^{45}$Ca by young fruit and leaves of tomato plants. Ann. Bot. 63, 281–288.

Ho L C, Belda R, Brown M, Andrews J and Adams P 1993 Uptake and transport of calcium and the possible causes of blossom-end rot in tomato. J. Exp. Bot. 44, 509–518.

Klein J D and Ferguson J B 1987 Effect of high temperature on calcium uptake by suspension-cultured pear fruit cells. Plant Physiol. 84, 153–156.

Knight M R, Campbell A K, Smith S M and Trewavas A J 1991 Transgenic plant aequorin reports the effects of touch and cold - shock and elicitors on cytoplasmic calcium. Nature 352, 524–526.

Minamide R T and Ho L C 1993 Deposition of calcium compounds in tomato fruit in relation to calcium transport. J. Hortic. Sci. 68, 755–762.

Palta J P 1990 Stress interactions at the cellular and membrane levels. Hort. Science 25, 1377–1381.

Poovaiah B W 1993 Biochemical and molecular aspects of calcium action. Acta Hortic. 326, 139–147.

Rengel Z 1992 The role of calcium in salt toxicity. Plant Cell Environ. 15, 625–632.

Santarius K A and Weis E 1988 Heat stress and membranes. *In* Plant membranes-structure, assembly and function. Eds. J L Harwood and T J Walton. pp 97–112. The Biochemical Society, London.

Schulte-Baukloh Ch and Fromm J 1993 The effect of calcium starvation on assimilate partitioning and mineral distribution in the phloem. J. Exp Bot. 44, 1703–1707.

Simon E W 1978 The symptoms of calcium deficiency in plants. New Phytol. 80, 1–15.

Starck Z and Cieśla E 1989 Possible role of growth regulators in adaptation to heat stress affecting partitioning of photosynthates in tomato plants. Acta Soc. Bot. Pol. 58, 71–84.

Starck Z and Witek-Czupryńska B 1987 Translocation of $^{45}$Ca in tomato plants treated with growth regulators. Zemljiste Biljka 36, 195–206.

Starck Z and Witek-Czupryńska B 1993 Diverse response of two cultivars tomato fruit explants to high temperature. Acta Soc. Bot. Pol. 62, 165–169.

Steponkus P L 1981 Responses to extreme temperatures. Cellular and sub-cellular bases. *In* Enc. Plant Physiol. 12A. Eds. O L Lange, P S Nobel, C B Osmond and H Ziegler. pp 372–402. Springer-Verlag, Berlin.

Tomala K and Dilley D R 1990 Some factors influencing the calcium level in apple fruits. Acta Hortic. 274, 481–487.

Toproven Y and Glinka Z 1976 Calcium ions protect beet root cell membranes against thermally induced changes. Physiol. Plant. 37, 131–134.

*F. Baluška et al. (eds.), Structure and Function of Roots*, 311–317, 1995.


# Safety zone as a barrier to root-shoot ice propagation

J. Zámećník, J. Bieblová and M. Grospietsch
*Department of Plant Physiology, Research Institute of Crop Production, Dmovská 507, Prague, CZ-161 06 Czech Republic*

*Key words:* hydraulic safety zone, ice propagation, winter cereals


## Abstract

Winter wheat, rye and triticale were sown at a depth of 12 cm to promote internode elongation. The root-shoot junction was separated by an internode about 4.5 cm long. The rate of ice propagation was measured separately across the root-internode and crown-internode at -4°C. Ice nucleation in seminal roots was experimentally induced by ice nucleating bacteria. The rate of movement of the ice front was measured by differential thermal analysis. The highest rate of ice propagation occurred in the internode. The rate of ice propagation across the root-internode section was slower than in the internode alone, but differences exist in the various genotypes. The rate of ice propagation between the seminal roots and the internode did not correlate with the number of roots with unsafe vessels, as detected by the pressurized air method. The rate of ice propagation between the internode and the crown was approximately one third of the rate in the internode. The discontinuity in internode-crown section reached nearly 100% as was observed by the pressurized air method. Anatomical studies supported the discontinuity of vessels evaluated by the pressurized air method. Experimental data support the view that the barriers in root-internode and in internode-crown section reduced the rate of ice propagation but these barriers are not correlated with either cold hardiness or number of safe roots.


*Abbreviations:* AN – adventitious nodal roots, AS – adventitious seminal roots, SS – secondary seminal roots

## Introduction

Roots of winter cereals may be torn by soil heaving, when the upper layer of soil repeatedly changes its volume. The broken roots are in direct contact with ice crystals in the soil. Those ice crystals can initiate ice formation which can propagate into the crowns. The crown meristems are the critical part for the survival of winter cereals after freezing.

Ice nucleation can be started by exogeneous and/or endogeneous nuclei. Ice nucleating microorganisms can act as exogeneous biogenic ice nuclei. Hoar frost, soil particles and soil ice can act as exogeneous non-biogenic ice nuclei. Plants have barriers to exogeneous air-born ice nuclei (Kaku, 1971), such as the hydrophobic wax coating on surfaces of leaves. However, little is known about the barriers to ice propagation in the root-shoot junction in annual plants.

The hydraulic safety zone is formed in the basal region of roots (Luxová, 1986, 1989) and in the junction of lateral roots (Luxová, 1990; North et al., 1992). The hydraulic safety zones were found in higher number in winter than in spring cereals. (Aloni and Griffith, 1991). This observation brings up the following questions: Does this difference in number of safety roots between winter and spring cereals also predominate in the winter cereals with various frost resistance? What is the role of hydraulic safety zones for the ice propagation in root-shoot plant continuum? The aim of this study is the detection of barriers to ice propagation in the root-shoot system and the investigation of the role of safety hydraulic zones in various species and genotypes differing in frost resistance to ice propagation. For this study we used contrast genotypes and species (winter rye, winter wheat and triticale) in cold tolerance. The plants were sown deeper than usual for winter cereals. Plant sprouting from deeper layers of soil developed internodes of various lengths, which separated the root system from the crown. This plant model provides a means of determining the localization of barriers to ice propagation between the root and crown.

## Materials and methods

Winter wheat (cv. Mironovskaya 808, Chlumecká 12, San Pastore) triticale (cv. Malno, Korm) and winter rye (cv. Musketeer) were sown at a depth of 12 cm in the field to promote elongation of the root-shoot internode. The plants were harvested for measurement in the middle of winter when they were in the 16th stage of the decimal code according to the scale of cereals growth stages (Nátrová and Jokeš, 1992) when adventitious nodal roots (AN) were absent, and in the late winter in the 35th stage of the decimal code when AN roots had developed.

*Pressurized air method*

The pressurized air method (Aloni and Griffith, 1991) was used for the evaluation of root-crown xylem continuum. An individual root or internode was sealed into a silicon tubing which was a little broader in diameter than the diameter of the root or internode. Pressurized air at 0.3 MPa was forced into the roots or internode. Higher pressure then 0.4 MPa caused rupture of the xylem. Bubbles from the opposite end of the sample were examined under water. The bubbling samples were designated as those without segmentation of root vessels from shoot vessels. After each observation, the nonbubbling plant parts were again cut near the sealing sites. The samples without bubbling after recutting were not included to eliminate samples wounded during preparation or with immature metaxylem. Data were collected separately for root-internode and internode-crown system.

*Microscopic observation*

The samples of root-shoot sections were trimmed into 2–3mm and fixed in formalin-aceto-alcohol fluid (FAA). The samples were cleared in butyl alcohol, embedded in paraffin and sectioned serially using a Multicut 2045 (Leica) microtome at 12 $\mu$m per section. Sections were stained using alcian blue (Polysciences) and dehydrated through alcohol to xylen and

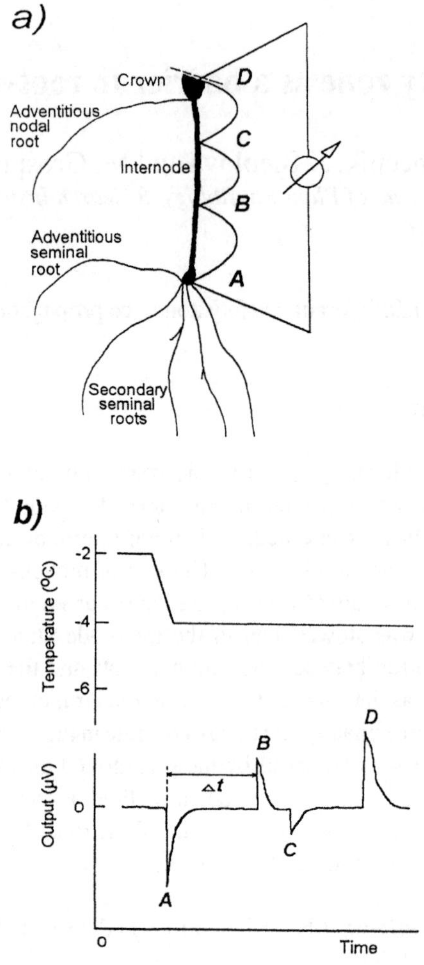


*Fig. 1.* **a**. Experimental design of thermocouple junction housing in root-crown ice propagation measurement. Note the equal distance between thermocouples. **b**. The time course of cooling rate and output from four thermocouples in differential connection. Δ t - time delay in ice propagation.

then mounted in Solacryl. Samples were examined and photographed using the Olympus Vanox-T Photomicroscope.

*Ice propagation method*

The leaves were detached from the crown near the meristematic tissue and only the 30 mm of roots tissue was left on the plant. The surface dried sample was attached by tape to polythene foil on copper plate to ensure a good thermal contact. The latent heat of crystallization in the tissue was measured using thermocouples. As the ice front propagates to the measurement point, the latent heat of crystallization increases the temperature in the surroundings of the measuring point. Four copper-constantan thermocouples were connected in series (ABCD - Fig. 1a) to detect the freezing exotherm.

The first thermocouple was situated just at the place where the seminal roots branch (Fig. 1a). Ice nucleation was initiated at the cut end of the roots using a 15 $\mu$L of ice nucleating bacteria *Pseudomonas syringae* CCM 4073 solution ($10^{10}$ cfu $mL^{-1}$). The 30 mm of root length represented a sufficient distance from the first thermocouple to avoid detecting latent heat of crystallization from drops containing ice nucleation bacteria. The second and third measuring junction were inserted between the internode and copper plate. The fourth thermocouple was inserted into the end of the cut crown, which was wetted by 10 $\mu$L of distilled water. The measuring junctions (A, B, C, D - Fig. 1a) were attached equidistantly along the plant. The distance between A-B, B-C and C-D differed according to the length of each plant internode. The plant on the copper plate was covered by a glass lid and sealed using a foam gasket. The whole measuring plate was inserted into the ethanol cooling bath (Ultra-Kryomat Lauda).

The plants were allowed to equilibralise at -2°C for 2 minutes (the ice nucleation bacteria were in the supercooling stage), the temperature was then lowered to -4°C at a cooling rate of 2.3°C $min^{-1}$. As the temperature was lowered, the bacteria started to nucleate ice. The release of heat was detected by differential thermal analysis at 0.2 second intervals using a multimeter connected to the PC computer. Sudden release of latent heat of crystallization manifests itself as a voltage peak from the thermobattery when the ice front reaches the next junction (Fig. 1b). The time difference between the onset of peaks was taken as the time of linear ice front propagation. To obtain reproductive data, the measuring points were situated at the same distance on each plant. The ice propagation was expressed as ice propagation rate per one mm of the measuring point distance, because of the differences between the internode length of each measured plant. The ice propagation rate was measured also on detached plant organ segments. The detached organ segments (i.e. root, internode, leaves) were measured in opposite cut sites by two thermocouple junctions in the same way.

Mean ice propagation rate was calculated by harmonic mean. The ice propagation rate in safety barriers was calculated from the formula of harmonic mean:

Ice propagation rate across the barrier =

$$\frac{r_{AB}\ d_B\ r_{BC}}{r_{BC}\ d_{AB} - (r_{AB}(d_{AB} - d_B))} \tag{1}$$

where r is the ice propagation rate, d is the length, the subscripts AB, BC are for the measuring points (Fig. 1a) and the subscript B is for the calculated length of the barrier.

## Results

The length of internodes does not differ greatly among genotypes (Table 1). The longest internode was found in cv. Musketeer (49±3.6 mm), the shortest cv. Mironovskaya 808 (28±2.0 mm).

### *Pressurized air*

In mid winter it was possible to distinguish the secondary seminal roots (SS) from adventitious seminal roots (AS) according to their colour. The number of safe connections in SS roots was smaller than the number in AS roots as shown from the results of the pressurized air method (Table 1). The smallest percentage of SS roots (43±10%) with safety zones was observed in Malno cultivar. The higher percentage (75±28) of SS roots with safety zone contrasts with the low percentage (20±13) of AS roots with safety zone in San Pastore cultivar. There were no observations of a continuity of individual AN roots to the crown in any cultivars except Mironovskaya 808 which showed safe systems only in 87±13% of the samples.

In late winter the visual distinction between the SS and AS roots was not apparent in the plants collected from the field. The measurements on both SS and AS roots using the pressurized air method indicated decreased percentage of roots with safety zones in late winter in comparison with mid winter (Table 1). The AN roots had safety zones in all observed cultivars again as in mid winter.

### *Microscopic observation*

The vessels in adventitious roots with safety zones terminating at the root-shoot junction as has been found in longitudinal sections of tissue. In the roots with safety zones, the ends of metaxylem elements were connected by tracheids to axial strand vessels (Figs. 2, 3a). In roots with unsafe systems, the direct connection of adventitious root vessels with axially strand vessels was observed (Fig. 3b).

*Table 1.* Mean values of number of seminal secondary (SS) and seminal adventitious (AS) roots and percentage of roots with safety zone of cereal cultivars determined by pressurized air method in mid and late winter. (n = 10 ± SE)

| Cultivar | Internode Length mm | Mid winter | | | | Late winter | |
|---|---|---|---|---|---|---|---|
| | | SS | | AS | | SS and AS | |
| | | No. | Safe % | No. | Safe % | No. | Safe % |
| Mironovskaya 808 | 28±2.0 | 3.5±0.38 | 51±5 | 0.8±0.28 | 100 | - | - |
| Chlumecká 12 | 45±1.4 | 4.0±0.41 | 58±17 | 1.0±0.48 | 100 | 3.8±0.48 | 29±9 |
| San Pastore | 37±3.4 | 4.3±0.48 | 75±28 | 5.0±0.50 | 20±13 | 5.0±0.55 | 63±7 |
| Malno | 38±2.7 | 3.3±0.26 | 43±10 | 0.9±0.20 | 88±10 | - | - |
| Korm | 34±1.7 | 7.6±0.51 | 61±14 | N[a] | N | - | - |
| Musketeer | 49±3.6 | 6.1±0.56 | 86±6 | N | N | 5.7±0.53 | 38±9 |

[a] Not developed.

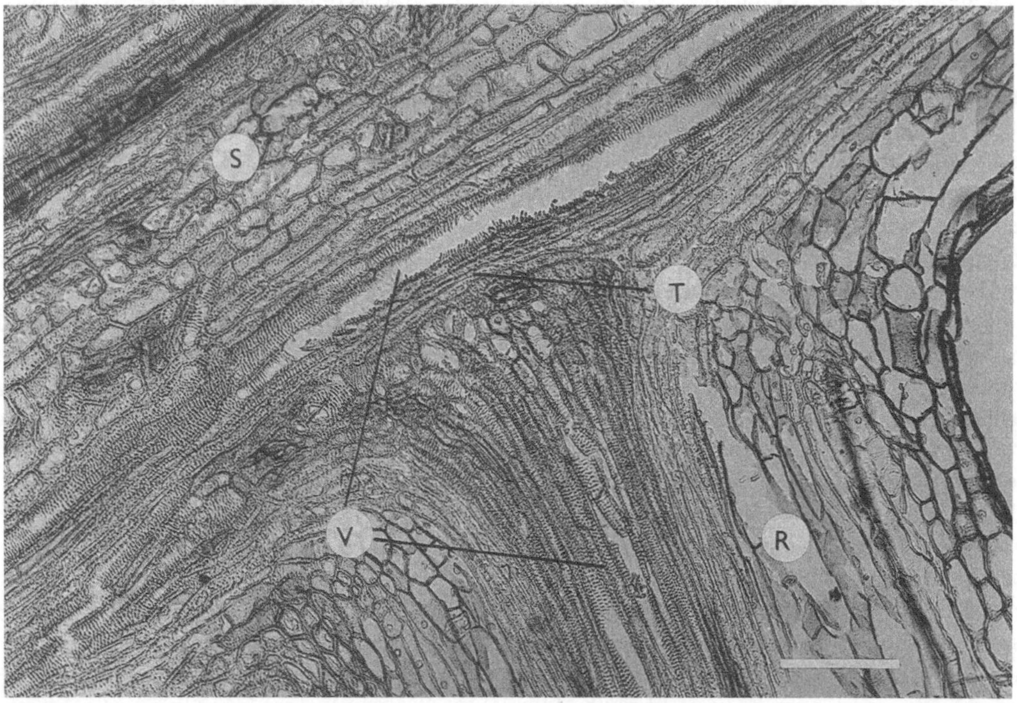


*Fig. 2.* Longitudinal section of safe root-shoot junction of winter rye cv. Musketeer shows the way of connection between vascular strands of adventitious root vessels and shoot vessels. The connection among metaxylem vessels of roots and those of shoots is not direct, but it is occurs through the tracheids. R-root, S-shoot, V-vessels, T-tracheids. The *bar* represents 100 $\mu$m.

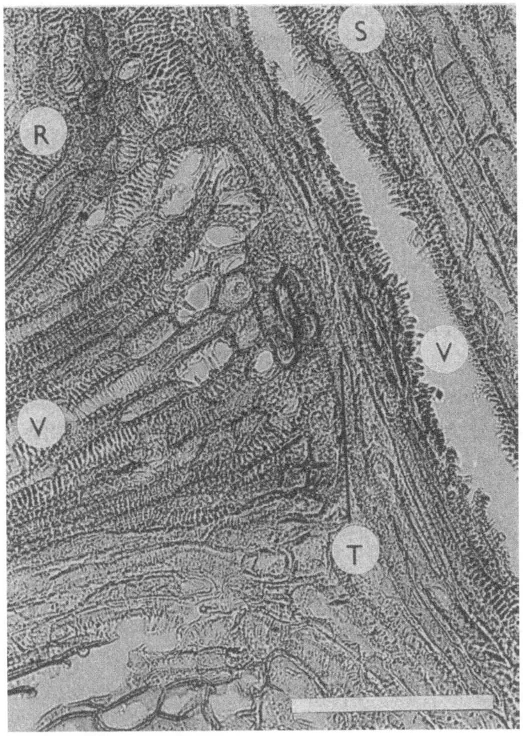


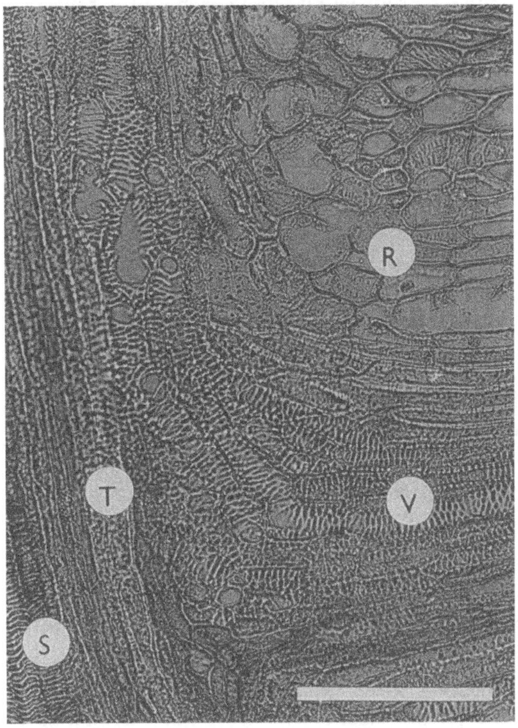


*Fig. 3.* **a.** An example of safe termination of root metaxylem vessels. The detail shows the lack of direct connection of the adventitious root vessels and vessels in the crown. **b.** An example of an unsafe termination of root vascular bundle. The majority of metaxylem vessels are terminated as in the previous case, but a few vessels on the upper part of the root vascular bundle are directly connected to the axial strand of the shoot. R-root, S-shoot, V-vessels, T-tracheids. The *bar* represents 100 $\mu$m.

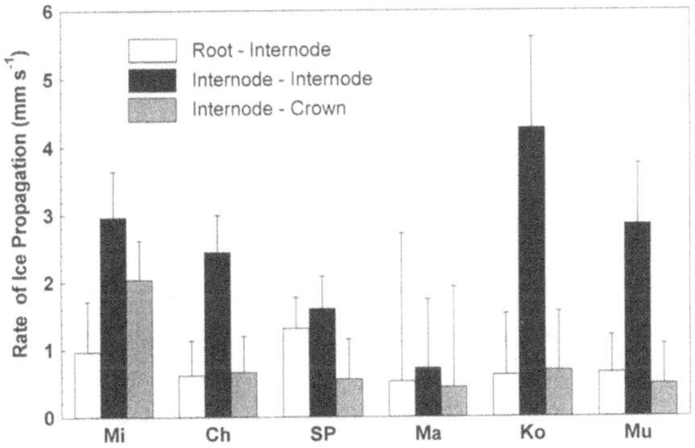


*Fig. 4.* Harmonic mean with standard error of ice propagation rate in the equal distance of root-internode-internode, internode-crown. Mi - Mironovskaya 808, Ch - Chlumecká 12, Sp - San Pastore, Ma - Malno, Ko - Korm, Mu - Musketeer.

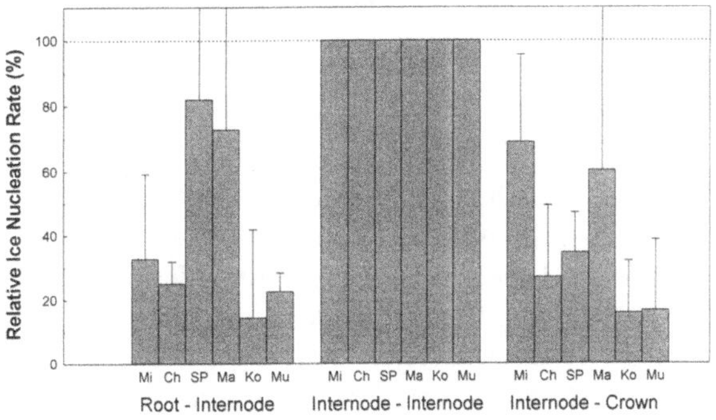


*Fig. 5.* Mean values with standard error of the relative ice propagation rate in the equal distance of root-internode, internode-internode, internode-crown of each plant. The ice propagation rate in internode-internode was taken as 100% for each cultivar. Mi - Mironovskaya 808, Ch - Chlumecká 12, Sp - San Pastore, Ma - Malno, Ko - Korm, Mu - Musketeer.

### *Ice propagation*

Endogeneous ice nucleation of plants was not detected at temperatures higher than -5.5°C. Therefore the bacteria strain with ice nucleation activity in a narrow range at -3°C was suitable for ice nucleation. The ice propagation rate between roots and the internode and between the internode and the crown was lower than the ice propagation rate in the internode alone (Fig. 4).

The ice propagation rate (Fig. 5) was expressed relative to the ice propagation rate in each cultivar internode. There were no differences in ice propagation rates among cultivars except in cv. Mironovskaya 808 which has higher ice propagation rate in the internode-crown section (2.04 mm $s^{-1}$) than in the other cultivars. In segments of individual AS roots ice propagation rate was 5.7±0.42 mm $s^{-1}$. For comparison the rate of ice propagation in leaf segments was 5.3±0.21 mm $s^{-1}$. The investigation of ice propagation in individual plant segments supports the findings on the whole plant.

The part of the internode between the measuring points AB and CD (Fig. 1a) is of decisive importance in higher ice propagation rate calculated in the root-internode and internode-crown sections. In a macroscopic view the length of the junction (2 mm) was approximately one fifth of distance between the A-B measuring points. The ice propagation rate in the 2 mm length of junction ($d_B$=2 mm) of cv. Musketeer was calculated from Equation (1), where $r_{AB}$=0.7 mm $s^{-1}$ and $r_{BC}$=2.9 mm $s^{-1}$ (Fig. 4). The average length of cv. Musketeer internode is 49 mm (Table 1), then the length between AB measuring points is $d_{AB}$=49/3=16.3 mm. The ice propagation rate from this calculation is 0.1 mm $s^{-1}$. From the cross longitudinal sections (Fig. 3a) it can be suggested in more detail that the safety zone is 50 – 200 $\mu$m thick. If 200 $\mu$m length is taken into account, then the ice propagation rate at the cellular level barrier calculated from the equation (1) is 0.01 mm $s^{-1}$ for cv. Musketeer.

## Discussion

The SS roots had a higher ratio of roots with unsafe zone than the AN roots (the ratio varied in different species and genotypes). The AN roots developed later than SS roots. The hydraulic safety roots are determined by the anatomical structure (Aloni and Griffith, 1991; North et al., 1992). Our anatomical studies were consistent with these observations and with those obtained by the pressurized air method. The higher number of AS roots with safety zones could be caused by the immature metaxylem at the junction. In spite of the diversity in the number of safety roots (Table 1) in low frost tolerant winter wheat genotypes, there seems to be no correlation between the number of safety roots and the frost tolerance (i.e. low frost tolerance cultivar San Pastore has higher number of safety roots than highest frost tolerant cv. Mironovskaya 808). The number of roots with safety zones as well as the ice propagation rate was not based on the genotype specificity of cold hardiness of the investigated cultivars. The correlations were not found either among the species (winter rye Musketeer has a high frost resistance Krol et al., 1984), or among the cultivars of one species (cv. Mironovskaya 808 belongs to hardy and cv. San Pastore to less hardy winter wheat, Prášil and Zámečník, 1991). These results did not support the working hypothesis about the highest number safety zones predetermined by highest frost tolerance of cereal species or genotypes.

From the comparison of the direct ice propagation measurements (Figs. 4, 5) and from comparison of the high ice propagation rate in detached internodes, in detached leaves and in detached roots, with the low ice propagation rate in root-internode and in internode-crown section, it is obvious that the ice propagation is lower in these parts. The barriers to ice propagation in cereal seedlings are located in the root-internode and internode-crown section as follows from these observations. Generally the barriers were located close to the meristems. The active meristems were not only in the crowns but also in the root-internode section as followed from low frequency of sprouting from root-internode section. The barriers of ice propagation of undetermined compositions are known in floral buds of several cold hardy species of woods. For example the barriers to ice propagation in *Azalea* buds appear to be of physical origin (Chalker-Scott, 1992). In spite of that the anatomical origin of barriers in cereals could not be excluded, because of the complicated anatomy of the root-internode and the internode-crown section.

No correlation was found between the number of safety roots in root-internode section and the rate of ice propagation in the tested genotypes. The ice propagation rate should not correlate with the number of roots with unsafe zones, because the ice can propagate through one root, irrespective of the number of other roots with safety zones.

The calculation of the ice propagation rate as defined by Equation (1) provides an approximation of decreased ice propagation rate through the barriers in root-internode and internode-crown sections. The calculation indicates that the ice propagation rate across the barrier can be six-times lower than if measured for the entire pathway. The value of the ice propagation rate can be further decreased to one sixtieth of our findings, if the barriers are assumed at cellular level. The barriers preventing the spread of an ice front were also found in the nodes of wheat stem (Single and Olien, 1967).

The ice propagation rates in the internode are comparable with the ice propagation rate in 10% sucrose (Lusena, 1955). These comparisons indicate no barriers exist in the internode to hinder ice propagation. The high rate of water freezing in the xylem of cereals may have a considerable effect on air bubbles released during freezing. The gas molecules have no time to diffuse away from rapidly moving ice/water interface and after thawing the gaseous bubbles can be formed. The gaseous bubbles imply xylem embolism (Sucoff, 1969). For example 40% of *Quercus ilex* twigs loss their hydraulic conductivity at -1.5°C (Lo Gullo and Salleo, 1993). The safety roots would be of advantage to plants exposed to xylem embolism not only during drought and phythopathogenic stresses but also during the freezing stress. The safety root may help to keep the hydraulic conductivity after air emboli seeding during the transition of water from the liquid to the ice phase.

## Acknowledgements

The first author thanks Dr Marilyn Griffith (University of Waterloo) for stimulating discussion and Dr G McLeod (Agriculture Canada, Swift Current, Sask.) for providing the winter rye seeds. We are grateful to Professor Mária Luxová for helpful comments on the manuscript draft. This paper is a part of the project 501/93/0329 supported by the Czech Grant Agency.

## References

Aloni R and Griffth M 1991 Functional xylem anatomy in root-shoot junctions of six cereal species. Planta 184, 123–129.

Chalker-Scott L 1992 Disruption of an ice-nucleation barrier in cold hardy *Azalea* buds by sublethal heat stress. Ann. Bot. 70, 409–418.

Kaku S 1971 A possible role of endodermis as a barrier for ice propagation in the freezing of pine needles. Plant Cell Physiol. 12, 941–948.

Krol M, Griffth M and Huner N P A 1984 An appropriate physiological control for environmental temperature studies: comparative growth kinetics of winter rye. Can. J. Bot. 62, 1062–1068.

Lo Gullo M A and Salleo S 1993 Different vulnerabilities of *Quercus ilex* L. to freeze and summer drought-induced xylem embolism: an ecological interpretation. Plant Cell Environ. 16, 511–519.

Lusena C V 1955 Ice propagation in system of biological interest. III. Effect of solutes on nucleation and growth of ice crystals. Arch. Biochem. Biophys. 57, 277–284.

Luxová M 1986 The hydraulic safety zone at the base of barley roots. Planta 169, 465–470.

Luxová M 1989 The vascular system in the roots of barley and its hydraulic aspects. *In* Structural and Functional Aspects of Transport in Roots. Eds. B C Loughman, O Gašparíková and J Kolek. pp 15–20. Kluwer Academic Publishers, Dordrecht, The Nederlands.

Luxová M 1990 Effect of lateral root formation on the vascular pattern of barley roots. Bot. Acta 103, 305–310.

Nátrová Z and Jokeš M 1992 A proposed scale using decimal code for classification of wheat ear development. Agriculture University of Nitra, CSFR.13 p. *(In Czech).*

North G B, Ewers F W and Nobel P S 1992 Main root-lateral root junctions of two desert succulents: Changes in axial and radial components of hydraulic conductivity during drying. Am. J. Bot 79, 1039–1050.

Prášil I and Záměčník J 1991 The effect of weather on the overwintering of winter crops. *In* Weather and Yield. Ed. J Petr. pp 141–152. Elsevier, Amsterdam.

Single W V and Olien C R 1967 Freezing processes in wheat stems. Aust. J. Biol. Sci. 20, 1025–1028.

Sucoff E 1969 Freezing of conifer xylem and the cohesion-tension theory. Physiol. Plant. 22, 424–431.

*Section editor: H Lambers*

*F. Baluška et al. (eds.), Structure and Function of Roots*, 319–326, 1995.


# The binding of zinc, but not cadmium, by phytic acid in roots of crop plants

R.F.M. van Steveninck, A. Babare, D.R. Fernando and M.E. van Steveninck[1]
*School of Agriculture and [1]Department of Botany, La Trobe University, Bundoora, Victoria, 3083, Australia*

*Key words:* cadmium, crop species, heavy metal binding, phytic acid, X-ray microanalysis, zinc


## Abstract

Plant species adapted to soils enriched with heavy metals often accumulate these metals in their above or below ground organs. In this study, electron probe microanalysis of fractured, quench-frozen root specimens of common crop species shows that an appreciable quantity of Zn can be bound as Zn phytate (*myo*-inositol *kis*-hexaphosphate) within small vacuoles of cells in the root elongation zone of lucerne, soybean, lupins, tomato, rapeseed, cabbage, radish, maize and wheat exposed to high levels of Zn (80–300 $\mu M$). Globular deposits of Zn phytate are most frequently observed in the endodermis of dicotyledonous species and in the pericycle of monocotyledonous species, but may also occur in the stele and inner cortex after prolonged exposure to toxic levels of Zn. The deposits could not be found in Zn-treated sunflower, field peas and Italian ryegrass. In three crop species, lucerne, soybean and maize, Zn-induced phytate globules were frequent, but exposure of roots to 30 $\mu M$ Cd did not induce the formation of Cd-containing globular deposits as observed in *Lemna minor* (Van Steveninck et al., 1990a, 1992). Simultaneous Zn and Cd treatment induced the formation of Zn phytate globules as effectively as Zn alone, and Cd was not detected in the deposits.


## Introduction

Accumulation of heavy metals such as Zn, Ni, Cu and Co in plants adapted to heavy metal-containing soils is a well-known phenomenon (Baker and Brooks, 1989). Plant species possessing this adaptation can be useful as geobotanical indicators in mineral exploration (Bollard and Butler, 1966; Reeves and Brooks, 1983). If accumulation occurs in parts of crops that are consumed, it may lead to the entry of toxic metals (e.g. Pb, Cd) into the food chain. However, the ability to hyperaccumulate metals may be an advantage if accumulation is restricted to plant parts that are not consumed, in particular roots of fodder plants and crops in which only the above ground parts are harvested for consumption. For example the introduction of metallothionein genes into tobacco (Maiti et al., 1989, 1991; Misra and Gedamu, 1990) aims to increase the retention of Cd in the roots and thus reduce Cd levels in the leaves. For similar reasons, lines of maize with and without the capacity to restrict Cd entry into the shoots are being compared (Florijn and Van Beusichem, 1993), and this work has shown that genetically determined differences occur in the degree of Cd-binding and in the distribution of Cd in roots and shoots of maize plants.

The accumulation of heavy metals may involve several processes, including metal binding in the apoplast, the formation of insoluble deposits in the cytoplasm and their secretion into vacuoles (Verkleij and Schat, 1990). When adapted species show cross tolerance or combined tolerance to several metals, the involvement of the same mechanism might be expected, especially in the case of Cd and Zn which exhibit close physical and chemical similarities and often occur together in metalliferous soils (Verkleij and Schat, 1990). However, Vázquez et al. (1992), although observing little difference between root and shoot in the degree of accumulation of Cd and Zn in the Cd/Zn hyper-accumulator *Thlaspi caerulescens*, found that Cd was accumulated mainly in the apoplast whereas Zn was principally found in vacuoles. Van Steveninck et al. (1990a) produced evidence for two distinct mechanisms of Zn and Cd binding in young fronds of the aquatic plant *Lemna minor*, but in mature frond tissue, Zn and Cd could both be bound by means of phytate (Van Steveninck et al., 1992). Considering that inner cortical cells in the root elongation zone of Zn-tolerant *Deschampsia caespitosa* are capable of binding Zn by means of phytate (Van Steveninck et al., 1987), and that this trait has now been shown to occur in the radicle of maize seeds (Mikus et al., 1992) and in the root endodermis (dicots) or pericycle (monocots) of a wide range of crop plants (Van Steveninck et al., 1993), it seemed worthwhile to check whether or not these plant species are capable of binding Cd by the same means.

## Materials and methods

### *Plant culture*

Crop species included in this investigation were radish (*Raphanus sativus* cv Mars), cabbage (*Brassica oleracea* cv Savoy King), rapeseed (*Brassica napus*), lucerne (*Medicago sativa* cv WL514), soybean (*Glycine max* cv Riddley), tomato (*Lycopersicon esculentum* cv Rouge de Marmand), sunflower (*Helianthus annuus* cv Hysun), wheat (*Triticum aestivum* cv Matong), lupins (*Lupinus angustifolius* cv Gungurre), field peas (*Pisum arvense* cv Dunn), maize (*Zea mays* cv SR103) and Italian ryegrass (*Lolium multiflorum*). Seeds were germinated on filter paper wet with distilled water and seedlings transferred to 2 L containers of aerated nutrient solution (after Godbold et al., 1984) containing 400 $\mu M$ $KNO_3$, 400 $\mu M$ $Ca(NO_3)_2$, 150$\mu M$ $MgSO_4$, 150 $\mu M$ $NaH_2PO_4$, 10$\mu M$ NaCl, 5 $\mu M$ $H_3BO_3$, 5 $\mu M$ FeNa $(EDTA)_2$, 1 $\mu M$ $MnSO_4$, 1 $\mu M$ $ZnSO_4$, 0.5 $\mu M$ $(NH_4)_6Mo_7O_{24}$, 0.1 $\mu M$ $CuSO_4$ and 0.02 $\mu M$ $CoCl_2$ or quarter strength Johnson solution (Johnson et al., 1957) containing 1500 $\mu M$ $KNO_3$, 500 $\mu M$ $Ca(NO_3)_2$, 250 $\mu M$ $NH_4H_2PO_4$, 150 $\mu M$ $MgSO_4$, 25 $\mu M$ FeK $(EDTA)_2$, 25 $\mu M$ $H_3BO_3$ 1 $\mu M$ $MnSO_4$, 1 $\mu M$ $ZnSO_4$, 0.25 $\mu M$ $CuSO_4$ and 0.25 $\mu M$ $H_2MoO_4$. The Johnson solution (Na-free) was included to check whether or not detectable quantities of Na in the plant tissue interfered with Zn $L_\alpha$ X-ray emissions through overlap with Na $K_\alpha$ X-rays. Containers of nutrient solution were topped with distilled water as necessary, and the solution was renewed every 5 days. Day/night temperatures and photoperiod were maintained at 25/20°C and 18/6 hours respectively with 220 $\mu$mol $m^{-2}$ $s^{-1}$ light intensity. Plants were supplied with an additional 80, 100, 160 or 300 $\mu M$ $ZnSO_4$ and/or 30 $\mu M$ $CdCl_2$ after 7–10 days growth in nutrient solution. Root tips approximately 5 mm from the apex were analyzed 7–14 days after addition of Zn and/or Cd to the nutrient solution.

### *Electron probe X-ray microanalysis*

Analytical results were obtained from root tips quench frozen in melting nitrogen (-210°C) using the methods described by Van Steveninck et al. (1990b). Briefly, root tips were collected from dominant roots (usually the primary root), quickly mounted in a carbon stub with the apical 3–5 mm protruding, quench frozen in melting nitrogen (-210°C), and fractured at the surface of the stub with a cold scalpel blade (-187°C). The specimen was then transferred to the cold stage (-187°C) of a JEOL JSM840 scanning electron microscope, equipped with a Tracor Northern TX3/47-6620 SMS X-ray detector and TN 5500 analyzer. Quench freezing, fracturing and transfer were all performed under vacuum using a Hexland CT1000 cryo-transfer system. Intracellular details of the fractured face were exposed by briefly raising the stage temperature to approximately -80°C (etching).

X-ray data were collected at 20 kV, with a probe current in the range of 0.1–0.3 nA giving detector dead times of approximately 25% for 100 sec of live time scans of 1 $\mu m^2$ area at 10,000 times magnification. In most cases Zn-P deposits were located by X-ray scanning programmed to detect Zn $K_\alpha$ X-ray emissions. Cd-containing deposits cannot be directly identified by means of X-ray scanning because Cd $L_\alpha$ emissions overlap with K $K_\alpha$ emissions. Therefore, detection of Cd deposits was attempted by X-ray mapping for P or S which may be associated with Cd deposits. X-ray data for each species were collected from several root specimens taken from different plants.

## Results

Symptoms of Zn and Cd toxicity generally appeared earlier and were more severe in crop species grown on Godbold's nutrient solution than in those grown on Johnson solution, but the occurrence of Zn-P containing deposits in root cells was unaffected by the composition and strength of the nutrient solution. Hence data obtained from plants grown on both types of solution have been combined.

The fracture faces of root specimens, which reveal a complete cross-section of the root elongation zone

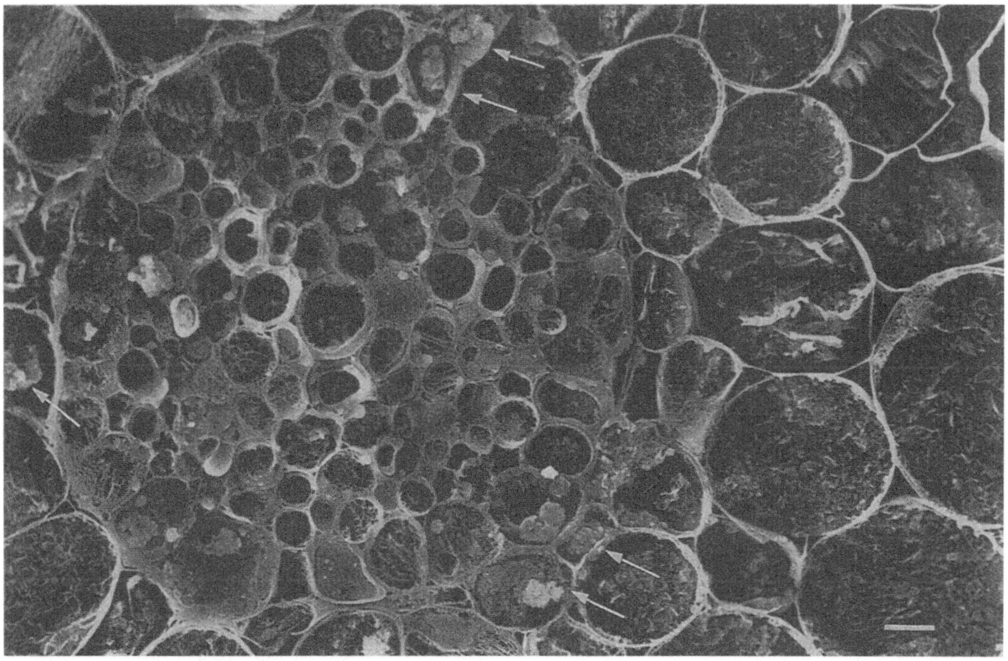

*Fig. 1.* Aggregates of globular deposits seen in the endodermis (arrows) in an etched, transversely fractured face 5mm from the apex of a lucerne root exposed to 100 $\mu M$ Zn in the culture solution. *Bar* represents 10 $\mu m$.

approximately 5 mm from the apex, were systematically X-ray scanned to determine the location of Zn-P deposits. Most deposits were well defined and occurred as distinct globules of 0.5–2 $\mu$m diameter, singly or in clusters. Their shape and apparent texture varied between species and to a lesser extent within species, perhaps due to the type or position of the fracture which could reveal either a stable, smooth deposit (globule) or one which is coated with or enveloped by some proteinaceous or membranous material. Clusters of Zn-containing deposits were mostly confined to cells of the endodermis in dicotyledonous species, and to cells of the pericycle in monocotyledonous species such as maize and wheat. Deposits were sometimes found in stelar or cortical cells immediately adjacent to the pericycle or endodermis, but these were often associated with some irregularity in the endodermis and may have been in endodermal cells (Fig. 1). A wider distribution beyond the endodermis and pericycle occurred especially in soybean and radish roots (inner cortex), and in maize roots (stelar parenchyma) that had been exposed to higher levels of Zn (300 $\mu M$ for maize) for more than one week. Also, the frequency of deposits in laterals of seminal roots of wheat and maize was higher than in the seminal roots themselves.

*Table 1.* Frequency of occurrence of phytate globules in the endodermis (dicots) or pericycle (monocots) in roots of crop plants exposed to high concentrations of Zn (80–160 $\mu M$) in the culture solution

| Frequent | Present | Rare | Absent |
|---|---|---|---|
| Lucerne | Lupins | Wheat | Sunflower |
| Soybean | Cabbage | | Field peas |
| Rapeseed | Radish | | Italian ryegrass |
| Tomato | Maize | | |

Table 1 lists the crop species surveyed and the observed frequency of the deposits. The absence of deposits is difficult to prove, and it is possible, as in the case of wheat, that on further investigation and exposure to higher levels of Zn, globules will be found.

The X-ray spectra collected from Zn-P deposits (Fig. 2a - f), in general, were characterised by high net counts and low X-ray fluorescence (background radiation) compared with spectra from regions immediately adjacent to these deposits which showed a relatively high X-ray fluorescence and low net counts for K, Cl, S and P (Fig. 2g). Low X-ray fluorescence indicates low water content and the spectra of Zn-P deposits (Fig. 2a

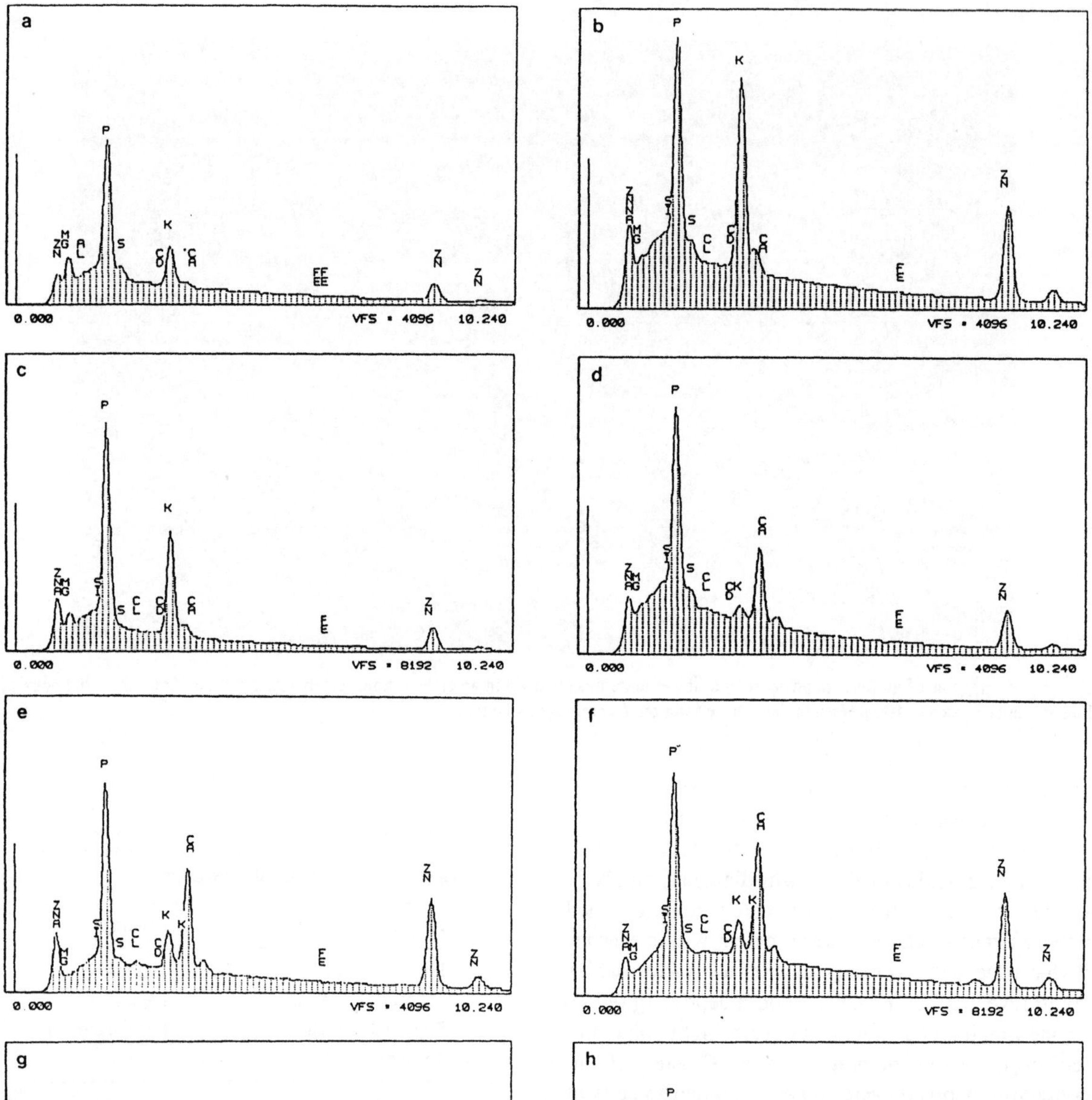


*Fig. 2.* Electron probe X-ray microanalysis spectra obtained at 10,000 × magnification with a reduced area scan of 1 $\mu m^2$ from: **a.** a conglomerate of small globules in the centre of an endodermal cell of a lupin root exposed to 100 $\mu M$ Zn, **b.** a large globule (approx 1 $\mu$m diameter) in the centre of an endodermal cell of a rapeseed root exposed to 160 $\mu M$ Zn, **c.** a large globule in a cortical cell immediately adjacent to the endodermis of a lucerne root exposed to 100 $\mu M$ Zn, **d.** a large globule in the endodermis of another lucerne root exposed to 100 $\mu M$ Zn, **e.** an area containing Zn-P deposits located in an edodermal cell by X-ray mapping before 'etching' the fracture face of a radish root exposed to 100 $\mu M$ Zn, **f.** a globular deposit in the same location as in e, but after 'etching', **g.** the cytoplasm of an endodermal cell immediately adjacent to globular deposits in a radish root exposed to 100 $\mu M$ Zn and **h.** a small crystal of K Mg phyphate (ex Sigma).

*Table 2*. Unadjusted net count ratios of Zn, K, Mg and Ca relative to P from Zn-P deposits in roots of various crops exposed to high concentrations of Zn (80–160 $\mu M$) in the culture solution

| | | | Net count ratio (mean ± SEM) | | | |
|---|---|---|---|---|---|---|
| | | | Zn/P | K/P | Mg/P | Ca/P |
| Radish | 80 $\mu M$ ZN | (n =17) | 0.25±0.01 | 0.10±0.01 | ND[a] | 0.18±0.01 |
| Cabbage | 100 $\mu M$ Zn | (n = 12) | 0.21±0.02 | 0.05±0.01 | 0.011±0.003 | 0.31±0.02 |
| Tomato | 100 $\mu M$ Zn | (n = 25) | 0.27±0.02 | 0.26±0.02 | ND | 0.54±0.03 |
| Soybean | 100 $\mu M$ Zn | (n = 11) | 0.20±0.01 | 0.61±0.03 | 0.042±0.007 | ND |
| Maize | 100 $\mu M$ Zn | (n = 14) | 0.23±0.03 | 0.50±0.07 | 0.051±0.006 | ND |
| Wheat | 160 $\mu M$ Zn | (n =3) | 0.12±0.01 | 0.79±0.03 | 0.062±0.007 | ND |
| Rapeseed | 160 $\mu M$Zn | (n =23) | | | | |
| -high | K deposits | (n = 6) | 0.25±0.02 | 0.93±0.06 | 0.013±0.002 | ND |
| -low | K deposits | (n = 17) | 0.26±0.02 | 0.53±0.07 | 0.062±0.004 | ND |
| Lucerne | 100$\mu M$ Zn | (n = 18) | | | | |
| -high | K deposits | (n = 13) | 0.20±0.02 | 0.39±0.05 | 0.036±0.006 | ND |
| -low | K deposits | (n =5) | 0.23±0.02 | 0.15±0.04 | 0.014±0.006 | 0.30±0.05 |
| Lupins | 100$\mu M$ Zn | (n = 20) | | | | |
| -high | K deposits | (n =14) | 0.10±0.01 | 0.86±0.05 | 0.077±0.006 | ND |
| -low | K deposits | (n = 6) | 0.10±0.01 | 0.34±0.02 | 0.16±0.001 | ND |

[a]Not detected.

- f) closely resemble those obtained from a solid sample of K Mg phytate (ex Sigma; Fig. 2h), confirming the locally high concentrations of P, Zn, Mg, Ca and K which were equally evident before and after etching (removal of ice by sublimation) (Fig. 2e, f).

Elemental net count ratios relative to P were calculated for each spectrum recorded. As is evident in Table 2, the Zn/P net count ratio is quite constant within and to a somewhat lesser extent between species with the exception of wheat and lupins. However, the relative contents of K, Mg and Ca fluctuated and depended very much on species and type of deposit. For instance, in radish and tomato roots, deposits contained K and Ca in constant proportions to P for each species, and no Mg was detected. In soybean, maize and wheat roots, deposits contained K and Mg, and no Ca was detected, while in cabbage root both Ca and Mg were present in deposits. In other species, two distinct types of deposit were evident; high K-containing deposits and low K-containing deposits. In rapeseed root, relatively high K deposits contained very little Mg, whereas Mg was relatively abundant in low K deposits (no Ca was detected in any of the rapeseed deposits). In lucerne root, Ca remained undetected in the relatively high K deposits but was abundant in relatively low K deposits. In lupins, as in rapeseed, high K deposits contained less Mg than in low K deposits, and again no Ca was detected.

In soybean and lucerne where Zn-P deposits were most frequent (Table 1), the addition of 30 $\mu M$ Cd instead of 100 $\mu M$ Zn caused severe toxic symptoms (water stressed foliage and stunted main roots with strongly promoted lateral root development). Hence, it was not possible to use equimolar concentrations of Cd or Zn to compare their effect on formation of globular deposits in the endodermal cells. Nevertheless, Zn-P deposits were frequent in root tips of plants treated with 100 $\mu M$ Zn (Fig. 3a) while similar Cd-P deposits were absent from root tips of plants treated with 30 $\mu M$ Cd. It was only after a prolonged search that some characteristic high density globular deposits were discovered in endodermal cells, 3 mm from the tip of a lateral root of soybean (Fig. 3b). However, it was shown by means of peak deconvolution of Cd $L_\alpha$ and K $K_\alpha$ X-ray emissions that these deposits did not contain Cd. Further extensive X-ray scanning for P revealed that generally 5–6 cells per cross section of endodermis contained high levels of both K and P (Fig. 3c), which could be a pre-requisite condition for phytate globule formation. Unfortunately, it was not possible to carry out similar X-ray scans for Cd due to the overlap of K $K_\alpha$ and Cd $L_\alpha$ X-ray emissions, K

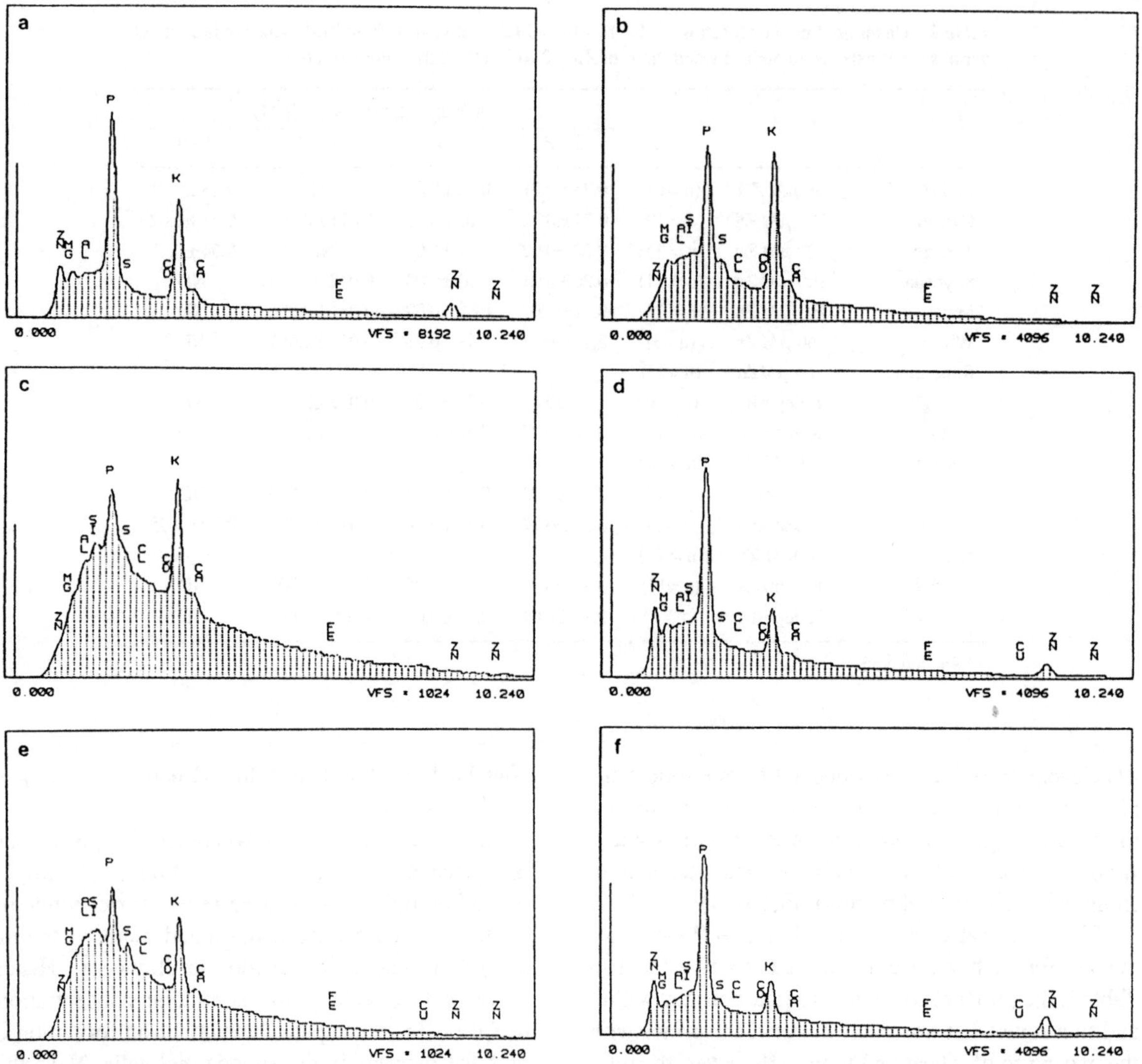


*Fig. 3.* Electron probe X-ray microanalysis spectra obtained at 10,000 × magnification with a reduced area scan of 1 $\mu m^2$ from: **a.** a large single globule (approx 1 $\mu$m diameter diameter) in an endodermal cell of the elongation zone (5 mm from apex) of a soybean root exposed to 100 $\mu M$ Zn, **b.** 2 medium sized globules (approx 0.5 $\mu$m diameter) in an edodermal cell of a soybean root exposed to 30 $\mu M$ Cd, **c.** an area with high K and P in one endodermal cell of a lucerne root tip (3 mm from apex) exposed to 30 $\mu M$ Cd, **d.** a conglomerate of small globules in a pericycle cell of a maize root tip (4 mm from apex) exposed to 300 $\mu M$ Zn, **e.** a deposit in the centre of a pericycle cell of a maize root exposed to 30 $\mu M$ Cd and **f.** a conglomerate of globular deposits in a pericycle cell of a maize root exposed to 300 $\mu M$ Zn + 30 $\mu M$ Cd.

being present at fairly high levels in most cells of the cross section of the root.

The results of the above experiments suggest that Cd might inhibit the formation of phytate and thus prevent formation of insoluble Cd-phytate deposits (Nolan et al., 1987). Hence, the effect of combined high levels of Zn and Cd on phytate globule formation was tested using maize. Raising the Zn level to 300 $\mu M$ greatly increased the frequency of Zn-P deposits compared with the 100 $\mu M$ Zn treatment. Again, phytate globules were absent in the 30 $\mu M$ Cd treatment (Fig. 3d), but Cd did not seem to prevent the formation of phytate deposits which appeared to be equally frequent in the Zn and Zn + Cd treatments (Fig. 3e, f). Further checks on the overlap of Cd $L_\alpha$ and K $K_\alpha$ X-ray emissions in 45 spectra of high density Zn-P globular deposits in the Zn + Cd treatment showed the absence of Cd. X-ray scans for the presence of S failed to act as an indicator for Cd detection.

## Discussion

In the majority of crop plants studied, globular deposits with a high Zn content occur, either singly or as aggregates, particularly in cells of the endodermis (dicotyledonous species) or pericycle (monocotyledonous species) in the elongation zone of roots exposed to high levels (80–300 *μM*) of Zn (Van Steveninck et al., 1993). In soybean, lucerne and maize supplied with 300 *μM* Zn, Zn-P deposits were particularly numerous and deposits were also found in the stele and inner cortex. However, Zn-P deposits could not be found in three crop species: sunflower, field peas and ryegrass. Absence of Zn-P deposits is difficult to prove and may depend on the degree of Zn-stress applied. For instance, Zn-P deposits occurred in roots of rapeseed supplied with 160 *μM* Zn but could not be found in roots supplied with 80 *μM* Zn, and the frequency of Zn-P deposits was greatly increased by exposing maize roots to 300 *μM* Zn instead of 100 *μM* Zn. Globular deposits were generally absent from most crop species exposed to normal levels of Zn (1 *μM*) in culture solutions. Hence, one might conclude that the formation of Zn-P deposits is induced by exposure of the roots to critical levels of Zn.

The aquatic plant, *Lemna minor* L., which has a known capacity for phytic acid synthesis (Roberts and Loewus, 1968) responds to high Zn or Cd levels with the production of Zn-P or Cd-P deposits respectively, but only in mature tissue. In immature tissue, two modes of Cd binding became evident, indicating a change from Cd-P binding to a mechanism depending on Cd binding by a S-containing compound, presumably phytochelatin. However, additional microprobe evidence indicated that Cd-sulphide crystallites may also occur (Van Steveninck et al., 1990a).

In soybean and lucerne root tips, where Zn-P deposits were particularly numerous, exposure to 30 *μM* Cd did not result in the detection of any Cd-P type deposits, and it was not possible to detect Cd associated with S. The failure to detect Cd by X-ray microanalysis suggests either absence or a homogeneous distribution of Cd over the apoplastic and symplastic space of the root tip elongation zone. Considering previous results obtained with mature fronds of *Lemna*, and the fact that both Zn and Cd can produce stable, insoluble, deposits with phytate in vitro (Nolan et al., 1987), it seemed possible that the absence of Cd-P deposits was due to a direct inhibitory effect of Cd on phytic acid synthesis. Further experiments, combining high levels of Zn and Cd, showed that the presence of Cd had no noticeable effect on the formation of Zn-P deposits in maize. However, Cd was not detected in any of these deposits. Keeping in mind that phytate should have a high affinity for Cd (Nolan et al., 1987), it appears that Cd might be spatially separated from the site of Zn-phytate accumulation. This phenomenon could be explained on the basis of a highly specifc compartmentation within plant cells.

Comparisons between crop species indicate that the kind of metal ion bound to phytic acid is highly specific and that even the relative proportions tend to be constant. For instance, Ca is present but Mg can not be detected in deposits in radish and tomato roots, whereas no Ca can be detected in soybean, maize, wheat, rapeseed and lupins, while in lucerne the presence of Ca depends on the proportion of K in the deposits. Such specificity in binding was also recorded in *Lemna* in which neither Zn nor Cd is bound by Ca-oxalate-containing raphide cells, while Sr is readily incorporated in mature raphide cells (Franceschi and Schueren, 1989), and Cd could be bound by phytate only in mature tissue (Van Steveninck et al., 1992). Considering the observed differences between crop species with respect to Ca and Mg, which are probably due to selective compartmentation, it appears likely that other plant species will be discovered where Cd is bound by phytate. Cd phytate binding is more likely to be found in typical metallophytes possessing a well-documented co-tolerance for Cd and Zn (e.g. *Thlaspi caerulescens* and *Silene vulgaris*).

At present, it appears that the binding of Zn in the endodermis or pericycle of roots by phytic acid is a Zn-specific process which occurs widely in the plant kingdom. It is suggested that the capacity to bind heavy metals by means of phytate in endodermal cells is genetically determined and may provide a strategy for minimising the heavy metal concentration in above ground parts. Recently we have observed binding of Zn by means of phytic acid in hairy roots of *Arabidopsis thaliana* transformed with *A. rhizogenes*. Genes that confer the ability to bind toxic ions could enable transformed roots to be used in the treatment of industrial wastes (Metzger et al., 1992).

## References


Baker A J M and Brooks R R 1989 Terrestial higher plants which hyper-accumulate metallic elements – a review of their distribution, ecology and phytochemistry. Biorecovery 1, 86–126.

Bollard E G and Butler G W 1966 Mineral nutrition of plants. Ann. Rev. Plant Physiol. 17, 77–112.

Florijn P J and Van Beusichem M L 1993 Cadmium distribution in maize inbred lines: effects of pH and level of Cd supply. Plant and Soil 153, 79–84.

Franceschi V R and Schueren A M 1989 Incorporation of strontium into plant calcium oxalate crystals. Protoplasma 130, 199–205.

Godbold D L, Horst W J, Collins J C, Thurman D A and Marschner H 1984 Accumulation of zinc and organic acids in roots of zinc tolerant and non-tolerant ecotypes of *Deschampsia caespitosa*. J. Plant Physiol. 116, 59–69.

Johnson C M, Stout P R, Broyer T C and Carlton A B 1957 Comparative chlorine requirements of different plant species. Plant and Soil 8, 337–353.

Maiti I B, Wagner G J, Yeargan R and Hunt A G 1989 Inheritance and expression of the mouse metallothionein gene in tobacco. Impact on Cd tolerance and tissue Cd distribution in seedlings. Plant Physiol. 91, 1020– 1024.

Maiti I B, Wagner G J and Hunt A G 1991 Light inducible and tissue specific expression of a chimeric mouse metallothionein cDNA gene in tobacco. Plant Science 76, 99–107.

Metzger L, Fouchault I, Glad C, Prost R and Tepfer D 1992 Estimation of cadmium availability using transformed roots. Plant and Soil 143, 249–257.

Mikus M, Bobák M and Lux A 1992 Structure of protein bodies and elemental composition of phytin from dry germ of maize (*Zea mays* L.) Bot. Acta 105, 26–33.

Misra S and Gedamu L 1990 Heavy metal resistance in transgenic plants expressing a human metallothionein gene. Plant Gene Transfer. pp 257–265 Alan R Liss, Inc.

Nolan K B, Duffin P A and McWeeny D J 1987 Effects of phytate in mineral bio availability. In vitro studies on $Mg^{2+}$, $Ca^{2+}$, $Fe^{3+}$, $Cu^{2+}$ and $Zn^{2+}$ (also $Cd^{2+}$) solubilities in the presence of phytate. J. Sci. Food Agric. 40, 79–85.

Reeves R D and Brooks R R 1983 European species of *Thlaspi* L. (Cruciferae) as indicators of nickel and zinc. J. Geochem. Explor. 18, 275–283.

Roberts R M and Loewus F 1968 Inositol metabolism in plants. VI. Conversion of myoinositol to phytic acid in *Wolffiella floridana*. Plant Physiol. 43, 1710–1716.

Van Steveninck R F M, Van Steveninck M E, Fernando D R, Horst W J and Marschner H 1987 Deposition of zinc phytate in globular bodies in roots of *Deschampsia caespitosa* ecotypes; a detoxification mechanism? J. Plant Physiol. 131, 247–257.

Van Steveninck R F M, Van Steveninck M E, Fernando D R, Edwards L B and Wells A J 1990a Electron probe X-ray microanalytical evidence of two distinct mechanisms of Zn and Cd binding in a Zn tolerant clone of *Lemna minor*. C. R. Acad. Sci. Paris 310, Serie III, 671–678.

Van Steveninck R F M, Van Steveninck M E, Wells A J and Fernando D R 1990b Zinc tolerance and the binding of zinc as zinc phytate in *Lemna minor*. X-ray microanalytical evidence. J. Plant Physiol. 137, 140–146.

Van Steveninck R F M, Van Steveninck M E and Fernando D R 1992 Heavy metal (Zn, Cd) tolerance in selected clones of duck weed (*Lemna minor*). Plant and Soil 146, 271–280.

Van Steveninck R F M, Babare A, Fernando D R and Van Steveninck M E 1993 The binding, of zinc in root cells of crop plants by phytic acid. Plant and Soil 155/156, 525–528.

Vázquez M D, Barcelo J, Poschenrieder Ch, Màdico J, Hatton P, Baker A J M and Cope G H 1992 Localization of zinc and cadmium in *Thlaspi caerulescens* (Brassicaceae), a metallophyte that can hyper-accumulate both metals. J. Plant Physiol. 140, 350–355.

Verkleij J A C and Schat H 1990 Mechanisms of metal tolerance in higher plants. *In* Evolutionary Aspects of Heavy-Metal Tolerance in Plants. Ed. J Shaw. pp 179–193. CRC Press Boca Raton, FL.

*F. Baluška et al. (eds.), Structure and Function of Roots*, 327–333, 1995.


# Differences in the metabolic responses of root tips of wheat and rye to aluminium stress

Jan J. Ślaski
*Plant Breeding and Acclimatization Institute - Radzików, Department of Plant Biochemistry and Physiology, Mineral Stress Laboratory, PO Box 1019, 00–950 Warszawa, Poland. Present address: Department of Botany, B-414 Biological Sciences Building, University of Alberta, Edmonton, Alberta, Canada T6G 2E9*

*Key words:* aluminium tolerance, enzymes, root tips, rye, wheat

## Abstract

The effects of aluminium (Al) ions on the metabolism of root apical meristems were examined in 4-day-old seedlings of two cereals which differed in their tolerance to Al: wheat cv. Grana (Al-sensitive) and rye cv. Dańkowskie Nowe (Al tolerant). During a 24 h incubation period in nutrient solutions containing 0.15 m*M* and 1.0 m*M* of Al for wheat and rye, respectively, the activity of first two enzymes in the pentose phosphate pathway (G-6-PDH and 6-PGDH) decreased in the sensitive cultivar. In the tolerant cultivar activities of these enzymes increased initially, then decreased slightly, and were at control levels after 24 h. In the Al-sensitive wheat cultivar a 50% reduction in the activity of 6-phosphogluconate dehydrogenase was observed in the presence of Al. Changes in enzyme activity were accompanied by changes in levels of G-6-P - the initial substrate in the pentose phosphate pathway. When wheat was exposed for 16 h to a nutrient solution containing aluminium, a 90% reduction in G-6-P concentration was observed. In the Al-tolerant rye cultivar, an increase and subsequently a slight decrease in G-6-P concentration was detected, and after 16 h of Al-stress the concentration of this substrate was still higher than in control plants. This dramatic Al-induced decrease in G-6-P concentration in the Al- sensitive wheat cultivar was associated with a decrease in both the concentration of glucose in the root tips as well as the activity of hexokinase, an enzyme which is responsible for phosphorylation of glucose to G-6-P. However, in the Al-tolerant rye cultivar, the activity of this enzyme remained at the level of control plants during Al-treatment, and the decrease in the concentration of glucose occurred at a much slower rate than in wheat. These results suggest that aluminium ions change cellular metabolism of both wheat and rye root tips. In the Al-sensitive wheat cultivar, irreversible disturbances induced by low doses of Al in the nutrient solution appear very quickly, whereas in the Al-tolerant rye cultivar, cellular metabolism, even under severe stress conditions, is maintained for a long time at a level which allows for root elongation to continue.

*Abbreviations:* G-6-PDH – glucose-6-phosphate dehydrogenase, 6-PGDH – 6-phosphogluconate dehydrogenase, G-6-P – glucose-6-phosphate, TEA – triethanolamine.

## Introduction

Aluminium (Al) toxicity is a major factor limiting the growth of plants in mineral acid soils with a pH below 5.0 (Foy, 1988). The toxic effect of this ion is primarily expressed as a drastic inhibition in root elongation, possibly a result of Al-induced disturbances of cell division in root apical meristems (Bennet and Breen, 1991; Eleftheriou et al., 1993; Taylor, 1988). Foliar symptoms of aluminium toxicity resemble those of calcium and phosphorus deficiency as aluminium acts as an antagonist of these two elements (Foy, 1992).

Cultivated plants differ significantly in their response to Al-toxicity. Among cereals, rye is considered to be the most Al-tolerant, whereas wheat is regarded as very Al-sensitive (Foy, 1988; Ponnamperuma, 1982; Ślaski, 1990). Although many mechanisms of Al-tolerance have been postulated, none of these can satisfactorily explain the observed differences among genotypes.

Our previous results (Ślaski, 1989, 1990) show that $NAD^+$ kinase, an enzyme which catalyses the only known biochemical reaction of $NADP^+$ synthesis, is involved in the response of cultivated plants to Al-stress. By regulating production of $NADP^+$, which is a rate-limiting factor in many enzymatic processes, $NAD^+$ kinase may control several biosynthetic pathways including pentose phosphate pathway, shikimic acid pathway, as well as pathways involved in the synthesis of fatty acids, amino acids and nucleic acids (Allan and Trewavas, 1987). Among these pathways, the pentose phosphate pathway may be of great importance in Al-tolerance mechanism, since this pathway provides a number of intermediates (pentoses, erythrose-4-phosphate, NADPH) which are involved in the synthesis of substances which are considered to play a role in root response to Al-stress. In addition, the first two enzymes of this pathway viz. glucose-6-phosphate dehydrogenase and 6- phosphogluconate dehydrogenase are $NADP^+$-dependent. Under Al-stress, $NAD^+$ kinase activity in an Al-tolerant rye cultivar was 4-times higher than in an Al-sensitive wheat cultivar (Ślaski, 1990), thus it appears that Al-tolerance may be connected with availability of $NADP^+$.

The aim of this work was to determine the effect of aluminium on the cellular metabolism of root tips from two cereals that differ in their tolerance to Al.

## Material and methods

### *Plant material*

Four-day-old seedlings of an Al-tolerant rye cultivar (*Secale cereale* cv. Dańkowskie Nowe) and an Al-sensitive wheat cultivar (*Triticum aestivum* L. cv. Grana) were used for experiments. Seeds were sterilized with a 0.1% solution (w/v) of $HgCl_2$ for 10 min., rinsed thoroughly with tap water and germinated overnight at room temperature on filter paper in Petri dishes. Plants were subsequently grown for 4 days under conditions previously described (Ślaski, 1992) with the following modifications. The nutrient solution (pH 4.50) contained: 0.4 m*M* $CaCl_2$, 0.65 m*M* $KNO_3$, 0.25 m*M* $MgCl_2$, 0.01 m*M* $(NH_4)_2SO_4$ and 0.04 m*M* $NH_4NO_3$. Phosphorus was omitted in the nutrient solution to avoid precipitation with aluminium. After 4 days, seedlings were transferred to fresh nutrient solution with aluminium added in the form of $AlCl_3 \times 6H_2O$, at a concentration of 0.15 m*M* and 1.0 m*M* for wheat and rye, respectively. Control plants were transferred to fresh nutrient solution without aluminium. Immediately after Al-treatment the activity of enzymes and concentration of substrates in root tips were assessed as described below. Besides about 30–40 seedlings from each combination were transferred to nutrient solution without aluminium for 48 h to determine the ability of roots to grow after removal of Al-stress. The root regrowth was determined by staining roots with a 0.1% solution (w/v) of Eriochrome Cyanine R for 10 min. (Aniol, 1984).

### *Extraction of the enzymes*

Root tips 3–4 mm long from 50–60 seedlings (200–300 mg fresh weight) were cut off from the root, pulverized in liquid nitrogen and extracted with 50 m*M* Tris buffer (1:10, w/v) (pH 7.4) containing 1 *M* KCl, 1 m*M* EDTA, 2 m*M* $MgCl_2$ and 2.5% (w/v) polyvinylpyrrolidone (MW 40 000). The mixture was shaken for 15 min. at 4°C and centrifuged at 22,000 *g* for 30 min. The pellet was discarded and the supernatant was used for assays of enzymes and metabolites.

### *Enzyme assays*

The activities of G-6-PDH, 6-GPDH and hexokinase were measured at 340 nm with a Beckman spectrophotometer, model 26 according to Deutch (1983), Bergmayer (1974a) and Bergmayer (1974b), respectively. The incubation mixture for G-6-PGDH assay contained: 150 $\mu$L 0.5 *M* Tris-HCl (pH 7.5), 100 $\mu$L 3.8 m*M* $NADP^+$, 100 $\mu$L 0.63 m*M* $MgCl_2$, 200 $\mu$L 33 m*M* G-6-P, 0.5 mg maleimide and 100 $\mu$L extract. The volume was adjusted to 100 $\mu$L with $H_2O$. The incubation mixture for 6-GPDH assay contained: 250 $\mu$L 0.5 *M* Tris-HCl (pH 7.5), 200 $\mu$L 31 m*M* gluconate-6-phosphate, 100 $\mu$L 0.63 *M* $MgCl_2$, 100 $\mu$L 3.8 m*M* $NADP^+$ and 100 $\mu$L extract. The volume was adjusted to 1000 $\mu$L with $H_2O$. For the hexokinase assay the incubation mixture contained: 500 $\mu$L 50 m*M* TEA buffer (pH 7.6), 500 $\mu$L 0.6 *M* glucose, 100 $\mu$L 0.1 *M* $MgCl_2$, 100 $\mu$L 4 m*M* $NADP^+$, 50 $\mu$L 1.6 m*M* ATP, 0.6 unit G-6-PDH and 100 $\mu$L extract. The final volume of the mixture was 1360 $\mu$L.

*Assay of metabolites*

Glucose and G-6-P were assayed in deproteinized samples at 340 nm with a Beckman spectrophotometer, model 26 according to Kunst et al. (1984) and Michal (1984), respectively. The concentration of reagents in 2 mL of the incubation mixture for glucose assay were as follow: 66 m*M* $NaH_2PO_4$, 3.8 m*M* $MgSO_4$, 1.5 m*M* ATP, 1.5 m*M* $NADP^+$. Subsequently 100 $\mu$L of extract (glucose) and finally enzyme solution consisted of hexokinase and G-6-PDH were added. The incubation mixture for G-6-P assay contained: 500 $\mu$L 0.4 *M* TEA buffer (pH 7.6), 100 $\mu$L 24 m*M* $NADP^+$, 100 $\mu$L 0.5 *M* $MgCl_2$, 500 $\mu$L extract and 150 units G-6-PDH.

The concentration of pentoses in root tips was determined according to Douglas (1981) as the difference in absorption at 552 and 510 nm of the reaction product of phosphoglycilol and xylose.

## Results

The first visual symptom of aluminium toxicity is inhibition of root elongation. A 24-h incubation of seedlings in nutrient solution containing Al (0.15 m*M*) caused irreversible damage to the root apical meristems in the Al-sensitive wheat cultivar. This damage was reflected by a lack of root regrowth, when seedlings were transferred to the medium without aluminium for 2 days (Fig. 1). In rye a 7-times higher concentration of Al in the nutrient solution (1.0 m*M*) had little effect on the ability of roots to continue the elongation growth. One would expect that Al-induced inhibition of root growth is preceded by changes in root cellular metabolism. Among other pathways, the pentose phosphate pathway was supposed to play an important role in the response of Al-tolerant genotypes to Al-stress, possibly by providing intermediates and NADPH for the synthesis of different compounds which may be involved in protecting the root against Al-induced injury (Ślaski, 1990).

Results of these studies indicate that in the Al-sensitive wheat cultivar, the activity of the first two enzymes of the pentose phosphate pathway i.e. glucose-6-phosphate dehydrogenase and 6-phosphogluconate dehydrogenase decreased under conditions of severe Al-stress (24h of Al treatment (Fig. 2A, B). The latter enzyme appeared to be more sensitive to aluminium than G-6-PDH as a 50% reduction of its activity was observed after 24-h stress (Fig. 2B). When the Al-tolerant rye cultivar was incubated

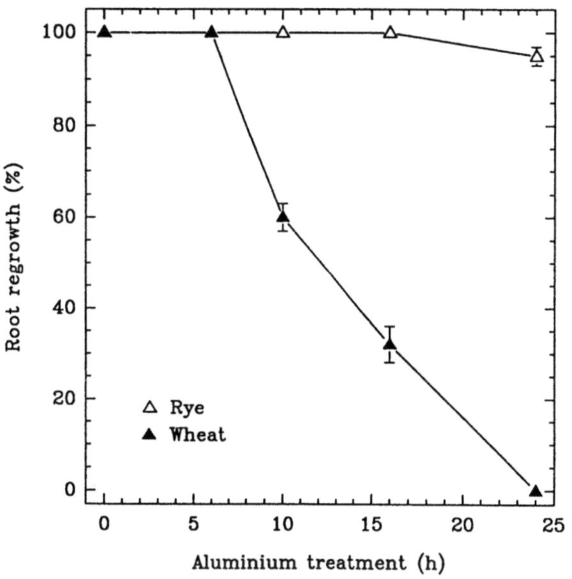


*Fig. 1.* Effect of aluminium treatment on regrowth of roots in cultivars of rye and wheat. About 40 four-day-old seedlings were incubated in a nutrient solution containing 0.15 m*M* and 1.0 m*M* $AlCl_3$ for wheat and rye, respectively. Data represents percent of seedlings which were able to continue to grow during the 48 h following treatment with aluminium. Values represent the mean of at least 3 replicates ± SE.

for 24 h in the presence of 1.0 m*M* Al, the activity of these enzymes increased initially, then decreased slightly, and after the Al-stress period the activity still remained at a level comparable to that of control plants (Fig. 2A, B).

The ability to maintain the activities of pentose phosphate pathway enzymes at relatively high levels in rye root tips under severe Al-stress may contribute to an increase of production of pentoses, which are one of the intermediates of this pathway (Fig. 3).

Changes in enzyme activities were also accompanied by changes the concentration of glucose-6-phosphate - a substrate which initiates the pentose phosphate pathway (Fig. 4).

When the Al-sensitive wheat cultivar was incubated for 16 h in nutrient solution containing 0.15 m*M* Al, a 90% reduction in G-6-P concentration was observed. However, in the Al-tolerant rye cultivar, an increase and subsequently a slow decrease in concentration of the G-6-P was detected, and after 16 h of Al-stress the concentration of this substrate was still higher than in control plants. This dramatic decrease in G-6-P concentration in the Al-sensitive wheat cultivar under Al-stress was connected with a decrease in both, the glucose concentration in the root tips and in hexokinase

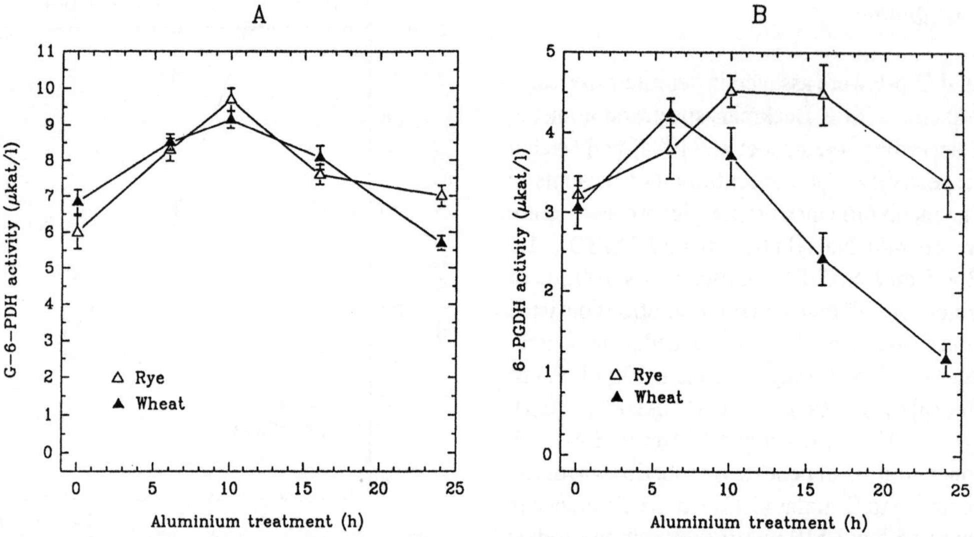


*Fig. 2.* Effect of aluminium treatment on activity of glucose-6-phosphate dehydrogenase (G-6-PDH) (**A**) and 6-phosphagluconate dehydrogenase (6-PGDH) (**B**) in root tips of wheat and rye. Plants were grown in a nutrient solution supplied with 0.15 m*M* and 1.0 m*M* of $AlCl_3$ for wheat and rye, respectively. The values indicated at time '0' represents levels of activity found in control plants which were not exposed to Al. Values represent the mean for at least 4 independent experiments ± SE.

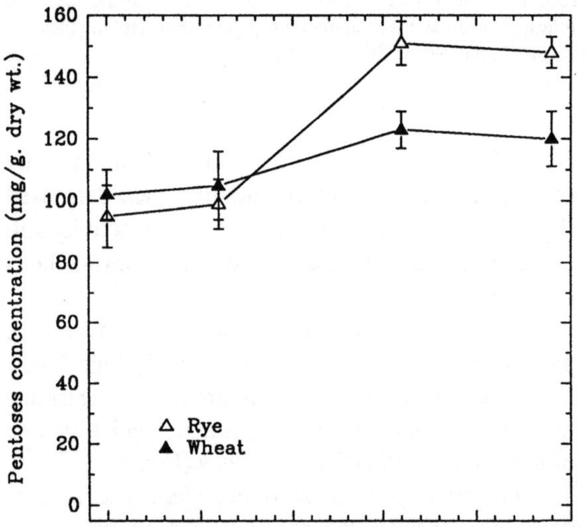


*Fig. 3.* Changes in the concentration of pentoses in root tips of wheat and rye in the presence of aluminium. Four-day-old plants were grown for 24 h in the presence of 0.15 m*M* and 1.0 m*M* $AlCl_3$ for wheat and rye, respectively. Values represent the mean of at least 3 replicates ± SE.

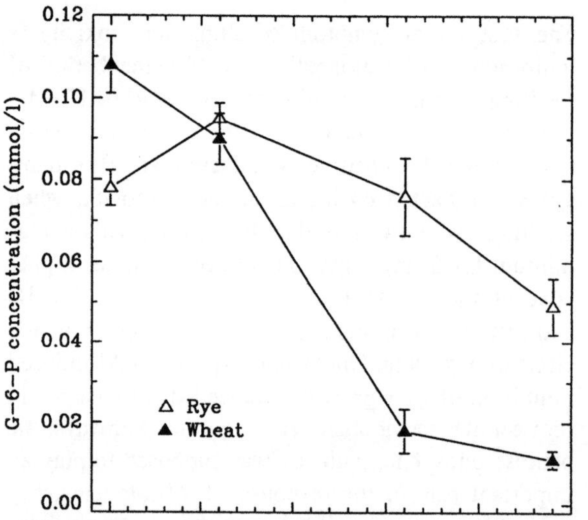


*Fig. 4.* Changes in the concentration of glucose-6-phosphate in root tips of wheat and rye exposed to aluminium. Plants were incubated in a nutrient solution containing 0.15 m*M* and 1.0 m*M* $AlCl_3$ for wheat and rye, respectively. Values represent the mean of at least 3 replicates ± SE.

activity, an enzyme which is responsible for phosphorylation of glucose to glucose-6-phosphate (Fig. 5A, B). However, in the Al-tolerant rye cultivar, activity of hexokinase remained at the control level during exposure to Al, and the decrease of glucose concentration occurred at a much slower rate than in wheat.

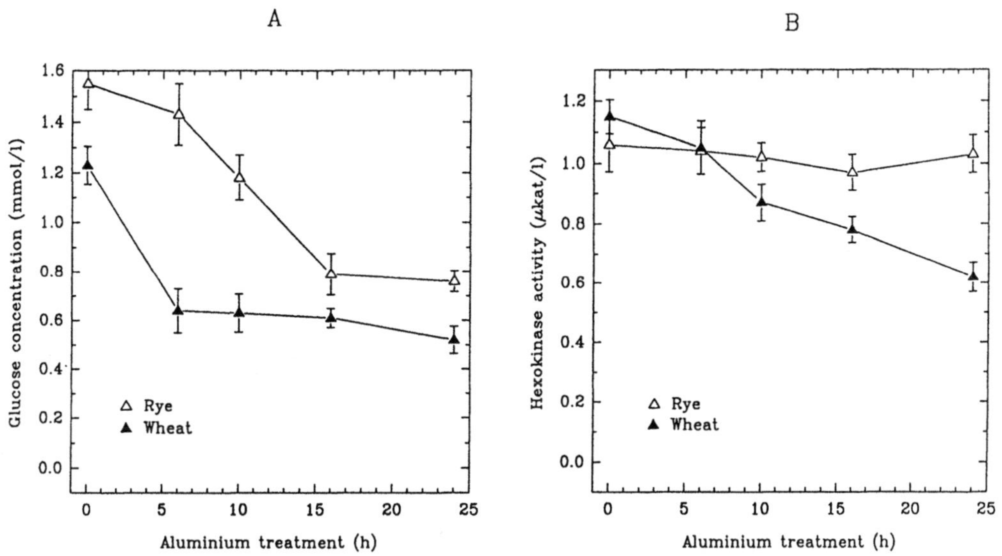


*Fig. 5.* Effect of aluminium treatment on glucose concentration (**A**) and hexokinase activity (**B**) in root tips of wheat and rye. Seedlings were grown for 24 h in the presence of 0.15 m*M* and 1.0 m*M* $AlCl_3$ for wheat and rye, respectively. Values represent the mean of at least 3 replicates ± SE.

## Discussion

During the past few years the effect of aluminium on biochemical processes in roots has been widely discussed and contradictory data concerning Al influence on cellular metabolism has been presented.

One group of evidences shows that in spite of Al-induced inhibition of root elongation, during the first few hours of Al-treatment these ions do not directly disturb intracellular metabolism, presumably because Al does not cross the plasma membrane (Horst et al., 1992; Kinraide, 1988). These results suggest that inhibition of root growth must be due to some kind of impairment of processes occurring in the root apoplast.

The other group of data shows that in plants exposed to relatively low level of Al-stress this ion influences the metabolic activity inside the cell, suggesting that inhibition of root growth is a consequence of metabolic disorders (Copeland and de Lima, 1992; Karataglis, 1986; Pfeffer et al., 1986; Ślaski, 1990). Many authors have also found genotypical differences in metabolic responses to Al between Al-tolerant and Al-sensitive cultivars (Foy et al., 1987; Gakuru and Lefebvre, 1991; Galvez et al., 1991; Moustakas et al., 1992; Ślaski, 1989).

Data presented in this paper seem to support the second opinion since after only 6 h incubation of roots in nutrient solution containing Al a significant change, in both enzyme activity and metabolite concentration were observed. A different response to Al-stress was observed between rye and wheat. In the Al-tolerant rye cultivar the activity of G-6-PDH and 6-PGDH did not decrease below control level even after 24 h incubation in nutrient solution containing 1.0 m*M* Al (Fig. 2A, B). The ability to maintain undiminished flux through the pentose phosphate pathway in the face of severe Al-stress seems to play an important role in the mechanism of Al-tolerance in rye. This pathway provides intermediates such as pentoses, erythrose-4-phosphate, and NADPH. These intermediates are involved in the synthesis of several compounds (for example, amino acids, nucleic acids, coenzymes) which may be involved in the response of plants to Al-stress. This suggestion is supported by the observed increase in concentration of pentoses in rye root tips treated with Al, whereas in wheat no significant changes in the level of pentoses was observed (Fig. 3). In the Al-sensitive wheat cultivar, severe Al-stress led to decrease in the activity of G-6-PDH and a 50% reduction of 6-PGDH activity (Fig. 2A, B). These decreases in activity could be explained by a direct interaction of Al with enzymes in wheat, which may have a higher affinity for aluminium than similar enzymes in rye. To verify this supposition in vitro experiments examining the influence of Al on enzymes from both cultivars are needed. Contradictory data was recently presented by Copeland and de Lima (1992) which showed that after 24 h and 48 h of treatment with 0.075 m*M* Al the activity of G-6-PDH in whole roots of wheat cv. Vulcan was not reduced. These results can be explained by the fact that this long-term Al-stress did not make irreversible changes in the root metabolism in this cultivar since even after 48 h of exposure to Al, a 25% of roots were still able to grow.

The activities of both enzymes are controlled by the ratio of NADPH to $NADP^+$ (Copeland and Turner, 1987; Hrazdina and Jensen, 1992). In some wheat cultivars the availability of $NADP^+$ is strongly reduced due to depression of $NAD^+$ kinase activity by Al ions (Ślaski, 1989). By disturbing other metabolic processes in which NADPH is used, Al could cause a reduction of the demand for this coenzyme and subsequent increase in concentration in root cells. Both these reasons may increase the $NADPH/NADP^+$ ratio and this in turn could lead to inhibition of enzymes. Another reason for the decrease in activity of pentose phosphate pathway enzymes may be a low level of glucose-6-phosphate – a substrate which is known to initiate this pathway (Lendzian, 1978; Turner and Turner, 1980). The 90% reduction in G-6-P concentration in wheat root tips during first the 16 h of Al-treatment (Fig. 4) suggests that besides the above mentioned reasons, the reduction of G-6-P availability may cause an addition reduction the activity of the pentose phosphate pathway. This hypothesis is supported by Hanson's (1991) work, which showed that in soybean Al-tolerant genotypes had higher concentrations of G-6-P than Al-sensitive selections.

Looking for a possible source of such a rapid decrease in G-6-P concentration in wheat, the substrate (glucose) and enzyme responsible for its phosphorylation (hexokinase) were examined (Fig. 5A, B). In the Al-sensitive wheat cultivar both factors were influenced by Al ions. The glucose level was drastically reduced by a 50% during the first 6 h of Al-stress. Since aluminium is known to be a strong inhibitor of hexokinase in pea seeds (Turner and Copeland, 1981), inhibition of hexokinase activity which can additionally disturb of G-6-P biosynthesis was expected. In fact, in the Al-sensitive wheat cultivar 50% reduction of hexokinase activity was observed, whereas this enzyme from rye root tips appeared to be completely insensitive to Al-toxicity, similarly to Copeland and de Lima (1992) results obtained for semi-tolerant wheat.

Taking into account the above presented results it may be concluded that Al changes the cellular metabolism in root tips of both wheat and rye. In the Al-sensitive wheat cultivar, irreversible disturbances induced by low doses of Al in the nutrient solution appeared very quickly, whereas in the Al-tolerant rye cultivar, cellular metabolism, even under severe stress conditions, was maintained for a long time at a level which allowed for root elongation to continue. In the future, studies are required to elucidate if the changes in metabolic activity observed in the presence of Al are connected with the concentration of aluminium at the site of enzyme action in the root symplast.

## Acknowledgment

This work was supported by the Polish-American Join Commission under the Maria Sklodowska-Curie Found, project no MR/USDA 92–93. The author thanks Julie L Stephens her editorial comments on the manuscript.

## References

Allan E and Trewavas A 1987 The role of calcium in metabolic control. *In* The Biochemistry of Plants. Ed. D D Davies. Vol 12, pp 117–149. Academic Press, San Diego.

Aniol A 1984 Induction of aluminum tolerance in wheat seedlings by low doses of aluminum in the nutrient solution. Plant Physiol. 75, 551–555.

Bennet R J and Breen C M 1991 The aluminum signal: New dimensions to mechanism of aluminum tolerance. *In* Plant-Soil Interactions at Low pH. Ed. R J Wright. pp 703–716. Kluwer Acad Publ., Dordrecht

Bergmayer H U 1974a 6-phosphogluconate dehydrogenase. *In* Methods of Enzymatic Analysis. Vol. 1. Academic Press, NY. 500p.

Bergmayer H U 1974b Hexokinase. *In* Methods of Enzymatic, Analysis. Vol 1. Academic Press, NY. 473p.

Copeland L and de Lima M 1992 The effect of aluminium on enzyme activities in wheat roots. J. Plant Physiol. 140, 641–645.

Copeland L and Turner J F 1987 The regulation of glycolysis and pentose phosphate pathway. *In* The Biochemistry of Plants. Vol 11, pp 107–127. Academic Press, San Diego.

Deutsch J 1984 Glucose-6-phosphate dehydrogenase. *In* Methods of Enzymatic Analysis. Ed. H U Bregmayer. Vol 3, pp 190–197. Verlag Chemie, Weinheim, Basel.

Douglas S G 1981 A rapid method for the determination of pentosans in wheat flour. Food Chem. 7, 139–145.

Eleftheriou E P, Moustakas M and Fragiskos N 1993 Aluminate-induced changes in morphology and ultrastructure of *Thinipyrum* roots. J. Exp. Bot. 44, 427–436.

Foy C D 1988 Plant adaptation to acid, aluminum toxic soils. Commun. Soil Sci. Plant Anal. 19, 959–987.

Foy C D 1992 Soil chemical factors limiting plant root growth. *In* Advances in Soil Science. Ed. B A Steward. Vol. 19, pp 97–149. Springer Verlag, NY.

Foy C D, Lee E H and Wilding S B 1987 Differential aluminum tolerances of two barley cultivars related to organic acids in their roots. J. Plant Nutr. 10, 1089–1101.

Gakuru S and Lefebvre C 1991 Acid phosphatase: Screening Zairian *Zea mays* varieties for aluminium tolerance. Cereal Res. Commun 19, 477–481.

Galvez L, Clark R B, Klepper L A and Hansen L 1991 Organic acid and free proline accumulation and nitrate reductase activity in sorghum genotypes differing in aluminum tolerance. *In* Plant - Soil Interactions at Low pH. Ed. R J Wright. pp 859–867. Kluwer Acad Publ., Dordrecht.

Hanson W D 1991 Root characteristics associated with divergent selection for seedling aluminum tolerance in soybean. Crop Sci. 31, 125–129.

Horst W J, Asher C J, Cakmak I, Szulkiewicz P and Wissemeier A H 1992 Short-term responses of soybean roots to aluminium. J. Plant Physiol. 140, 174–178.

Hrazdina G and Jensen R A 1992 Spatial organization of enzymes in plant metabolic pathways. Annu. Rev. Plant Physiol. Molec. Biol. 43, 241–267.

Karataglis S 1986 Aluminium toxicity in *Avena sativa* cv. Kassandra and a comparison with the toxicity caused by some other metals. Phyton 27, 1–14.

Kinraide T B 1988 Proton extrusion by wheat roots exhibiting severe aluminum toxicity symptoms. Plant Physiol. 88, 418–423.

Kunst A, Draeger B and Ziegenhorn 1984 D-glucose. *In* Methods of Enzymatic Analysis. Ed. H U Bregmayer. Vol 6, pp 163–172. Verlag Chemie, Weinheim, Basel.

Michal G 1984 D-Glucose-6-phosphate and D-Fructose-6-phosphate. *In* Methods of Enzymatic Analysis. Ed. H U Bregmayer. Vol 6, pp 191–198. Verlag Chemie, Weinheim, Basel.

Lendzian K 1978 Interactions between magnesium ions, pH, glucose-6-phosphate and NADPH/NADP$^+$ ratios in the modulation of chloroplast glucose-6-phosphate dehydrogenase in vitro. Planta 141, 105–110.

Moustakas M, Yupsanis T, Symeonidis L and Karataglis S 1992 Aluminum toxicity effects on durum wheat cultivars. J. Plant Nutr. 15, 627–638.

Pfeffer P E, Tu S, Gerasimowicz W V and Cavanaugh J R 1986 In vivo $^{31}$P studies of corn root tissue and its uptake of toxic metals. Plant Physiol. 80, 77–84.

Ponnamperuma F N 1982 Genotype adaptability as a substitute for amendments on toxic and nutrient deficient soils. *In* Proc. Ninth Int. Plant Nutr. Colloq. Ed. A Scaife. Vol 2, pp 467–473. Commonwealth Bureau, Slough, UK.

Ślaski J J 1989 Effect of aluminium on calmodulin-dependent and calmodulin-independent NAD kinase activity in wheat (*Triticum aestivum* L.) root tips. J. Plant Physiol. 133, 696–701.

Ślaski J J 1990 Response of calmodulin-dependent and calmodulin-independent NAD kinase to aluminium in root tips from various cultivated plants. J. Plant Physiol. 136, 40–44.

Ślaski J J 1992 Physiological and genetical aspects of the tolerance of cereals to soil acidity and to toxic effects of aluminium ions. Bull. IHAR 183, 37–45.

Taylor G J 1988 The physiology of aluminium tolerance in higher plants. Commun. Soil Sci. Plant Anal. 19, 1179–1194.

Turner J F and Copeland L 1981 Hexokinase II of pea seeds. Plant Physiol. 68, 1123–1127.

Turner J F and Turner D H 1980 The regulation of glycolysis and the pentose phosphate pathway. *In* The Biochemistry of Plants. Ed. D D Davies. Vol 2, pp 279–316. Academic Press, NY.

F. Baluška et al. (eds.), Structure and Function of Roots, 335–339, 1995.


# Phytohormonal levels in spruce roots under aluminium stress

R. Čížková
*Department of Biology, Pedagogical Faculty, Masaryk University, Mendlovo náměsti 19, cz-603 00 Brno, The Czech Republic*

*Key words:* aluminium, phytohormones, spruce


## Abstract

Aluminium induces morphological changes of spruce (*Picea abies* (L.) Karst.) seedling roots. Changes in root length are recorded as parameters sensitive to aluminium stress.

The possibility of Al-induced increase of cytokinin-like substances and enhanced ethylene evolution has been established as being dependent on the concentration and period of Al exposure in our previous studies. In the present study, 0.74 and 1.85 m*M* Al in nutrient solution increased levels of cytokinin- and gibberellin-like substances after 4 months of stress. Predominantly, the increase of the indole-acetic acid content was highly significant (Experiment 1). Higher external Al concentrations, than the above mentioned, or longer periods of exposure, were responsible for inhibition of all phytohormones (Experiment 2). In the Experiment 3, 5 h exposure to 5.56 m*M* Al decreased the level of cytokinin-like substances, 40 h exposure to Al evoked their increase. The levels of gibberellin-like substances were stimulated during both intervals. The increase in indole-acetic acid content was significant after 5 and 40 h. Changes in phytohormonal levels induced by Al are undoubtedly a reason for the typical morphological changes in spruce roots under aluminium stress.


*Abbreviations:* IAA – indole-acetic acid, $GA_3$ – gibberellic acid, BAP – benzylaminopurine

## Introduction

It has been postulated that one of the main causes of forest tree decline and dieback is the high level of aluminium in soil solution induced by acidic deposition (Ulrich, 1989). Forest trees suffer Al stress especially at early stages of their development (Hutchinson et al., 1986; Shaedle et al., 1989). Aluminum toxicity is manifested primarily in damage to roots. Investigations of Al stress on tree seedlings have centered on the deleterious effects on root growth, nutrient uptake, morphological changes, including their anatomy and accumulation of Al (Andersson, 1988; Godbold and Kettner, 1991; Hutchinson et al., 1986; Raynal et al., 1990). In connection with a decrease in the vitality of forest trees, interest in the effect of aluminium on tree physiology has increased as well. In the present study, phytohormonal levels were studied in spruce (*Picea abies* (L.) Karst.) seedlings as a possible reaction to Al stress. Time of exposure and Al concentrations were varied.

## Material and methods

Pregerminated seedlings of Norway spruce (*Picea abies* L. Karst.) were grown in hydroponic culture at increasing concentrations of total aluminium, supplied as $Al_2(SO_4)_3$ identically to a previous study (Čižková, 1992). Plants were grown under controlled climatic conditions at 42 W $m^{-2}$ fluorescent irradiance, 16 h light period and air temperature 23°C ± 2°C. The pH of all nutrient solutions was checked frequently and, if necessary, adjusted to the present value of 3.8.

In Experiment 1, two-week-old spruce seedlings were grown at total aluminium concentrations of 0, 0.74, 1.85, 5.56 m*M* for 4 months. There were ninety seedlings per Al treatment.

In Experiment 2, two-week-old spruce seedlings were exposed to an Al concentration of 2.59 m*M* for 6.5 and 8.5 months. There were thirty seedlings per treatment.

In Experiment 3, spruce seedlings, precultivated for 8.5 months in nutrient solution without Al, were exposed to an aluminium concentration of 5.56 m*M* for 5 and 40 hours. There were thirty seedlings per treatment.

After the above-mentioned intervals of aluminium stress, phytohormonal levels were determined in the roots. Cytokinin-like substances were analysed by the *Amaranthus*-betacyanin bioassay after root homogenization, ethanol extraction (24 h at 5°C) and TLC chromatography on Silufol using N-butanol : ammonium hydroxide (4:1, v/v) as the mobile phase (Čížková, 1990; Rodrigues and Houwen, 1972). BAP standards of 0.01 mg $dm^{-3}$ and 0.1 mg $dm^{-3}$ increased the betacyanin content by 143.2 and 180.3 %, respectively. Gibberellin-like substances were analysed by the *Lactuca*-hypocotyl bioassay (Frankland and Wareing, 1961) after root homogenization, ethanol extraction (24 h at 5°C) and TLC chromatography on Silufol using chloroform : acetic acid : ethylacetate (60:5:40, v/v) as the mobile phase (Sembdner et al., 1962). $GA_3$ standards of 0.01 mg $dm^{-3}$ and 0.1 mg $dm^{-3}$ increased the hypocotyl length by 143.5% and 205.3%, respectively. Indole-acetic acid (auxin) content was analysed after root homogenization and extraction with redistilled ethanol (24 h at 5°C). The extract was filtered and the filtrate was evaporated to the water residue. The water residues were subjected to purification according to Knegt and Bruinsma (1973). Free IAA was estimated according to $\alpha$-pyrone fluorescence (Mousdale et al., 1978) using a Shimadzu RF-540 spectrofluorimeter. Each analysis was repeated three times and represented extracts of ten seedlings. If statistically significant, differences in IAA content in spruce roots between seedlings cultivated with or without aluminium were evaluated by the G-test (Fryer, 1957).

## Results

After 4 months, the indole-acetic acid content at 0.74 and 1.85 m*M* Al in nutrient solution was significantly increased in comparison to that of seedlings grown in the absence of Al (Fig. 1). Concentrations of 0.74 and 1.85 m*M* Al were responsible for an increase in levels of cytokinin- and gibberellin-like substances, while those levels were decreased at 5.56 m*M* Al (Figs. 2,3).

A concentration of 2.59 m*M* Al did not significantly affect the IAA concentration after 6.5 or 8.5 months (Fig. 4). Levels of cytokinin- and gibberellin-like substances were negatively influenced by Al (Figs. 5,6).

In experiment 3 the IAA content was highly significant and significant increased after 5 and 40 h expo-

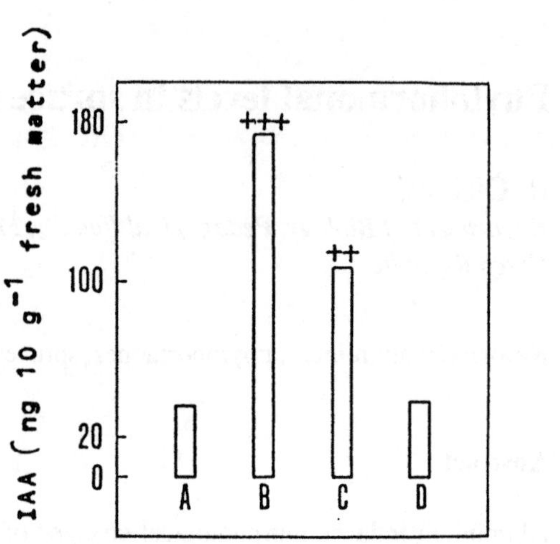


*Fig. 1.* Indole-acetic acid content (ng 10 $g^{-1}$ fresh matter) in the roots of spruce seedlings grown in hydroponic culture at 0 (**A**), 0.74 (**B**), 1.85 (**C**) and 5.56 (**D**) m*M* Al for 4 months. Significance of differences in variants B, C, D as related to variant A is indicated by ++ at $p = 0.01$, +++ at $p = 0.001$ (n = 3). Experiment 1.

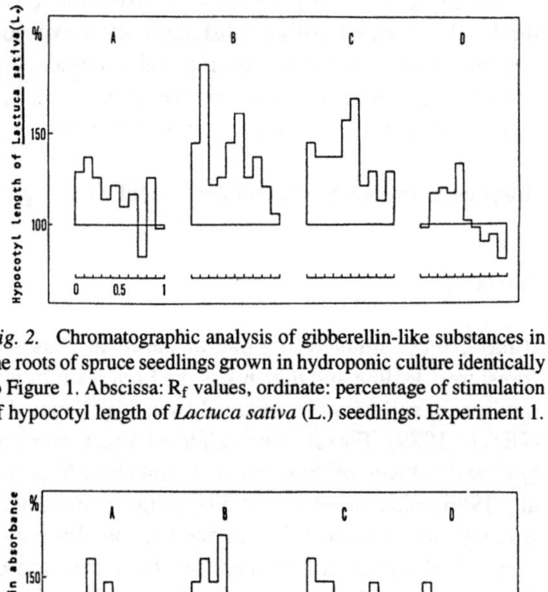


*Fig. 2.* Chromatographic analysis of gibberellin-like substances in the roots of spruce seedlings grown in hydroponic culture identically to Figure 1. Abscissa: $R_f$ values, ordinate: percentage of stimulation of hypocotyl length of *Lactuca sativa* (L.) seedlings. Experiment 1.

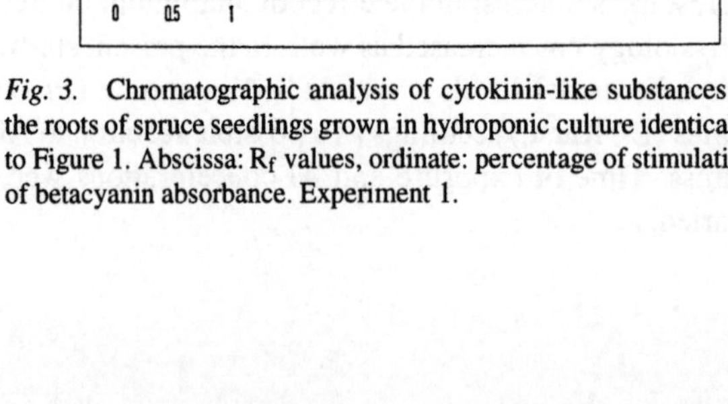


*Fig. 3.* Chromatographic analysis of cytokinin-like substances in the roots of spruce seedlings grown in hydroponic culture identically to Figure 1. Abscissa: $R_f$ values, ordinate: percentage of stimulation of betacyanin absorbance. Experiment 1.

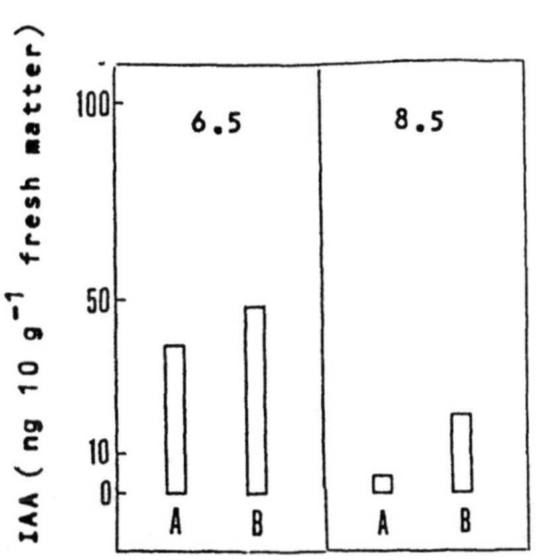


*Fig. 4.* Indole-acetic acid content (ng 10 $g^{-1}$ fresh matter) in the roots of spruce seedlings grown in hydroponic culture at 0 (**A**), and 2.59 (**B**) m*M* Al for 6.5 and 8.5 months. Experiment 2.

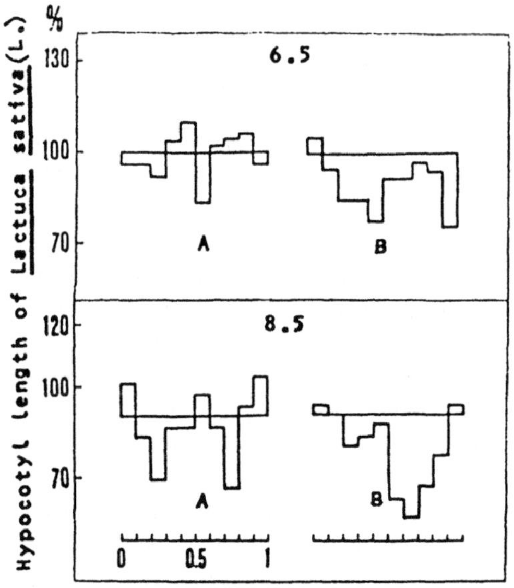


*Fig. 5.* Chromatographic analysis of gibberellin-like substances in the roots of spruce seedlings grown in hydroponic culture identically to Figure 4. Abscissa: $R_f$ values, ordinate: percentage of stimulation of hypocotyl length of *Lactuca sativa* (L.) seedlings. Experiment 2.

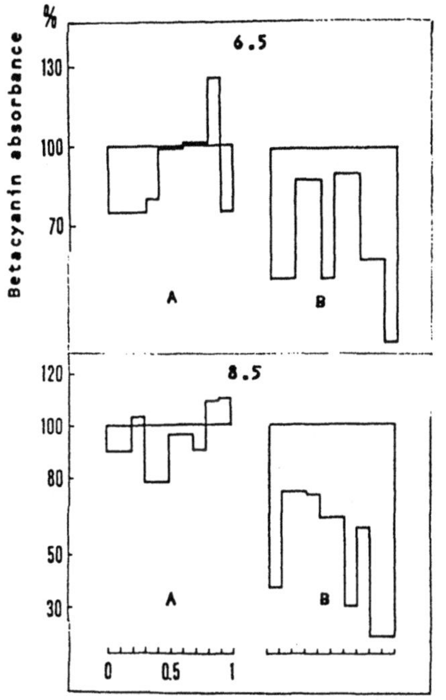


*Fig. 6.* Chromatographic analysis of cytokinin-like substances in the roots of spruce seedlings grown in hydroponic culture identically to Figure 4. Abscissa: $R_f$ values, ordinate: percentage of stimulation of betacyanin absorbance. Experiment 2.

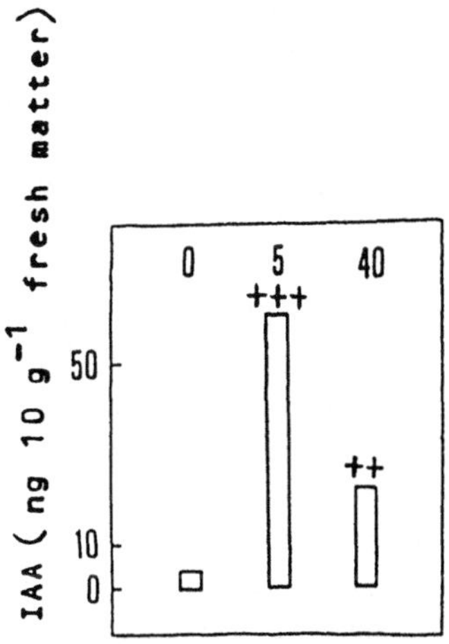


*Fig. 7.* Indole-acetic acid content (ng 10 $g^{-1}$ fresh matter) in the roots of spruce seedlings exposed to 5.56 m*M* Al for 0, 5 and 40 hours. Significance of differences in IAA after 5 and 40 h, as related to 0 h, is indicated by ++ at $p = 0.01$, +++ at $p = 0.001$ (n = 3). Experiment 3.

sure to 5.56 m*M* Al, respectively (Fig. 7). Levels of gibberellin-like substances were increased after both intervals of Al exposure (Fig. 8). After 5 hours exposure to Al, levels of cytokinin-like substances were decreased. After 40 h this decrease was less (Fig. 9).

## Discussion

There are many recent studies aimed at evaluating the aluminium sensitivity of trees. Coniferous trees are rather less sensitive than herbaceous plants, therefore many authors use high Al concentrations in their experiments. Schier (1985) and Thornton et al. (1987), for example, used Al concentrations as high as 7400 and 2000 $\mu M$, respectively. It coresponded with the

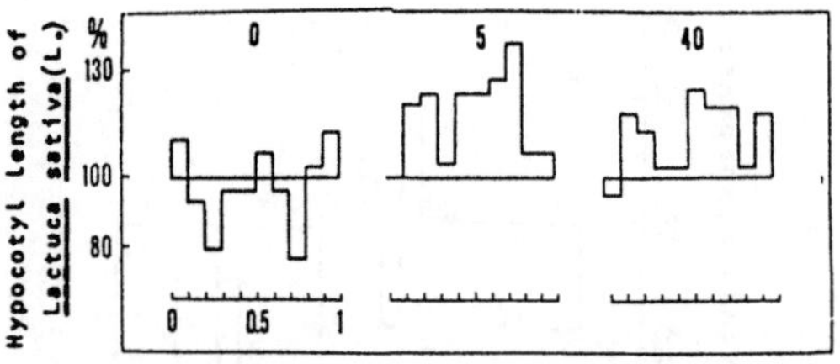


*Fig. 8.* Chromatographic analysis of gibberellin-like subtances in the roots of spruce seedlings treated by aluminium identically to Figure 7. Abscissa: $R_f$ values, ordinate: percentage of stimulation of hypocotyl length of *Lactuca sativa* (L) seedlings. Experiment 3.

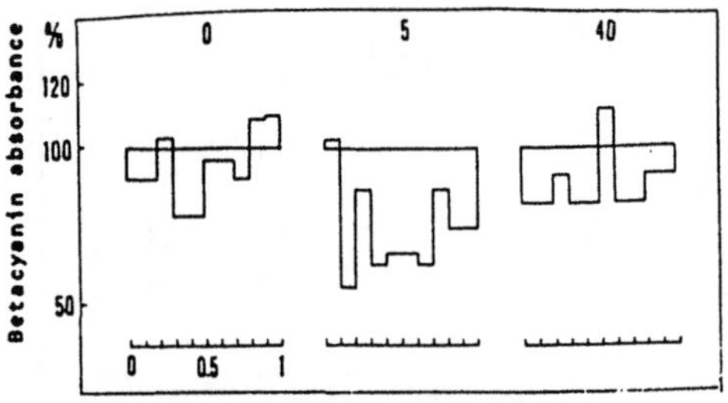


*Fig. 9.* Chromatographic analysis of cytokinin-like substances in the roots of spruce seedlings treated by aluminium identically to Figure 7. Absissa: $R_f$ values, ordinate: percentage of stimulation of betacyanin absorbance. Experiment 3.

arrangement of our experiments. Aluminium phytotoxicity is modified by ionic strength, specific ion effects, and possibly the nature of the charge-balancing anions. The present study only compares the spruce seedlings cultivated in nutrient solutions without or with aluminium supplied as $Al_2(SO_4)_3$.

In general, roots are the most frequently investigated subject, as related to aluminium toxicity. With respect to morphological changes, limitation of the elongation of the main root is often observed. Reduction of root length is associated with an enlargement of root diameter. Likewise, anomalies in the development of lateral roots are characteristic for some cases of aluminium toxicity (Andersson, 1988). In our previous experiments, we investigated Al effect on root morphology of spruce seedlings affected by Al identically as in the present study. This allows comparison between Al-induced changes in root morphology and Al-induced changes in internal phytohormone levels. Al mostly reduced both the length of primary and secondary roots of spruce seedlings (Čížková, 1989, 1992). Nevertheless, at the lowest Al concentrations under investigation the average length of the secondary root was stimulated, while the average length of the primary root was limited. In addition, secondary roots were initiated closer to the apex of the primary root. Obviously, low Al concentrations were already effective in decreasing the apical dominance of the primary root, but they were not toxic enough to inhibit the growth of lateral roots. Probably Al disturbs the apical dominance which prevents the development of lateral roots next to the tip. It is obvious, that apical dominance is regulated by phytohormones. Intensification of apical dominance of roots is supported by cytokinins, but reduced by auxins (Burström and Svensson, 1972; Šebánek, 1983). No investigation of a possible phytohormonal gradient along roots were made in the present experiments. Total phytohormonal levels were estimated in the whole root system. Nevertheless, it seems to be likely that Al-induced increase of the IAA content, and Al-induced morphological changes, in roots of spruce seedlings are interrelated.

It appears to be rather controversial that levels of both cytokinin- and gibberellin-like substances will be increased at lower Al concentrations, as observed in Experiment 1 and after short-term Al exposure (Experiment 3). Contrary to auxins, cytokinins should be responsible for intensifying apical dominance in roots. Furthermore, phytohormonal levels in the root system could include defense reactions of spruce seedlings to Al stress. Especially the stimulation of cytokinins can be attributed to a defense reaction against unfavourable environmental factors (Čížková, 1990). In Experiment 2, concentrations and period of Al exposure were obviously too high and aluminium reduced the levels of all phytohormones under investigation.

Another morphological change in roots of spruce seedlings induced by Al (at concentrations above 2.0 m*M*) described in previous studies was the swelling of root apices. Root swelling as a result of Al excess resembled ethylene effects on barley (Crosset and Campbell, 1971). Based on findings that Al induced ethylene evolution by spruce seedling (Čížková and Reinöhl, 1992), it is obvious that an Al-induced morphological anomaly might be a consequence of ethylene production. Data cited in the literature give no precise insight into the mechanism by which Al inhibits root elongation. Inhibition in mitotic activity by Al was established (Tepper et al., 1989). It was reported that Al acts also indirectly on mitosis through integrated control function of the root as well (Roy et al., 1988). Godbold et al. (1988) stated that displacement of Ca by Al in the cell wall may inhibit root elongation. Our data suggest that Al-induced phytohormonal changes possibly also participate in Al-induced root morphological changes.

## References


Andersson M 1988 Toxicity and tolerance of aluminium in vascular plants. Water, Air, Soil Pollut. 39, 439–462.

Burström H G and Svensson S B 1972 Hormonal regulation of root growth and development. *In* Hormonal Regulation in Plant Growth and Development. Eds. H Kaldewey and Y Vardar. pp 125–136. Verlag Chemie, Weinheim, Germany,

Crossett R N and Campbell J D 1971 The effect of ethylene on the development of the barley system. Annual Report, Letcombe Lab. Arc, UK. 173 p.

Čížková R 1989 Changes in root system morphology of spruce (*Picea abies* (L.) Karst.) owing to aluminium. *In* Proc. Vth Days of Plant Physiology. pp 195. Ed. J E Purkyně. Univ. Fac. Nat. Sci. Brno, The Czech Republic. (*In Czech*).

Čížková R 1990 Acidification stress of root environment as related to endogenous cytokinins and gibberellins in oak seedlings. Biol Plant. 32, 97–103.

Čížková R 1992 Effect of Aluminium on Growth and Phytohormones in Norway Spruce Seedlings. Acta Sci. Nat. Brno 26, The Czech Rebublic. 51 p.

Čížková R and Reinöhl V 1992 Effect of aluminium in root environment on ethylene production spruce (*Picea abies* (L.) Karst.) seedlings. Biol. Plant. (Suppl.) 34, 514.

Frankland B and Wareing PF 1961 Effect of gibberellic acid on hypocotyl growth of lettuce seedlings. Nature 185, 255–256.

Fryer H C 1957 Elements of Statistics. New York, USA. 127 p.

Godbold D L, Fritz E and Hüttermann A 1988 Aluminium toxicity and forest decline. Proc. Nat. Acad. Sci. USA 85, 3888–3892.

Godbold D L and Kettner C 1991 Use of root elongation studies to determine aluminium and lead toxicity in *Picea abies* seedlings. J. Plant Physiol. 138, 231–235.

Hutchinson T C, Bozic L and Munoz-Vega G 1986 Responses of five species of conifer seedlings to aluminum stress. Water, Air, Soil Pollut. 31, 283–294.

Knegt E and Bruisma J 1973 A rapid sensitive and accurate determination of indolyl-3-acetic acid. Phytochemistry 12, 753–756.

Mousdale D M A, Butcher D N and Powell R G 1978 Spectrophotofluorimetric methods of determining indole-3-acetic acid. *In* Isolation of Plant Growth Substances. Ed. J R Hillman. pp 27–39. Cambridge University Press, Cambridge, UK.

Raynal D J, Joslin J D, Thornton F C, Schaedle M and Henderson G S 1990 Sensitivity of tree seedlings to aluminum: III. Red spruce and loblolly pine. J. Environ. Qual. 19, 180–187.

Rodrigues P A S and Houwen P S W 1972 Cytokinins and bacterial symbiosis *Ardisia* species. Z. Pflanzenphysiol. 68, 170–171.

Roy A K, Sharma A and Talukder G 1988 Some aspects of aluminium toxicity in plants. Bot. Rev. 54, 145–178.

Schaedle M, Thornton F C, Raynal D J and Tepper H B 1989 Response of tree seedlings to aluminium. Tree Physiol. 5, 337–356.

Schier G A 1985 Response of red spruce and balsam fir seedlings to aluminum toxicity in nutrient solution. Can. J. For. Res. 15, 29–33.

Sembdner G, Gross R and Schreiber K 1962 Die Dünnschicht-chromatographie von Gibberellinen. Experimentia 18, 584–585.

Šebánek J 1983 Plant Physiology. SZN Praha, The Czech Republic. 558 p. (*In Czech*).

Tepper H B, Yang C S and Schaedle M 1989 Effect of aluminium on growth of root tips of honey locust and loblolly pine. Environ. Exp. Bot. 29, 165–173.

Thornton F C, Schaedle M and Raynal D J 1987 Effects of aluminium on red spruce seedlings in solution culture. Environ. Exp. Bot. 27, 489–498.

Ulrich B 1989 Effects of acid deposition on forest ecosystems in Europe. *In* Acid Precipitation. Adv. Environ. Sci. Eds. D C Adriano and A H Johnson. pp 189–272. Springer Verlag, New York, USA.

*F. Baluška et al. (eds.), Structure and Function of Roots,* 341–347, 1995.


# Distribution of root biomass and length in *Picea abies* ecosystem under different immission regimes

Milan Kodrík
*Institute of Forest Ecology, Slovak Academy of Sciences, Štúrova 2, 960 53 Zvolen, Slovak Republic*

*Key words:* air pollution, forest ecosystem, *Picea abies*, root biomass, root length

**Abstract**

This study reports the results of root biomass research in an 80-year-old spruce monoculture where four monitoring plots were established. Three of them were exposed to various influences of immissions; the fourth one was a control plot. The investigation of the below-ground tree biomass and its length was done by means of destructive analysis of sample trees. Twelve complete spruce root systems were processed.

Individual stands achieved values of the below-ground biomass from 54.5 t $ha^{-1}$ to 72.6 t $ha^{-1}$ in terms of dry weight. The largest total below-ground biomass was found on the control plot. Total root length varied from 430 km $ha^{-1}$ to 990 km $ha^{-1}$. Thin roots ($\oslash \leq 0.5$ cm) formed the most substantial share of the total root length, about 880 km $ha^{-1}$ on the control plot.

## Introduction

Research of root biomass represents a part of the basic information that is important for understanding the dynamics of forest ecosystems. Aspects such as growth, productivity or energy cycling can be characterized through monitoring changes in biomass from one sampling occasion to another. Investigations on tree biomass distribution in forests are necessary in order to quantify the description of the ecosystems, and to answer questions about the effects of human impact on them.

In natural forest ecosystems the biomass of wood species forms approximately 98 per cent of the total ecosystem biomass. However, from a methodological point of view it is laborious to determine the tree biomass, due to the dimensions and weight of above-ground and below-ground tree parts.

Historically, research on the below-ground biomass can be divided into three different stages. During the first stage it was directed towards the anatomical and morphological description of roots. Later on, activity was first directed towards ecological factors influencing root growth, and finally towards physiological ones. Many papers from the first two stages can be found in surveys by Karizumi and Tsutsumi (1958), Köstler et al. (1968), Röhring (1966), Sutton (1969).

Results from the third research stage have been summarized by Ovington (1962), Santantonio et al. (1977) and Persson (1993). Results of a large number of studies dealing with woody species were furthermore published through IUFRO in issues devoted to the study of forest biomass (e.g., Young, 1973). Many new techniques of root investigation have been described by Böhm (1979), Hendrik and Pregitser (1992), and Vogt and Persson (1991).

In spite of problems with extraction of root systems many root investigations have been carried out in Slovakia, among them investigations on black locust (Benčat, 1988, 1989), Scots pine and European beech (Kodrík, 1992a; Oszlányi, 1979, 1986), Norway spruce (Kodrík, 1992b; Oszlányi, 1986) and the oak-hornbeam ecosystem (Šimonovič, 1978). Other related studies can be found in Grier et al. (1981), Harris et al. (1977), Jones and Raynal (1986), Nihlgård (1972) and Persson (1978, 1983, 1990).

The aim of the present article is to gain data on the division and the total quantity of the below-ground biomass in Norway spruce stands under varying pollution regimes. Measurements were furthermore carried

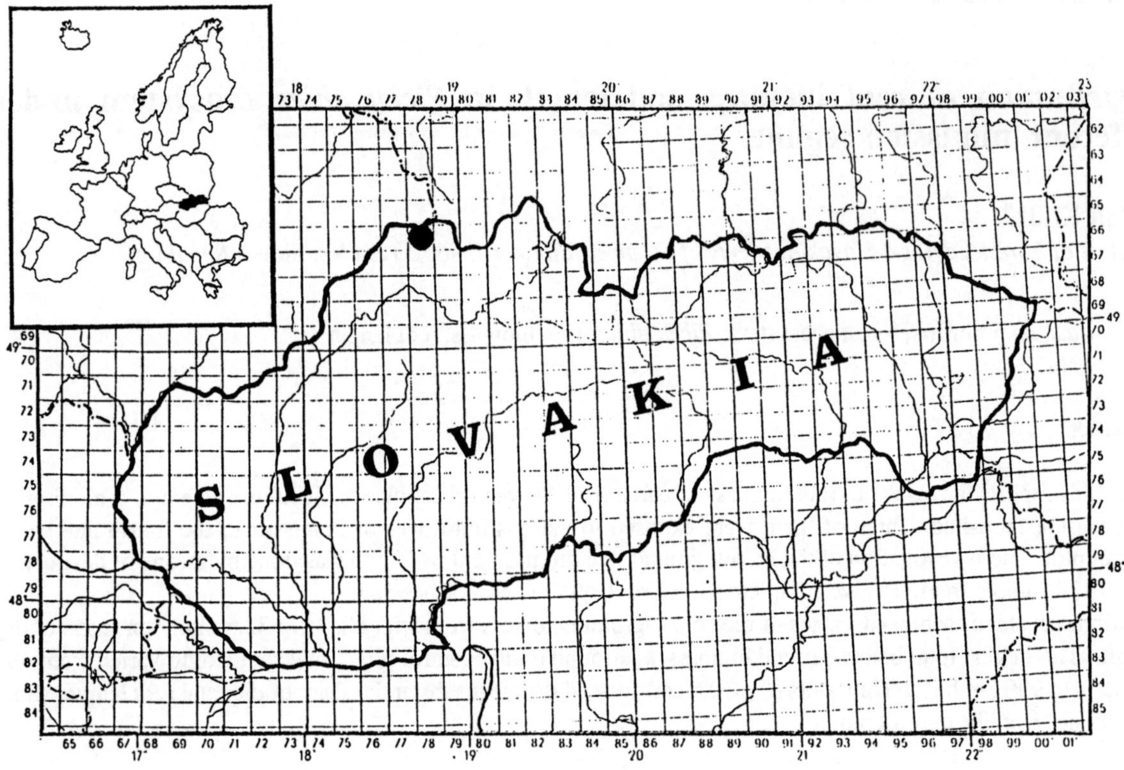


*Fig. 1.* Location of the monitoring plots $MP_1$-$MP_4$.

*Table 1.* Stand characteristics of four monitoring plots in the area of Moravian-Silesian Beskides ($MP_{1-4}$)

| Plot | Altitude (m) | Age (years) | Height (m) | Diameter (cm) |
|---|---|---|---|---|
| $MP_1$ | 900 | 80 | 23 | 26 |
| $MP_2$ | 920 | 75 | 22 | 28 |
| $MP_3$ | 910 | 75 | 22 | 28 |
| $MP_4$ | 830 | 70 | 22 | 25 |

*Table 2a.* Immission situation in the Čadca area — average values in 1980–1990 (Kontrisova 1988, 1990)

| Dust | Trace elements in dust ($mg\ kg^{-1}\ year^{-1}$) | | | | | |
|---|---|---|---|---|---|---|
| ($g\ m^{-2}\ month^{-1}$) | Cd | Cu | Cr | Mn | Ni | Pb |
| 8.67 | 9.6 | 72.1 | 34.9 | 440.4 | 94.5 | 462.6 |

*Table 2b.* $SO_2$ and $NO_2$ content ($\mu g\ m^{-3}$) the air in the Žilina region (28 km away from the monitoring plots) in 1986–1990

| | | Arithmetic Mean | Min. | Max. |
|---|---|---|---|---|
| $SO_2$ | Winter | 138.8 | 11.0 | 696.7 |
| | Summer | 62.3 | 0.4 | 153.9 |
| $NO_2$ | Winter | 126.2 | 28.0 | 898.0 |
| | Summer | 79.7 | 26.2 | 237.8 |

out in order to obtain the length of the below-ground biomass fractions.

## Materials and methods

In the area of Moravian-Silesian Beskides four monitoring plots ($MP_1$-$MP_4$) were established. In all cases the stand was formed by Norway spruce monocultures. The soil substrate consisted of an iron podzol. All plots were situated on the northwest exposure with an incli-

*Table 3.* Root biomass of Norway spruce (*Picea abies* (L. Karst.) in different diameter classes in terms of dry weight kg ha$^{-1}$ and percentage share in every diameter class (for $MP_1$-$MP_4$ see Materials and methods)

| Plot | Root-size classes (cm) | | | | | | | Total |
|---|---|---|---|---|---|---|---|---|
| | ≤0.5 | 0.6–2.0 | 2.1–5.0 | 5.1–7.0 | 7.1–10.0 | >10.0 | Stump | |
| $MP_1$ | 740 | 2810 | 5960 | 3960 | 3880 | 19420 | 18370 | 55140 |
| | 1 | 5 | 11 | 7 | 7 | 36 | 33 | 100 |
| $MP_2$ | 880 | 4500 | 7570 | 3630 | 3600 | 19850 | 23500 | 63530 |
| | 1.5 | 7 | 12 | 5.5 | 6 | 31 | 37 | 100 |
| $MP_3$ | 1390 | 2340 | 4480 | 2990 | 2360 | 18960 | 22010 | 54530 |
| | 2.5 | 4 | 8 | 5 | 4.5 | 35 | 41 | 100 |
| $MP_4$ | 2130 | 3500 | 6840 | 3910 | 4470 | 32160 | 19600 | 72610 |
| | 3 | 5 | 10 | 5 | 6 | 44 | 27 | 100 |

nation of 10–15°. More details can be found in Table 1 (see also Kodrík 1992b).

Pollution regimes were taken into consideration when the plots were chosen. The first three plots were characterized by pollution loadings of different intensities (expressed by 50–60 % loss of needles), while the fourth plot was a control plot which was loaded with the minimum amount of immissions. The content of immission loading comes from industrial enterprises in the Ostrava and Karvina regions, as well as from more distant sources in Poland. Concrete values of immission loading on the investigated monitoring plots are given in Tables 2a and b (Kontrišová 1988, 1990).

We have gained all data on below-ground biomass by destructive sampling of trees. To get a complete picture of the position of the trees as the individuals, repeated stock-takings of trees assorted each tree into Kraft's classification scale (Vyskot et al., 1971). This scale takes the relative altitudinal position of the tree and formation of the crown into account. Kraft (Vyskot et al., 1971) distinguishes the following classes:

- *Dominant trees* with extremely strongly developed crowns;
- *Codominant trees* with relatively well developed crowns constituting, as a rule, the main stand;
- *Partially codominant trees* with still sufficiently normal but relatively weakly developed and oppressed crowns;
- *Undertopping trees* with a more or less shortened unilaterally developed crown oppressed from several sides;
- *Fully shaded trees* - typically undertopping trees;

The selection resulted from the calculated mensurational tree variables, separately for each Kraft class (Oszláni 1975).

There is a linear dependency between the basal area at breast height and the tree weight (Ovington et al., 1968). In the same way the dependency between the above-ground and below-ground tree biomass was proved (Harris et al., 1977).

On the basis of these results as well as our own research in previous period we found that the major representatives the above-ground biomass (including the below-ground biomass) are the trees from the Kraft's tree class 1, 2 or (Kodrík, 1992c). The above-ground biomass of the next two classes is negligible compared to the trees of the former classes. For this reason and in view of the difficult technical feasibility of the root system extraction, our selection was directed to the sample trees of dominant, codominant and partially codominant trees. We processed one sample tree out of the first three classes on each research plot. The tree root system was elevated by means of the archeological method (Kodrík, 1992b). The whole root system as gradually uncovered using shovels, hoes and brushes. Furthermore, a tractor with a winch and a powersaw was used. The below-ground biomass was processed immediately after it was uncovered. The fresh weight

*Table 4.* Average root length (m) in different diameter classes and tree classes

| Root-size class cm | Trees | | |
|---|---|---|---|
| | Dominant | Codominant | Partially codominant |
| ≤0.5 | 1 369.9 | 1 090.4 | 880.8 |
| 0.6–2.0 | 180.6 | 146.5 | 97.2 |
| 2.1–5.0 | 48.8 | 26.3 | 22.1 |
| 5.1–7.0 | 9.8 | 5.4 | 3.6 |
| 7.1–10.0 | 4.1 | 3.5 | 2.2 |
| >10.0 | 7.4 | 3.9 | 2.5 |

was determined in the field on scales with an accuracy of 0.05 kg. A sample from every diameter fraction was taken for further laboratory processing.

## Results

### *Below-ground biomass*

The Norway spruce below-ground biomass on individual monitoring plots is shown according to diameter classes in Table 3, which also gives the total below-ground biomass as well as the percentage of individual diameter classes from the total below-ground biomass. It is evident from the data that the most substantial share of the below-ground biomass is on $MP_4$ - 72.6 t $ha^{-1}$ in terms of dry weight.

In the below-ground biomass structure there are no substantial differences except in the first diameter category. The most substantial share of the biomass in this diameter category is on $MP_4$, where it represents 3 % of the total below-ground biomass. The plots were subject to various pollution loads (see Materials and methods). This means that $MP_4$ is a control plot to be compared to $MP_1$, $MP_2$ and $MP_3$.

It is obvious from the biomass distribution (Table 3) that almost all diameter categories on the control plot ($MP_4$) obtained higher estimates. Also, the total biomass is 72. 6 t DW $ha^{-1}$ on $MP_4$. The lowest total biomass on the polluted plots was found on $MP_3$ - 54. 5 t DW $ha^{-1}$, which is equal to 75 % of the total below-ground biomass on the control plot. The differences in biomass on the other polluted plots were not so big.

For a better comparison of differences in biomass on individual plots the investigated sample trees were directly compared. In this way effects caused by varying stand density were avoided. On the whole, it can be presumed that the figures for below-ground biomass on the control sample trees should be higher than on the plots with high immissions. This is to be explicitly claimed for the sample trees of class 3 (partially codominant trees). There was a slight difference in the data of partially codominant trees in some diameter classes compared to the immission plot. The tree class 1 (dominant trees) the greatest differences were found between the control tree ($MP_4$) and the others. The estimated total biomass of the dominant sample tree on $MP_4$ was not the heaviest biomass fraction (Fig. 2).

The roots with a diameter >10 cm represent the most substantial share of the below-ground biomass in terms of dry weight (more than one third). They are followed by the stump biomass - almost one third. All other diameter fractions represent about 32 % of the total root biomass.

### *Below-ground length*

A survey of average root length of sample trees in individual diameter classes according to tree classes is shown in Table 4. It is obvious from these data that he most substantial root length was in the first diameter class. For the codominant and dominant trees this means root lengths of more than 1,000 m for a single tree. The length of the root decreases with the raising diameter and depends on the tree class. The lowest root length was estimated for partially codominant trees.

A survey of below-ground root length on the individual monitoring plots according to diameter in different classes is shown in Table 5. The first two diameter classes have the highest values. When comparing immission plots to the control plot, we can only see differences in the first diameter category. In all other categories there were no differences. The situation concerning below-ground biomass on the immission plots is found in Table 5. It is obvious that immission has an important effect on fine roots. The monitoring plots differ from each other in the total below-ground length biomass of the root, as well as in terms of dry weight. In $MP_4$ the highest root length was achieved: 990 km $ha^{-1}$. The highest total root length on the polluted plots was found on $MP_3$ (648 km $ha^{-1}$). The estimate from the control plot was around 50 % higher than that.

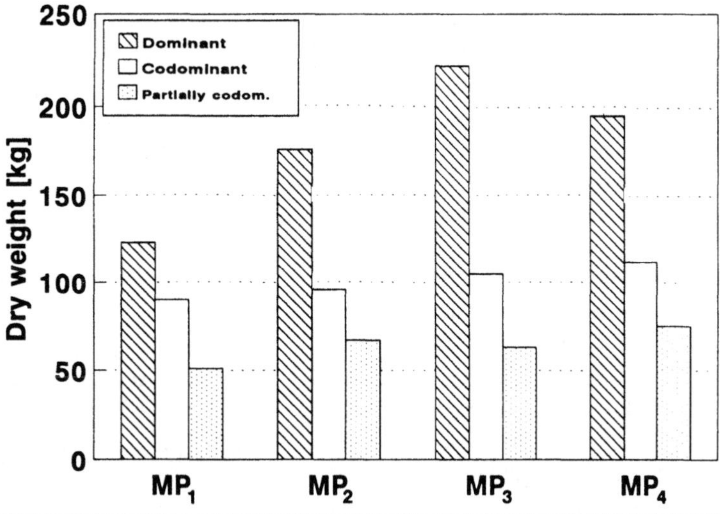


| Trees | $MP_1$ | $MP_2$ | $MP_3$ | $MP_4$ |
|---|---|---|---|---|
| Dominant | 123 | 176 | 222 | 195 |
| Codominant | 90 | 96 | 105 | 112 |
| Partially codom. | 51 | 67 | 63 | 75 |

**Sample trees**

*Fig. 2.* Total below-ground biomass review of sample trees in terms of dry weight (in kilograms) according to their biosociological position in forest stand.

*Table 5.* Root length in different diameter classes in km ha$^{-1}$ (for $MP_1$-$MP_4$ see Materials and methods)

| Plot | Root-diameter classes (cm) | | | | | | Total |
|---|---|---|---|---|---|---|---|
| | ≤0.5 | 0.6–2. | 2.1–5. | 5.1–7. | 7.1–10. | >10. | |
| $MP_1$ | 437.3 | 60.7 | 16.6 | 3.5 | 2.1 | 1.7 | 521.9 |
| $MP_2$ | 311.8 | 93.7 | 16.5 | 2.5 | 1.5 | 1.8 | 427.8 |
| $MP_3$ | 575.2 | 54.0 | 12.9 | 2.6 | 1.0 | 2.0 | 647.7 |
| $MP_4$ | 880.0 | 80.7 | 19.8 | 3.6 | 2.2 | 3.6 | 989.9 |

## Discussion and conclusion

On the investigated plots 12 Norway spruce root systems were excavated. The plots were selected in such a way that they were as similar as possible. The thickest roots (diameter > 10 cm) represented the largest share of biomass. This is about 40% of the total below-ground biomass. The stump represented 25–30% of the below-ground biomass of a Norway spruce stand. The total below-ground biomass of the roots on the $MP_1$-$MP_4$ plots achieved values from 114.5 t ha$^{-1}$ to 168.0 t ha$^{-1}$ in terms of fresh weight. Manakov (1961) estimated 180 t ha$^{-1}$ for a 120-year-old Norway spruce stand in agreement with our results. Remezov (1959, in Rodin and Vasilievich 1967) reported the below-ground biomass in a 72-year-old spruce of 65 t DW ha$^{-1}$ and in an 83-year-old spruce of 78 t DW ha$^{-1}$. Nihlgård (1972) estimated the below-ground biomass in a 55-year-old Norway spruce stand to be 59 t DW ha$^{-1}$. On the other hand, Parshevnikov (1975) estimated the root biomass of 66 t DW $ha^{-1}$ in a 110-year-old Norway spruce stand. It may be concluded with regard to $MP_1$-$MP_4$ that the stand density is higher and the growth conditions are different. Our data are furthermore confirmed by results from Oszlányi (1986), who estimated the root biomass in terms of below-ground fresh weight in a 60-year-old Norway spruce stand at 115 t $ha^{-1}$.

Data on the Norway spruce below-ground biomass length are known from various publications. Nobbe (1876) found 253 roots in a young Norway spruce stand that were 1.9 m long. Bernbeck (1917) found that spruce roots went down mostly to a depth of 1.5 m and only very rarely reached a depth of 3 m. This corresponds with our own data. The depth reached by roots is influenced by soil depth and environmental conditions. The results often cannot be compared, owing to the different growth conditions. Material achieved in this way can be used in the future for calculation of rootage, root surface area, and other characteristics of Norway spruce.

## References


Benčat' T 1988 Root system and belowground biomass of Black Locust (*Robinia Pseudacacia* L.) in Southern Slovakia. Lesnictví 34, 51–60 (*In Slovak*).

Benčat' T 1989 Black Locust biomass production in Southern Slovakia. Acta Dendrobiologica, Veda Bratislava, Slovakia, 192 p.

Bernbeck 0 1917 The wind as a plantpatology factor. Englers bot. Jahrb. 45 (*In German*).

Böhm W 1979 Methods of studying Root Systems. Springer-Verlag, Berlin, Heidelberg, New York. 188 p.

Grier CH C, Vogt K A and Keyes M R 1981 Biomass distribution and above- and below-ground production in young and mature *Abies amabilis* zone ecosystems of the Washington Cascades. Can. J. For. Res. 11, 117–124.

Harris W, Kinerson R and Edwards N 1977 Comparison of below-ground biomass of natural deciduous forest and Loblloly pine plantations. Pedobilogia 17, 369–381.

Hendrik R L and Pregitser K S 1992 The demography of fine roots in a northern hardwood forest. Ecology 73, 1094–1104.

Jones R H and Raynal D J 1986 Spatial distribution and development of root sprouts in *Fagus grandifolia* (Fagaceae). Am. J . Bot. 73, 1723–1731.

Karizumi N and Tsutsumi T 1958 A reference of the root systems on the forest trees (to 1955). J. Jap. Forest. Soc. 40, 202–223.

Kodrík M 1992a Beech in below-ground biomass. *In* Ecological and Economic Significance of Beech. pp 103–109. Technical University Press, Zvolen, Slovakia.

Kodrík M 1992b Below-ground biomass distribution of Norway spruce. *In* Forest-Wood-Ecology, Internat . Sci . Conf., pp 151–157. Technical University Press, Zvolen, Slovakia.

Kodrík M 1992c Root biomass distribution of Norway spruce of older forest ecosystem. *In* Root Ecology and its Practical Application. Eds. L Kutschera, E Lichtenegger, H Persson and M Sobotik. pp 534–536 Verein für Wurzelforschung, A-9020 Klagenfurt, Austria.

Kontrišová O 1988 Air quality on the monitoring plots. Report of Institute of Forest Ecology SAS, Slovakia. 13 p (*In Slovak*).

Kontrišová O 1990 Air quality on the monitoring plots. Final report, Institute of Forest Ecology SAS, Slovakia. 27 p (*In Slovak*).

Köstler J N, Brückner E and Bibelriether H 1968 Roots of forest trees, Hamburg und Berlin. 412 p (*In German*).

Manakov K N 1961 Adsorption of mineral elements and nitrogen by plants from soil in forests of Kola peninsula. Pochvovedenie 8, 34–41 (*In Russian*).

Nihlgård B 1972 Plant biomass, primary production and distribution of chemical elements in a beech and planted spruce forest in South Sweden. Oikos, 23, 69–81.

Nobbe F 1876 Seed production booklet. Berlin, Germany. 118 p (*In German*).

Oszlányi J 1975 Characteristics of ecological structure of the growing stock at the research site of Báb. *In* Research Project Báb IBP Progress Report II. Ed. V Biskupský. pp 61–70. Publish. House of the Slovak Acad. Sci., Bratislava, Slovakia.

Oszlányi J 1979 Biomass energetic value of different biosociological position trees. Lesnícky Časopis 23, 177–188 (*In Slovak*).

Oszlányi J 1986 Analysis into biomass production and into its energy equivalent of the tree layer in five forest ecosystems. Biologické Prace 32, 1, Veda Bratislava, 1–157 (*In Slovak*).

Ovington J D 1962 Quantitative ecology and the woodland ecosystems concept. Advert. Ecol. Res. 1, 103–192.

Ovington J D, Forrest W and Armstrong J 1968 Tree biomass estimation. *In* Symposium on primary production and mineral cycling of natural ecosystems. Ed. H Young. pp 69–81. Univ. Maine, Orono.

Parshevnikov A L 1975 Productivity and turnover of chemical elements in northern phytocoenoses. Nauka, Leningrad, USSR. 128 p (*In Russian*).

Persson H 1978 Root dynamics in a young Scot pine stand in Central Sweden. Oikos 30, 3, 508–519.

Persson H 1983 The distribution and productivity of fine roots in boreal forest. Plant and Soil 71, 87–101.

Persson H 1990 The effects of forest liming on fertilization on fine root growth. Water, Air, Soil Pollut. 21, 516–527.

Persson H 1993 Factors affecting fine root dynamics of trees. Suo 43, 163–172.

Remezov N P, Bykova L N and Smirnova K M 1959 Uptake and turnover of nitrogen and ash elements in forests of European part of USSR. Izd. MGU, Moscow, USSR. 282 p (*In Russian*).

Rodin L and Vasilievich N 1967 Production and Mineral Cycling in Terrestrial Vegetation. Oliver and Boyd, Edinburgh and London.

Röhring E 1966 Development of the forest wood species roots in relationship to ecological conditions. Forstarchiv 37, 237–249 (*In German*).

Santantonio D, Hermann R K and Overton W S 1977 Root biomass studies in forest ecosystems. Pedobiologia 17, 1–31.

Sutton R F 1969 Form and development of conifer root systems. Technic. Communic. of 7th Commonwealth Forestry Buroev. Oxford, UK. 46 p.

Svoboda P 1952 Life of the forest. Brazda Praha, Czechoslovakia, 895 p (*In Czech*).

Šimonovič V 1978 Study of the root biomass in the herb layer of an oak-hornbeam forest. Biológia (Bratislava) 28, 11–22.

Vogt C and Persson H 1991 Measuring growth and development of roots. *In* Techniques and Approaches in Forest Tree Ecophysiology. Eds. J P Lassoie and T M Hincley. pp 477–501. CRS Press, Boca Raton, FL.

Vyskot M, Doležal B, Jurča J, Korf V, Korpel' Š, Macháč D, Polák L, Priesol A, Řehák J and Wolf J 1971 Bases of growth and production of forest, State Agricultural Publishing house Praha Czechoslovakia. 440 p (*In Czech*).

Young H E 1973 IUFRO Biomass studies. Univ. Maine, Orono, USA. 92 p.

*Section editor: H Lambers*

*F. Baluška et al. (eds.), Structure and Function of Roots*, 349–352, 1995.


# Sporocarp succession of mycorrhizal fungi in the Norway spruce plantations in formerly agricultural land

Ján Gáper and Pavel Lizoň
*Institute of Forest Ecology, Slovak Academy of Sciences, Sturova 2, SK-960 53 Zvolen, Slovakia and Plant Pathology Herbarium, Cornell University, 401 Plant Science Bldg., Ithaca, NY 14853-4203, USA*

*Key words: Picea abies*, sporocarp succession, spruce plantations

## Abstract

Successional processes in ectomycorrhizal macrofungi associated with Norway spruce plantations on former agricultural land were studied as reflected by sporocarps. Abundance and temporal distribution of sporocarps were recorded in permanent observation plots within stands of various ages in 1989–1992.

The succession started mainly with *Laccaria laccata* (Scop.:Fr.) Berk. and Br. s.l. *Cortinarius* spp. and *Dermocybe* spp., *Hebeloma crustuliniforme* (Bull.:Fr.) Quél. and *Chalciporus piperatus* (Bull.:Fr.) Bataille dominated in later stages. *Amanita rubescens* (Pers.:Fr.) S.F.Gray, *Hygrophorus pustulatus* (Pers.:Fr.) Fr. and some others species were the most abundant species in the 24th – 27th years after plantation.

## Introduction

Temporal distribution of ectomycorrhizal macrofungi is due the host associations (Last et al., 1983) and microenvironmental factors. The succession of epigeous ectomycorrhizal fungi as reflected from sporocarps was studied among others in *Betula pendula* and *B. pubescens* (Mason et al., 1982; Last et al., 1984), in *Pseudotsuga menziensii* (Ammirati et al., 1987; O'Dell et al., 1992; Vogt et al., 1992), *Pinus patula*, *P. radiata* and *P. sylvestris* (Chu-Chou, 1979; Last et al., 1981; Termoshuizen and Schaffers, 1987, 1989). Observations of sporocarp production of selected species in relation to the age of young birch trees resulted in a suggested division of mycorrhizal fungi into "early stage" and "late stage" fungi (Mason et al., 1982). The classification, adopted for fungi associated with woods established on agricultural soils, has many limitations (Fleming et al., 1986; Newton, 1991). The main difference between early and late stage fungi was defined in their ability to colonize root systems from spores and vegetative mycelium respectively (Newton, 1992). Thus some late stage species can better infect a host tree than early stage fungi (Read, 1991).

The process of colonization of forested farmland is different as in forest soils which have a high mycorrhizal inoculum potential (Newton, 1992). There is naturally a lack of true ectomycorrhizal populations in old soils exploited for agriculture production. Forestation by ectotrophic trees predestinate colonization of these soils by ectomycorrhizal fungi. The aim of the present study was to follow the process of ectomycorrhizal colonization during the first ca. 30 years of development of planted Norway spruce stands as reflected by production of epigeous sporocarps.

## Materials and methods

The research site is located on slopes of the hill Vrch Dobroc (740–917 m a.s.l.) close to the village of Detvianska Huta, Central Slovakia and is managed by the State Forests Company and the Institute for Forest Research. Stands of deciduous and coniferous trees were planted on former agricultural land starting in the 1960s and later they were incorporated to the IUFRO research net. The mean yearly air temperature is 6.0°C, becoming 12.0°C in the growing season. Annual precipitation averages 900 mm, including 300 mm of precipitation in the growing season. Bedrock is formed by crystalline rocks covered mostly by loamy sandy brown forest soils. Both organic and mineral horizons are acid (pH 4.2–6.8). Potatoes, rye and oat were the main crops in this area during extensive farming for

*Table 1.* Characteristics of permanent plots

| Characteristic | Permanent plot | | | |
|---|---|---|---|---|
| | A (lopped) | B (unlopped) | C | D |
| Altitude (m a.s.l.) | 820 - 835 | 820 - 835 | 810 - 815 | 860 |
| Exposure | SW | SW | SW | ±plain |
| Size (m) | 125 × 10 | 125 × 10 | 41.66 × 30 | 35.4 × 35.3 |
| Year of plantation | 1965 | 1965 | 1982 | 1989 |
| Seedling spacing (m) | 1.4 × 1.4 | 1.4 × 1.4 | 1.4 × 1.4 | 1.9 × 1.2 |
| Soil | typical cambisols | typical cambisols | typical cambisols | typical cambisols |

a few hundred years (M. Kamenský, 1994, personal communication).

Four permanent plots (each 1250 $m^2$) were established inside larger Norway spruce (*Picea abies* (L.) Karst.) stands of various age (Table 1). The maximum distance between individual plots is ca. 1.5 km. Plots A and B are divided only by a 2–2.5 m wide clear-cut strip and both are without an herbaceous layer. Soil is covered by needle litter. Crown thinning was used in plot A: all trees, except a few, were lopped and all wood removed. Plot B was treated by low thinning; trees were unlopped and logged trees were left lying in the stand. Soils are the same typical cambisols in all of the permanent plots (J. Kukla, 1992, personal communication) but presence of soil inoculum and conditions for a spontaneous mycorrhizal infection were completely different at the time of plantation for a particular stand. The most unfavorable conditions were in the 1960s when the forestation started. At that time no mycorrhizal populations were available in these soils. On the other hand, the site of the youngest studied stand (plot D), when planted, was surrounded by ca. 30 years-old Norway spruce stands and stands of another ectotrophic trees which were fully colonized. The plot C was established inside Norway spruce stands of the same age. Unfortunately, there are no data on the provenance and age of the seedlings.

According to the preliminary plant sociological investigations (J Uhlířová, 1992, personal communication) the site of plot C was originally occupied by the association *Anthoxantho-Agrostietum tenuis* Sill. 1933 (subass. typicum Jurko 1971, alliance *Cynosurion cristati* R.Tx. 1947, suballiance *Polygalo-Cynosurenion* Jurko 1974). Herbaceous grass communities of plot D are heterogeneous. More oligotrophic and less mesophytic species are present due to extensive farming in the past. No cleaning was done both in plot C and D.

Sporocarp presence and abundance of epigeous ectomycorrhizal macrofungi were determined in each plot. Abundance of sporocarps was recorded in June-July and September–October and 11 visits include investigations in plots A, B, C in 1989–1992 and in plot D in 1990–1992 (Table 2). August is the dryest summer month in this area (Petrovic, 1972) and therefore no study was done in this month. Except for voucher specimens no sporocarps were removed from the plots.

*Thelephora terrestris* and species of *Ramaria* are growing usually in clusters and it is nearly impossible to distinguish individual sporocarps and they were not counted for this study.

## Results

Sporocarps of 27 ectomycorrhizal species in 20 genera was been recorded in 1989–1992 (Table 2). The entire number of sporocarps ranged from 327 in plot B to 3977 in plot C. At least 12 species were recorded only either in plot A or in plot B. It is hard, of course, to conclude whether the forest treatment (plots A and B) influenced sporocarp producing species, since there are no replicates. However, the process of sporocarp succession during the first 27 years of development of planted Norway spruce stands is clearly shown in Table 2.

*Table 2.* Abundance of fruitbodies of ectomycorrhizal macrofungi in Norway spruce plantations (Vrch Dobroc, Central Slovakia). Data in parentheses refer to the lopped plot A. Abundance of *Thelephora* and *Ramaria* species entered only as present (+) or absent (0)

| Recognized fungi | Years after plantation | | | | | | | | | | |
|---|---|---|---|---|---|---|---|---|---|---|---|
| | 1st | 2nd | 3rd | 7th | 8th | 9th | 10th | 24th | 25th | 26th | 27th |
| *Amanita excelsa* (Fr.) Bertillon | 0 | 0 | 0 | 0 | 0 | 0 | 0 | 0(0) | 0(0) | 0(0) | 0(4) |
| *Amanita muscaria* (L.:Fr.) Hook | 1 | 1 | 0 | 74 | 34 | 10 | 22 | 4(3) | 0(0) | 2(9) | 0(0) |
| *Amanita regalis* (Fr.) Michael | 0 | 0 | 0 | 0 | 0 | 0 | 0 | 0(0) | 3(1) | 0(0) | 0(0) |
| *Amanita rubescens* (Pers.:Fr.) S.F.Gray | 0 | 0 | 0 | 0 | 0 | 1 | 0 | 23(48) | 0(0) | 0(1) | 0(0) |
| *Boletus edulis* Bull.:Fr. | 0 | 0 | 0 | 0 | 0 | 0 | 0 | 0(1) | 0(2) | 0(0) | 0(0) |
| *Cantharellus cibarius* Fr. | 0 | 0 | 0 | 0 | 0 | 0 | 0 | 1(0) | 0(0) | 0(0) | 0(0) |
| *Chalciporus piperatus* (Bull.:Fr.) Bataille | 0 | 1 | 0 | 3 | 94 | 72 | 74 | 0(0) | 2(0) | 0(0) | 0(0) |
| *Clitopilus prunulus* (Scop.:Fr.) Kummer | 1 | 0 | 1 | 0 | 0 | 0 | 0 | 22(1) | 12(4) | 0(0) | 1(8) |
| *Cortinarius* spp. and *Dermocybe* spp. | 0 | 8 | 0 | 277 | 1789 | 497 | 6 | 32(33) | 13(9) | 33(153) | 0(0) |
| *Gomphidius glutinosus* (Schaeff.) Fr. | 0 | 0 | 0 | 0 | 0 | 0 | 0 | 0(1) | 1(0) | 7(4) | 0(0) |
| *Hebeloma crustuliniforme* /Bull.:Fr.) Quél. | 0 | 0 | 0 | 190 | 277 | 2 | 7 | 0(0) | 0(6) | 0(0) | 0(1) |
| *Hebeloma perpallidum* Mos. | 0 | 0 | 0 | 0 | 0 | 75 | 0 | 0(0) | 0(0) | 0(0) | 0(0) |
| *Hebeloma sinapizans* (Paulet:Fr.) Gill. | 0 | 0 | 0 | 0 | 16 | 2 | 0 | 0(0) | 0(0) | 0(0) | 0(0) |
| *Hygrophorus olivaceoalbus* (Fr.:Fr.) Fr. | 0 | 0 | 0 | 0 | 0 | 0 | 0 | 0(0) | 2(0) | 0(0) | 0(0) |
| *Hygrophorus pustulatus* (Pers.:Fr.) Fr. | 0 | 0 | 0 | 0 | 0 | 0 | 0 | 0(13) | 29(17) | 64(0) | 0(0) |
| *Inocybe* sp. | 1 | 0 | 0 | 0 | 0 | 0 | 0 | 0(0) | 0(0) | 0(0) | 0(0) |
| *Inocybe lacera* (Fr.) Kummer | 0 | 0 | 0 | 1 | 75 | 17 | 0 | 3(8) | 1(4) | 8(1) | 0(0) |
| *Laccaria laccata* (Scop.:Fr.) Berk. and Br.s.l. | 74 | 246 | 6 | 0 | 6 | 82 | 77 | 11(2) | 5(1) | 18(37) | 0(0) |
| *Lactarius deterrimus* Gröger | 0 | 0 | 0 | 0 | 188 | 2 | 7 | 0(0) | 0(0) | 0(0) | 0(0) |
| *Lactarius* cf. *rufus* (Scop.:Fr.) Fr. | 0 | 0 | 0 | 0 | 0 | 0 | 0 | 0(0) | 6(0) | 15(0) | 0(0) |
| *Leccinum piceinum* Pilát et Dermek | 0 | 0 | 0 | 0 | 0 | 0 | 0 | 0(0) | 2(0) | 2(0) | 0(0) |
| *Paxillus involutus* (Batsch:Fr.) Fr. | 1 | 0 | 0 | 0 | 0 | 0 | 0 | 0(1) | 0(0) | 0(0) | 0(0) |
| *Ramaria* sp. | 0 | 0 | 0 | 0 | 0 | 0 | 0 | 0(0) | 0(0) | +(0) | 0(0) |
| *Russula aeruginea* Lindbl.in Fr. | 0 | 0 | 0 | 0 | 0 | 0 | 0 | 0(0) | 0(0) | 0(0) | 0(1) |
| *Russula cyanoxantha* (Schaeff.) Fr. | 0 | 0 | 0 | 0 | 0 | 0 | 0 | 1(4) | 0(0) | 0(2) | 0(0) |
| *Russula turci* Bres. | 0 | 0 | 0 | 0 | 0 | 0 | 0 | 0(0) | 0(2) | 0(5) | 0(0) |
| *Thelephora terrestris* (Ehrh.:Willd.) Fr. | + | 0 | + | + | + | + | 0 | 0(0) | 0(0) | 0(0) | 0(0) |
| *Tricholoma* sp. | 0 | 0 | 0 | 0 | 0 | 0 | 0 | 0(0) | 0(0) | 0(4) | 0(0) |
| *Tricholoma vaccinum* (Pers.:Fr.) Kummer | 0 | 0 | 0 | 0 | 0 | 0 | 0 | 1(1) | 0(0) | 0(0) | 0(0) |
| *Xerocomus badius* (Fr.:Fr.) Gilb. | 0 | 0 | 0 | 0 | 0 | 0 | 0 | 0(0) | 0(1) | 2(1) | 0(0) |
| *Xerocomus chrysenteron* (Bull.) Quél | 0 | 0 | 0 | 0 | 0 | 0 | 0 | 1(0) | 0(0) | 0(0) | 0(0) |

## Discussion

Generally, the studies regarding the primary (and secondary) macrofungal succession are associated with many difficulties. As pointed out by Jansen and De Nie (1988), species diversity of sporocarps is reflected by the number of mycorrhiza but the correlation between sporocarp density and mycorrhizal density is not very high. The presence of sporocarps probably expresses only the carbohydrate draining ability of a specific species. Further, mycorrhizae and sporocarp production have been suggested to reach their highest number at the time of canopy closure (Jansen and De Nie, 1988). However, this correlates very well with our records since the canopy has closed in plot C where abundance of the species of *Cortinarius* and *Hebeloma* was the highest. Biomass of sporocarps, of course, should be better reported as dry matter because of their interspecific variability. Unfortunately, such data were not available in the introductory stage of this project.

Few fungi in the initial stage of mycorrhizal succession, as *Thelephora terrestris* and *Laccaria laccata* s.l., were likely transferred with the seedlings from nurseries. It is even possible that all species occurring in the first three years of the colonization of farmland were transferred from nurseries.

In spite of difficulties in evaluation of primary mycorrhizal succession, recorded data document very well the process of the sporocarp succession during the macrofungal colonization of farmland.

## Acknowledgement

This project was partially supported by Grant 2/023/93 from the Slovak Grant Agency for Sciences.

## References

Ammirati S, Ammirati J F and Bledsoe C 1987 Spatial and temporal distributions of ectomycorrhizal fungi in a Douglas-fir plantation. *In* Proc. 7th NACOM: Mycorrhizae in the Next Decade. Eds. D M Sylvia et al. pp 81. Gaiensville, USA.

Arnolds E 1991 Decline of ectomycorrhizal fungi in Europe. Agric. Ecosyst. Environ. 35, 209–244.

Chu-Chou M 1979 Mycorrhizal fungi of *Pinus radiata* in New Zealand. Soil Biol. Biochem. 11, 557–562.

Dighton J and Mason P 1985 Mycorrhizal dynamics during forest tree development. *In* Developmental Biology of Higher Fungi. Eds. D More et al. pp 117–139. Cambridge University Press, Cambridge, UK.

Fleming L V, Deacon J V and Last F T 1986 Ectomycorrhizal succession in a Scottisch birch wood. *In* Physiological and Genetical Aspects of Mycorrhizae. Eds. V Gianinazzi-Pearson and S Gianinazzi. pp 259–264. INRA, Paris, France.

Harper J E and Webster J 1964 An experimental analysis of the coprophilous fungus succession. Trans. Brit. Mycol. Soc. 47, 511–530.

Jansen A E and De Nie H W 1988 Relations between mycorrhizas and fruitbodies of mycorrhizal fungi in Douglas fir plantations in The Netherlands. Acta Bot. Neerl. 37, 243–249.

Last F T et al. 1981 Factors affecting the production of fruitbodies of *Amanita muscaria* in plantations of *Pinus patula.* Proc. Indian Acad. Sci. (Plant Sci.) 90, 91–98.

Last F T, Mason P A and Wilson J 1983 Fine roots and sheathing mycorrhizas: their formation, function and dynamics. Plant and Soil 71, 9–21.

Last F T et al. 1984 Succession of fruitbodies of sheathing mycorrhizal fungi associated with *Betula pendula.* For. Ecol. Manage. 9, 229–234.

Mason P A et al. 1982 Ecology of some fungi associated with an ageing stand of birches (*Betula pendula* and *B. pubescens*). For. Ecol. Manage. 4, 19–39.

Molina R, Massicotte H B and Trappe J M 1992 Specificity phenomena in mycorrhizal symbioses: Community-ecological consequences and practical implication. *In* Mycorrhizal Functioning: An Interactive Plant-Fungal Process. Ed. M F Allen. pp 357–423. New York, USA.

Newton A C 1991 Mineral nutrition and mycorrhizal infection of seedling oak and birch. III. Epidemiological aspects of ectomycorrhizal infection and the relationship to seedling growth. New Phytol, 117, 53–60.

Newton A C 1992 Towards a functional classification of ectomycorrhizal fungi. Mycorrhiza 2, 75–79.

O'Dell T E, Luoma D L and Molina R J 1992 Ectomycorrhizal fungal communities in young, managed and old-growth Douglas-fir stands. North. Environ. J. 8, 166–168.

Petrovic S 1972 Climatic and phenological conditions of the Central Slovakia region. HMV, Bratislava, Slovakia. 432 p. (*In Slovak*).

Read D J 1991 Mycorrhizas in ecosystems - nature's response to the "law of the minimum". *In* Frontiers in Mycology. Ed D L Hawksworth. pp 101–130. CAB International, Wallingford, Oxon, UK.

Termoshuizen A J and Schaffers A P 1987 Occurrence of carpophores of ectomycorrhizal fungi in selected stands *Pinus sylvestris* in the Netherlands in relation to stand vitality and air pollution. Plant and Soil 104, 209–217.

Termoshuizen A J and Schaffers A P 1989 Succession of mycorrhizal fungi in stands of *Pinus sylvestris* in the Netherlands. Agric. Ecosyst. Environ. 28, 503–507.

Vogt K A et al. 1992 Sporocarp production by Basidiomycetes, with emphasis on forest ecosystems. *In* The Fungal Community, its Organisation and Role in the Ecosystems. Eds. G C Carroll and D T Wicklow. pp 563–581. Marcel Dekker Inc., New York, Basel, Hong Kong, USA.

# Developments in Plant and Soil Sciences

1. J. Monteith and C. Webb (eds.): *Soil Water and Nitrogen in Mediterranean-type Environments.* 1981 ISBN 90-247-2406-6
2. J. C. Brogan (ed.): *Nitrogen Losses and Surface Run-off from Landspreading of Manures.* 1981 ISBN 90-247-2471-6
3. J. D. Bewley (ed.): *Nitrogen and Carbon Metabolism.* 1981 ISBN 90-247-2472-4
4. R. Brouwer, I. Gašparíková, J. Kolek and B. C. Loughman (eds.): *Structure and Function of Plant Roots.* 1981 ISBN 90-247-2510-0
5. Y. R. Dommergues and H. G. Diem (eds.): *Microbiology of Tropical Soils and Plant Productivity.* 1982 ISBN 90-247-2624-7
6. G. P. Robertson, R. Herrara and T. Rosswall (eds.): *Nitrogen Cycling in Ecosystems of Latin America and the Caribbean.* 1982 ISBN 90-247-2719-7
7. D. Atkinson, K. K. S. Bhat, M. P. Coutts, P. A. Mason and D. J. Read (eds.): *Tree Root Systems and Their Mycorrhizas.* 1983 ISBN 90-247-2821-5
8. M. R. Sarić and B. C. Loughman (eds.): *Genetic Aspects of Plant Nutrition.* 1983 ISBN 90-247-2822-3
9. J. R. Freney and J. R. Simpson (eds.): *Gaseous Loss of Nitrogen from Plant-Soil Systems.* 1983 ISBN 90-247-2820-7
10. United Nations Economic Commission for Europe (ed.): *Efficient Use of Fertilizers in Agriculture.* 1983 ISBN 90-247-2866-5
11. J. Tinsley and J. F. Darbyshire (eds.): *Biological Processes and Soil Fertility.* 1984 ISBN 90-247-2902-5
12. A. D. L. Akkermans, D. Baker, K. Huss-Danell and J. D. Tjepkema (eds.): Frankia *Symbioses.* 1984 ISBN 90-247-2967-X
13. W. S. Silver and E. C. Schröder (eds.): *Practical Application of* Azolla *for Rice Production.* 1984 ISBN 90-247-3068-6
14. P. G. L. Vlek (ed.): *Micronutrients in Tropical Food Crop Production.* 1985 ISBN 90-247-3085-6
15. T. P. Hignett (ed.): *Fertilizer Manual.* 1985 ISBN 90-247-3122-4
16. D. Vaughan and R. E. Malcolm (eds.): *Soil Organic Matter and Biological Activity.* 1985 ISBN 90-247-3154-2
17. D. Pasternak and A. San Pietro (eds.): *Biosalinity in Action.* Bioproduction with Saline Water. 1985 ISBN 90-247-3159-3
18. M. Lalonde, C. Camiré and J. O. Dawson (eds.): Frankia *and Actinorhizal Plants.* 1985 ISBN 90-247-3214-X
19. H. Lambers, J. J. Neeteson and I. Stulen (eds.): *Fundamental, Ecological and Agricultural Aspects of Nitrogen Metabolism in Higher Plants.* 1986 ISBN 90-247-3258-1
20. M. B. Jackson (ed.): *New Root Formation in Plants and Cuttings.* 1986 ISBN 90-247-3260-3
21. F. A. Skinner and P. Uomala (eds.): *Nitrogen Fixation with Non-Legumes* (Proceedings of the 3rd Symposium, Helsinki, 1984). 1986 ISBN 90-247-3283-2
22. A. Alexander (ed.): *Foliar Fertilization.* 1986 ISBN 90-247-3288-3
23. H. G. v.d. Meer, J. C. Ryden and G. C. Ennik (eds.): *Nitrogen Fluxes in Intensive Grassland Systems.* 1986 ISBN 90-247-3309-X
24. A. U. Mokwunye and P. L. G. Vlek (eds.): *Management of Nitrogen and Phosphorus Fertilizers in Sub-Saharan Africa.* 1986 ISBN 90-247-3312-X
25. Y. Chen and Y. Avnimelech (eds.): *The Role of Organic Matter in Modern Agriculture.* 1986 ISBN 90-247-3360-X
26. S. K. De Datta and W. H. Patrick Jr. (eds.): *Nitrogen Economy of Flooded Rice Soils.* 1986 ISBN 90-247-3361-8
27. W. H. Gabelman and B. C. Loughman (eds.): *Genetic Aspects of Plant Mineral Nutrition.* 1987 ISBN 90-247-3494-0
28. A. van Diest (ed.): *Plant and Soil: Interfaces and Interactions.* 1987 ISBN 90-247-3535-1

# Developments in Plant and Soil Sciences

29. United Nations Economic Commission for Europe and FAO (eds.): *The Utilization of Secondary and Trace Elements in Agriculture.* 1987 ISBN 90-247-3546-7
30. H. G. v.d. Meer, R. J. Unwin, T. A. van Dijk and G. C. Ennik (eds.): *Animal Manure on Grassland and Fodder Crops.* Fertilizer or Waste? 1987 ISBN 90-247-3568-8
31. N. J. Barrow: *Reactions with Variable-Charge Soils.* 1987 ISBN 90-247-3589-0
32. D. P. Beck and L. A. Materon (eds.): *Nitrogen Fixation by Legumes in Mediterranean Agriculture.* 1988 ISBN 90-247-3624-2
33. R. D. Graham, R. J. Hannam and N. C. Uren (eds.): *Manganese in Soils and Plants.* 1988 ISBN 90-247-3758-3
34. J. G. Torrey and J. L. Winship (eds.): *Applications of Continuous and Steady-State Methods to Root Biology.* 1989 ISBN 0-7923-0024-6
35. F. A. Skinner, R. M. Boddey and I. Fendrik (eds.): *Nitrogen Fixation with Non-Legumes* (Proceedings of the 4th Symposium, Rio de Janeiro, 1987). 1989 ISBN 0-7923-0059-9
36. B. C. Loughman, O. Gašparíková and J. Kolek (eds.): *Structural and Functional Aspects of Transport in Roots.* 1989 ISBN 0-7923-0060-2; Pb 0-7923-0061-0
37. P. Plancquaert and R. Haggar (eds.): *Legumes in Farming Systems.* 1990 ISBN 0-7923-0134-X
38. A. E. Osman, M. M. Ibrahim and M. A. Jones (eds.): *The Role of Legumes in the Farming Systems of the Mediterranean Areas.* 1990 ISBN 0-7923-0419-5
39. M. Clarholm and L. Bergström (eds.): *Ecology of Arable Land – Perspectives and Challenges.* 1989 ISBN 0-7923-0424-1
40. J. Vos, C. D. van Loon and G. J. Bollen (eds.): *Effects of Crop Rotation on Potato Production in the Temperate Zones.* 1989 ISBN 0-7923-0495-0
41. M. L. van Beusichem (ed.): *Plant Nutrition – Physiology and Applications.* 1990 ISBN 0-7923-0740-2
42. N. El Bassam, M. Dambroth and B.C. Loughman (eds.): *Genetic Aspects of Plant Mineral Nutrition.* 1990 ISBN 0-7923-0785-2
43. Y. Chen and Y. Hadar (eds.): *Iron Nutrition and Interactions in Plants.* 1991 ISBN 0-7923-1095-0
44. J. J. R. Groot, P. de Willigen and E. L. J. Verberne (eds.): *Nitrogen Turnover in the Soil-Crop System.* 1991 ISBN 0-7923-1107-8
45. R. J. Wright, V.C. Baligar and R. P. Murrmann (eds.): *Plant-Soil Interactions at Low pH.* 1991 ISBN 0-7923-1105-1
46. J. Kolek and V. Kozinka (eds.): *Physiology of the Plant Root System.* 1992 ISBN 0-7923-1205-8
47. A. U. Mokwunye (ed.): *Alleviating Soil Fertility Constraints to Increased Crop Production in West Africa.* 1991 ISBN 0-7923-1221-X; Pb 0-7923-1222-8
48. M. Polsinelli, R. Materassi and M. Vincenzini (eds.): *Nitrogen Fixation* (Proceedings of the 5th Symposium, Florence, 1990). 1991 ISBN 0-7923-1410-7
49. J.K. Ladha, T. George and B.B. Bohlool (eds.): *Biological Nitrogen Fixation for Sustainable Agriculture.* 1992 ISBN 0-7923-1774-2
50. P.J. Randall, E. Delhaze, R.A. Richards and R. Munns (eds.): *Genetic Aspects of Plant Mineral Nutrition.* 1993 ISBN 0-7923-2118-9
51. K.S. Kumarasinghe and D.L. Eskew (eds.): *Isotopic Studies of Azolla and Nitrogen Fertilization of Rice.* 1993 ISBN 0-7923-2274-6
52. F.A. Bliss and G. Hardarson (eds.): *Enhancement of Biological Nitrogen Fixation of Common Baen in Latin America.* 1993 ISBN 0-7923-2451-X
53. M.A.C. Fragoso and M.L. van Beusichem (eds.): *Optimization of Plant Nutrition.* 1993 ISBN 0-7923-2519-2
54. N.J. Barrow (ed.) *Plant Nutrition - From Genetic Engineering to Field Practice.* 1993 ISBN 0-7923-2540-0
55. A.D. Robson (ed.): *Zinc in Soils and Plants.* 1993 ISBN 0-7923-2631-8
56. A.D. Robson, L.K. Abbott and N. Malajczuk (eds.): *Management of Mycorrhizas in Agriculture, Horticulture and Forestry.* 1994 ISBN 0-7923-2700-4

# Developments in Plant and Soil Sciences

57. P.H. Graham, M.J. Sadowsky and C.P. Vance (eds.): *Symbiotic Nitrogen Fixation*. 1994 ISBN 0-7923-2781-0
58. F. Baluška, M. Čiamporová, O. Gašparíkova and P.W. Barlow (eds.): *Structure and Function of Roots*. 1995 ISBN 0-7923-2832-9
59. J. Abadía (ed.): *Iron Nutrition in Soils and Plants*. 1995 ISBN 0-7923-2900-7
60. P.S. Curtis, E.G. O'Neill, J.A. Teeri, D.R. Zak and K.S. Pregitzer (eds.): *Belowground Responses to Rising Atmospheric $CO_2$*. Implications for Plants, Soil Biota, and Ecosystem Processes. 1994 ISBN 0-7923-2901-5
61. P.C. Struik, W.J. Vredenberg, J.A. Renkema and J.E. Parlevliet (eds.): *Plant Production on the Threshold of a New Century*. 1994 ISBN 0-7923-2903-1
62. L.O. Nilsson, R.F. Hüttl and U.T. Johansson (eds.): *Nutrient Uptake and Cycling in Forest Ecosystems*. CEC/IUFRO Symposium, Halmstad, Sweden, June 7–10, 1993. 1995 ISBN 0-7923-3030-7

Kluwer Academic Publishers – Dordrecht / Boston / London

GPSR Compliance

*The European Union's (EU) General Product Safety Regulation (GPSR) is a set of rules that requires consumer products to be safe and our obligations to ensure this.*

*If you have any concerns about our products, you can contact us on ProductSafety@springernature.com*

In case Publisher is established outside the EU, the EU authorized representative is:

Springer Nature Customer Service Center GmbH
Europaplatz 3
69115 Heidelberg, Germany

**Batch number: 10762213**

Printed by Printforce, the Netherlands